SNOW ENGINEERING
RECENT ADVANCES AND DEVELOPMENTS

PROCEEDINGS OF THE FOURTH INTERNATIONAL CONFERENCE ON SNOW ENGINEERING
TRONDHEIM/NORWAY/19 – 21 JUNE, 2000

Snow Engineering

Recent Advances and Developments

Edited by

Erik Hjorth-Hansen
Norwegian University of Science and Technology, Trondheim, Norway

Ivar Holand (1924 – 2000)
SINTEF Civil and Environmental Engineering, Trondheim, Norway

Sveinung Løset
Norwegian University of Science and Technology, Trondheim, Norway

Harald Norem
Norwegian University of Science and Technology, Trondheim, Norway

A.A.BALKEMA/ROTTERDAM/BROOKFIELD/2000

Cover illustration courtesy of Truls Mœlmann

Published by
A.A. Balkema, P.O. Box 1675, 3000 BR Rotterdam, Netherlands
Fax: +31.10.413.5947; E-mail: balkema@balkema.nl; Internet site: www.balkema.nl
A.A. Balkema Publishers, Old Post Road, Brookfield, VT 05036-9704, USA
Fax: 802.276.3837; E-mail: info@ashgate.com

ISBN 90 5809 148 1
© 2000 A.A. Balkema, Rotterdam
Printed in the Netherlands

Snow Engineering, Hjorth-Hansen, Holand, Løset & Norem (eds) © 2000 Balkema, Rotterdam, ISBN 90 5809 148 1

Table of contents

3 Structural engineering

4 Housing and residential planning

5 Transportation

Snow Engineering, Hjorth-Hansen, Holand, Løset & Norem (eds) © 2000 Balkema, Rotterdam, ISBN 90 5809 148 1

Foreword

Since 1988, three Snow Engineering Conferences have been held at the following venues: Santa Barbara, USA 1988 & 1992; Sendai, Japan 1996. Each of these Snow Engineering Conferences have been a forum for discussions and the exchange of research results and field experience. They have become premier events for the international exchange of information on developments in snow engineering.

The 4th International Conference on Snow Engineering was organized by the Norwegian University of Science and Technology (NTNU), Department of Structural Engineering and Department of Road and Railway Engineering, and arranged in Trondheim, Norway, 19 – 21 June, 2000. Further details about the conference may be found on the web site www/bygg.ntnu.ktek/snow2000/.

The Proceedings from the conference have been arranged in four main chapters as follows:

Snow technology and science: Snow drift, avalanches, friction, fundamentals and measurements of snow properties, snow on ground;

Structural engineering: Snow loads on structures, shape factors, avalanche impact loads, snow-creeping loads;

Housing and residential planning: Snow accumulation related to the location and design of houses and structures, residential planning for operation and maintenance;

Transportation: Road and railway engineering in snowy areas, operation and maintenance of roads, railways and airfields, friction on roads and airfields.

A European snow load research project was completed in 1999. The work was related to the new European standard (CEN standard) on snow loads, which is now under preparation. Many of the papers in the chapter on 'Structural Engineering' describe the detailed results from this project. Consequently, this chapter may also function as 'Justification Notes' or a 'Background Document' for the new CEN standard.

Since the whole field of 'Snow Engineering' is a continuum, the individual papers are not easily arranged in the chapters specified above, and some papers may relate to two or more chapters. To help the readers find the specific topic they are looking for, a subject index is included which gives page number references.

We feel it is imperative that the proceedings are published by an international publishing house and are available for distribution at arrival of the participants at the conference venue. These requirements have necessitated a very strict deadline for the complete camera-ready manuscript to be delivered to the publishers. As a result, we have not been able to give manuscripts that arrive late all of the desired formal revisions before printing.

We wish to express our thanks to all authors for their contributions to this volume and to the publishers for their excellent cooperation during the work.

Trondheim, March 2000
Ivar Holand (1924 – 2000)
Chairman of the Organizing Commitee

Snow Engineering, Hjorth-Hansen, Holand, Løset & Norem (eds)© 2000 Balkema, Rotterdam, ISBN 90 5809 148 1

Postscript

Ivar Holand (1924-2000)

The Foreword on the previous page was written 3 1/2 months before the conference, at a time when Professor emeritus Ivar Holand had only had a month to adjust to the realities of a cancer diagnosis, and to turn over his responsibilities as Chairman to his Trondheim colleagues in the Organizing Committee. The anticipated slow development of his condition soon proved too optimistic, and he died on 28 April 2000.

We, who have taken over his task, commend his outstanding leadership during the almost two years of planning for the conference, and we miss him greatly.

Erik Hjorth-Hansen
Sveinung Løset
Harald Norem

An almost complete biography of Ivar Holand is found as the opening article of the book honouring Ivar Holand at his 70th anniversary:
Bell, K. (editor), 1994. *From Finite Elements to the Troll Platform*
Department of Structural Engineering
The Norwegian Institute of Technology
Trondheim, Norway

1 Invited paper

Snow Engineering, Hjorth-Hansen, Holand, Løset & Norem (eds)© 2000 Balkema, Rotterdam, ISBN 90 5809 148 1

Snow loads on roofs in Europe: Research and standardization

L. Sanpaolesi
Department of Structural Engineering, University of Pisa, Italy

ABSTRACT: This note deals with structural engineering and, at the same time, links research and standardization. Nowadays the perfecting of design is correlated to the streamlining of structural knowledge, which in turn can only result from research and, in particular from research addressed to the definition of actions on structures, which is most beneficial to realizations in terms of cost reduction and improvement of safety standards. In 1996 the European Commission DGIII/D-3 sponsored a wide pre-normative research aimed at improving the knowledge in the field of snow loads on structures. At the same time, in 1998, the Commission stated to prepare, through CEN, the final EN 1991-1-3 Snow Loads standard text. In the present note are given information about the links between the European research project and in the CEN normative activity in the field of snow loading, which were both leaded by the Author.

1 INTRODUCTION

The title of this note makes it clear that we are discussing structural engineering and, at the same time, correctly links research and standardization.

Actually, this link has always existed but perhaps it has increased recently, also due to project complexities towards which we are heading, abandoning oversimplified design criteria. I mean to say that the perfecting of design is correlated to the streamlining of structural knowledge, which in turn can only result from research.

I am convinced that, with current knowledge standards, we can aim at refining structural projects further in terms of cost reduction and higher safety standards by working more on actions than on structural computation techniques. That is to say, it is worth intensifying research on actions because they are most beneficial to realizations.

One of the major aspect that have by now permeated both research and standardization is the probabilistic evaluation of all phenomena and this has helped to compound problems.

In recent years I dealt with a major European research program and here I shall illustrate some of the main aspects that will be examined in the numerous papers on the subject addressed to our Conference.

In 1996 the European Commission General Direction III sponsored a pre-normative research aimed at improving scientific knowledge and models to determine snow loads on buildings by producing a sound common scientific basis acceptable to all European countries involved in the drafting of new European standards.

In fact the main part of existing studies about snow loads on structures were carried out in cold climatic areas; studies conducted in continental and Mediterranean Europe therefore present a major and new contribution to scientific knowledge in the snow loading field.

The study, started in December 1996 and concluded in June 1999, was carried out by the following research Institutes from six different European countries:

1. Building Research establishment Ltd, Construction Division (United Kingdom)
2. CSTB, Centre de Recherche de Nantes (France)
3. Ecole Polytechnique Fédérale de Lausanne, (Switzerland)
4. IMES, Structural Engineering Department (Italy)
5. Joint Research Centre, ISIS (EU)
6. SINTEF, Civil and Environmental Engineering (Norway)
7. University of Leipzig, Institute of Concrete Design (Germany)

8. University of Pisa, Department of Structural Engineering (Italy)

The research program is in two consecutive phases. Phase I provides methods and techniques to determine *ordinary* and *exceptional* snow loads on the ground in order to produce a new European ground snow load map. Phase II provides methods and techniques to determine snow loads on roofs and define appropriate criteria to determine serviceability loads on such roofs.

At the same time, since 1998, the European Commission has started preparing, through the CEN (European Normalization Committee), the final EN-1991-1-3 Snow Loads standard text, to be applied right across Europe, specifically in the nineteen CEN member states.

The Snow Loads on buildings standard forms part of the important Eurocode program, structural engineering design standards to be progressively adopted throughout Europe, replacing current national sets of standards and covering all the various aspects: the Basis of Design, Actions, Design Rules for Reinforced Concrete, for Steel Constructions, for Masonry, etc.

Here too, I had the fortune to be involved with other Colleagues as Researchers, with fresh results made available to us, and as Regulators at the same time. To tell the truth, this was not always an "easy" position to be in due to the numerous evident uncertainties in the field of research and on account of the way we are conditioned by the body of legislators.

Nonetheless, we are trying to do our best and we hope to be able to help develop knowledge in our field.

However, before examining more technical subjects, I would like to comment on Standardization, particularly in Europe.

First and foremost, research is universal, free and unites Researches worldwide.

Standardization is highly conditioned on the following accounts, for example;

- the legal validity of Standards which varies greatly in various European Countries;
- regulative traditions;
- various habits and local typologies in various countries;
- the stickiness manifested in many countries by National Competent Authorities with each regulative change.

On the other hand, Europe is steeped in tradition and history and it is not easy, not even in an apparently neutral sector such as Action Standards, to standardize different, well-established opinions and guidelines.

That is why the European Commission makes slow progress in the regulative sector, trying to get members to agree rather than forcing its ideas and accepting in between stages in approaching a common standard.

A last remark on the recently approved standard ISO/4355, Determination of Snow Loads on Roofs. It is undoubtedly an excellent, up-to date set of standards but amongst the European Code Study Group the prevalent opinion was to set a special standard better suited to the European situation, which ranges from temperate Mediterranean climates to North European climates such as the Scandinavian countries and Iceland, as the ISO standard appears to be slightly too complex and more fitting for North European countries. However, they still intended to acquire major parts of the ISO/4355 standard.

Having started these general remarks, I would now like to pass on to more technical subjects.

2 THE EUROPEAN GROUND SNOW LOADS MAP

The Research work reviewed current practice in eighteen European countries and, in consultation with the appropriate National Meteorological offices, identified statistical techniques and data that were both available and essential in determining characteristic snow load values.

The analyzed database consist of 2600 weather stations, where records have usually been kept for 20 to 40 winters, though at some stations over 90 winters.

Similarities and differences between individual national approaches were identified, creating the need to develop a reference model for statistical analysis of ground snow loads, to be adopted all across Europe. This statistical model was defined according the values of the best correlation coefficients, obtained during interpolation of the sampled snow data, based on the following three different cumulative distribution functions: Gumbel (Type I), Lognormal and Weibull.

In order to avoid inconsistencies it was agreed to use, as reference procedure, the Gumbel (TypeI) distribution, determining its parameters by means of the least square method.

Climatic Regions

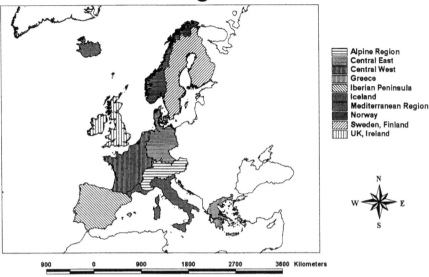

Figure 1. European climatic regions.

Having obtained 50-years characteristic ground snow loads at each weather station, the next step is to investigate the geographic variation in loads through regionalisation of available data.

Ten different homogeneous climatic regions were identified on the basis of geographic and climatic consideration.

These regions are shown in figure 1.

Within a distinct climatic region, snow load is considered to be a function of site altitude, yet affected by random deviation (including topographical effects). This allows a mean altitude function to be defined for each climatic region, except for those areas which show a poor correlation with altitude (e.g. Norway).

A spatial interpolation analysis was performed with GIS techniques (Geographical Information System) to obtain the spatial representation of the variation of snow loads on the ground, reduced at sea level, right across the examined climatic area.

Particular attention was paid to setting GIS interpolation phase parameters and to reduce inconsistencies at borderlines between climatic regions.

Figure 2 shows, as an example, the altitude function for the climatic region of Sweden and Finland.

Figure 3 shows the ground snow map of Sweden and Finland and figure 4 indicates the location of Weather Stations.

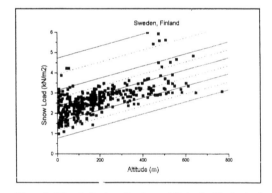

Figure 2. Climatic Region: Sweden and Finland: altitude - load relationship.

Figure 3. Climatic Region: Sweden and Finland: ground snow load map.

Sweden, Finland: Snow Load at Sea Level

Figure 4. Climatic Region: Sweden and Finland: location of weather stations

Several checks ware done to finalize maps. Two of them are particularly relevant:
- a comparison of values calculated using maps with values obtained directly at each weather station as result of the statistical procedure and values obtained from the ENV 1991-2-3 Snow Loads (preliminary code);
- the verification of snow load steps encountered at borderlines between climatic regions.

Both of the above checks gave good result, showing that the map obtained from studies gives a good representation of the variation in snow loads across Europe according to the updated snow database retrieved and that differences at borderlines between defined climatic regions are sufficiently limited.

However, a map thus obtained as a final Research product cannot be introduced as such into the European code, as annex, for numerous reasons:

- the map obtained is physical, whereas standard wise administrative boundaries have to be taken into account;
- local situations and requirements better known to each country must be considered;
- there must be consent and agreement with National Competent Authorities.

Therefore, it was decided to place the physical European map in an "Informative Annex" and to introduce National Competent Authorities' update maps, based on the physical European map, into the code.

This is evidently a compromise that enables a progressive approach of national maps to the common European map and numerous European countries have already declared that they will provide maps of their territory consistent with or identical to the European map.

That is all that can be obtained for the moment.

3 EXCEPTIONAL SNOW LOADS

In some isolated climatic regions very heavy snow falls have been recorded resulting in abnormally large snow loads. Such snow-falls significantly disturb the statistical processing of more regular snowfall data and clearly didn't fit in with the distribution calculated for the remaining values. These snow loads are referred to as 'exceptional snow loads'.

Before the present research study only the French code in Europe gave information on how to identify and treat these exceptional values in ultimate limit states verifications.

"Exceptional" values were encountered in many areas of Europe, mainly in maritime and coastal areas and in mild climates where snow is intermittent and generally short lasting snowfalls are registered. These values were defined numerically and places featuring such exceptional values were localized in Europe, in order to find out geographical, meteorological and any other source of influence which should have led to register exceptional snowfalls.

Figure 5 shows an example of exceptional snow load according to the given definition: *if the ratio of the largest load value to the characteristic load determined without the inclusion of that value is greater than 1,5 then the largest load value shall be treated as an exceptional value.*

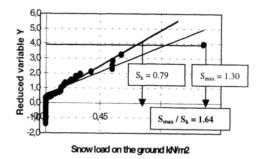

Figure 5. Exceptional Snow load at Pistoia (Italy 58. m.a.m.s.l.)

When encountered, exceptional values were discarded from samples to calculate the characteristic ground snow load, used to define the map.

Out of 2600 weather stations analyzed, approximately 160 of them feature exceptional snow loads. The location of such stations is shown in figure 6.

In European code undergoing definition, the issue has been addressed.

With the aforementioned data available, it is impossible to draw up a specific European map with values of exceptional loads expected, especially as there is no chance of achieving statistical reliability and with numbers insufficient to be able to draw any map whatsoever.

Figure 6. Location of the European weather station where exceptional snow load were registered.

What can be asserted following on experimental research is that these exceptional loads exist in all Countries with a European and Mediterranean continental climate and they cannot be included amongst ordinary data and treated as such.

The proposed regulative solution is based on the following points:

- to leave local Authorities to define the areas of their territory where exceptional events could occur (in my opinion, all countries with a Mediterranean continental climate could be included);
- to consider these events as "accidental" as defined in the EN-1990-1-1 Basis of design code;
- to use as "accidental" load A_d the characteristic snow loads on the ground value s_k, multiplied by a k coefficient k $\approx$ 2, i.e. $A_d = 2S_k$.

However, the issue is still being defined.

4 COMBINATION COEFFICIENTS FOR SERVICEABILITY LIMIT STATES VERIFICATIONS

In the Research studies the serviceability load combination factors were investigated, starting from the available daily snow data and following different statistical procedures, in order to define the optimal ψ coefficients for ULS and SLS verifications. Important and interesting results were obtained, particularly with regard to the differentiation in ψ values within European climates.

The combination factor ψ_0 is applied to the snow load effect when the dominating load effect is due to some other external load, such as wind. Accordingly, a derivation of this combination factor strictly requires a refined modeling of both snow and wind including the modeling of their variation in time. However, procedure based on such a refined model are typically time-consuming both with regard to retrieval of input data, numerical algorithms and computation time. As a consequence, simplified procedure, and in particular, the following three different approaches, as described in the Eurocode 1 Basis of Design (ENC 1994) and ISO 2394 (ISO 1998) background documents, were used:
(i)   Simplified methods: Turkstra's rule and the Design Value Method;
(ii)   the Borghes-Castanheta model;
(iii)   the upcrossing-rate analysis model.

These were applied to snow and wind data coming from several weather stations in Germany and Norway, leading to the following main conclusions:
- the factors obtained from the Borghes-Castanheta and upcrossing rate models agree fairly well;
- the values obtained from the simplified method gives results that are generally on the safe side than those obtained from other methods.

Due to major difficulties in getting refined and top quality combined data sets for snow and wind data at each European weather station, it was decided to calculate the combination coefficient ψ_0 by means of the Design Value Method (which gives good and slightly conservative results in comparison with Turkstra's rule) and extreme value distribution Type I.

One of the most interesting achievements of this task is undoubtedly the calibration of the combination coefficient ψ_0 values in the various climatic regions as shown in table 1.

The obtained range for the values stresses the need to differentiate the combination coefficient as recommended in Eurocode 1, whereas, at the moment only one value, equal to 0,6 is given for the whole CEN area.

For the studies of ψ_1 and ψ_2, snow load on the ground (or on a roof) may be considered as a process in time. Introducing an appropriate resolution of the time scale (e.g. daily measurements), observed load value may be taken as a sample, allowing an empirical distribution of the (daily) snow load to be obtained. For some climatic regions, the likelihood of high snow loads is not constant during the whole winter season due to the snow loading not being stationary process (e.g. mountain or continental climate). Combining all load values measured during several winters in one single histogram, irrespective of the date of occurrence means simplifying these types of climate.

The corresponding cumulative distribution function of daily snow loads makes it possible to obtain the load level with a certain probability P of not being exceeded. The ψ_1 and ψ_2 coefficients are the relation between these fractiles of P probability and the characteristic value of the snow load. The probability P may be interpreted as part of the entire time, during which the snow load is equal or less than the given load level.

Investigations were based directly on the empirical cumulative distribution representing the observed short-term snow loads or on an appropriate theoretical distribution, adjusted by choosing the best fitting parameters.

ψ_1 and ψ_2 value have been investigated for different climatic regions. The frequent and quasi-permanent values of the snow load are based on time series of daily snow loads, whereas the derivation of the characteristic value was performed in the first phase of the present research program and based on annual maximum of the snow load (by using the statistics of extremes).

The frequent value is determined such that fractile of time during which it is exceeded is chosen to be 0,05 (a base case).

The quasi-permanent value is determined so that the time during which it is exceeded forms a considerable part of the reference period of time. The time during which it is exceeded is be set as 0,5 of the reference period.

The above described procedures for the calculation of coefficients ψ_1 and ψ_2 were applied to 63 weather stations across the whole CEN members area in Europe, where daily snow data was available. The stations were chosen to represent as far as possible all climatic regions in the CEN area (including the different levels of altitude). The results obtained are summarized in table 1.

In the standard undergoing definition, the summary of results obtained has been introduced; this summary is present in the table and is important to applications because it gives rise to ψ_0, ψ_1 and ψ_2, which differ widely in various European countries.

5 SHAPE COEFFICIENTS

In the majority of the current codes the roof snow load is normally calculated from the ground snow load by multiplying by one or more conversion factors, which accounts for the roof shape, thermal characteristics, exposure and other possible influencing factors, such as the presence of glass parts in the roof.

Table 1. ψ coefficients for the European climatic regions

Region	ψ_0	ψ_1	ψ_2	Notes
Alpine	0,6	0,45	0,10	Altitude: > 1000m
	0,5	0,30	0,00	<= 1000m
UK and Eire	0,4	0,04	0,00	
Iberian	0,5	-	-	Altitude: > 500m
	0,4	0,00	0,00	<= 500m
Mediterranean	0,5	0,10	0,00	
Central East	0,55	0,40	0,10	Altitude: > 500m
	0,4	0,20	0,00	<= 500m
Central West	0,4	0,05	0,00	
Greece	0,5	0,00	0,00	
Norway	0,7	0,50	0,20	Altitude: > 300m
	0,6	0,40	0,20	< 300m
Finland-Sweden	0.65	0,60	0,20	
Iceland	0,6	0,20	0,00	Area: South-west
	0,6	0,40	0,10	North-East

Experimental investigations and theoretical modeling of snow deposits on roofs was mainly developed in cold countries, such as Canada, USA, Russia and Norway. No relevant information was available for the European climatic area.

Within this research, an attempt was made to cover the lack of scientific background in determining roof shape coefficients in Europe.

As a first step it was reviewed the state of the art about the theoretical modeling of snow deposits on roofs, with particular reference to snow drifting, to metamorphism and to ablation models.

The above investigations provided a great deal of important information, aimed at setting up an experimental measurement to retrieve in-situ measurements to calculate shape coefficients.

The results of the experimental measurements campaign, which actually lasted for two winters, are to be considered a starting point for the assessment of shape coefficients commonly adopted in the Eurocode.

Studies carried out in the research project combined the results of theoretical investigations with experimental results obtained during two winters in-situ measurements performed in the UK, Germany, Switzerland and Italy out of a total of 81 roofs: 26 flat and 55 gabled.

In addition to the field measurement program, a large experimental campaign was undertaken in the Climatic Wind Tunnel at CSTB, Nantes. Experimental models are calibrated using date available from in situ measurements.

The reduction of snow load on glass roof was also weighed up.

Following on data retrieval both in the field and in the wind tunnel, a probabilistic method was used to analyze data from both these studies and the drift and depletion model.

Based on retrieve data, a multiple regression analysis for the different parameters was performed and a proposal for the normative use of a new set of shape coefficients for gabled roof and for exposure coefficients was formulated.

In particular, despite the limited set of roof shapes, research has led to numerous indications.

For instance, for gabled roofs with slope equal to 30°, the leeward side shape coefficient does not reach as high a value as 1,2 as it is indicated in several codes.

Further evidence is that on the windward side of a gabled roof a considerable reduction in the shape coefficient is observed for roof angles greater than 15°-30°, in accordance with current Eurocode provisions.

It was also possible to evaluate the exposure coefficient, which is normally set to 1,0 in the Eurocode. Measurements revealed the following range of values: from 0,7 in windswept areas up to 1,1 in sheltered sites.

As initial consequences of these Research result, standard wise a proposal pursuant to fig 7 is under review.

Other shape coefficients improvements are under review.

6 CONCLUSIONS

The illustration of Research on snow loads and the normative study in progress are the best demonstration of how Research and Standardization can produce major interlinked result fostered in the case in question by the almost contemporaneousness of the conclusion of Research and of the study of the normative text and by the fact that the people committed on both fronts are mainly the same.

However, undoubtedly Research problems never end and each time some further contribution can only help to deepen knowledge, while normative problems often appear to be conditioned by external, legal, bureaucratic constraints and by a certain resistance to change.

However, the development of both sectors in undoubtedly continuous and in progress.

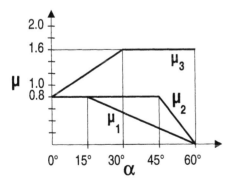

Figure 7 Shape coefficient for a duo-pitched roof.

2 Snow technology and science

Snow Engineering, Hjorth-Hansen, Holand, Løset & Norem (eds) © 2000 Balkema, Rotterdam, ISBN 90 5809 148 1

Program *Haefeli* – Two-dimensional numerical simulation of the creeping deformation and temperature distribution in a phase changing snowpack

Perry Bartelt & Marc Christen
Swiss Federal Institute for Snow and Avalanche Research, Davos Dorf, Switzerland

Stefan Wittwer
Swiss Federal Institute for Forest, Snow and Landscape Research, Birmensdorf, Switzerland

ABSTRACT: The two-dimensional finite element program *Haefeli* is presented. This program has been especially developed at our institute to solve heat transfer and creep problems in phase changing snowpacks. Snow is treated as a multi-component (ice, water, air) porous medium. The granular microstructure of the ice lattice is modeled using metamorphism algorithms that track grain growth and decay. The instationary heat conduction equation is solved to find the bulk temperature of the snowpack given a complex set of meteorological boundary conditions. Micro-plane material models have been introduced to find the time-dependent visco-elastic deformation of the snowpack. Constitutive laws for heat conductivity and snow viscosity are formulated using both continuum and micro-structure based models. Mass conserving subsurface melting and refreezing routines have been implemented. Two snowpack simulations are described and discussed: (1) a temperature drop imposed on a snow block is modeled using four different thermal conductivity models and (2) snow forces exerted on avalanche defense structures are calculated and the results are compared to the Swiss Guidelines.

1 INTRODUCTION

Snowpack modeling has become an integral part of snow and avalanche engineering. A one-dimensional snowpack model is now used on a day-to-day basis to predict the temperature change and mechanical deformation of the snowpack at over 50 locations within the Swiss Alps (Lehning et al. 1998, Lehning et al. 1999). The model calculations are used by avalanche warning experts to supplement measurements from a sophisticated system of automatic weather stations. The model predicts the layered structure, temperature distribution and settlement rate of the snowcover. Snowpack modeling is also important in long term avalanche prevention. The creep forces exerted on avalanche defence structures positioned in the starting zones of avalanches must be accurately determined. These calculations determine the spacing between the rows (Figure 1) and thus the financial cost of the protective measures.

The physical modeling of the snowpack is difficult. The snowpack consists of snow layers with different grain morphology. The size and shape of the grains change constantly due to inter-grain, intra-grain and inter-layer water vapour transport processes which are driven by temperature gradients within the snowpack. The temperature gradients, in turn, are driven by meteorological energy exchanges at the snowpack surface. Sudden changes in air temperature, wind speed, solar radiation and rainfall can change the state of the snowpack dramatically and rapidly. Phase changes at the snowpack surface and within the interior of the snowpack are commonplace. The movement of meltwater and meltwater refreezing involve both energy and mass transport processes.

However, for avalanche experts the temperature and mass-movements are not the primary parameters of interest, rather the state of stress, strain and defor-

Figure 1: Avalanche defense structures in starting zones prevent spontaneous avalanche release (Foto: Schiahorn, Switzerland).

mation rate which determine the stability of the snowpack. Snow is a visco-elastic material that undergoes large irreversible volumetric straining during it's lifetime. Triaxial tests with snow have clearly shown that snow viscosity is a strain-rate and temperature dependent non-Newtonian material again highly dependent on the microstructure (Bartelt & von Moos 2000).

A complete physical modeling of the snowpack therefore involves solving a system of highly instationary, nonlinear and coupled differential equations which impose mass, momentum and energy conservation at a *macro*-level. Since the mechanical (e.g. viscosity), thermal (e.g. conductivity), optical (e.g. albedo) and hydrological properties (e.g. matrix potential) are all a function of the snow morphology the governing system of equations would idealy employ constitutive laws based on snow *micro*-structure.

In this paper we will present a two-dimensional finite element model – *Haefeli* [1]-- which models the visco-elastic deformation and instationary heat transfer in snow using both empirical and micro-structure based constitutive laws. Two-dimensional finite element simulations have been used previously to determine the state of stress of mountain snowpacks (Smith & Sommerfeld 1971, Curtis & Smith 1974, Lang & Sommerfeld 1977, Bader et al. 1989). These finite element models, however, have treated snow as a small deformation, linear visco-elastic material. They have also ignored snow microstructure and the temperature dependency of snow. Simpler one-dimensional numerical models such as SNTHERM (Jordan 1991), CROCUS (Brun 1989) and DAISY (Bader 1992) have been applied to model the snowpack. These models, however, can only predict a one-dimensional state of stress.

2 SNOWPACK MODEL OF *HAEFELI*

The model domain is defined in a two-dimensional Lagrangian coordinate system $\vec{x} = (x, y)$. The primary problem unknowns are the temperature field $T(x, y, t)$ and the creep deformations in the x and y directions, $u(t)$ and $v(t)$. The deformations define the densification rate of the snowpack. The temperature and displacements are found by solving partial differential equations governing energy and momentum conservation. Because we are using a finite element model in which the displacements are continuous in the model domain, mass conservation is automatically fulfilled when solving the momentum equation. Mass conservation of the ice matrix does not have to be invoked separately.

---

[1] Prof. Haefeli was the first Swiss snow researcher to investigate snow creep, see literature list.

Snow is parameterized according to *both* macroscopic and microscopic parameters. Both parameter sets are used in *constitutive* laws. *Metamorphism* laws describe the changing properties of the ice matrix. These laws can only be implemented within the framework of a Lagrangian coordinate system where both the macroscopic and microscopic material properties can be tracked over time.

2.1 Macroscopic Parameters

Snow consists of ice, water and air. We will always assume that the air is fully saturated with water vapor. A volume V of snow therefore contains the sum of the component volumes V_ϕ:

$$V = \sum_\phi V_\phi . \tag{1}$$

Here ϕ is the component index and stands for ice, water and air. A forth component can also exists in permafrost problems: soil (Phillips et al. 2000). Each component ϕ occupies a part V_ϕ of the total volume V which is the volumetric content θ_ϕ of component ϕ:

$$\theta_\phi = \frac{V_\phi}{V} \tag{2}$$

The sum of the volumetric components is

$$\sum_\phi \theta_\phi = 1 \tag{3}$$

This requirement is fulfilled at all times during the calculations – for both phase change operations (exchange between the water and ice components) and densification (exchange between the air and ice components). The *total snow density* ρ is the fraction of the total element mass to the total element volume:

$$\rho = \frac{m}{V} = \sum_\phi \theta_\phi \rho_\phi \tag{4}$$

2.2 Microstructure Parameters

Microstructure parameters describe the granular properties of the ice matrix. They are divided into *primary* parameters r_g, r_b, S_g and D_g which are the grain radius, bond radius, grain sphericity (grain shape, rounded or edged) and grain dendricity (new or old snow). These parameters are termed primary because metamorphism laws have been experimentally formulated to describe the rates of growth and decay, $\dot{r}$, $\dot{r}_b$, $\dot{S}_g$ and $\dot{D}_g$.

The metamorphism laws are based on laboratory experiments (Marbouty 1980) and are function of temperature and temperature gradient. The ice matrix is also described by *secondary* parameters such as coordination number N_3, the total number of ice grains per unit volume n and the grain and bond lengths l_g and l_b. These parameters are termed secondary because they are calculated directly from the primary parameters.

2.3 Heat Transfer (Energy Conservation)

The heat transfer equation expresses the conservation of thermal energy and is given by

$$\rho c \frac{\partial T}{\partial t} + \nabla \left(k_{\text{eff}} \cdot \nabla T \right) = Q \tag{5}$$

where

ρ	total snow density according Eq. (4)
c	specific heat capacity of snow
$T = T(x, t)$	snow Temperature (function of location x and time t)
k_{eff}	thermal conductivity
Q	heat sinks and sources (see below).

This equation is solved for the unknown temperature $T(x,t)$ using a standard finite element method for heat transfer (Bathe 1982). Three-node triangular elements are employed to discretize the model domain. A fully implicit time integration method is used to solve the finite element system of equations. The temperature T is the bulk temperature, that is, all three components of the snow have the same temperature. No temperature differences exist; the heat exchanges between the ice, water and air phases occur rapidly in comparison to the heat conduction within the snowpack.

2.3.1 Effective Specific Heat Capacity
The specific heat capacity of a multi-component porous medium can be expressed as:

$$c = \frac{1}{m} \sum_{\phi} m_{\phi} c_{\phi} \tag{6}$$

The above equation can be rewritten in terms of the volumetric contents of the components:

$$c = \frac{1}{\rho} \sum_{\phi} \rho_{\phi} \theta_{\phi} c_{\phi} \tag{7}$$

Tab. 1 shows the constant settings of the specific heat capacities of Haefeli.

Table 1: Specific heat capacity of the ice, water and air components.

c_{ϕ}	Value [Jkg^{-1}K^{-1}]
c_{ice}	2100
c_{water}	4182
c_{air}	1005

2.3.2 Effective, Empirical and Microstructure-based Thermal Conductivity Laws
The effective thermal conductivity is defined using the thermal conductivities of the ice, water and air phases:

$$k_{\text{eff}} = \sum_{\phi} \theta_{\phi} k_{\phi} \tag{8}$$

Tab. 2 presents the values used by Haefeli:

Table 2: Thermal conductivities of ice, water and air.

k_{ϕ}	Value [Wm^{-1}K^{-1}]
k_{ice}	2.2
k_{water}	0.598
k_{air}	0.026

Several empirical thermal conductivity laws have also been implemented in Haefeli:

Table 3: Empirical thermal conductivity laws.

Author (Year)	Thermal Conductivity k [Wm^{-1}K^{-1}]
Jansson (1901)	$k_{\text{eff}} = 0.0209 + 0.000795\rho$ $+ 2.511e^{-12}\rho^4$
Devaux (1933)	$k_{\text{eff}} = 0.0293 + 2.93\left(\dfrac{\rho}{1000}\right)^2$
Sturm et al. (1997) (for $\rho < 156$ kgm^{-3})	$k_{\text{eff}} = 0.023 + 0.234\left(\dfrac{\rho}{1000}\right)$
Sturm et al. (1997) (for $\rho \geq 156$ kgm^{-3})	$k_{\text{eff}} = 0.138 - \left(\dfrac{1.01\rho}{1000}\right) + 3.233\left(\dfrac{\rho}{1000}\right)^2$

2.3.3 Microstructure-based Thermal Conductivity
The microstructural thermal conductivity law is based on the work of (Adams & Sato 1993), (Sato et al. 1994) and (Edens et al. 1997). It models the snow as a system of ice grains and pores in series and parallel with each other. Some of the heat conduction is trough the ice, where the effect of the constriction induced by bonds and necks is taken into account. Another portion of the heat conduction is due to the transfer of heat from one ice grain across a pore space and then into another ice grain. The third portion of the transfer is associated with heat transfer within a pore which is large enough that the upper and lower bounding ice grain can be ignored. The effective thermal conductivity is given by

$$k_{\text{eff}} = \frac{n_{ca}}{n_{cl}} \left[\frac{\pi^2 r_b k_i N_3}{32} + \frac{k_a k_{ap} A_{ip}}{l_{gs} k_{ap} + l_{ps} k_i} + \frac{k_a A_p}{l_p} \right], \tag{9}$$

15

where n_{cl} is the cube root of the number of grains per unit volume and n_{ca} is the square of n_{cl}. The lengths l_{gs}, l_{ps} and l_p are mean pore and grain lengths obtained from stereological analysis. A_p and A_{ps} are the mean cross-sectional areas of the pores and grains. k_{ap}, k_a and k_i are thermal conductivity of pore space due to combined direct conductivity and transport of latent heat, thermal conductivity of air and ice. The above equation implies that thermal conduction increases with density, bond size (r_b), temperature and coordination number (N_3).

2.3.4 Thermal Energy Sinks and Sources

The term Q denotes the energy rate at which the snowpack absorbs or loses energy; it is composed of three contributions:

$$Q(\bar{x},t) = q + S(\bar{x},t) + Q_L(\bar{x},t) \qquad (10)$$

$$q = q_s + q_l + q_{lw} \qquad (11)$$

where

q	power per unit area due to energy fluxes at the boundaries, primarily the snowpack surface.
S	power per unit volume generated by the absorbed shortwave intensity
Q_L	power per unit volume generated (> 0) or consumed (< 0) by phase changes.

The ground boundary of the snowpack is usually held at constant temperature (a Dirichlet boundary condition). The top surface boundary is open for the energy exchange with the atmosphere and the net exchange rate q is made up of convective and radiative heat exchanges (Neumann boundary conditions), which will be described in the following.

2.3.5 Sensible and Latent Heat

The sensible q_s and latent heat q_l exchanges are implemented in *Haefeli* according to (Bader & Weilenmann 1992):

$$q_s = \kappa_s(t)[T_A(t) - T_s(t)] \qquad (12)$$

with

$\kappa_s(t)$	convection coefficient; $\kappa_s(t)$ is a function of the wind speed, w.

and

$$q_L = \kappa_l(t)[rh \cdot e_A(T_A) - e_s(T_s)] \qquad (13)$$

where

$\kappa_l(t)$	exchange coefficient; $\kappa_l(t)$ is a function of the wind speed, w.
rh	relative humidity
e_A	vapor pressure air
e_s	vapor pressure snow.

The program thus requires at least three meteorological input parameters; the air temperature $T_A(t)$, the wind speed, $w(t)$ and the relative humidity $rh(t)$. The air space at the snowpack surface is assumed to be fully saturated at temperature T_S.

2.3.6 Net Longwave Radiation

The net longwave radiation intensity q_{lw} is given by the difference between the incoming longwave atmospheric and emitted snowcover radiation:

$$q_{lw} = \sigma[\varepsilon_A(t) \cdot T_A^4(t) - \varepsilon_s(t) \cdot T_s^4(t)] \qquad (14)$$

with

σ	Stefan-Boltzmann constant [$Wm^{-2}K^{-4}$]
ε_A	atmospheric emissivity, cloudcover
ε_s	snow emissivity = 1.0

Haefeli treats snow in the longwave range as pure black body ($\varepsilon_s = 1$).

2.3.7 Absorption of the Shortwave Radiation

The term S in Eq. (10) is the radiation power in the shortwave range absorbed by a snowpack element. The extinction coefficient is given by (Warren 1982):

$$\frac{dI(z)}{I(z)} = -Edz \qquad (15)$$

With $I(z)$ is the angle-integrated shortwave (solar) intensity left in the depth z below the surface. *Haefeli* integrates Eq. (15) over the triangular area of the finite mesh elements. The extinction coefficient E depends on the density and on the grain size of the snow (Mellor 1977). At present, E is a constant between $20 \ m^{-1} < E < 150 \ m^{-1}$.

2.4 Phase Change

Meltwater refreezing and subsurface melting are treated at the finite element level as volumetric heat sources and sinks, respectively, that correspond to changes in an element's volumetric ice and water contents. The procedure is mass conserving.

Subsurface melting is treated as a constraint on the temperature field: the temperature $T(x,t)$ can never be greater than $0°$ C. When the numerical heat transfer solution calculates temperatures which are greater than $T(x,t) > 0°$ C, the "excess" energy is used to melt the ice matrix and not raise the temperature of the snowpack to some unphysical value. Let ΔT be the difference between the calculated snow temperature T and the melting temperature $T_m=0°C$,

$$\Delta T = T - T_m .\qquad(16)$$

For melting to occur $\Delta T > 0° C$. The mass of meltwater Δm_w that can be produced is

$$\Delta m_w = \frac{cm_i\Delta T}{L_f}\qquad(17)$$

where c is the effective specific heat of snow; m_i is the ice mass and L_f is the latent heat of fusion. Since the volumetric ice and water contents are used throughout the computations to describe the macroscopic state of the snow matrix, we can also write that the change in volumetric water content is

$$\Delta\theta_w = \frac{c\theta_i\rho\,\Delta T}{L_f\rho_w} .\qquad(18)$$

Since mass is conserved at all times, $\Delta m_w = -\Delta m_i$. The change in the volumetric ice content is then

$$\Delta\theta_i = \frac{\rho_w\theta_w}{\rho_i} .\qquad(19)$$

The latent heat energy per unit volume required for the phase change is

$$Q_L = L_f\Delta\theta_i\rho_i .\qquad(20)$$

In the case of melting $\Delta\theta_i$ is negative and Q_L is consequently a heat sink. Meltwater refreezing is treated similarily to subsurface melting. For meltwater refreezing to occur two conditions are required. Firstly, meltwater is present ($\theta_w > 0$) and secondly the calculated temperature of the snow is below $0° C$.

2.5 Instationary Snow Creep

In our model for snow creep and fracture, the constitutive laws are defined on planes of various orientations within the material. (Bazant & Oh 1985) called these planes "microplanes" because the inelastic material behaviour of each plane is based on the mechanics of the material microstructure. He introduced the prefix "micro" in order to emphasize the fact that for many materials the mechanical behaviour, especially fracture and damage, is best characterized on weak planes that are found in the material microstructure, specifically the interaggregate contact planes which, for the case of snow, are the ice grain bonds and necks. The behaviour of the individual microplanes is described by simple one-dimensional rheological material laws relating the normal stresses and strains in a single direction. We

postulate that the mechanical behaviour and strength of the ice grain chains is more realistically and more simply described by the N-directional one-dimensional laws (Figure 2) than complicated tensorial constitutive laws with many parameters.

In a first approximation shear strains in the microplane directions are unopposed. However, this does not imply that the material as a whole has no shear stiffness: shear deformations are resisted by the normal stiffness of each microplane. The total material response is found by superimposing the contributions from all directions.

Viscous behaviour (creep) is introduced into the N-directional material model by assuming that the total strain in the n-th direction is composed of elastic and viscous parts, i.e.,

$$\varepsilon_n = \varepsilon_n^e + \varepsilon_n^c\qquad(21)$$

Also, the total volumetric strain is likewise decomposed into elastic and viscous parts of:

$$\gamma = \gamma^e + \gamma^c .\qquad(22)$$

The N-directional material law is given by

$$\sigma_n = \alpha\varepsilon_n^e + \beta\gamma^e\qquad(23)$$

where α and β are material constants with the values

$$\alpha = \frac{8\mu}{N} = \frac{4C}{N(1+\nu)}\qquad(24)$$

and

$$\beta = \frac{2\lambda-\mu}{N} = \frac{C}{N}\frac{(4\nu-1)}{(1+\nu)(1-2\nu)}\qquad(25)$$

The superscript e has been introduced to denote that the quantities ε_n and γ refer to the elastic N-directional and volumetric strains. The symbol C denotes the modulus of elasticity and ν Poisson's ratio.

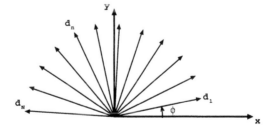

Figure 2: The direction vectors d_n for $n = 1$ to N. As a first approximation, we assume that snow is an isotropic material and that the direction vectors are equally distributed.

A simple Maxwell model is used to describe viscoelastic material behaviour. Of course, more complicated models are possible. The Maxwell model defines the viscous strain rate according to

$$\dot{\varepsilon}_n^c = \frac{\sigma_n}{\eta} . \qquad (26)$$

The parameter η is termed the snow viscosity (with dimensions [Pa s]). Usually, the viscosity is defined as a function of the snow density and temperature $\eta = f(\rho, T)$. Several of the different laws are listed in Table 4.

Table 4: Empirical viscosity laws, implemented in *Haefeli*. In the formulas T and ρ are to be substituted in [K] and [kgm^{-3}].

Author (Year)	Viscosity [Pa s]
Kojima (1974)	$\eta = 8.64(10^6)e^{0.021\rho}$
Mellor (1975)	$\eta = 5.0(10^7)e^{0.022\rho}$
Claus (1978)	$\eta = 6.57(10^7)e^{0.014\rho}$
Gubler (1994)	$\eta = 1.86(10^{-6})e^{0.02\rho + 8100/T}$
Morris (1994)	$\eta = 5.38(10^{-3})e^{0.024\rho + 6042/T}$
Loth & Graph (1993)	$\eta = 3.70(10^7)e^{8.10(10^{-2})(273.15-T)}e^{2.10(10^{-2})\rho}$

A detailed description of the two-dimensional plane strain finite element creep solution can be found in (Bartelt & Christen 1999).

3 SIMULATIONS

In this section two snowpack simulations are presented. In the first example we back-calculate the temperature response of a snow sample subjected to a sudden temperature drop; in the second example we compare the Swiss Guideline calculation procedures for creep forces acting on defense structures to finite element simulations. The guidelines treat snow as a Newtonian fluid; *Haefeli* treats snow as a highly non-linear and instationary creeping material.

3.1 *Example 1*

In March 1998 Dr. Martin Schneebeli from our Institute and Prof. Dr. Bob Brown, Montana State University, began an experiment to study snow metamorphism in the cold rooms of the Weissfluhjoch, Davos. They produced snow samples by sieving freshly deposited new snow into 120 x 50 x 24 cm bins (length x width x height). The bin was placed in a cold room on top of a heating element that kept the bottom of the sample at a constant temperature of -2°C. The cold room itself had a constant temperature of -11°C (temperature gradient 36°C/m). Thermocouples were inserted into the snow to record the temperature at different heights of the block throughout the experiment. Several other data was recorded, such as penetrometer measurements, grain-size distribution, density distribution and grain morphology. In this example, however, only the temperature measurements will be used for the calculations.

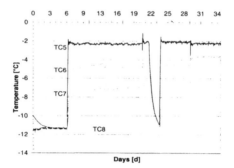

Figure 3: Temperature measurements of Schneebeli and Brown's snow block. Thermocouple (TC) positions from the top of sample: top (TC5), 8cm (TC6), 16cm (TC7), 24cm (TC8). The initial density of the snow was 175 [kgm^{-3}].

Figure 3 shows the temperature measurements at different heights in the block over the entire four week testing period. According to Schneebeli and Brown, the heating plate was accidently turned off between 31.3.98 13:40 and 2.4.98 13:40, causing the temperature of the sample to plunge. Afterwards the heating element was switched back on. This temperature history provides us with an ideal test to investigate different thermal conductivity laws.

In this example we tried to calculate the temperature drop (from -2°C to -10°C within one day), taking into account different thermal conductivity laws from literature (see Table 3).
According to the scientists, the heating plate was switched off shortly after a sample was removed from the test for a grain morphology analysis on 30.3.98. The initial density of the snow sample was 213 kgm^{-3} (top); 227 kgm^{-3} (middle) and 242 kgm^{-3} (bottom). The initial grain size of the new snow was 0.75 mm throughout the sample.

3.1.1 *Finite Element Simulations*
The two-dimensional finite element mesh consisted of 205 nodes and 351 elements. The mesh was refined to contain 760 nodes and 1404 elements. The temperature boundary conditions at the top and bottom of the mesh were prescribed to the temperature measurements (sensors TC5 and TC8).
Only the 3.5 day period of the temperature drop was simulated using time integration steps of 30 minutes (168 steps). The snow was assumed to be rigid. Only the instationary heat transfer solution was em-

ployed since the snowpack did not settle significantly during this period.

3.1.2 *Results*
The results of the simulations are depicted in Figures 4-7. A comparison is made between measured and simulated temperatures.

We note a very good agreement between the measurements and simulation results for the law of Jansson. The worst result is the one using the thermal conductivity law of Sturm. The results of Devaux and Adams are also very good, but not as good as Jansson's during the warming phase of the experiment, after the heat element has been turned back on.

3.1.3 *Conclusions*
All four constitutive formulations could be applied to back-calculate both the temperature drop and subsequent temperature rise after the heating element was turned back on. We believe that an exact simulation of temperature can only be achieved when the conductivity law is based on snow microstructure (grain and bond size, coordination number). Therefore, the Adam's conductivity law yields very good results especially because it is a physical model. The fact that the oldest empirical conductivity law of (Jansson 1901) provides better results than the newest conductivity law of (Sturm 1997) is due to the fact that for this particular experiment the snow microstructure in the sample was probably more similar to the microstructure of the tests of (Jansson 1901). It does not indicate a better constitutive formulation.

3.2 *Example 2*
During the extreme avalanche winter of 1999 defense structures in avalanche starting zones were responsible for protecting many communities from extreme avalanche damage. These structures must be strong enough to withstand the snow forces caused by the interruption of the natural glide and creep processes of the snowpack. Defense structures are dimensioned according to the Swiss Guidelines for Avalanche Control in Starting Zones.
The Guidelines apply analytical calculation procedures based on the work of (Haefeli 1942) and (Salm 1977).
The forces exerted on avalanche defense structures are calculated using *Haefeli* and the results are compared to the Swiss Guidelines. Both the horizontal S_N and vertical S_Q force components are calculated.

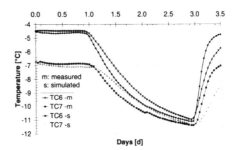

Figure 4: Simulation using Sturm's conductivity law.

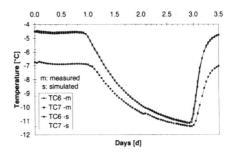

Figure 5: Simulation using Jansson's conductivity law.

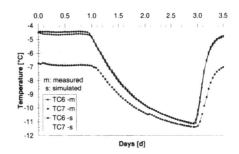

Figure 6: Simulation using Devaux's conductivity law.

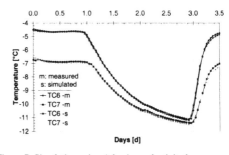

Figure 7: Simulation using Adam's conductivity law

19

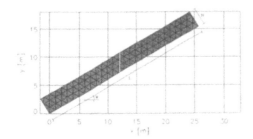

Figure 8: Unrefined finite element mesh depicting ψ, L and H.

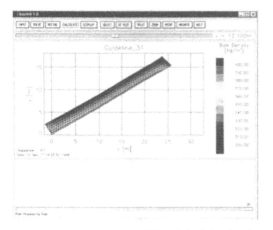

Figure 9: Deformed mesh after 20 days of simulation time (refined mesh).

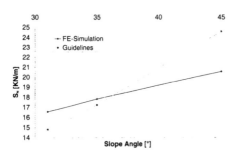

Figure 10: Snow force S_N parallel to the slope. The figure shows that the guidelines predict larger forces on steeper slopes.

A comparison between the finite element results and the guidelines is useful in order to delimit the application boundaries of the guidelines.

In the following simulations the viscous material model of (Mellor 1975) has been employed,

$$\eta = 5.0(10^7)e^{0.022\rho}.$$

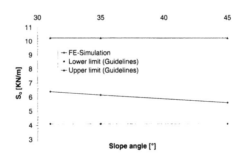

Figure 11: Snow force S_Q perpendicular to the slope. The figure shows that the finite element solution is between the upper and lower guideline limits for S_Q.

In our triaxial testing we have found that this law is valid for strain rates of the order $\dot{\varepsilon} \approx 10^{-8}$, which is the correct order of magnitude for natural snowpacks (Bartelt & von Moos, 2000). For higher strain rates it over-estimates the viscosity.

3.2.1 Finite Element Simulations

In the finite element simulations, three different slope angles were modeled, ψ=31°, ψ=35° and ψ=45°. The snow height was the same in all three cases, D=3m measured perpendicular to the slope. The snow density was ρ=300 kgm⁻³. The Swiss Guideline creep factor was K=0.76sinψ. The distance L between two rows of supporting structures was (according to the Swiss Guidelines): L=30.5m, L= 27.2m and L=21.2m, for each slope angle. Snow gliding at the ground was neglected.

For each slope angle, two calculations were made, one without horizontal friction at the supporting structure (no calculation of S_Q necessary) and one with a rough supporting structure (calculation of S_Q necessary).

Figure 8 shows Mesh No. 1 and depicts L, H and ψ. The unrefined finite element mesh consists of 110 nodes and 159 elements. The number of nodes/elements depends on the slope angle and the distance L between two rows of supporting structures.

A total of 20 days were simulated using a time step of 0.05 days. The temperature of the snow remained constant at T=-10°C. No instationary heat transfer solution was therefore performed. The initial input mesh was automatically refined, resulting in a finite element mesh with 378 nodes and 636 elements (shown in Figure 9). We modeled snow using only 10 directions (N=10). More directions did not improve the results.

3.2.2 Results

After 20 days, the snowpack crept down the slope as indicated in Figure 9. This figure depicts the density distribution after 20 days. Note that the density has increased considerably indicating large strains. Figures 10 and 11 compare the analytical guideline and simulation results for the snow force parallel to the slope, S_N, as well as for the perpendicular snow force, S_Q.

Figure 10 shows that the guideline-calculations predict larger forces on steeper slopes. At slope angles less than 36°, the finite element solution predicts higher S_N forces. Figure 11 shows that for S_Q the finite element solution is between the upper and lower guideline limits.

4 CONCLUSIONS

A highly non-linear, large strain, visco-elastic finite element model has been used to verify the Swiss Guidelines for Avalanche Control in Starting Zones. The finite element simulations reveal the following:

1. The simulation results are within the range of the Swiss Guidelines.
2. The simulation results show that the Swiss Guidelines predict higher creep forces on slopes steeper than 36°. On slopes less than 36°, the numerical calculations predict slightly higher forces.
3. The viscosity law of Mellor was applied in the simulations and provided realistic results. The newly developed micro-plane material model could thus be validated.

The finite element models are based on snowpacks of constant density and snow height. Both these assumptions can be dropped in order to back-calculate actual pressure measurements on defense structures. Snow layers with different densities could be modeled. Furthermore, newly developed micro-structure based, strain-rate dependent material models based on triaxial testing can then be used to simulate the behaviour of the snow layers.

It should be pointed out that it is not possible to specify a sliding friction law at the base of the snowpack that could be used to investigate the influence of snow gliding on the snowpack stress distribution. It is, however, possible to specify ground oscillations which would allow studying the influence of surface roughness on the simulation results.

Finally, the constitutive models do not include material failure. In the next research step, snow failure will be introduced into program *Haefeli*.

Further validations of the visco-elastic constitutive formulation based on field and laboratory experiments are presented (Bartelt & Christen 1999).

REFERENCES

Adams, E. E. & Sato, A. 1993. Model for effective conductivity of a dry snow cover composed of uniform ice spheres, *Annals of Glaciology*, 18, 300-304.

Bader, H.P., Gubler, H.U. & Salm, B. 1989. Distributions of stresses and strain-rates in snowpacks, *Proceedings of the Conference: Numerical Methods in Geomechanics*, edited by Swoboda, Innsbruck 1988.

Bader, H.P. & Weilenmann, P. 1992. Modeling temperature distribution, energy and mass flow in a (phase-changing) snowpack. I. Model and case studies, *Cold Regs. Sci. and Technol.* 20, 157-181.

Bartelt, P. & Christen, M. 1999. A Computational Procedure for Instationary Temperature-Dependent Snow Creep, *Advances in cold region thermal engineering and sciences: Technological, environmental and climatological impact.* Proceedings of the 6th international symposium held in Darmstadt, Germany, 22-25 August 1999. Kulumban Hutter, Y. Wang and H. Beer (eds.), (Lecture notes in physics; Vol. 533), 367-386.

Bartelt, P. & von Moos, M. 2000. Triaxial Creep Tests on Snow and a New Microstructure-Based Snow Viscosity Law, *In these Proceedings*.

Bathe, K. 1982. Finite Element Procedures in Engineering Analysis, *Prentice-Hall, Inc., Englewood Cliffs*, New Jersey, USA.

Bazant, Z. & Oh, B. 1985. Microplane Model for Progressive Fracture of Concrete and Rock, *ASCE Journal of Engineering Mechanics*, 111, No. 4, 559-582.

Brun, E., Martin, E., Simon, V., Gendre, C. & Coleou, C. 1989. An energy and mass model of snow cover suitable for operational avalanche forecasting, *J. Glaciol.* 35, 333-342.

Claus, B. 1978. Compactive viscosity of snow from settlement gauge measurements, *Interner Bericht Nr. 565, Eidgenössisches Institut für Schnee- und Lawinenforschung, Davos.*

Curtis, J. & Smith, F. 1974. Material property and boundary condition effects in avalanche snowpacks, *Journal of Glaciology*, 13, No. 67, 99-108.

Devaux, J. 1933. L'économie radio-thermique des champs de neige et des glaciers, *Ann. Chim. Phys.*, 20, 5-67.

Edens, M. Q., Brown, R. L. & Barber, M. 1997. Surface segmentation and measurements of surface sections of snow, *Cold Regions Science and Technology*,

Gubler, H.U. 1994. Physik von Schnee, *Interne Herausgebung, Eidgenössisches Institut für Schnee- und Lawinenforschung, Davos.*

Haefeli, R. 1942. Spannungs- und Platizitätserscheinungen in der Schneedecke. *Mitteilungen aus der Versuchsanstalt für Wasserbau an der Eidg. Technische Hochschule (Zürich)*, Nr. 2.

Jansson, M. 1901. Ueber die Wärmeleitfähigkeit des Schnees, *Ofversigt af Kongl. Vetenskaps-Akademiens Forhandlinger*, 58, 207-222.

Jordan, R. 1991. A one-dimensional temperature model for a snow cover, *Cold Regions Research and Engineering Laboratory Special Report*, 91-16, 49.

Koh, G. & Jordan, R. 1995. Sub-surface melting in a seasonal snow cover, *J. Glaciol.* 41, No. 139, 474-482.

Kojima, K. 1974. A field experiment on the rate of densification of natural snow layers under low stress, *IAHS-AISH Publication No. 114, Snow Mechanics Symposium, held at Grindelwald, Switzerland*, 298-308.

Lang, T. & Sommerfeld, R. 1977. The modeling and measurement of the deformation of a sloping snowpack, *Journal of Glaciology*, 19, No. 81, 153-163.

Lehning, M., Bartelt, P. & Brown, R. 1998. Operational use of a snowpack model for the avalanche warning service in Switzerland, *Proceedings of the NGI Anniversary Conference, Voss, Norway, 14-16 May 1998*, Oslo, Norges Geotekniske Institutt, NGI Report 203, 169-174.

Lehning, M., Bartelt, P., Brown, B., Russi, T., Stöckli, U. & Zimmerli, M. 1999. SNOWPACK model calculations for avalanche warning based upon a new network of weather and snow stations, *Cold Regions Science and Technology*, 30, 145-157.

Loth, B. & Graf, H.F. 1998. Snow Cover Model for Global Climate Simulations, *Journal of Geophysical Research*, 98, No. D6, 10451-10464.

Marbouty, D. 1980. An experimental study of temperature gradient-metamorphism, Journal of Glaciology, 26, No. 123, 303-312.

Mellor, M. 1975. A review of basic snow mechanics, *IAHS-AISH Publication No. 114, Snow Mechanics Symposium, held at Grindelwald, Switzerland*, 251-291.

Mellor, M. 1977. Engineering properties of snow, *J. Glaciol.* 19, No. 81, 15-66.

Morris, E.M. 1994. Modeling mass and energy exchange over polar snow using the DAISY model, *IAHS Publication*, No. 223, 53-60.

Phillips, M., Bartelt, P. & Christen, M. 2000. Anchor temperatures of avalanche defense snow supporting structures in frozen ground, *In these Proceedings*.

Salm, B. 1977. Snow Forces, *Journal of Glaciology*, 19, No. 81, 67-100.

Sato, A., Adams, E. E. & Brown, R. L. 1994. Effect of microstructure on heat and vapor transport in snow composed of uniform fine ice spheres, *Proceedings of Int. Snow Science Workshop*, 176-184

Smith, F. & Sommerfeld, R. 1971. Finite-element stress analysis of avalanche snowpacks, Short note, *Journal of Glaciology*, 10, No. 60, 401-405.

Sturm, M., Holmgren, J., König, M. & Morris, K. 1997. The thermal conductivity of seasonal snow, *Journal of Glaciology*, 43, No. 143, 26-41.

Warren, S.G. 1982. Optical properties of snow, *Reviews of Geophysics and Space Physics*, 20, No.1, 67-89.

Snow Engineering, Hjorth-Hansen, Holand, Løset & Norem (eds)© 2000 Balkema, Rotterdam, ISBN 90 5809 148 1

Triaxial creep tests on snow and a new microstructure-based snow viscosity law

Perry Bartelt
Swiss Federal Institute for Snow and Avalanche Research, Davos Dorf, Switzerland

Markus Von Moos
Institute of Geotechnical Engineering (IGT), Swiss Federal Institute of Technology, ETH-Hoenggerberg, Zürich, Switzerland

ABSTRACT: Snow is a highly non-linear visco-elastic material characterised by a complex microstructure. In this paper we present a constitutive model for snow that is based on the axial and inelastic micro-structural deformations of the ice lattice. We determine the average micro-structural stress in the grain necks and then apply a strain-rate and temperature dependent viscosity law of pure ice to describe the deformation of the solid ice lattice. The law is valid for strain-rates $\dot{\varepsilon}$ between 10^{-7}s^{-1} and 10^{-5}s^{-1}. The constitutive model is validated using mechanical tests performed with a specially designed deformation controlled triaxial apparatus. Stereological thin sections were taken of the experimental samples to determine the grain and bond size as well as co-ordination number. In a final step, we attempt to apply our micro-structural theory to determine a micro-structure based elasticity law, but without success. We conclude that the elastic and viscous behaviour of snow are governed by different micro-structural deformation processes.

1 INTRODUCTION

A constitutive law to describe the mechanical deformation of snow has two fundamental applications in snow and avalanche engineering:

- It is applied in numerical models to predict the deformation rate and stress state of multi-layered snowpacks. These models are now being applied operationally on a day-to-day basis to predict avalanche hazard in Switzeland (Lehning et. al. 1998), (Bartelt & Christen 2000).
- It is used to predict the creep forces that are exerted on avalanche defense structures. During the extreme avalanche winter of 98/99, these structures prevented the spontaneous release of avalanches and were primarily responsible for the safety of many mountain communities in Switzerland.

Both these important tasks require a constitutive theory for snow that is valid for large strains (volumetric strains of 100 percent) and strain rates between 10^{-8}s^{-1} and 10^{-3}s^{-1}. These rates of deformation occur in naturally deforming snowpacks -- as opposed to snowpacks subjected to skier or vehicle loadings, where the strain rates are much higher than 10^{-2}s^{-1}.

At the continuum level snow is classified as an isotropic, non-linear visco-elastic material. The snowpack is clearly anisotropic because it consists of layers with strongly varying densities and micro-structural properties. Previous triaxial tests with snow have validated the isotropy assumption for a single layer, at least for a thick, well-metamorphosed layer with rounded grains (Salm 1977).

For a visco-elastic material, the total state of strain, ε, is given by the sum of the elastic ε_e and viscous parts ε_v,

$$\varepsilon = \varepsilon_e + \varepsilon_v \qquad (1)$$

or, in rate form,

$$\dot{\varepsilon} = \dot{\varepsilon}_e + \dot{\varepsilon}_v . \qquad (2)$$

The viscous strain rate, $\dot{\varepsilon}_v$, is related to the applied stress by the viscosity, η,

$$\dot{\varepsilon}_v = \frac{\sigma_s}{\eta_s} . \qquad (3)$$

The parameter η_s is termed the snow viscosity (with dimensions Pa s). We have subscripted the stress σ_s and viscosity η_s with the letter s to denote that these quantities represent macroscopic or continuum snow values. Usually, the viscosity is defined as a function of the snow density and temperature $\eta_s = f(\rho, T)$. Several different viscosity laws are listed in the Table 1.

Table 1: Empirical viscosity laws. In the formulas T and ρ are to be substituted in [K] and [kgm^{-3}].

Author (Year)	Viscosity [Pa s]
Kojima (1974)	$\eta_s = 8.64(10^6)e^{0.021\rho}$
Mellor (1975)	$\eta_s = 5.0(10^7)e^{0.022\rho}$
Claus (1978)	$\eta_s = 6.57(10^7)e^{0.014\rho}$
Gubler (1994)	$\eta_s = 1.86(10^{-6})e^{0.02\rho + 8100/T}$
Morris (1994)	$\eta_s = 5.38(10^{-3})e^{0.024\rho + 6042/T}$
Loth (1993)	$\eta_s = 3.70(10^7)e^{8.10(10^{-2})(273.15-T)}e^{2.10(10^{-2})\rho}$

An analysis of the table shows the highly non-linear nature of snow: all the proposed laws are exponential. These laws have been based primarily on field observations. Our experience with them is that they are inadequate to model the snowpack in general, see (Lehning et. al. 1998) or (Bartelt & Christen, 1999).

Also note that for our applications, the irreversible viscous deformations are larger than the elastic deformations since we want to calculate the long-term creep deformations of the snowpack. Moreover,

$$\varepsilon_v \gg \varepsilon_e. \qquad (4)$$

The state of stress is given by the snow elasticity, E_s,

$$\sigma_s = E_s \varepsilon_e. \qquad (5)$$

The problem with this continuum analysis is that the state of macroscopic stress, σ_s, poorly describes the stress state of the necks or bonds which is responsible for the overall material behaviour of snow. Because of the complex interaction of the alpine snowpack with the environment, snow undergoes a variety of metamorphic processes such as grain growth and decay, sintering and bonding of grains

and melt/freeze processes. These processes occur essentially at the same time scale as the creep deformations and subsequently their influence on the mechanical behaviour cannot be excluded from a constitutive theory. Material failure is also based on the micro-structural stress state. Thus, the macroscopic state of stress and strain are not ideal parameters to formulate constitutive theories for snow.

Mahajan and Brown (1993) have attempted to relate the deformation of snow to the micro-structural properties of the ice-lattice. Using a model in which snow is modelled as a collection of spherical particles joined by areas of much smaller cross-section called bonds or necks, they proposed two basic micro-structural deformation mechanisms:

- **Inelastic Deformation of the Necks** : When the magnitude of the neck stress is less than 4kPa, then insignificant neck fracturing is assumed and inelastic straining of the necks is the primary deformation mechanism. Because the necks will not all have the same orientation with respect to the applied load it is expected that the neck stresses will vary: some necks will have higher stresses and thus fracture or undergo superplastic straining. However, the majority of the necks will be subjected to low stress and thus the inelastic deformation of the necks is the dominant mechanism.

- **Interparticle Sliding** : When the stress in the necks is large, the bonds fracture causing the grains to slide against each other. Sliding friction between the ice grains, not the straining of the ice bonds, is the primary mechanism of mechanical resistance. Mahajan and Brown (1993) postulated that "the first crack in a neck occurs when the lateral strain equals the strain for tensile fracture at the same instantaneous strain rate". The failure strain thus depends on the strain rate.

In order to apply the theory of Mahajan and Brown (1993) a problem must be solved: the state of macroscopic stress must be related to the stress in the bonds of the ice-lattice, σ_i. They proposed the following factor, which we have denoted β_η and termed the *neck stress factor for viscous deformation*,

$$\beta_\eta = \frac{N_3}{4\pi}\left(\frac{r_b}{r_g}\right)^2 \qquad (6)$$

In the above equation N_3 is the snow co-ordination number; r_b and r_g are the bond and grain radius. The parameter β_η relates the applied macroscopic snow stress, σ_s, to the stress state of the ice bonds or necks, σ_i, according to

$$\sigma_s = \beta_\eta \frac{\rho_s}{\rho_i} \sigma_i \qquad (7)$$

where ρ_i is the density of ice.

In this paper we experimentally show that the parameter β_η accurately predicts the state of stress in the bonds of the ice lattice. Our analysis is based on a series of experiments using a newly developed triaxial apparatus. In a follow-up step, we also show that it is possible to formulate a viscosity law for snow according to

$$\eta_s = \beta_\eta \frac{\rho_s}{\rho i} \eta_i = \frac{N_3}{4\pi} \left(\frac{r_b}{r_g} \right)^2 \frac{\rho_s}{\rho i} \eta_i \qquad (8)$$

where η_i is the viscosity of ice. We will limit ourselves to snow composed of well rounded ice grains. We will also show why the existing viscosity laws of Table 1 cannot be used to model snow.

The paper proceeds in four parts:

- First, we describe our newly developed triaxial apparatus that we have used to perform deformation controlled creep tests (applied strain rates $\dot{\varepsilon}$ between $10^{-7}s^{-1}$ and $10^{-5}s^{-1}$) on snow (densities between 180 kgm^{-3} and 440 kgm^{-3}) with well-rounded ice grains(grain radii between 0.25mm and 1.0mm).
- Second, we analyse the test results according to the micro-structure theory of Mahajan and Brown (1993) and show that their model accurately predicts the state of micro-structural stress in the ice lattice.
- Third, we formulate a one-dimensional micro-structure-based constitutive law for snow viscosity which assumes that the primary mechanism of snow deformation is the straining of the ice bonds or necks within the snow ice-lattice. For this law we apply a viscous strain-rate dependent power law for pure ice.
- Forth, if it is possible to accurately predict the state of micro-structural stress, then it should also be possible to define an elasticity law based on the deformation of the ice bonds. In the final part of this paper, we show why this is not possible.

2 THE TRIAXIAL APPARATUS

The primary components of the triaxial apparatus are depicted in Fig. 1.

In a triaxial test a snow cylinder with a height of 126 mm and diameter of 58 mm is placed between two loading plates. A lateral pressure and axial deformation is applied to the sample. The probe size was chosen such that it's volume is exactly one-third of a litre and the height to width ratio is larger than 2 to 1. This ensures a homogeneous stress state in the sample. The diameter defines the smallest snow-pack layer that can be tested, approximately 6 cm.

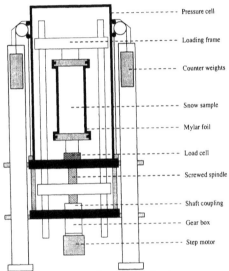

Figure 1: The mechanical system.

The cylinder is enclosed in a 12 μm thick gas-tight Mylar foil, which is glued at the top and bottom of the sample to two PVC rings. This guarantees the separation between the sample and the applied lateral pressure. The PVC rings are attached to the loading plates when the sample is installed in the testing device.

The deformation-controlled testing apparatus is driven by a step motor which propels a geared screwed spindle over a shaft coupling. The loading frame can be moved up or down at different axial strain rates between $\dot{\varepsilon}$ between $2(10^{-7})s^{-1}$ and $2(10^{-5})s^{-1}$. One step of the motor is equivalent to 1 μm feed. The loading frame and the load cell are both inside the pressure cell in order to avoid measuring any friction in the loading system. Presently the load cell can measure up to 2kN (750 kPa); however, the cell can be easily be exchanged if higher or lower axial pressures are to be measured.

The triaxial cell was specifically designed to test the snow samples in deformation-controlled mode -- as opposed to load-controlled mode -- since constitutive models are based on determining a state of stress (which we measure) from a state of strain (which we apply). The relative displacement between the top and bottom of the loading plates is recorded and defines the axial deformation of the sample.

The triaxial apparatus contains two air systems: the pore air system within the probe and the lateral pressure system. Both are entirely separate, see Fig. 2.

The lateral pressure is produced by an air-pump that delivers a maximum of 30kPa. The lateral pressure is with air because a fluid with a sub-zero freezing point would produce a hydrostatic pressure gradient from the top to the bottom of the sample, thus creating a non-homogeneous stress state in the probe.

To control the air pressure in the two systems, three pressure transducers are used, see Fig. 2. The first measures the difference between the lateral pressure, p_l, and the pressure in the pore air, p_p. The second transducer measures the difference between the lateral pressure and the atmospheric pressure, p_a. The third transducer measures the absolute pore air pressure, p_p. The pore air system is closed before the experiment begins. Subsequently, at the beginning of the experiment $p_p = p_a$. The atmospheric pressure can of course change during the duration of the experiment.

The volume change over time is measured by capturing the displaced pore air when the sample is loaded. Each loading plate contains a filter which allows the pore air to escape freely from the sample. When the probe is deformed, air flows out an pushes a drop of antifreeze to the left or right of two light barriers. At the beginning of the experiment, the drop is positioned such that both barriers are interrupted. As soon as a barrier is free, a piston is activated to bring the drop back to it's original position. The position of the piston, which is driven by a step motor, corresponds to the volume change of the sample.

Finally, the temperature inside the pressure cell is recorded using three temperature sensors: one in the top load plate; one in the bottom load plate and one in the pressure cell in the air that surrounds the sample.

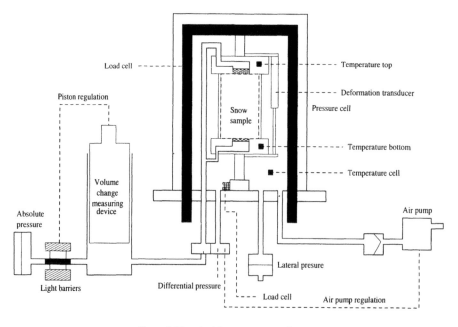

Figure 2: The triaxial apparatus control system.

3 EXPERIMENTAL DETERMINATION OF THE MICRO-STRUCTURAL STRESS STATE

In order to determine the state of micro-structural stress, we first define the dimensionless parameter α_η:

$$\alpha_\eta = \frac{\sigma_s}{\sigma_i}\frac{\rho_i}{\rho_s} \qquad (9)$$

where σ_s is the applied snow stress; ρ_s and ρ_i are the density of snow and ice, respectively and σ_i is the ice stress which we define, in general, according to any constitutive law for ice, for example,

$$\sigma_i = \eta_i \dot{\varepsilon}_i \qquad (10)$$

In our present analysis we will use the viscosity law of Barnes, see (Sanderson 1988),

$$\dot{\varepsilon}_i = A_i \exp\left(\frac{Q_i}{RT}\right)\sigma_i^3 \qquad (11)$$

with the constants $A_i = 4.1(10)^{-10}$ Pa^{-3}s^{-1} and $Q_i = 78$ kJmol^{-1} for T < -8° C and $A_i = 7.8(10)^{-2}$ Pa^{-3}s^{-1} and $Q_i = 120$ kJmol^{-1} for T > -8° C. R is the universal gas constant. The ice viscosity is thus,

$$\eta_i = \left(A_i \exp\left(\frac{Q_i}{RT}\right)\dot{\varepsilon}_i^2\right)^{-\frac{1}{3}} \qquad (12)$$

Physically, the parameter α_η indicates how much ice is resisting viscous deformation. If $\alpha_\eta = 1$ then all available mass with the ice lattice is being stressed and contributing to resistance. For this reason we have termed α_η the *lattice viscous-load carrying factor*.

Assuming that the strain rate of the ice lattice, $\dot{\varepsilon}_i$, is the same as the applied strain rate, $\dot{\varepsilon}_s$, then α_η can be determined for every experiment, since σ_s, $\dot{\varepsilon}_i$ and ρ_s are known. The results are shown in Fig. 3.

The results show that $0.02 < \alpha_\eta < 0.20$. In other words, in our experiments from 2 to 20 percent of the total ice in the crystal lattice contributes to the mechanical resistance. Moreover, for snow with higher densities, the applied force is being carried by 1/5 of the total available ice mass; for lower density snow, the applied force is being carried by only 1/50 of the total available ice mass. This implies that the stress on the ice bonds must between

5 and 50 times higher than the applied macro-stress.

In the next step in our analysis we made a direct comparison between the *experimental* α_η and the *theoretical* β_η values. We have first fitted the experimental results with the exponential curve $\alpha_\eta = 0.0028 \exp(.0080 \rho_s)$. A plot containing the fitted α_η curve superimposed with β_η values for different r_b/r_g ratios is shown in Fig. 4. We observe a very good agreement between α_η and β_η. The reason why α_η increases with snow density is that the r_b/r_g ratios are different for every experiment and increase in general with snow density. It reflects the fact that the snow samples have some metamorphic history and different micro-structural states. Note also that the experimental scatter increases with density because higher density snow has a larger micro-structural scatter.

We measured the micro-structural parameters (r_b, r_g N_2) using stereological thin sections and thus could compare the experimental β_η values directly with α_η, see Fig. 5. A typical thin section of one of our snow samples is shown in Fig. 6. Note that three-dimensional co-ordination number N_3 was determined from the two-dimensional co-ordination number N_2 according to the theory of (Alley 1986),

$$N_3 = \frac{\pi}{8}\frac{r_b}{r_g}\left[1+\left(1+\left(\frac{r_b}{r_g}\right)^2\right)^{\frac{1}{2}}\right]N_2 \qquad (13)$$

The results show that the parameter β_η is an effective way to approximate the neck stress, or, the governing stress in the ice-lattice, especially for snow densities $\rho_s < 350$ kgm^{-3}.

4. A MICRO-STRUCTURE BASED VISCOSITY LAW

In the preceding section we showed that

$$\beta_\eta \approx \alpha_\eta = \frac{\sigma_s \rho_i}{\sigma_i \rho_s} \qquad (14)$$

Substituting $\sigma_s = \eta_s \dot{\varepsilon}_s$ and $\sigma_i = \eta_i \dot{\varepsilon}_i$ into the above equation and rearranging to find the snow viscosity, we find,

$$\eta_s = \alpha_\eta \frac{\rho_s}{\rho_i}\eta_i = \alpha_\eta \frac{\rho_s}{\rho_i}\left(A_i \exp\left(\frac{Q_i}{RT}\right)\dot{\varepsilon}_s^2\right)^{-\frac{1}{3}} \qquad (15)$$

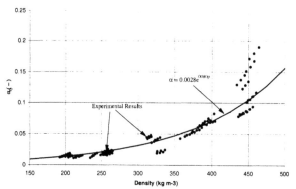

Figure 3: The parameter α_η indicates which percentage of the ice lattice is resisting deformation. The thick curved line is an exponential fit to the test results: $\alpha_\eta = 0.0028\exp(.0080\,\rho_s)$. The experimental results indicate α_η lies between 2 and 20 percent.

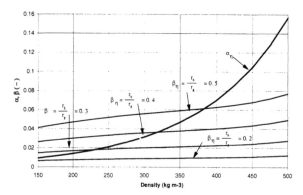

Figure 4: A comparison between α_η and β_η. The straight lines are the theoretical predictions of Mahajan and Brown for constant r_b/r_g ratios. The lower line $r_b/r_g =0.2$; upper line $r_b/r_g =0.5$. Note the good agreement between theory and experiment for low density snow.

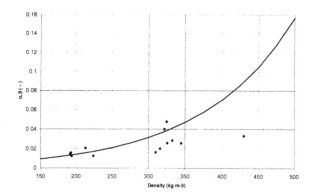

Figure 5: A comparison between α_η (curve) and β_η (solid dots). The β_η values were determined using stereological thin sections, see Fig. 6.

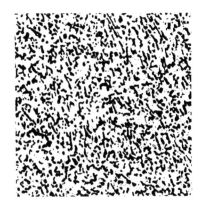

Figure 6: A stereological thin section of a snow sample. Snow density ρ_s =320 kgm^{-3}, r_g=0.094mm, r_b=0.047mm, N_3 =2.34, β_η =0.04655.

In Fig. 7 the constitutive model results are superimposed on the experimental results. The agreement between the experiments and Eq. 15 is excellent, especially with respect to the strain rate dependency of the results. The results are based on three important assumptions:

- We assume *axial* straining in the ice bonds. That is, we assume that the macro- and microstructural stress states are defined according to the compactive viscosities and strain-rates.
- We assume that the strain rate in the bonds, $\dot{\varepsilon}_i$, is equal to the macroscopic strain rate $\dot{\varepsilon}_s$. Thus, we assume that the state of strain is homogeneous over the snow sample volume.
- In order to fit the experimental results we have determined an α_η= 0.0028exp(.0080 ρ_s) based on

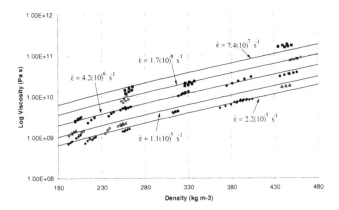

Figure 7: Comparison between experimentally determined viscosity and constitutive model. Note the good agreement for all strain rates. Experimentally applied strain-rates: diamonds $\dot{\varepsilon}$ = 2.2(10)$^{-5}$s^{-1}; triangles $\dot{\varepsilon}$ = 1.1(10)$^{-5}$s^{-1}; circles $\dot{\varepsilon}$ = 4.2(10)$^{-6}$s^{-1}; crosses $\dot{\varepsilon}$ = 1.7(10)$^{-6}$ s^{-1}; squares $\dot{\varepsilon}$ = 7.4(10)$^{-7}$ s^{-1}.

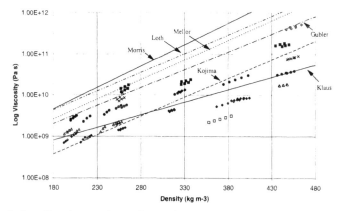

Figure 8: Existing viscosity laws (lines) in comparison to the experimental results (dots). The laws are defined in Table 1. Note the inability of all the laws to predict the correct strain rate.

an average of all experiments. However, in the above analysis we have shown that $\alpha_\eta \approx \beta_\eta$. Therefore, we rewrite the viscosity law as

$$\eta_s = \frac{N_3}{4\pi} \left(\frac{r_b}{r_g} \right)^2 \frac{\rho_s}{\rho_i} \left(A_i \exp\left(\frac{Q_i}{RT} \right) \dot{\varepsilon}_s^2 \right)^{-\frac{1}{3}} \quad (16)$$

- We are applying the viscosity law of Barnes. Other constitutive models for ice could be applied. Furthermore, all the experiments were performed at -12° C and therefore the temperature dependency of our constitutive model could not be validated. The temperature dependency is treated only in the η_i law, meaning that β_η is a temperature-independent factor. In future tests at different temperature -- with snow samples of similar micro-structure -- can be used to validate the constitutive model. For example, the calculated α_η values depicted in Fig. 3 should not change when the temperature of the tests change.

6 ANALYSIS OF EXISTING VISCOSITY LAWS

A comparison between existing viscosity laws and our experimental results is shown in Fig. 8. The viscosity laws are listed in Table 1.

We note that the viscosity laws of Morris, Loth and Mellor all over-predict the viscosity for the considered density range. We conclude that their formulations are valid for either (1) very low strain rates (say $10^{-8}s^{-1}$ or $10^{-9}s^{-1}$), (2) much colder temperatures or (3) snow with non-rounded grains. These laws also appear to mix several different strain rates since the slope(log viscosity / density) of the curves does not match the experimental results. In fact, an interesting feature of the experimental results is that the slope of the experimental results for a particular strain rate are all similar. None of the existing viscosity laws matches this experimental result well.

The laws of Kojima and Claus are based on field experiments with new snow. It is therefore hardly surprising that their formulations match the experimental results for low density snow at high strain rates.

7 SNOW ELASTICITY

Our constitutive formulation for snow viscosity is based on the idea that the axial deformation of the ice-chains is responsible for the macroscopic mate-

rial behaviour. Theoretically, it should thus be possible to formulate a similar model for the snow elasticity:

$$E_s = \alpha_e \frac{\rho_s}{\rho_i} E_i = \beta_e \frac{\rho_s}{\rho_i} E_i \quad (17)$$

where α_e and β_e are the experimental lattice load carrying factor and the micro-structural factor for elastic deformation, respectively. We determined the E_s from our experimental tests; our results are depicted as a function of snow density in Fig 9. Knowing that the elasticity of ice is (Sanderson 1988)

$$E_i = 9.5 \text{ GPa} \quad (18)$$

it is possible to determine α_e. These results are shown in Fig. 10.

A comparison with Fig. 3 (α_η) reveals that

$$\alpha_\eta \gg \alpha_e \quad (19)$$

We emphasise that this result is based purely on an analysis of the experimental results. No material model has been employed. Both Fig. 3 and Fig. 10 show that the results scatter with increasing density. This is expected since the micro-structural parameters r_g, r_b and N_3 will vary more with increasing density. This result again reflects the fact that the snow samples have some metamorphic history.
From this result we conclude that the elastic deformation of snow is governed by a different micro-structural deformation process, perhaps, for example, the bending and not the axial deformation of the ice-chains.

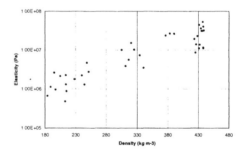

Figure 9: Experimentally determined snow elasticity values, E_s. The results showed no strain-rate dependency, similar to pure ice.

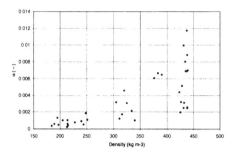

Figure 10: Experimentally determined values for the parameter α_e

CONCLUSIONS

In this paper we have formulated a snow viscosity law which is based on three factors:

- β_η: a snow microstructure factor that relates the macroscopic and microscopic states of stress. For well rounded snow, we have found that $0.02 < \beta_\eta < 0.2$, implying that the stress in the ice bonds is between 5 and 50 times higher than the applied macroscopic stress.
- ρ_s/ρ_i: a factor which accounts for snow density.
- η_i: a temperature and strain rate dependent viscosity law.

The parameter β_η has been derived by assuming that the axial and inelastic deformation of the solid ice chains determines the micro-structural stress state.

The proposed model has been validated by a series of triaxial tests that have been performed at different strain-rates. Finally, we attempt to apply the same theory to find the micro-structural based snow elasticity law. We have shown that $\alpha_\eta \gg \alpha_e$. This is an important result since it shows that perhaps the elastic and viscous response of snow are governed by different micro-structural deformation processes.

REFERENCES

Alley, R. 1986, Three dimensional coordination number from two-dimensional measurements: A new method,, *Journal of Glaciology*, 32, No. 112.

Bartelt, P. & Christen, M. 1999, A Computational Procedure for Instationary Temperature-Dependent Snow Creep, *Advances in cold region thermal engineering and sciences: Technological, environmental and climatological impact.* Proceedings of the 6th international symposium held in Darmstadt, Germany, 22-25 August 1999. Kulumban Hutter, Y. Wang and H. Beer (eds.), (Lecture notes in physics; Vol. 533), 367-386.

Bartelt, P. & Christen, M. 2000. Program *Haefeli* – Two-Dimensional Numerical Simulation of the Creeping Deformation and Temperature Distribution in a Phase Changing Snowpack, *In these Proceedings.*

Claus, B. 1978. Compactive viscosity of snow from settlement gauge measurements, *Interner Bericht Nr. 565, Eidgenössisches Institut für Schnee- und Lawinenforschung, Davos.*

Gubler, H.U. 1994. Physik von Schnee, *Interne Herausgebung, Eidgenössisches Institut für Schnee- und Lawinenforschung, Davos.*

Kojima, K. 1974. A field experiment on the rate of densification of natural snow layers under low stress, *IAHS-AISH Publication No. 114, Snow Mechanics Symposium, held at Grindelwald, Switzerland*, 298-308.

Lehning, M., Bartelt, P. and Brown, R. 1998. Operational Use of a Snowpack Model for the Avalanche Warning Service in Switzerland: Model Development and First Experiences, *In Proceedings of the NGI Anniversary Conference, Voss, Norway, 14--16 May 1998.* Oslo, Norges Geotekniske Institutt, 169-174. (NGI Report 203.)

Loth, B. & Graf, H.F. 1998, Snow Cover Model for Global Climate Simulations, *Journal of Geophysical Research*, 98, No. D6, 10451-10464.

Mahajan, P. & Brown R. 1993. A microstructure-based constitutive law for snow, *Annals of Glaciology*, Vol. 18, 287-294.

Mellor, M. 1975. A review of basic snow mechanics, *IAHS-AISH Publication No. 114, Snow Mechanics Symposium, held at Grindelwald, Switzerland*, 251-291.

Morris, E.M. 1994. Modeling mass and energy exchange over polar snow using the DAISY model, *IAHS Publication*, No. 223, 53-60.

Salm, B. 1977. Snow Forces, *Journal of Glaciology*, 19, No. 81, 67-100.

Sanderson, T. 1988. *Ice Mechanics*, London, Graham and Trotman Limited, Publishers.

Snow Engineering, Hjorth-Hansen, Holand, Løset & Norem (eds) © 2000 Balkema, Rotterdam, ISBN 90 5809 148 1

Detection and analysis of avalanches using wave measurements

Bjarni Bessason, Gunnar I. Baldvinsson & Ódinn Thórarinsson
Earthquake Engineering Research Centre, Engineering Research Institute, University of Iceland, Iceland

Gísli Eiríksson
Public Roads Administration, Ísafördur, Iceland

ABSTRACT: The road along the Óshlíd-hillside is one of the most hazardous roads in Iceland due to rock-fall, debris flows and avalanches. There are 23 known avalanche tracks in the hillside. Below four of the tracks with the highest avalanche frequencies concrete avalanche sheds have been constructed to protect the road. When there is an avalanche threat in the Óshlíd-hillside often many avalanches fall from different tracks in a short time and therefore it is likely that avalanches will hit the concrete sheds. Based on this knowledge a research project was initiated in February 1996 with the objective of developing a system to detect and analyse avalanches that hit one of these sheds and to send a warning to a control station. Later the project was extended and two more tracks with no avalanche sheds were included. This paper describes the system and presents the first results of the research project.

1 INTRODUCTION

1.1 Background

Vestfirdir in Northwest Iceland (see figure 1) are known as an avalanche prone area. The landscape consists of mountains and fjords with limited amount of flat land. In 1995 two major snow avalanches in the area struck the small villages, Súdavík and Flateyri, and caused a number of deaths and injuries.

Road construction is in general quite difficult in the area and many roads are exposed to avalanches. The 6 km long road section along the Óshlíd-hillside between Ísafjördur and Bolungarvík (see figure 1 and figure 2) is one of the most hazardous roads in Iceland due to rock-fall, debris flows and avalanches. At the top of the approximately 700 m high hillside there are powerful scars. Below there are steep slopes, consisting of rock, gravel, sand and soil, which extend all the way down to sea level. The road along the hillside is at a low height or about 20 to 50 m above sea level. The road was first opened as a track in 1949, rebuilt in the seventies and again in the early eighties.

Observations carried out by the Icelandic Public Roads Administration have shown that snow avalanches are frequent in 23 tracks or paths in Óshlíd, see figure 2. In an 11 years period from 1976 to 1986 a total of 948 snow avalanches that caused road closure were registered in these paths.

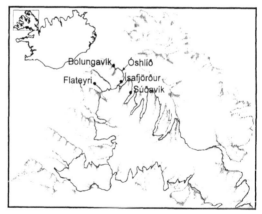

Figure 1. Vestfirdir in Northwest Iceland

The distribution of the avalanches is shown in figure 3. As can be seen from the figure four of the paths are more active than the others and there concrete avalanche sheds have been constructed to protect the road. The first shed was built in path 15 (Steinsófæra) in 1986 and the last one in 1994. When there is an avalanche threat in the Óshlíd-hillside often many avalanches fall from different paths in a short time and therefore it is likely that avalanches will hit the concrete sheds. Therefore by detecting avalanches that hit these sheds it is possible to send out an avalanche warning for the road

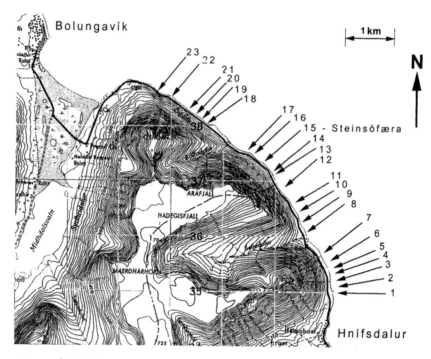

Figure 2. The road along the Óshlíð-hillside showing 23 known avalanche paths.

and the surrounding area.

When avalanches fall they cause wave motion on the surface of the earth. The basic idea is to utilise this motion to detect the avalanches.

1.2 Objective

The main objective of the research project, presented in this paper, is to develop a system to continuously monitor a given avalanche path in order to be able to automatically detect when an avalanche has fallen. Further the extent of the avalanche will be estimated and an alert signal transmitted to a control station where decisions regarding actions will be made.

1.3 Methodology

To accomplish the objective above a measurement system using a similar approach as used when measuring seismic signals was developed. Further it was necessary to formulate a processing method to analyse and assess the resulting seismic data. When developing the processing method it is essential to support the measured data with visual inspections at the site to identify the type and extent of the measured seismic events. Finally, the measured events have to be correlated with meteorological data.

1.4 Benefits

The benefits of an automatic detection and alerting system that estimates the type and extent of a recorded event can be summarised as follows:

- It assists in traffic control during bad weather on mountain roads or other roads were avalanches are frequent and where it may be necessary to close roads for safety reasons. This can be implemented either by sending alert signals to a control station or directly alerting traffic by traffic lights or sound signals.
- By installing a network of detection systems at selected sites which are known to be active with respect to avalanches it may be possible to alert inhabitants in the area when avalanches start to fall. Sites with different avalanche sensitivity with respect to weather conditions (wind- and precipitation direction) should be selected.
- The detection system can be used for avalanche research. With the system it is possible to time exactly when avalanches fall. By studying weather data and correlating it to the time of avalanche releases it may be easier to understand when an avalanche threat develops.

Several systems for recording rock-fall and avalanches that are in motion have been tested. In Norway and also earlier in Óshlíð geophones have been used to measure ground vibrations. If peak value ex-

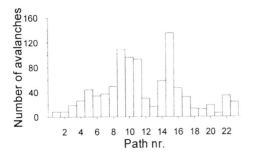

Figure 3. Distribution of avalanches in paths 1 to 23 in Óshlíð in the period 1976-1986, causing closure of the road.

ceeded a predefined level traffic was alerted by light and sound signals (see Norwegian Public Roads Administration, 1994). There were problems in choosing a reasonable sensitivity level for the geophones, to avoid too many false warnings.

A literature survey only showed a few papers describing systems where wave motion is used to detect avalanches. Leprettre et al. (1996) (the year this project started) shows first results from a similar system but in a different environment.

2 MEASUREMENT SYSTEM

2.1 *Measurement system #1*

The first measurement system was installed in a 90 m long avalanche concrete shed at path 15 (Steinsófæra) in Óshlíð in February 1996. A three-axial FBA-23 accelerometer from *Kinemetrics Inc.* was located at the top of the inner wall of the shed, i.e. at the hillside side. The accelerometer was connected to a SSA-1 data acquisition system (DAQ) also from *Kinemetrics Inc.* The SSA-1 had a predefined threshold and only recorded events when the motion exceeded this threshold. The sampling frequency was 200 Hz and the measurement range was adjusted so that it was from 0.001 to 2.5 m/s². Each event had a 10 s preevent and 10 s postevent. The SSA-1 was connected to a modem and measured events could be transferred through the telephone system from Óshlíð to Reykjavik where the system was observed from.

2.2 *Measurement system #2*

In November 1998 a new system was installed in Óshlíð. The SSA-1 was replaced by a K2 data acquisition system from *Kinemetrics Inc.* The K2 is capable of automatically calling a central station in Reykjavik when an event has been recorded. Secondly, it is possible to use 6 or even 12 channels in the K2 compared to only 4 in the SSA-1. Finally, the resolution is higher, 19 bits, in the K2 compared to

12 bits in the SSA-1. When changing the data acquisition system the three-axial accelerometer was replaced by an uni-axial FBA-11 accelerometer from *Kinemetrics Inc.* The new accelerometer was located at the same place as before in the shed and measures in a horizontal direction perpendicular to the wall. Further to obtain experience with sensors located on rock outcrop in avalanche tracks, it was decided to add tracks 14 and 16 to the system. A second uniaxial FBA-11 accelerometer was therefore installed in path 14, approximately 120 m south of the avalanche shed in Steinsófæra. The accelerometer was bolted on rock and shielded by a steel box. The third uniaxial accelerometer was then installed in path 16 about 210 m north of the avalanche shed. Also beside the accelerometer in path 16 a SM-6 geophone (natural frequency 4.5 Hz) from *Sensor Netherlands* was located. This is a low-cost geophone, which is much cheaper than the accelerometers. A total of four channels are used in measurement system #2. The measurement range in this new system is 0.04 mm/s² to 9.8 m/s² for the accelerometers and 0.33 µm/s to 86.6 mm/s for the geophone.

2.3 *Measurement system #3*

In August 1999 two geophones were added to the system one beside the accelerometer in path 14 and one beside the accelerometer in path 15.

3 DATA PROCESSING - PARAMETERS

For each recorded event characteristic parameters are determined. By studying these parameters and when possible correlating them with visual inspections data for events that have been confirmed and estimated by field inspection it is possible by means of statistics and experience to use them to identify and assess events.

Prior to determining the characteristic parameters raw data have to be corrected for instrument bias. The data is also filtered with a 50 Hz cut-off frequency low-pass filter and with a 2 Hz cut-off frequency high-pass filter. The characteristic parameters are determined both from time series data and power spectrum data.

3.1 *Peak acceleration and peak velocity*

Peak acceleration and peak velocity are the peak values of measured acceleration or velocity data. Alternatively peak velocity can be found by numerical integration of the acceleration signal and correspondingly peak acceleration can be found from the velocity signal by numerical differentiation.

3.2 Characteristic frequency, f_{char}

Characteristic frequency, f_{char}, is determined from the power spectral density, $S(f)$, of the acceleration time series as:

$$f_{char} = \sqrt{\frac{\int_0^\infty f^2 S(f)\, df}{\int_0^\infty S(f)\, df}} \qquad (1)$$

The characteristic frequency is an indication of the dominating frequency of the power spectrum.

3.3 Half-power bandwidth, B_e

Half-power bandwidth, B_e, of the power spectrum is defined as the width of the frequency band which contains half of the power in the spectrum. It is based on the following equations:

$$I_0 = \int_0^\infty S(f)\, df \qquad (2)$$

$$\int_0^{f_1} S(f)\, df = 0.25 \cdot I_0 \qquad (3)$$

$$\int_0^{f_2} S(f)\, df = 0.75 \cdot I_0 \qquad (4)$$

$$B_e = f_2 - f_1 \qquad (5)$$

3.4 Duration, T_d

Duration, T_d is a measure of the time length of an event. Duration can be defined in many ways, see for instance Vanmarcke (1980). Here it is defined as the sum of 0.2 s long time segments, t_i, in the time series where standard deviation is greater than two times the standard deviation of background vibration noise at the respective site:

$$T_d = t_1 + \ldots + t_i + \ldots + t_N \qquad (6)$$

The standard deviation of background noise is determined from a one second long segment at the beginning of the 10 second long preevent segment. For a signal that contains a lot of pulses there may be quiet parts in between that do not contribute to the duration.

3.5 Energy, E

Energy, E, is found by integrating the squared signal of all segments that contribute to the duration of each event, that is:

$$E = \int_{t_1} a^2(t)dt + \ldots + \int_{t_i} a^2(t)dt + \ldots + \int_{t_N} a^2(t)dt \qquad (7)$$

In many cases there is good correlation between the peak value of a time series and its energy. It is however clear that transient and stationary processes may have similar peak values although the energy is quite different.

3.6 Peak factor, S_{factor}

Peak factor, S_{factor}, is defined as the ratio of the peak value to the root-mean-square value of the signal. High peak factor indicates a transient for instance caused by rock-fall while a lower value indicates more continuos flow.

4 SUMMARY OF MEASUREMENTS

Monitoring at the site started in February 1996. Most of the acquired data were recorded by measurement system #1 and covers 2½ year. Measurement system #2 covers 1 year and measurement system #3 has been operative since August 1999.

A total of 323 events were recorded on measurement system #1. In the beginning the threshold was set at 2.5 mm/s² and a lot of recorded events were caused by car traffic. By increasing the threshold to 20 mm/s² it was possible to get rid of these traffic events. Analysis of the collected data showed that the sensor measuring in a horizontal direction perpendicular to the wall gave the largest amplitudes and little or no more information was obtained from the other two sensors. Interpretation and analysis of the data was therefor only done for data from the horizontal perpendicular sensor. In Bessason et al. (1999) there is a table that shows all events recorded by system #1 where 5 mm/s² peak acceleration level was exceeded. Only a few snow avalanches were recorded in the period, but a lot of events were caused by rock-fall and debris flows. Weather conditions have been a bit special during the measurement period, with few snowstorms from Southwest or West, which are the most hazardous wind-directions for the Óshlíð-hillside.

More than 1000 events were recorded on measurement system #2. Although some experience had been obtained for the avalanche shed in path 15 this experience could not be used for the rock outcrops in paths 14 and 16. The rock response is much lower than the response at an avalanche shed wall. Some avalanches in path 14 and 16 were not recorded the first winter (1998-1999) because the threshold was too high. After the first winter it could be deduced that the threshold should be set to a level around 0.5 mm/s² for the rock-outcrop in tracks 14 and 16.

Events measured by measurement system #3 have not been systematically processed so far since this system has only been operative from August 1999.

In the following only results from interpretation of data from the shed in avalanche track 15 are presented.

5 INTERPRETATION OF RECORDED EVENTS IN PATH 15

The recorded events are quite diverse with respect to the characteristic parameters. In figure 4 peak acceleration is plotted against characteristic frequency, f_{char}, and in figure 5 energy, E, is plotted against the characteristic frequency. As shown in the figures peak acceleration varies from 3 mm/s^2 up to around 2000 mm/s^2 and energy from 4 to 400,000 mm^2/s^3. In figure 6 time series and corresponding power spectrum for three different events are shown. Figure 6a shows a small avalanche event, which was confirmed by field investigation, figure 6b shows the biggest debris flow that has been recorded so far and finally figure 6c shows vibrations from a passing car. If plots of time series and corresponding power spectra are inspected along with the characteristic parameters it is possible in most cases to classify the events recorded in path 15.

5.1 Traffic

By setting the threshold at 5 mm/s^2 vibrations caused by small cars do not trigger the system. Trucks however easily trigger the system. In most cases peak acceleration for trucks is below 30 mm/s^2. In rare cases the peak value can be higher or up to about 50 mm/s^2. The characteristic frequency for events caused by trucks is generally lower than for other events or typically between 13 and 15 Hz. Bandwidth, B_e, is also typically smaller or around 2 Hz for these events. This can clearly be seen from figure 6c. Small trucks and large cars are not capable of exciting the structure enough to create a "tonal" response in the structure. The results are that events caused by these vehicles have higher characteristic frequencies and greater bandwidth than the events caused by large trucks. Duration of events created by cars is typically 7 to 8 s and generally shorter than 15 s.

5.2 Debris flows and avalanches

Events that are not generated by vehicles are caused by rock-fall, debris flows and snow avalanches. So far only a few recorded events have been confirmed to be snow avalanches. The main reason for that is that the frequency of snow avalanches in the Óshlíð-hillside has been less in the past years compared to the frequency in the late seventies and the eighties.

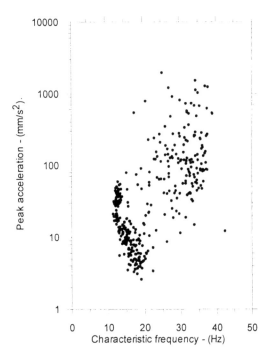

Figure 4. Peak acceleration versus characteristic frequency for all recorded events in path 15, where peak acceleration is greater than 3 mm/s^2.

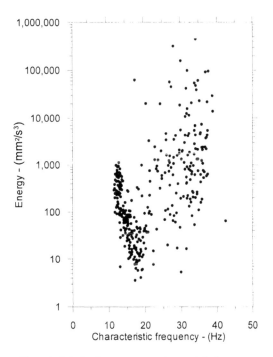

Figure 5. Calculated energy versus characteristic frequency for all recorded events in path 15, where peak acceleration is greater than 3 mm/s^2.

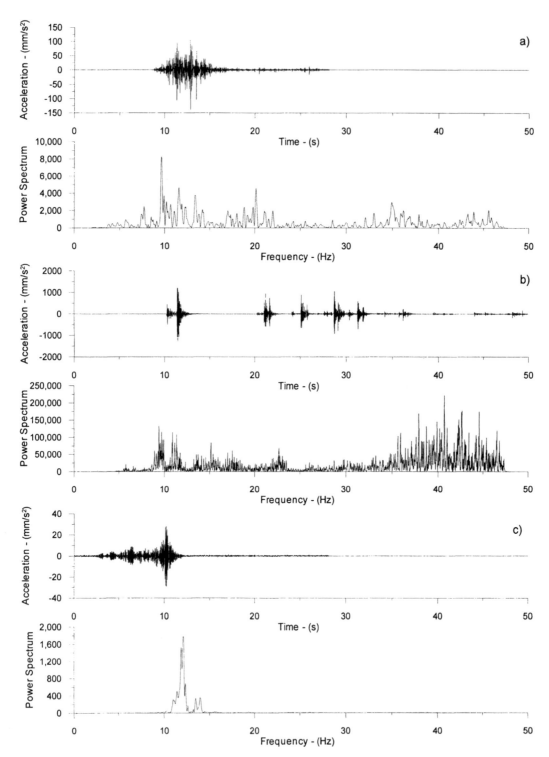

Figure 6. Time series and below corresponding power spectrum for three mesured events: a) Snow avalanche from February 22. 1997, b) Debris flow from September 22. 1998 and c) Passing truck.

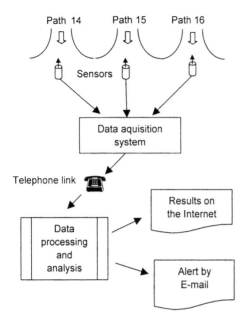

Figure 7. Snow avalanche detection and alert system.

Table 1. Statistical distribution of characteristic parameters in path 15.

Cumulative frequency	Peak acceleration (mm/s²)	Peak velocity (μm/s)	Duration (s)	Energy (mm²/s³)
10%	12.7	158	2.2	68.3
30%	27.1	282	6.2	164
70%	106	725	10.8	980
90%	278	2317	17.8	7001
100%	1964	18613	63.4	453334

Typically recorded snow avalanches have the characteristics shown in figure 6a, that is a smooth envelope. There are also examples of more transients like time series, probably because of rock and ice blocks mixed into the avalanche mass. Debris flows generally show more transients as for example shown in figure 6b. From the existing data it is difficult to see any significant difference between the characteristic parameters for avalanches and debris flows. Taking into account that snow avalanches usually fall in snowstorms or during special weather conditions it is in most cases easy to distinguish between avalanches and debris flows. Use of a digital camera is another alternative in distinguishing between avalanches and other types of events.

5.3 Statistical distribution of parameters

The measurements have now been carried out for three and a half-year. During this time the threshold has been varied several times. In table 1 statistical distributions of parameters for the recorded events in path 15 are shown. The table shows 10%, 30%, 70%, 90% and 100% cumulative frequency for the parameters.

The table shows the wide range of these parameters. It is worth mentioning that the second largest event is of similar size as the largest one.

6 ALERT SYSTEM

The basic idea behind the alert system that includes paths 14, 15 and 16 is shown in figure 7. Data acquisition system located in the avalanche shed records events when the threshold is exceeded. The system then contacts via modem and telephone link a computer located in Reykjavik where the data are processed and analysed and the size and type of event is assessed. The results of the detection process are then published in a home page on the Internet and an alert message is sent out via E-mail. The whole process is completely automatic.

On the Internet the time and location of event is presented. In addition the numerical value of three of the characteristic parameters are shown, that is peak acceleration, duration and energy. In addition to giving the numerical value of these parameters they are also described verbally, i.e. *small, medium or large,* based on the statistical distribution in table 1. Finally, there is one column showing the estimated type, i.e. vehicle, avalanche, rock-fall or debris flow.

7 CONCLUSIONS AND SUMMARY

The research project was initiated in February 1996. In the beginning measurements were only made in one avalanche track, i.e. path 15, where the road is protected by a concrete avalanche shed. The sensors are bolted on the inner wall of the shed. To obtain experience with avalanche tracks were no protecting constructions are installed two new tracks, i.e. paths 14 and 16, were included in the research project in November 1998. In these paths the sensors are bolted on rock outcrops. At present the main conclusions and status of the research project are as follows:

- The operation of measurement equipment is satisfactory and stable. There have been some problems with data from the accelerometers related to electrical noise while the geophones have been much more stable.
- A lot of data has been obtained for path 15. There the sensors are bolted on a concrete wall of the avalanche shed and the generated vibrations are greatly affected by the dynamic characteristics of the structure. It is easy to distinguish between events caused by vehicles and events caused by avalanches and debris flows. Many events have been recorded that are caused

by rock-fall and debris flows. Still relatively few events have been recorded as a result of snow avalanches.

- In paths 14 and 16 more experience has to be obtained. The sensors in these paths are bolted on rock-outcrop. The monitoring has only been active for a short period of time and to start with the threshold was to high and some events including avalanches were lost. Roughly the threshold in these paths should be 1/10 of the threshold in the avalanche shed. This means a threshold of 0.5 mm/s^2 for the accelerometers and 5 μm/s for the geophones.
- The communication link between the data acquisition system in Óshlíð and the central computer in Reykjavik works as expected. Whenever an event is recorded the system automatically calls the central computer and transfers the data.
- Data processing is automatic and results are directly (within one to two minutes) written to a home page, which can be accessed through the Internet.
- Parallel to putting the results on the Internet an alert message is sent out via E-mail.

The next step in the research project is to obtain more experience for paths 14 and 16. It is also necessary to confirm more recorded events by field investigations. Furthermore, data processing routines have to be improved in order to be able to determine more reliably the type of event.

8 REFERENCES

Bessason, B., Baldvinsson, G. & Thórarinsson, Ó. 1999. Skynjun og greining snjóflóða með bylgjumælingum, Engineering Research Institute, University of Iceland, report no. 99002, (written in Icelandic).

Leprettre, B.J.P., Navarre, J.P. & Taillefer, A. 1996. First results from a pre-operational system for automatic detection and recognition of seismic signals associated with avalanches, *Journal of Glaciology*, Vol. 42. No. 141, pp. 352-363.

Norwegian Public Roads Administration, 1994. *Snow Engineering for Roads – About Snow Avalanches and Drifting Snow*, Handbook, Serial No. 174.

Vanmarcke E.H. & Shih-Sheng, P.L. 1980. Strong-Motion Duration and RMS Amplitude of Earthquake Records, *Bull. Seism. Soc. Am.*, Vol. 70. No. 4, pp. 1293-1307.

Snow Engineering, Hjorth-Hansen, Holand, Løset & Norem (eds) © 2000 Balkema, Rotterdam, ISBN 90 5809 148 1

The kinetic friction of polyethylen on snow, comparison between laboratory and field measurements

D. Buhl, M. Fauve & H. Rhyner
Swiss Federal Institute for Snow and Avalanche Research (SFISAR), Davos Dorf, Switzerland

ABSTRACT: Experiments were done in the freezing chamber with polyethylen specimens sliding on ice using a tribometer. The coefficient of friction was measured as a function of the surrounding temperature (varyied between –25°C and 0°C) and the pressure on the slider (varyied between 50 MPa and 300 Mpa).

The coefficient of friction is strongly depending on the temperature of the ice surface. It is on its minimum around –3°C. In the temperature range between –3°C and 0°C the coefficient of friction is increasing with increasing temperature. In the temperature range between –3°C and –25°C the coefficient of friction is increasing with decreasing temperature.

In the low temperature range the coefficient of friction is depending on the pressure on the slider. The higher the pressure, the lower is the coefficient of friction. The influence of the pressure is decreasing for an increasing temperature and above –6°C there is no difference visible between different loads anymore.

The temperature of the ice surface is higher than the temperature of the surrounding temperature. Especially for low temperatures the temperature of the ice surface is significantly increased by frictional heating.

The results correspond well to the idea of three different components forming the overall kinetic friction. At low temperatures dry friction dominates, whereas with increasing temperature a more lubricated friction decreases the coefficient of friction and for even higher temperatures capillary drag effects increase the friction again. The results are confirmed by extended field measurements of gliding with skis on snow under racing conditions.

1 INTRODUCTION

It is well known, that the low kinetic friction of snow is due to a thin water film that is produced by frictional heating [Bowden]. It is well accepted, that the kinetic friction is the sum of dry friction, wet friction and capillary drag effects together with resistance caused by displacing and compressing the snow [Glenne, Colbeck]. In different weather and snow conditions a different type of friction dominates and the total friction can be found by summing the individual friction components [Lehtovaara].

This splitting of the total friction into individual components is related to the formation of the water film between the glider and the snow surface. The thickness of the water film mainly determines the distribution of the different friction mechanisms. Solid-to solid interactions take place when the thickness of the water film is insufficient to prevent direct contact of snow grains and the glider. As the water film thickens and lubricates the whole contact area the slider is only slowed down by the viscous resistance of the water film and the overall resistance is

on its minimum. The production of more water increases extrusion and shearing and results in capillary suction that drastically increases the resistance [Colbeck].

The thickness of the water film is believed to be the crucial parameter for the kinetic friction. Since this water film is formed by frictional heating of the contact area, there are a lot of parameters that contribute to the system. These parameters can be roughly divided into two regions: ski and snow. From the ski, there are parameters like roughness, hardness, wetability, thermal conductivity, pressure distribution, vibration and flexion that influence the friction. The snow is determined by temperature, density, water content, grain size, grain shape, thermal conductivity, roughness and hardness. Besides these sets, the speed and the load contribute to the system , too.

Focussing on snow there has been done some investigations on different parameters and their influence on the friction. Spring found that the coefficient of friction is increasing with increasing the speed.

Ericksson showed that friction is increased for smaller grain size. Shimbo tried to determine the influence of roughness and hardness on the friction. However, the most important snow parameter, the snow temperature, has not been the subject of intense studies, so far. It is crucial to determine the influence of the snow temperature on the friction in order to be able to optimize the system ski/snow for reduced friction. Therefore, the aim of this work is the investigation of the influence of the snow temperature on the friction in laboratory measurements and in gliding tests on the field. It is believed, that this knowledge enables to contribute the individual friction components to different temperature ranges, which is the key for optimizing the ski parameters in order to reach the best gliding performance.

2 EXPERIMENTAL

A tribometer was placed in a freezing chamber to measure the coefficient of friction (c.o.f.) of a specimen of polyethylen on snow and ice. The tribometer consists of a turning disc (diameter 15 cm), i.e. a mandrel filled with ice/snow, and a stamp that was pressed on the turning surface. The force trying to move the fixed stamp was measured and related to the weight of the stamp. Thus, the quotient of the friction force and the gravity determines the c.o.f.

Special emphasis had to be laid on producing a completely flat surface in order to avoid any jumping of the specimen. The mandrel rotated at maximum possible speed in order to reach at least a comparable speed to field tests. However, the reached 5 m/s are still low compared to field measurements where a velocity up to 30 m/s was reached. The second drawback is the fact, that the specimen glides always in the same track. Nevertheless, the experiment showed that even with these restrictions some clear results were obtained.

The temperature of the surroundings was controlled between 0°C and –25°C. The load on the specimen was varied between 5N and 30N. The temperature of the snow was determined during the experiment by measuring the infrared radiation of the surface. The surrounding temperature was measured by ordinary thermocouples.

Field measurements were performed on a gliding slope under racing conditions. More than 2000 runs on different skis contributed to the data. On each run a large set of parameters (ski, weather conditions, snow conditions, test results) was collected. In this study, special emphasis is laid on the influence of the snow temperature on the gliding performance without any further classification of ski or snow.

3 RESULTS

The coefficient of friction (c.o.f) for polyethylen on ice/snow in general is much lower than any c.o.f. on other materials. The measured values may reach a minimum value of about 0.02, which is about one order of magnitude lower than on steel (0.2), for example. However, the c.o.f. is not constant under the varyied conditions during the experiments, but strongly depends on the temperature of the snow surface and on the load.

3.1. Influence of the temperature (figure 1)

The c.o.f. reaches a minimum of 0.02 at a temperature of about –3°C on the snow surface. When increasing the temperature from –3°C to 0°C, the c.o.f. significantly increases up to 0.05. Decreasing the temperature from –3°C to –20°C increases the c.o.f. to values up to 0.18.

3.2. Influence of the load (figures 1 and 2)

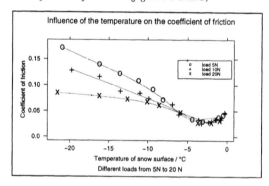

Figure 1. Influence of the snow temperature on the coefficient of friction. Higher friction for low temperatures and close to 0°C. Minimum friction at around –3°C. The friction is depending on the load only for low temperatures.

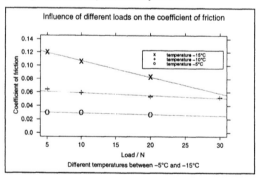

Figure 2. The load influences the coefficient of friction only at low temperatures. For temperatures higher than –10°C, there is no difference between the friction under different loads, whereas at low temperatures the coefficient of friction is lower for higher loads.

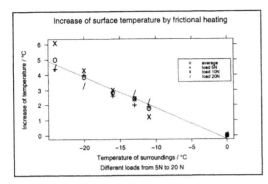

Figure 3. Increase of the snow surface temperature during running because of frictional heating. At lower surrounding temperatures the increase is more pronounced. A correlation between the load ant the heating could not be found.

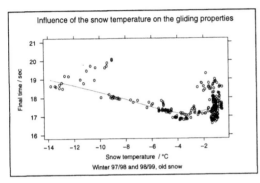

Figure 4. Influence of the snow temperature on the gliding properties of downhill skis. The gliding was performed under racing conditions (maximum speed 115 km/h) on different skis but always on the same slope.

The load influences the c.o.f. only in the low temperature region. The correlation between the temperature and the c.o.f. remains the same for any load (highest c.o.f. for lowest temperature, minimum at −3°C, increasing c.o.f. for increasing temperature). As soon as the temperature is higher than −5°C, any difference concerning different loads vanishes and the c.o.f. is the same. However, at cold temperatures below −10°C. the c.o.f. is significantly higher for lower loads. For example, at a temperature of −20°C, the c.o.f. under a load of 20N is 0.09, whereas the c.o.f. under a load of 5N is 0.18 (factor 2). The decrease of the c.o.f. is direct proportional to the increase of the load, but the factor depends on the temperature (the factor is 0.01 per 5N at −15°C). Above around −5°C the factor is 0 which means, that any change of the load produces no shift in the c.o.f. anymore.

3.3. Heating of the surface (figure 3)

The surface of the is heated by the repeated passing of the specimen. Thus, the surrounding temperature may be several degrees lower than the actual temperature of the snow surface. The increase of the temperature of the snow surface is proportional to the surrounding temperature. At a surrounding temperature −20°C, the surface temperature is increased during the experiment to −16°C. At higher temperature this increase is less pronounced but still present.

3.4. Field tests (figure 4)

The results of the gliding tests of skis shows a similar dependence of the c.o.f. on the temperature as found in the laboratory. The gliding was performed on the same slope all the time and the data of about 2000 runs were collected. The correlation between the final time and the temperature of the snow surface exhibits the same curvature as shown above. The fastest conditions are found at a temperature around −3°C. The conditions are slower when increasing the temperature from −3°C to 0°C. On the other hand, the conditions are slower when decreasing the temperature from −3°C to −15°C, as well. The variation of the times (especially for conditions close to 0°C) is due to the fact, that there is a huge amount of further parameters that influence the gliding properties, too. These parameters are examined elsewhere [Buhl].

4 DISCUSSION AND CONCLUSIONS

Most of the results can be explained by the formation of a water film between the gilding specimen and the snow surface. The generation of this meltwater is controlled by the combination of weight, speed, friction and the energy balance at the interface. There are a lot of parameters from the snow and the surrounding conditions that contribute to this energy balance, such as the snow temperature. The rate of meltwater production decreases as the temperature drops and the thickness of the water film decreases with decreasing temperature. Thus, the composition of the total friction changes with decreasing or increasing the temperature. The total friction is a composition of dry friction, lubricated friction and capillary drag and the distribution of the single components depend on the thickness of the water film.

It is difficult to quantify these effects because the friction, the produced heat, the temperature and the load are depending on each other. E.g., at a low temperature the friction increases because the produced heat is insufficient for a complete lubrication of the tribological interface. Thus, the amount of dry friction increases and therefore the c.o.f. increases. On the other hand, the frictional heating increases for a higher c.o.f. which increases the amount of heat in the contact area and enables to melt more snow. This delicate equilibrium of the components of the friction and the heat production has not been described quantitatively, so far. There exist only some theoretical approaches.

We suggest that in the low temperature region the production of frictional heating is insufficient to form a complete meltwater film. Thus, the amount of dry friction increases and the overall friction increases , as well. In this temperature range, the load contributes to the system: the higher the load, the more heat is produced by the friction. But an increased amount of heat influences shifts the proportion of dry to lubricated friction towards the last one, which results in a decreased overall friction.

As soon as the temperature is increased, the difference of the c.o.f. for varying loads vanishes more and more. We believe that in the temperature region where only lubricated friction contributes to the total friction, the load has no influence on the tribological system anymore. This is the case at around –4°C to –3°C. In that temperature region, the coefficient of friction is at its minimum. The thickness of the meltwater film is sufficient for a complete lubrication of the contact area and the dry friction is eliminated.

However, a further increase of the temperature that is always coupled to an increased frictional heat rate increases the c.o.f. again. In this high temperature region between –3°C and 0°C, to much water is produced which results in capillary drag effects. Again, this tribological phenomenon is independent of the load.

The linear dependence of the c.o.f. on the load in the low temperature region corresponds well to the theory of the total friction as a sum of dry friction and lubricated friction. A higher load produces more frictional heating especially by dry friction. Thus, more heat is available for the formation of a meltwater film and the part of lubricated friction is increased which reduces the overall friction.

Increasing the temperature reduces the dry friction that is mainly contributing to the production of frictional heating. Therefore, the heating of the snow surface is most pronounced in the low temperature region. In addition, the heating has to increase when the surrounding temperature closes up to the melting temperature. However, according to theory, the load should also influence the heating the way that more heat is produced for higher loads. This correlation could not be confirmed by our experimental results.

The results of the field tests show a good correlation to the experimental results in the laboratory. The curvature starts with higher c.o.f. for low temperatures and reaches a minimum when increasing the temperature range –4°C to –3°C. A further increase of the temperature towards 0°C increases the c.o.f. again[1]. Besides this overall correlation between the c.o.f. (or the gliding properties) and the temperature, the results show that there are some temperature regions, where the gliding properties are strongly influenced by some other parameters, especially in the high temperature region close to 0°C. The knowledge of these parameters and their influence on the system ski/ snow is crucial in order to optimize the gliding properties of the skis.

5 ACKNOWLEDGEMENTS

This project is supported by the KTI (Komission für Technologie und Innovation, CH) and several industrial partners (Stöckli AG, Toko AG, Fritschi AG, Nidecker SA, IMS Kunststoffe AG).

REFERENCES

Bowden F.P., Hughes T.P., The mechanics of sliding on ice and snow. Proc. Roy. Soc., London, Ser. A217 (1939), pp. 280-298.

Buhl D., Bruderer C., Fauve M., Rhyner H., Schlussbericht KTI-Projekt 3781.1. Internal report (2000).

Colbeck S.C., The kinetic friction of snow. J. of Glaciology, Vol 34, No. 116 (1988), pp. 78-86.

Eriksson R., Friction of runners on snow and ice. SIPRE Report TL 44, Meddelande 34/35 (1955), pp. 1-63.

Glenne B., Sliding friction and boundary lubrication of snow. ASME J. of Tribology, Vol 109 (1987), pp. 614-617.

Lehtovaara A., Kinetic friction between ski and snow. Acta Polytechnica Scandinavica, Mech. Eng. Ser. No. 93, Helsinki (1989).

Shimbo M., Mechanism of sliding on snow. General Assembly of Helsinki, Publ. No. 54, Int. Ass. of Hydrological Sciences, Gentbrugge, Belgium (1961), pp. 101-106.

Spring E., A method for testing the gliding quality of skis. Tribologia, Vol 7. No.1 (1988), pp. 9-14.

---

[1] We note, that this curvature was found for old snow. Field tests in new snow conditions showed a different influence of the temperature on the c.o.f [Buhl].

Snow Engineering, Hjorth-Hansen, Holand, Løset & Norem (eds)© 2000 Balkema, Rotterdam, ISBN 90 5809 148 1

Snow pillows: Use and verification

R.V. Engeset, H.K. Sorteberg & H.C. Udnæs
Hydrology Department, Norwegian Water Resources and Energy Directorate, Oslo, Norway

ABSTRACT: Snowmelt yields a substantial contribution to spring floods in Norway. The most severe floods, such as in southeastern Norway in 1995, are fed from snowmelt over extensive snow-covered high-mountain areas. To monitor the temporal evolution of the snow mass during winter and spring, a network of 23 snow pressure pillows has been established in Norway, covering 58°N–71°N, 6°E–28°E, and 30–1400 m a.s.l. Hourly data are supplied twice daily to national authorities. To investigate the performance of the snow pillows, manual snow surveys, a snow accumulation-ablation model and nearby meteorological data were analysed for the 1998-99 winter and spring periods. The results suggest that snow pillows and the snow model represent the accumulation period well. During the melt period, both the snow pillows and the snow model produced data in agreement with manual surveys, but at a lower accuracy than observed during the accumulation period. Problems with ice/crust layers and internal forces were pronounced during freeze-melt cycles. Reduced performance was also observed when the amount of snow was either very high or very low, due to increased disturbance from wind, solar radiation, local topography and internal forces in the snowpack. Snow melt and refreezing were simulated fairly well, but factors not accounted for in the model reduced its accuracy during the melt period. The model could be used to identify periods when the snow pillow showed artefacts related to wind transport of snow and changes in internal forces caused by freeze-melt cycles.

1 INTRODUCTION

The environment in most Norwegian river-systems is strongly influenced by the snow conditions since snowmelt runoff often gives a substantial contribution to floods in many regions. Droughts and power production shortages are also possible results from extreme snow situations. Knowledge of the snow-conditions is also useful for hydropower companies, water suppliers and studies of climatic changes. Other areas of application of snow pillow data are snow avalanche forecasting and research.

NVE is responsible for national flood warning and runoff forecasting in Norway. To forecast the runoff, the HBV-model (Bergström 1976, Bergström 1992) is used. The model simulates snow accumulation from observed precipitation and temperature, and the melting by a degree-day method. This method has proved useful in runoff-simulations in Sweden (Lindström et al. 1996).

Observations of temperature and precipitation are limited, especially in mountainous areas. Furthermore, observed precipitation is usually less than true precipitation due to wind-effects (Førland et al.

1996). For this reason a network of 23 snow pillows (Sorteberg 1998) has been established to get more accurate information about the snow water equivalent (SWE).

Earlier studies show that snow pillows give an accurate point-estimate of the SWE (e.g., Beaumont 1965, Kerr 1976). Some problems with the technique have been reported (Tollan 1970, Tveit 1971, Andersen 1981), mostly related to ice and crust formation.

The objective of this work was to evaluate whether the new snow pillows are reliably measuring the SWE under different climatic conditions and locations, and to investigate whether a snow model could simulate the temporal variation in SWE at each pillow.

2 STUDY AREA AND DATA SETS

NVE has a regional network of snow pillows, in which 19 stations were operated during the 1998-99 winter. To determine the evolution and decay of SWE during the winter and spring, the stations are placed at elevations and locations in the country,

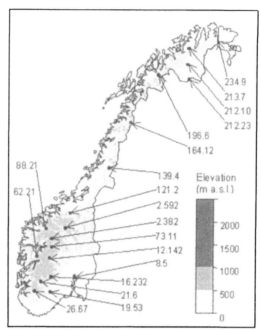

Figure 1. Snow pillows operated by NVE during the winter 1998-99. Three elevation intervals (<500, 500-1000 and >1000 m a.s.l.) are shown.

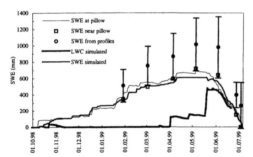

Figure 2. Sognefjellshytta. Observed SWE and simulated SWE and LWC. Standard deviation of SWE surveyed manually from profiles are shown with error bars. Note that LWC has been scaled by a factor of 10.

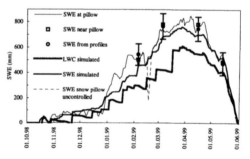

Figure 4. Duge Værstasjon. Observed SWE and simulated SWE and LWC. Standard deviation of SWE surveyed manually from profiles are shown with error bars. Note that LWC has been scaled by a factor of 10.

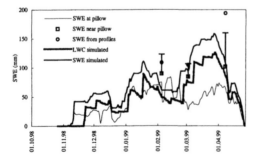

Figure 3. Fjalestad. Observed SWE and simulated SWE and LWC. Standard deviation of SWE surveyed manually from profiles are shown with error bars. Note that LWC has been scaled by a factor of 10.

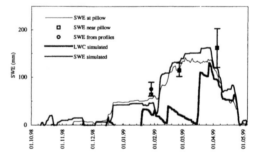

Figure 5. Maurhaugen-Oppdal. Observed SWE and simulated SWE and LWC. Standard deviation of SWE surveyed manually from profiles are shown with error bars. Note that LWC has been scaled by a factor of 10.

which are of interest to the national flood forecasting service. The stations are deployed between 58°N–71°N and 6°E–28°E, and cover the elevation interval 30–1400 m a.s.l. (Fig. 1).

The first snow pillow in Norway was deployed in 1967 at Kyrkjestølane, during the International Hydrological Decade (Andersen 1981, Furmyr & Tollan 1975, Tollan 1970, Tollan 1971). Subsequently, stations were installed at Groset in Telemark (1971) and Brunkollen in Bærum (1983), (Atterås 1991, Myrabø 1994). Additional 11 snow pillows were deployed in 1997 and 5 in 1998 (Tab. 1).

2.1 Snow pillow observations

The type of snow pillow operated by NVE is equipped with plastic bags filled with anti-freeze liquid. The new pillows have a diameter of 2 m and register the overburden pressure from the snow using a pressure sensor. The old pillows use a floating device and have a diameter of about 3.7 m (stations 8.5 and 16.232), see Sorteberg (1998, 1999) and Sundøen (1997). Snow pillow data are sampled every hour and automatically transferred to NVE twice a day.

2.2 Meteorological observations

Meteorological data from weather stations operated by the Norwegian Meteorological Institute (DNMI) are used for the snow simulations (Tab. 1).

During the winter 1998-99, temperatures were around 0°C at low elevations in parts of southeastern and western Norway. At high elevations of southeastern Norway snow condition were normal, with more snow than usual in the southwestern mountains. Troms, Nordland and Trøndelag had less snow than usual. Finnmark in northern Norway had a long and very cold period. For example, Karasjok observed a temperature of −51.2°C on 28 January (the minimum value for the period from 1888 to present is −51.4°C). In most parts of the country, a long and mild period occurred in April and gave an early start to the snowmelt. Due to the early start the snowmelt period was longer than normal, which reduced flood risk.

3 METHODS

3.1 Snow measurement

Manual snow surveys were conducted at each snow pillow site on a monthly basis from January 1999. The observations included:

- four snow depth samples around each pillow
- twenty snow depth samples along two crossing profiles of 50 m
- vertical snow density profile near the pillow

- vertical profile of snow moisture, stratigraphy and ice layers
- ground frost
- snow covered area

3.2 Snow model

The snow model simulated accumulation, using wind-catch and dislocation corrected precipitation observed at the nearest synoptic weather station. The model separated precipitation as snow from rain using observed air temperature. Snowmelt was modeled using a degree-day approach (Bergström 1976, Bergström 1992, Engeset & Schjødt-Osmo 1997), as data for the parameters required by energy-balance models were not available. Liquid water and refreezing were also simulated. The accumulation-ablation model used precipitation and air temperature as input variables. Internal variables were used for separating rain from snow using a fixed threshold, and fixed temperature-dependent thresholds were used to identify snowmelt and refreezing. Snowmelt intensity was specified by a time-varying variable, and refreezing intensity was fixed. The state variables described snow water equivalent and snow liquid water content, and were updated on a daily basis. The model also simulated water yield from snowmelt and rain. Studies have indicated that the model performance may improve by introducing other climatic parameters such as radiation (e.g., Kustas et al. 1994) and air humidity (Lindström et al. 1996).

4 RESULTS AND DISCUSSION

A total of 19 snow pillows were investigated. For a comprehensive project report see Engeset et al. (2000).

The results from five pillows are presented in Figures 2-6, which include snow pillow observations, manual snow surveys and results from modelling of snow accumulation, ablation and liquid water content. These graphs represent snow and climate variability at the investigated sites. Figures 2 and 4 represent areas with deep snow (maximum SWE higher than 500 mm) and Figures 3, 5 and 6 represent areas with shallow snow (maximum SWE less than 150 mm). Relatively stable below freezing conditions prevailed during the accumulation period at the pillows shown in Figure 2 (much snow) and Figure 6 (little snow), while a series of melt-refreeze cycles occurred at the pillows in Figure 3 (little snow) and Figure 4 (much snow). Figure 6 shows the effects of wind in the accumulation period and the effect of increased melt due to high solar energy input in the spring.

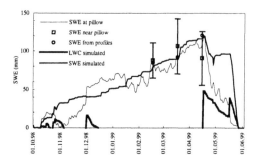

Figure 6. Siccajavre. Observed SWE and simulated SWE and LWC. Standard deviation of SWE surveyed manually from profiles, are shown with error bars. Note that LWC has been scaled by a factor of 10.

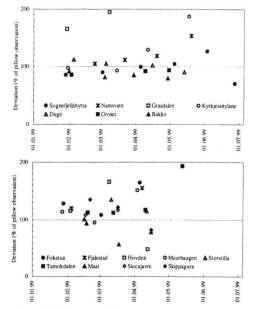

Figure 7. SWE observed from manual snow surveys near snow pillow relative to SWE observed by the pillow. Top panel shows pillows with maximum SWE above 300 mm during the winter, and bottom panel shows pillows with maximum SWE below 300 mm.

4.1 Accumulation

Good correlation was found between measurements of SWE near the pillows, model simulations of SWE and SWE at some pillows as illustrated in Figures 2 and 4. The relative deviations between SWE from manual snow surveys near pillows and SWE from pillows are presented in Figure 7.

In the beginning of the winter period the correlation between SWE observed near the pillow and by the pillow was good for most stations, especially in areas where few, if any changes in the air temperature around 0°C occurred, e.g., Sognefjelshytta,

Maurhaugen-Oppdal and Tamokdalen (Fig. 7). Under such conditions, a relatively homogenous snow pack was developed, without ice- and crust-layers. The modelling also indicated good correlation with pillow values in under these conditions.

Observed ice- and crust layers did not always influence the registrations of snow accumulation at the pillows, e.g., Duge Værstasjon (Figs 4 and 7). At Duge, both modelling and the measurements agreed fairly well with the pillow observations.

Under some circumstances the values differs a lot between models, pillows and snow measurements. In areas with changes in the air temperature around 0°C, ice- and crust layers forms frequently, and density measurements were thus difficult to conduct. Also, at some stations snow depth and density to the ground were difficult to measure, since thick ice-layers existed (e.g., Hovden and Kyrkjestølane). Problems were also encountered during measurements as a result of ice- and crust layers. This measurement error is well known and several investigations and descriptions of the topic are recorded, e.g. Bader et al. (1954), Goodison et al. (1981), Ramsli (1981), Kuusisto (1984).

The manual surveys indicated that in areas with continually shifts in air temperature around 0°C, the freeze-melt cycles induced large changes in the snow pack. The snow pack was in periods wet and ice-/cruster layers were formed. At Fjalestad (Fig. 3), the snow cover was thin and the snow pack was influenced by temperature variations. Results from Fjalestad (Fig. 3), Hovden and Grasdalen indicated that large deviation between snow pillow and manual observations occurred under such condition (Fig. 7). The explanation could be that the water in the snow pack froze and ice-/crust layers were formed. These layers could influence the snow at the pillow by relief of the weight, like as a bridge. In such a case, the pillows will underestimate the real snow water equivalent. Earlier studies describe similar problems (Furmyr et al. 1975, Tollan 1970, Tollan 1971, Tveit 1971, Tveit 1975). A decrease in observed SWE was observed at several pillows at temperatures far below 0°C. This normally occurred when periods of snowmelt were succeeded by a cold period. An example of this is shown for Duge Værstasjon from 15 to 17 January 1999 (Fig. 4), where a reduction in SWE of 300 mm was observed. These artefacts were not simulated in the model.

Wind has been a climatic factor that influenced snow pillow values, in particular for pillows elevated above the ground (similar problems are described by Furmyr et al. 1975, Tollan 1970). Wind led to snowdrift and may have caused additional snow accumulation or removed snow from the pillow. It is not unlikely that snowdrift has caused sudden changes in addition to melting or precipitation. It was for instance registered a decrease of about 20 mm in SWE at Siccajavre between 15 and 16 January, which the model did not simulate (Fig. 6). Sudden increases were observed 1 December 1998 (20 mm) and 14 February 1999 (30 mm) probably due to accumulation by snowdrift. Strong wind was observed by the weather station in the area and the low temperatures suggest that the snow was dry. Differences in observations at pillows and model simulations are also caused by catch loss at the precipitation gauge. This is in particular pronounced during strong wind and snow precipitation, when gauge observations are known to underestimate the actual amount of precipitation (Furmyr & Tollan 1975, Killingtveit & Sælthun 1995, Førland & Harstveit 1996).

In some situations, rain has been interpreted as snow or vice versa in the model, mainly due to incorrect threshold temperature for snow-rain classification The threshold temperature is a constant value in the model but will in fact vary under different atmospheric conditions (Skaugen 1998). At Fjalestad, an overestimated accumulation in the beginning of the winter resulted in an overestimation of the simulated values the entire winter.

4.2 Melting

Only a few snow measurements were carried out during the melting period. Also the melt season is considerably shorter than the accumulation season. Earlier investigations indicated large changes in the snow during the melting period (Male & Gray 1981, Bengtsson 1982, Killingtveit & Sælthun 1995). Manual surveys at a higher temporal frequency are required for assessing the snow pillow and model

performance properly. This is a very important period to monitor for flood forecasting.

In general, snow surveys observed higher SWE values than recorded by the pillows. Relative deviation is large by several of the pillows during the melting period compared to the winter period (see Fig. 7). During the melting, density measurements at some stations were taken in very wet snow, under circumstances that probably were unrepresentative for the snow wetness and density at the pillow. This could apply to all sites, and in particular uncertain results were recorded at Hovden and Kyrkjestølane.

In general, the model results show good correlation with the pillow observations (e.g., Figs 4 and 6). Nevertheless, at some stations, the models show too little or a slower melting compared to the pillow. Observed melting may be more rapid than that simulated, due to high solar radiation not readily described in a degree-day model. Including a radiation component may in these cases improve the model performance (Bengtsson 1976, Kustas et al. 1994). Under conditions with high insolation and old snow with a low albedo, high melt intensity is expected (Male & Gray 1981, Ramsli 1981) and such effects are not simulated well in the degree-day model, as it simulates snow melt as a function of temperature only. This effect is probably highest at pillows at high elevations and in northern Norway. In these areas melting commences relatively late and high insolation values are experienced on clear days. This effect was seen at Bakko, Groset, Kyrkjestølane, Masi, Siccajavre, Øvre Leirbotn and Skippagurra. The pillow values at Hovden and Grasdalen show slow melting, and the snow measurements and model results correlate poorly with pillow observations. The reason for the late melt could be a thick ice layer covering the pillows.

4.3 Liquid water content and refreezing

Simulations of liquid water in the snow pack were in agreement with the qualitative manual observations at several snow pillows, see for example Sognefjellshytta (Fig. 2) and Fokstua.

Occasionally, liquid water was simulates when observations showed that the snow was dry. An incorrect temperature elevation-correction could explain this, e.g. see Grasdalen (Fig. 5), Namsvatn tunnel Vekteren and Maurhaugen-Oppdal (Fig. 6).

In areas with great variations in the temperature, liquid water was observed to decrease slowly by during periods when the temperature fell. This was seen at Fjalestad (Fig. 3), Hovden and Reimegrend. The refreezing appears to have been underestimated in the model. However, no quantitative observations were available for a detailed analysis of the model's performance in terms of liquid water simulations.

Table 1. Snow pillow and weather station details.

NVE snow pillow (m a.s.l.)	Obs. period	DNMI weather station (m a.s.l.)	Distance pillow-weather station (km)
2.382 Sognefjellshytta 1435	1998-	55290 Sognefjell 1413	0
		15730 Bråtå 664	40
2.592 Fokstua 1000	1997-	16610 Fokstua 972	1
8.5.0 Brunkollen Klima 370	1983-	18960 Tryvasshøgda 528	10
12.142 Bakko 1020	1998-	25590 Geilo 810	20
16.232.14 Groset 990	1971-	31620 Møsstrand 977	10
19.53 Fjalestad 330	1998-	37230 Tveitsund 252	15
21.6 Hovden 855	1997-9	40880 Hovden 836	0
26.67 Duge Værstasjon 760	1997-	42920 Sirdal-Tjørhom 500	25
62.21 Reimegrend 595	1997-	51590 Voss 125	15
		51800 Mjølfjell 695	5
73.11 Kyrkjestølane 1000	1967-	54120 Lærdal 24	35
88.21 Grasdalen 935	1997-	60500 Tafjord 15	30
121.2 Maurhaugen-Oppdal 660	1998-	66770 Oppdal-Maurhaug. 668	0
139.4 Namsvatn tunnel Vekteren 460	1997-	77420 Majavatn 339	30
		77750 Susendal 265	65
		77550 Fiplingvatn 370	40
164.12 Storstilla nf Balvatn 565	1997-	81680 Saltdal 81	30
196.6 Tamokdalen 230	1997-	91300 Oteren 12	20
		89350 Bardufoss 76	45
212.10 Masi 272	1997-	93300 Solovomi 374	20
212.23 Siccajavre 385	1997-	93900 Siccajavre 382	0
213.7 Øvre Leirbotn 190	1998-	93140 Alta lufthavn 3	20
234.9 Skippagura 35	1997-	96800 Rustefjelbma 9	30

5 ASSESSMENT OF SNOW PILLOWS

The manual snow surveys indicated no malfunctions at the snow pillows when the snow was dry and there were no icing and crusting. Icing and crusting lead as expected to problems at some of the snow pillows. These problems were particularly pronounced at temperatures fluctuating around 0°C when subsequent freeze-melt cycles occurred. The snow model was useful to detect these sorts of problems and it seemed clear that temperature measurements are required to assess the quality of the snow pillow observations, e.g., Duge (Fig. 4).

When the snow was wet during the melt period, large differences were seen between the manual measurements and the snow pillow observations. This indicated that the snow density or depth varied in the area around the pillow. The spatial variability of the snow density was not measured, but this will be of high priority in the following studies of the snow pillows. At most of the snow pillows, the manual snow surveys indicated that the snow pillows were representative for the nearby area in the accumulation period and less representative in the melting period, e.g., Maurhaugen-Oppdal (Fig. 5). These problems could in many cases be related to the local topography. Local variations in solar exposition would give similar variations in melt intensity. At pillows with small amounts of snow, high relative difference between measurements and pillow values were identified during the winter season (Fig. 7). Thus the pillows give an uncertain point estimate of the snow amount under such conditions.

The average SWE from snow surveys along the profiles correlated well with SWE measured manually near the pillow. However, great differences were observed at some places, e.g., Sognefjellshytta (Fig. 2). To achieve good drainage this pillow was situated too high and did obviously not represent the amount of snow in the nearby area. Another pillow, Hovden, was situated too low compared to the surrounding ground, which reduced drainage from the pillow during the most intense melting.

Wind-drift seemed to be a problem at some of the pillows, e.g., Siccajavre (Fig. 6). At this pillow, and other pillows situated in dry areas, the wind drift at occasions gave relatively large fluctuation in the observed data during the accumulation period when the snow was dry. Although relevant, redistribution of snow under different wind-speeds and directions were not assessed during this project.

Other problems concerning the reliability of the snow pillows were human activity and extreme high amounts of snow (e.g., Grasdalen).

6 CONCLUSIONS

Daily observations of the evolution and decay of the snow cover, observed as the snow water equivalent by 19 snow pillows, have been assessed using manual snow surveys on a monthly basis and daily simulations from a snow model. The results suggest that both snow pillows and the snow model represent the accumulation well. During the melt period, both the snow pillows and the snow model produced data in agreement with manual surveys, but at a lower accuracy than observed during the accumulation period. However, the changes in SWE during the melt period occurred during a rather short time-span, and manual snow samples should be performed more frequently and from more points in the vicinity of the pillows to better assess the performance during the spring. In areas where the air temperature was frequently oscillating around 0°C, problems with ice- and crust layers within the snow pack were evident and thus the pillows underestimated the SWE at certain periods. Furthermore, pillow reliability was reduced when the amount of snow was either very high or very low, due to increased disturbances from wind, solar insulation, local topography and internal forces in the snowpack.

The snow model proved to model accumulation very well. Snow melt and refreezing were also simulated well, but the results suggest that factors not accounted for in the model reduced the accuracy of the model results during the melt period. The model could be used to identify periods when the snow pillow results showed artefacts related to such effects as wind transport of snow and changes in internal forces caused by freeze-melt cycles.

ACKNOWLEDGEMENTS

We would like to thank the people that have contributed during the fieldwork: Lom Fjellstyre, Laurits Sønstebø, Torleiv Fjalestad, Torgeir Reime, Lill-Johanne Myrhaug, Odd Fossmo, Agnar Johnsen, Einar Pettersen, Øst-Telemarkens Brukseierforening, Otra Kraft DA, Sira-Kvina Kraftselskap, Salten Kraftsamband AS, Nord-Trøndelag Elektrisitetsverk and Oslo Energi Produksjon AS.

The Norwegian Water Resources and Energy Directorate financed this project by *Vassdragsmiljøprogrammet 1997-2001*, a research programme on river environment management.

REFERENCES

Andersen, T. 1981. Automatisk registrering av snøens vannekvivalent ved snøputer. *Norsk hydrologisk komitè*, rapport nr 6.

Atterås, G. 1991. Hydrologiske undersøkelser i Grosetfeltet., *Norges vassdrags- og energidirektorat*. Dokument HB notat Nr 15/91.

Bader, H., Haefeli, R., Bucher, E., Neher, J., Eckel, O., Thams, C. & Niggli, P. 1954. Snow and its Metamorphism. *Snow, ice and permafrost research establishment.* Corps of Engineers, U.S. Army, Wilmette, Illinois.

Beaumont, R.T. 1965. Mt Hood pressure pillow snow gage. *33rd Western Snow Conference:* 29-35

Bengtsson, L. 1976. Snowmelt Estimated from Energy Budget Studies. *Nordic Hydrology,* 7.

Bengtsson, L. 1982. Avrinning vid snösmältning, Observationer och analys. *Teknisk rapport, Högskolen i Luleå.* WREL, Serie A nr 106.

Bergström, S. 1976. Development and application of a conceptional runoff model for Scandinavian catchments. *Departement of Water Resources Engineering. Lund Institute of Technology.* Bulletin Series A No. 52.

Bergström, S. 1992. The HBV-model - its structure and applicatoins. *SMHI RH no. 4.* Norrköping, Sverige.

Engeset, R., & Schjødt-Osmo, O., 1997. Retrieval of snow information using passive microwave data in norway. Proc. EARSL Workshop Remote Sensing of Land Ice and Snow, 17-18 April 1997, Freiburg, Germany.

Engeset, V.R., Sorteberg, K.H & Udnæs, H-C. 2000. Nosit - utvikling ac NVEs operasjonelle snøinformasjonstjeneste. *Norges vassdrag- og energidirektorat.* Dokument nr 1/2000.

Furmyr, S. & Tollan, A. 1975. Resultater og erfaringer av snøundersøkelser i Filefjell representative område 1967-1974. *Den Norske IHD – komitè.* Oslo.

Førland, E.J., Harstveit, K. & Lystad, S.L. 1996. Estimering av snøakkumulering og snøsmelting. *DNMI Klima.* Rapport nr 12/96.

Goodison, E.B., Ferguson, L.H. & McKay, G.A. 1981. Measurement and data analysis, in Gray D.M, Male D.H. *Handbook of snow. Principles, Management & Use.* Pergamon Press: 191-274.

Kerr, W.E. 1976. Snow pillow experience in a prairie (Alberta) environment. Poc. 44th Annu. Meet. West. Snow Conf.: 39-47.

Killingtveit, Å. & Sælthun, N.R. 1995. Hydrology, Volume no 7 in Hydropower development. *Norwegian Institute of technology, Trondheim.*

Kuusisto, E. 1984. Snow accumulation and snowmelt in Finland. *Publications of the water research institute 55.*

Kustas, P.W. & Rango, A. 1994. A simple energy budget algorithm for the snowmelt runoff model. *Water resources research.* Vol. 30. No.5: 1515-1527.

Lindström, G., Gardelin, M., Johansson, B., Persson, M. & Bergström, S. 1996. HBV-96. -En areellt fördelad modell för vattenkrafthydrologin. *SMHI RH.* NR 12.

Male, D.H & Gray. 1981. Snowcover ablation and runoff, in Gray D.M, Male D.H. *Handbook of snow. Principles, Management & Use.* Pergamon Press: 360-430.

Myrabø, S. 1994. Sæterbekken forsøksfelt. *Norges vassdrag- og energidirektorat.* Rapport 2/1994.

Ramsli, G. 1981. Snø og snøskred. *Universitetsforlaget Oslo-Bergen-Tromsø.*

Skaugen, T. 1998. Studie av skilletemperatur for snø ved hjelp av samlokalisert snøpute, nedbør- og temperaturdata. *Norges vassdrag- og energidirektorat.* Rapport 11/1998.

Sorteberg, K.H. 1998. NVEs snøputer vinteren 1997-98. *Norges vassdrag- og energidirektorat.* Rapport 27/1998.

Sorteberg, K.H. 1999. Snø, Informasjon om snøputene. *Hydrologisk månedsoversikt Oktober 1999, Norges vassdrags- og energidirektorat:* 29-32.

Sundøen, K. 1997. Snøputa, en sensor for måling av snølagets vekt. *Hydrologisk månedsoversikt Mars 1997, Norges vassdrags- og energidirektorat:* 20-21.

Tollan, A. 1970. Experiences with snow pillows in Norway. *Bulletin of the International Association of Scientific Hydrology.* XV, 2-6/1970: 113-120.

Tollan, A. 1971. Determinations of areal values of the water equivalent of snow in a representative basin. *Nordisk Hydrologisk Konferens Stocholm 1970.* Vol. II: 97-107.

Tveit, J. 1971. Temperatureffekter ved ei snøpute. *Nordic IHD.* Report No1: 49-65.

Tveit, J. 1975. Anvendt Hydrologi. *Institutt for vassbygning, Norges tekniske høgskole.*

Snow Engineering, Hjorth-Hansen, Holand, Løset & Norem (eds) © 2000 Balkema, Rotterdam, ISBN 90 5809 148 1

Snow melting system using a conducting pavement and an aquifer penetrating bore-hole

Teruyuki Fukuhara & Laurel Goodrich
Fukui University, Japan

Hiroshi Watanabe
Yamada Technology Institute Company Limited, Fukui, Japan

Kazuma Moriyama
Misawa Environmental Technology Company Limited, Hiroshima, Japan

ABSTRACT: This paper describes the performance of an experimental pavement snow-melting system in which ground heat provided by a Bore-hole Heat Exchange System (BHES) constructed in an alluvial soil is supplied to a serpentine coil placed in a high thermal conductivity pavement. The measurements show that the inflow of cold water returning to the bore-hole during snow melting operational cycles leads to a rapid drop in ground temperature in the vicinity of the bore-hole, but the temperature drop is noticeably less at those depths containing aquifers than at those with no water flow. The data also indicate that the ground temperature recovery after shutting off the circulation pumps is much more rapid at levels with aquifers than at those without. The energy extracted from the piping system was calculated for two types of heated roadway: (As$_1$), whose surface asphalt layer was made using a high thermal conductivity aggregate and (As$_2$), made using a normal thermal conductivity material. A significant difference was observed in the rate of snow melting between the two test sections. The section overlain with a high thermal conductivity asphalt pavement, As$_1$, performed substantially better than that made of regular asphalt concrete. In addition, the total energy extracted from the piping system for section As$_1$ was about 30% greater than for the normal section, As$_2$.

1 INTRODUCTION

Snow removal from roads is a significant problem in many regions of the world and appropriate countermeasures often depend on specific local conditions. Fukui is situated on the Japan Sea side (NW side) of the main island of Japan and, although temperatures are mild, this region experiences heavy snowfalls. Mean total snow fall over a winter is about 300cm while the mean daily minimum temperature is just above 0°C.

Solar energy could be an attractive natural energy source for snow melting but, unfortunately, the duration of sunshine in Fukui is extremely short and unreliable in winter. The potential for significant artificial storage of solar energy is low. However, since Fukui is in the southern part of the heavy snow region, even at shallow depths, the ground temperature is fairly warm throughout the year. In the Fukui Valley, temperatures at 3m below the ground surface are never lower than 12°C. It follows that, for regions like this at least, using ground heat may be more effective than using solar energy directly.

The Bore-hole Heat Exchange System (BHES), shown in Figure 1, was built to prevent icing and snow accumulation at a car park on the campus of Fukui University. The BHES consists of a long bore-hole and roadway pavement incorporating a heat exchange pipe. A propylene glycol-water antifreeze mixture serves as the heat transfer fluid and is circulated in a closed loop circuit linking the

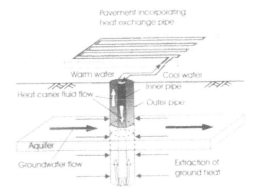

Fig.1 Bore-hole Heat Exchange System

pavement and the bore-hole by a 0.3KW line pump. In summer, the pavement acts as a solar collector while the bore-hole serves as a heat sink. The net heat transfer direction is reversed in winter.

BHES technology was originally developed primarily for thermal energy storage in bedrock but many big cities in Japan are constructed on alluvial aquifers having a shallow groundwater table. Since ground heat has a low energy density, acceptable snow-melting performance by a BHES hinges on being able to effectively extract ground heat through the bore-hole and transfer it efficiently to the road surface. The effectiveness of a heated pavement for snow melting also depends on achieving sufficiently uniform pavement surface temperatures. In addition, the potential for flowing groundwater to improve the thermal performance of bore hole heat extraction systems is significant in principle. Several BHES snow melting installations already exist in Japan but no information has yet been made available concerning the effects of groundwater flow on their snow-melting performance.

This paper describes the performance of an experimental closed-loop hydronic pavement snow-melting system constructed at Fukui University in which ground heat provided by a BHES constructed in an alluvial soil is supplied to a serpentine coil placed in a specially constructed high thermal conductivity pavement.

2 DESIGN OF THE BHES

2.1 Structure of the BHES and description of heat flows

As shown in Figure 1, the BHES is a closed-loop hydronic system consisting of a roadway pavement containing a serpentine connected to a concentric pipe placed in a bore-hole that penetrates an aquifer.

(1) *Hydronic pavement:* The experimental heated roadway (5m x 12m) is divided into two sections: (As$_1$), whose surface layer (0.03m thick) was made using a high thermal conductivity silica aggregate bituminous concrete (asphalt), and (As$_2$), whose surface layer was made from normal thermal conductivity asphalt. Comparisons were also made with an existing non-heated control pavement (As$_3$) constructed according to standard practice. Underneath the asphalt surface layer, both sections As$_1$ and As$_2$ included a 0.16m thick, factory made, Portland cement concrete base panel that was also made using a silica aggregate to maximise thermal conductivity, and the serpentine heat exchange piping system embedded in it was made of steel to improve heat transfer. The thermal conductivity of the asphalt layer of section As$_1$ (2.4W/mK) was 1.7 times higher than that of section As$_2$ (1.4W/mK). The thermal conductivity of the concrete base layer

was 2.3W/mK, nearly double that of normal concrete. The pipe spacing was 0.1m.

(2) *Bore-hole*: The bore-hole heat exchanger was 70m in length and comprised a co-axial double pipe made of PVC. The inner pipe was 56mm diameter with a wall thickness of 3mm while the outer was 90mm diameter with 4mm wall thickness. The bore-hole penetrated two aquifers. The uppermost consisted of a gravel layer lying between 30m and 44m deep while the lower one was a sand layer located between 46m and 52m below the ground surface

In operation, fluid comes out of the roadway pavement serpentine, enters the inner bore-hole tube and proceeds down to the bottom of the pipe. Subsequently the fluid flows upwards in the annular space between the inner and outer pipes and returns to the pavement again. During winter, the heat extracted from the ground via the bore-hole is used as a heat source for snow-melting. During summer, the heat that the pavement absorbs from the atmosphere is injected into the ground by the circulation of the fluid in the bore-hole. As a result, the BHES suppresses the excessive rise in asphalt pavement temperatures that usually occur and that lead to accelerated rutting damage on road surfaces in many regions of Japan.

3 EXPERIMENTAL RESULTS AND DISCUSSION

3.1 Thermal interaction between fluid in bore-hole and ground

Figure 2 shows spatial distributions of fluid temperature, Tw, along the inner and outer pipes during snow-melting operations. Positions 0m, 70m, and 140m on the x-axis correspond to the inlet, bottom, and outlet of the bore-hole, respectively. The BHES was shut down for a full month just prior to the start of snow-melting operations at 05:00 hours, 1998/01/01. The initial fluid temperatures were therefore steady and in thermodynamic equilibrium. Pavement base layer temperatures at start-up were in the range of 16°C to 16.5°C. As soon as the pump was started and snow-melting operations began, Tw dropped rapidly over the whole length of the bore-hole and subsequently rose again slowly while the fluid was warmed by the ground around the bore hole.

Figure 3 shows the fluid temperature recovery response after shut off of the circulation pump following the snow melting operational cycle described in Figure 2. A rapid rise in Tw is observed at the depths corresponding to the two aquifers, especially to the lower one. A possible explanation is that groundwater flow differences affected the rise in Tw because there is no significant difference in the ambient ground temperatures at the two depths corresponding to the aquifers. This behaviour was mirrored in other fluid temperature recovery responses.

The solid lines in Figures 2 and 3 represent the fluid temperature response calculated using a numerical model (Ohki et al. 1997, Tanimoto et al. 1999, Fukuhara et al. 1999) and the computed results are seen to agree well with the experimentally measured values.

Figure 4 shows the corresponding time variations of the fluid temperatures, Tw, in the outer pipe from January 1 to January 3, 1998 at depths of 20m, 40m, and 50m. There is no groundwater flow between the surface and 20m depth (aquitard). The 40m level corresponds approximately to the centre of the upper aquifer and 50m the centre of the lower one. The notations, "on" and "off" at the top of Figure 4 refer to operation of the fluid circulation pump during snow-melting operations. A rapid drop in Tw is observed at all depths but the drop is noticeably less at the 40m and 50m depths than at 20m. The rapid drop in Tw at the 50m depth is smaller than that at the 40m depth. The fluid temperature recovery is also more rapid at greater depth and follows the order 50m, 40m and 20m.

Figure 5 shows the corresponding time variations of the ground temperatures in the vicinity of the bore-hole at the same three depths as Figure 4. The measurements indicate that the inflow of cold water returning to the bore-hole during snow melting operational cycles leads to a rapid drop in ground temperature in the vicinity of the bore-hole at all depths, but the temperature drop is noticeably less at the 40m and 50m depths than at the 20m depth. The temperature drop during snow melting is also seen to be smaller at the 50m depth than at the 40m depth. In addition, the data indicate that the ground temperature recovery after shutting off the line pump following a snow-melting operation cycle is much more rapid at the 40m and 50m depths than at the 20m depth.

3.2 Effect of high thermal conductivity pavement on snow-melting performance

Figure 6 is a photograph to illustrate the improved snow melting performance achieved with the high thermal conductivity experimental pavement built at Fukui University. The time variations of the snow depth, S_D, and snowfall rate, h, on the day the photos were taken are indicated in Figure 7. All three pavement surface types are visible in the photograph. The middle segment of the photograph is section As_1, the high conductivity asphalt closed-loop hydronic pavement, while the normal conductivity asphalt hydronic section, As_2, is the section on the far left. The hydronically heated zones are surrounded by the unheated pavement of

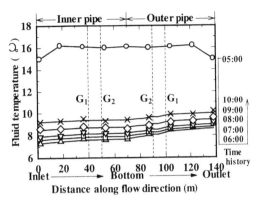

Fig.2 Fluid temperature along bore-hole during snow-melting operation

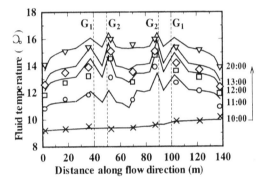

Fig.3 Fluid temperature recovery response after the shut off of the circulation pump

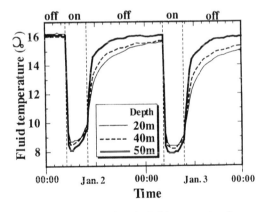

Fig.4 Changes over time in fluid temperature in outer pile

standard construction, labelled As_3 in the figure. At 07:00, the snow depth on the As_3 pavement was 18cm but the snow on section As_1 had already disappeared completely, while a slushy snow layer

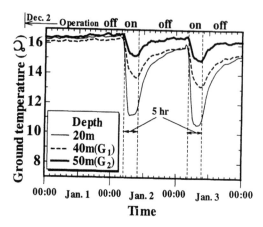

Fig.5 Changes over time in fluid temperature in the vicinity of the bore-hole

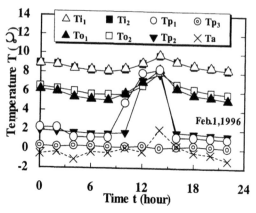

Fig.8 BHES fluid, roadway pavement and air temperatures variations with time

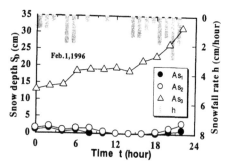

Fig.6 Snow-melting of the car park using the BHES and conducting pavement

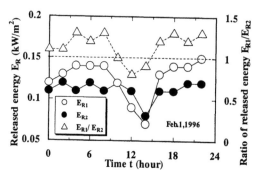

Fig.9 Comparison of energy flux released from the piping system with the high thermal conductivity asphalt pavement layer to that with normal conductivity asphalt

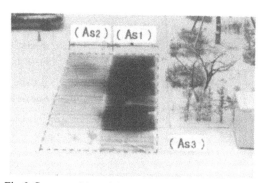

Fig.7 Time variations of snow depth and snowfall intensity

still remained on section As_2. As seen in Figure 7 it took until 12:00 for the snow to completely disappear on section As_2. After 14:00 the snowfall rate, h, remained steady at about 2cm/h for six hours and, by 22:00, the snow depth on the unheated section As_3 had increased to as much as 32cm. During the same time, the snow depth on section As_1

did not exceed 1cm while that on section As_2 built up to about 5cm.

Figure 8 shows, for 1996/01/01, hourly values of the inlet (return) fluid temperatures Ti_1 and Ti_2 of the heat exchange pipes in the two heated pavement sections, As_1 and As_2 , the corresponding outlet (supply) temperatures, To_1 and To_2, and the asphalt temperatures 0.01m below the surface of all three pavement sections, Tp_1, Tp_2, and Tp_3. Instantaneous air temperature, Ta is also indicated. It is seen that Ta was continually below 0°C except around 14:00 while Tp_3 was approximately 0°C over the course of the whole day because of the effects of the snow cover. The heated section pavement temperatures, Tp_1 and Tp_2, remained near 1°C at night and then begin to rise at about 09:00 and 10:00, respectively. This indicates that the snow melting was faster for section As_1 than for section As_2. Tp_1 and Tp_2 both reached a maximum (8.2°C) at 14:00 after snow melting was complete and the black asphalt pavement surfaces were exposed to similar micrometeorological conditions. After that, however, new snowfall occurred, and this lowered both Tp_1 and Tp_2 back to about 1°C again.

Regarding the fluid temperatures, for both heated sections, To was always lower than Ti, and this implies that the ground heat transported by the fluid was released to the pavement. The data also show that, at night, To_1 was lower than To_2 even though the inlet temperatures Ti_1 and Ti_2 were similar. This is indicative of the better heat transfer characteristics of the high thermal conductivity section.

Figure 9 shows the time variation of the heat flux, E_R, per unit area of pavement released from the piping system. The heat fluxes for the two sections, E_{R1} and E_{R2}, exhibit a diurnal variation and values are larger at night. They also increase during snowfall events. It is significant that the total energy extracted from the piping system for section As_1 was about 30% greater than for section As_2 during night-time as well as during snowfall. This result may explain the clear difference in snow-melting rate between the high thermal conductivity asphalt pavement section As_1 and the normal conductivity asphalt section As_2 seen in Figure 6.

4 CONCLUSION

The snow-melting performance of a high thermal conductivity roadway pavement combined with a Bore-hole Heat Exchange System (BHES) in an alluvial soil was evaluated. In addition, the thermal interaction between the ground and the bore-hole fluid was compared at levels containing aquifers and at those with no water flow (aquitards).

The following conclusions were drawn:

(1) The temperature drop of the ground around the bore-hole that occurs when the system is operated for snow-melting is significantly reduced within aquifer layers as compared to aquitards.

(2) Recovery of the ground temperatures in the vicinity of the bore-hole when the circulation pumps are shut off following a snow-melting operation cycle is much more rapid at levels with aquifers than at those without.

(3) Sensible heat associated with groundwater flow in aquifer layers contributes substantially to stable extraction of ground heat by the BHES.

(4) The total energy extracted from the serpentine piping system in the Portland cement concrete sub-layer was about 30% greater than normal when the surface asphalt layer had a relatively high thermal conductivity.

(5) It is concluded that a high thermal conductivity surface pavement can improve the energy performance of hydronic snow melting systems, particularly those using low density energy sources such as ground heat.

5 REFERENCES

Ohki et al., 1997, Control of Pavement Temperature by Bore Hole Heat Exchange System, *Proceedings of 7th International Conference on Thermal Energy Storage*, Sapporo, Japan, pp. 127-132

Tanimoto et al., 1999, Influence of Groundwater Flow on Extraction of Ground Heat by Bore Hole Heat Exchange System, *Proceedings of Japan Society for Snow Engineering*, Vol.16, pp. 89-90

Fukuhara T. and Sakamoto N., 1999, Snow Melting and Cooling of Pavement Using Ground Heat at a "*Michi no Eki*" Service Area on Route 8, *Journal of Snow Engineering of Japan*, Vol. 15, No.4, pp. 31-33

Snow Engineering, Hjorth-Hansen, Holand, Løset & Norem (eds) © 2000 Balkema, Rotterdam, ISBN 90 5809 148 1

Investigations on mechanism of landslide in the snowy regions

Takeshi Ito
Department of Civil and Environmental Engineering, Akita National College of Technology, Japan

ABSTRACT: Seasonal snow thawing, snow load, snow glide and sudden rain fall at the end of winter those are considered to be main influence factors for snow region landslide. This paper treats first statistics of snow induced landslide disasters, then the mechanism of landslide which will be discussed with changes of daily snow cover depth and snow melt water through the field observation. From the result of observation, it is pointed out that the most of large scale landslides often take place when snow melt water reached more than 20 mm/day especially at period of a couple of weeks to snow melt away. Model analyses for snow load on the slope by R. Haefeli, quantity of snow melt water by degree-day method are applied to assist the explanation of landslide mechanism in the snowy regions.

1. INTRODUCTION

In the snowy regions, a landslide frequently occurs when the inherent factor is triggered by inducing phenomena. Landslide behavior in the snowy regions where are not well known as its observation is difficult in winter. However, in Japan, a number of large scale landslides have often been taken placed in snow period especially at places facing the Sea of Japan. These areas experience some peculiar snowfall phenomena and is known as a world-famous heavy snow fall region even though it is located at relatively low latitudes. Because of heavy snow, several snow induced unstable ground conditions are produced, for example, snow melting, snow load, snow glide, and freezing and thawing phenomena are main influence factors for snowy region's landslides.

In addition to the above, sudden May storm is considered as a fatal influence factor for landslide. On the other sides, comparatively steep slopes and weathered soft rocks such as weak green tuff, mud stone, shale etc. exist redundantly on whole over the areas along the Sea of Japan. These common topographic, geological and climatical conditions prompt to make easily soil mass movement and ground failures. Heavy snow yields large quantity of snow melt water that makes an equivalent to heavy rain fall particularly at the end of spring.

In this paper, its quantity is calculated by the use of degree-day method. Snow load is also one of inducing factors, so that the estimated safety

factors are represented by a method proposed by R. Haefeli（1965）, next soil behavior will be analyzed. These influence factors will be discussed with using field observation data.

2. LANDSLIDE DISASTERS IN SNOWY REGIONS

From the data of the Ministry of Construction of Japan, 11,288 potential landslide sites were selected in areas under his jurisdiction as of July 1, 1998, in which 3,291 areas were designated as high risk landslide threatened areas.

Figure 1 shows potential landslide sites by each prefecture. As is seen from the figure, a lot of potential landslide sites exist in Niigata, Nagano, and Ishikawa where locate all at heavy snowy regions. Nagasaki has also many but few snow, however, locates in green tuff areas. Similarly, Wakayama, Tokushima, and Ehime have also many, and all is running along the central fault zone. Although snow induced landslide disaster is not high rate compare with other factors, however, each landslide scale is quite large.

Figure 2 shows a relationship between snow cover depth and landslide frequency which was investigated in 1987/1988 winter. It can be pointed out that a huge landslide not always appears at extreme heavy snow regions, it appears prefer in areas at snow cover depth between 100 cm to 300 cm.

3. SNOW PHENOMENA RELATE TO LANDSLIDE

Large scale landslides in the snowy regions prone to appear in March to May, and its disaster areas occupy approximate 70% to all landslide disasters. This kind of disasters depend highly on various snow phenomena especially snow melting water and snow load. We have experienced a huge landslide disaster on May 11, 1997 so called Hachimantai Sumikawa hot spa landslide, northern part of Honshu island of Japan (Ito, 1998).

Figure 3 shows bird's view of the landslide on May 12, 1997, the next day of the event. Soil block of the slide is roughly 850 m in length to south-north, width is roughly 350 m in length to east-west. The suffered site sectional area is 35 ha, and estimated sediment discharge volume is 2.5×10^6 m³ . Crown site of the landslide locates at the vicinity of 1000 m from the sea level, and tongue site have felt down. Its position just starts from snow line. Naturally, snow cover depth in areas is at least more than 300 cm every year.

According to the records of daily snow cover depth that were recorded at Sumikawa geothermal electric power station (elevation: 1050 m) shown in the right upper of the photo, its peak snow cover depth was observed as 360 cm in 1996/1997 winter, and its value is rather lower than its average value.

Figure 4 is a diagram of weather situations in those days. It is noted that snow melting would make a good progress from April, and last 110 cm snow pack had thawed away within 10 days at the beginning of May. Sum of rain fall and calculated snow melt water by degree-day method using the coefficient of snow melting 3.8 mm/℃ · days was drawn in this diagram. At snow melting period, it is a general case that snow melt water is more than 10 mm/day, if it becomes more than 20 to 40 mm/day, a large scale landslide probably be occurred, particularly it will be accelerated by sudden heavy rain fall. As is seen in the figure, May storm brought a heavy rain fall as 129 mm/day on May 8, 1997 just before 3 days of the event. This kind of sudden additional heavy rain fall in the snow melting period is an important key factor for every landslide disaster (Ito, 1999).

4. SNOW LOAD EFFECT

One more another reason to cause landslide at snowy season, snow load might be worked to change of shear strength of landslide soil layer as an effective stress of soil mass movement. For example, if snow load become larger, normal stress in the soil become large, it would make high pore

water pressure and it saturates soil continuously then soil shear strength decreases, while under ground water absorbent gets lower from outside. This will make a reservoir curtain like an underground dam which may produce following high pore water pressure particularly at late snow season. On the other hands, sudden heavy snow causes soil displacement in case of shallow landslide because it acts directly on a shallow slip plane forming a circular arc, accompanying soil consolidation. Weight of landslide mass will change accordance with snow pack, therefore, thrust force against landslide increases then it gets large and soil mass will move down gradually. Landslide might be occurred with these conditions.

Figure 5 (a) and (b) is a supportive picture

Figure 1. 11,288 potential landslide sites by prefectures under the jurisdiction of the Ministry of Construction as of July 1, 1998.

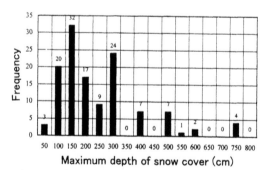

Figure 2. Relationship between snow cover depth and landslide frequency in 1987/1988 winter (data source: Ministry of Construction and Forestry Agency of Japan).

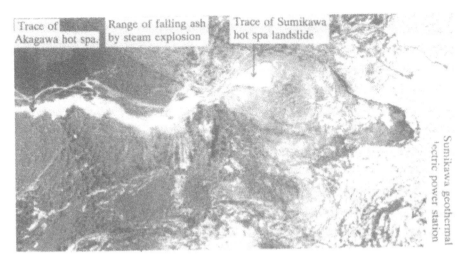

Figure 3. Sumikawa hot spa landslide occurred on May 11, 1997.

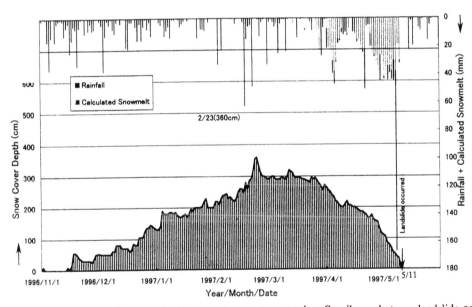

Sumikawa Geothermal Electric Power Station

Figure 4.   Weather situations in 1996/1997 snow season when Sumikawa hot spa landslide occurred.

which describes snow load acting against landslide. At the end of snow season, snow load is decreasing and contrary snow melt water would be supplied that help soil mass will be saturated. Thus pore water pressure is increasing and ground water level is rising. Therefore, landslide soil mass movement is gradually accelerated by the rate of rising of ground water table. Pore water also acts as lifting pressure to the landslide soil mass, thus finally these forces lead to flow it downward. This progress might be described geotechnically in the Figure 5 (c). In this figure, e is as void ratio and SF is landslide safety factor.

It has been reported that several landslides appeared in the earlier snow fall season. For example, a landslide at Yamagata prefecture

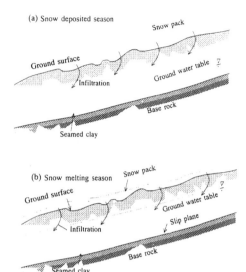

(a) Snow deposited season

(b) Snow melting season

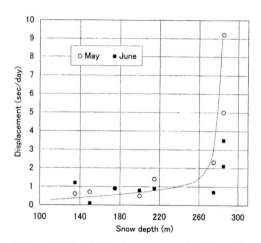

Figure 6. Landslide displacement in May and June against snow cover depth.

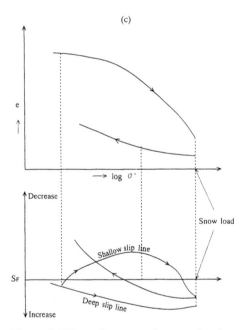

Figure 5. Effect of snow on the ground and geotechnical relationship between effective stress (log σ'), void ratio (e) and landslide safety (S_F).

occurred when heavy snow fall was recorded as 100 cm within two days, at which maximum snow cover depth was 200 cm. One more landslide example occurred when snow cover depth had reached to 400 cm. The reason is considered to be a snow load because of sudden heavy snow fall, it might be made an additional soil mass load.

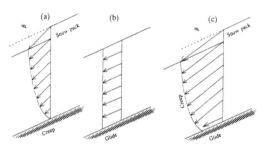

Figure 7. Models of snow creeping and gliding.

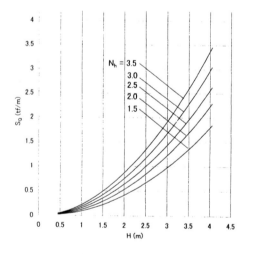

Figure 8. Relationship between snow cover depth (H), coeeficient of glide, and snow load (S_Q) on sloped ground.

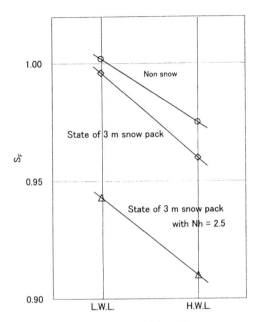

Figure 9. Landslide safety (S_F) taking account of snow load, S_Q.

Figure 6 is a typical example of certain landslide displacement which was observed at heavy snow region of the northern part of Honshu island, Japan. From the records which had been observed extend over several years, the soil mass movement becomes remarkable while the snow cover depth reached around 250 cm, rapid landslide movement had been appeared snd its displacement was about 2 radian second per day. Although average snow cover depth in those areas is approximate 300 cm, so that 250 cm appears generally at the end of January or at the beginning of February.

5. CREEPING AND GLIDING OF SNOW PACK

Snow pack on the slope acts, not just effect of distribution load but also has a character as snow creeping. In this situation, snow works as if like a sailor mast and landslide soil mass is just like a sailor. According to Haefeli's theory, snow pack effects on slope at right angle of static snow pressure (Ss) is,

$$Ss = (\gamma_t \cdot H^2/2)(1- 2\tan i \cdot \tan \beta) \cos^2 i \quad (1)$$

in which, γ_t: snow density, H: snow cover depth, i: slope angle, and β: vector toward deformation. And snow pressure on the slope as,

$$Sk = \gamma_t \cdot H^2 \cdot K \cdot Nh / 2 \quad (2)$$

in which, K: coefficient of creep of snow, Nh: coefficient of glide, and they are represented as,

$$K = (\sin 2i / 3) \sqrt{2/(\tan i \cdot \tan \beta)} \quad (3)$$
$$Nh = \sqrt{1+3n} \quad (4)$$

Total snow pressure is,

$$S = (\gamma_t \cdot H^2/2)(1- 2\tan i \cdot \tan \beta) \cos^2 i$$
$$+ (\sin 2i / 3) \sqrt{2/(\tan i \cdot \tan \beta)} Nh \quad (5)$$

This situation is described in the Figure 7. Application of this principal will be calculated in the following section.

6. SNOW LOAD AND LANDSLIDE SAFETY

Owing to investigate landslide safety, a sample was cited from a heavy snow region because of its deformation was remarkable. This landslide had four blocks and main moving block's depth was approximate 15 m. Soil mass was consisted mainly from weathered tuffs including 10 cm thick of soft clay at the slip plane whose N value is fairy low but permeable because parting angular stones. The slope was long and irregular forming, so that we used Janbu's non-circular method to examine safety.

Figure 8 shows the effect of snow load against snow cover depth. Following Figure 9 is a calculated safety in cases of no snow and 3 m snow cover depth. In this calculation, we used Nh=2.5 and snow density as γ =0.35tf/cm³.

According to the results, landslide safety was affected whether taking account of coefficient of glide or not, that is, landslide safety decreased 7% when considering coefficient of glide for the case of 3 m snow pack. In this way, our investigation proved that landslide safety is affected by snow glide and snow creep phenomena especially more than 1 m depth of snow pack.

7. CONCLUSIONS

We examined conduct of sloped snow load based on field data by using Haefeli's theory. As the results, we recognized that snow load is one of complicated conduct of snow and one of important factor to cause landslide, nevertheless, its mechanism is not known enough. Throughout winter observation, it is essential to clear up relationship between snow and landslide based on accumulating observation and analytical data.

REFERENCES

R.Haefeli 1965. Creep and progressive failure in snow, soil, rock and ice, *Proc. 6th int'l conference. SMFE,3,pp134-148.*

T.Ito 1998. Meteorological conditions and clay characteristics in the Hachimantai Sumikawa hot spa landslide/mud flow disasters, *J.Japan Landslide Society.* 35-2: 77-85.

T.Ito 1999. Snow induced landslides in Japan, *Slope Stability Engineering, Vol.2, 1223-1228*, Balkema.

Snow Engineering, Hjorth-Hansen, Holand, Løset & Norem (eds) © 2000 Balkema, Rotterdam, ISBN 90 5809 148 1

Drifting snow in complex terrain – Comparison of measured snow distribution and simulated wind field

Christian Jaedicke – *The University Courses on Svalbard, Norway*

Thomas Thiis – *Narvik Institute of Technology, The University Courses on Svalbard, Norway*

Anne Dagrunn Sandvik – *University of Bergen, Norway*

Yngvar Gjessing – *The University Courses on Svalbard, Norway*

ABSTRACT: Snowdrift in complex terrain can cause significant problems for maintenance of road and railway systems during winter and is a major factor for snow avalanche releases. The location of accumulation and erosion areas is highly connected to the local wind field. The snow distribution in an area of 3 x 4 kilometers is measured by ground penetrating radar. The study area is located at 78° North on Spitsbergen and features elevations from 200 m to 900 m above sea level. Snow distribution measurements are done twice during the field season in winter 1999 to monitor the spatial changes in snow distribution caused by snowstorms. In addition, the wind speed and wind direction is measured at four stations in the study area. The local wind field is simulated using a non-hydrostatic numerical mesoscale model. It can be seen that there is a relationship between snow distribution and wind climate in the area. High wind speeds correspond well to erosion areas while low wind speeds correspond well to accumulation areas. The achieved data can be used in future projects to validate distributed snowdrift models for complex terrain both for risk analysis and hydrological applications.

1 INTRODUCTION

Wind transported snow is a phenomena of the atmospheric boundary layer which is often observed in regions with little vegetation and a seasonal snow cover.

The wind redistributes the precipitated snow particles in the terrain. This process causes many problems for mobility in the affected areas, such as decreased visibility and accumulated snowdrifts on traffic lines. The redistribution affects the melting and runoff in spring as an important factor for hydroelectric power plants and flooding. Furthermore, drifting snow is a major input to snow avalanche release areas. In complex terrain the snowdrift pattern has a very high local variability, changing the con-

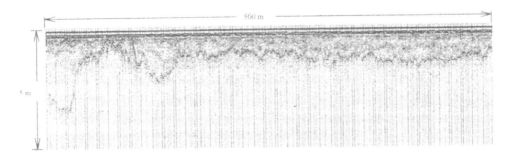

Figure 1. Radar profile from the study area. The length is 500 m and the depth 5 m. The interface between the snow and underlaying material can easily be seen.

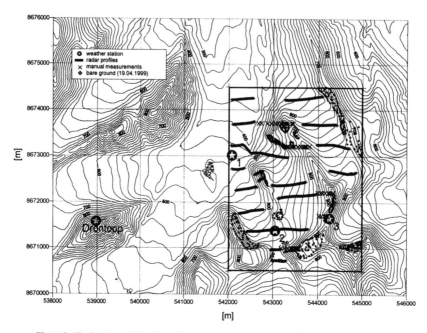

Figure 2. The location at Drønbreen. The study area is shown by a black frame.

ditions completely along short distances of only a few hundred meters. This effect makes the prediction of drifting snow difficult and requires high resolution of numerical drifting snow models. Some of the effects of drifting snow can be handled by the application of snow fences (Tabler 1988), but it is necessary to know the areas of strong snow fluxes before the construction of traffic lanes and buildings. Therefore, knowledge of the snow accumulation process in the terrain and the dominating wind field should provide the needed information to optimize the location of new structures.

The effect of complex terrain was early recognized by Andreé (1883). Norem (1974) considered the effects of the terrain during his experiments. However, field studies were mainly done on homogenous fetches (Budd et al. 1966, Mellor 1965). The snow transport over a mountain crest in the Swiss Alps is described by Föhn (1980) and Föhn & Meister (1983) They showed the modified wind and snow flux profiles on the crest due to the speed up effect and the resulting snow distribution. The snow transportation can be divided into three modes, creep, saltation and suspension. Creep is generally not considered in flux modeling due to its very small contribution to the total flux. The difference between the total mass transported by saltation and suspension is discussed by Dyunin & Kotlyakov (1980). They consider saltation to be the major contributor to the snow mass transported in complex terrain. Iversen (1980) modelled the effect of highway bridges and the surrounding terrain in wind tunnel experiments.

However, three-dimensional distributed snowdrift models for complex terrain are still under development. Pomeroy et al. (1997) presented a distributed snow model for the Canadian prairies. It uses a statistical method to determine the wind field in the terrain and works with standard meteorological observations of wind speed, temperature and humidity. Liston & Sturm (1998) developed a model in which saltation, suspension and sublimation were handled separately. The wind field was generated by interpolation of observed wind data in connection with empirical wind-topography relationships.

Gauer (1998) applied a hydrodynamic flow model for the wind field and coupled it to a drift model. The results were tested on a single snowstorm in the Swiss Alps and proved to reproduce the snow distribution well. A rule-based model was presented by Purves et al. (1998), which gives the flux from cell to cell using meteorological and snow surface properties.

For validation of these models, high quality distributed data of snow depth in complex terrain is needed. In this study a snow distribution data set from the high Arctic is provided and is compared with results from a high-resolution numerical wind model.

2 THEORY

The theoretical background on snowdrift was recently described by Pomeroy (1997), Liston & Sturm (1998) and Gauer (1998).

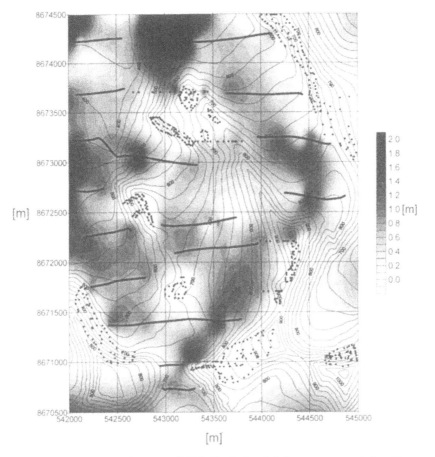

Figure 3. Snow distribution 19. April 1999. The depth scale is in metre water equivalent. Lines and dots show the data coverage.

The transported amount of snow is dependent on the applied shear stress on the surface $\tau = \rho_a u_*^2$, where τ is the shear stress, ρ_a is the air density. u_* is the friction velocity, which can be found from measurements assuming a logarithmic wind profile:

$$u(z) = \frac{u_*}{k} \log\left(\frac{z}{z_0}\right) \tag{1}$$

where $u(z)$ is a the wind speed at a reference height, z_0 is the roughness length and k is the von Karman constant.

Saltation is started when the friction velocity reaches a certain threshold value u_{*t}, which is dependent on snow-surface properties. Following Pomeroy & Gray (1990), this process can be described by

$$Q_{salt} = \frac{0.68 \rho_a u_{*t}}{u_* g}\left(u_*^2 - u_{*t}^2\right) \tag{3}$$

where Q_{salt} is the flux of saltating snow, and g the acceleration due to gravity. At higher wind speeds particles are lifted far above the ground and are transported over large distances in suspension. The snow flux by suspension Q_{susp} is given by

$$Q_{susu} = \int_{z_0}^{\infty} n(z) u(z) \, dz \tag{2}$$

with the snow drift density n as

$$n = n_{z_0}\left(\frac{z}{z_0}\right)^{\frac{-\omega}{k u_*}} \tag{3}$$

Here ω is the fall velocity of the snow particles. Partly the transported snow mass is reduced by sublimation of the particles during the transport in the air stream.

3 STUDY AREA

The study area is located at 78° 16' North in the high Arctic on the Island of Spitsbergen. The glacier Drönbreen is a 16-km long glacier with several side

67

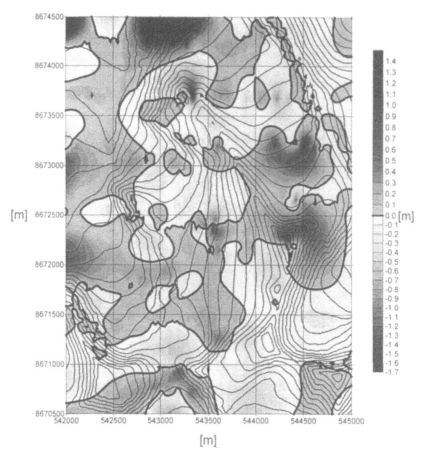

Figure 4. Losses and gains from 23. March – 19. April 1999 in meter water equivalent. Isoline show the topography.

arms. The study area itself is bare of vegetation and partly glaciated. It features elevations from 200 m to 900 m over sea level with slopes up to 45°. The mean annual air temperature in the area is around -5°C. The mean temperature during the season from February to May was -14°C with a maximum of +1.5°C and a minimum of -26°C. The prevailing wind direction in the area is from east. Westerly winds occur less frequent and bring higher air temperature and precipitation with them. Precipitation is not measured near the study area. Estimations and indirect measurements (Hagen et al. 1993) give an estimated value of 0.5 m/a for the Drönbreen area.

4 FIELD MEASUREMENTS

In an area of three times four kilometers the snow depth was surveyed on two occasions by manual measurements, ground penetrating radar (GPR) and visual observations. Ground penetrating radar is a powerful tool to investigate subsurface features. The system sends electromagnetic waves into the ground, which are partly reflected by layers in the subsurface. Sigurdsson (1998) gives a detailed description of the applied system. The boundary between snow and ice or bedrock can easily be seen as a strong reflector (Fig. 1). The system measures only the speed of the electromagnetic waves from the sender back to the receiver. Therefore the speed of the waves in the media must be found by careful calibration work. GPR has been used for snow depth surveys by Sand & Bruland (1998) and Annan et al. (1996), who also give a formulation to calculate water equivalent from calibrated speed and depth.

In the GPR measurements a 900 MHz system was applied, triggered every 0.5 m by an odometerwheel. The resolution at this frequency is around 0.12 m, depending on the speed on the radar wave in snow. The radar system was rigged on a sled and pulled by a snowmobile along the profiles. The profiles were spaced with a distance of 500 m in east west direction, depending on the topography. The profiles

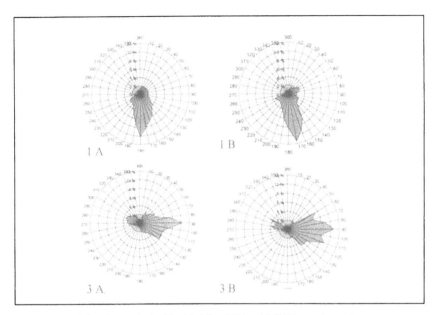

Figure 5: Wind directions at Station No. 1 & 3 for all (A) and drift (B) occasions only

were marked with bamboo sticks to allow an accurate repetition of the first survey. In areas, not accessible with the radar system, some of the profiles where continued, marked and measured manually with a snow avalanche probe. During this work, the area was observed from different locations by binoculars to detect areas with clearly visible stones. These areas where marked in a map and later noted as bare ground areas. The position of the radar-profiles, manual measurements and areas of bare ground are shown in Figure 2. Together about 6500 m of radar profiles 100 manual measuring points and large areas of bare ground covered the study area. After the last survey the profiles where positioned with a differential GPS system. For calibration of the radar system, the snow depth was manually measured between the radar antennas at selected locations and several snow pits were dug in connection with the radar measurements.

Four automatic weather stations (AWS) were placed in the study area of which one is working all year round on a near by mountaintop at 1020 m elevation. The purpose of this station is to give an indication of the geostrophic wind directions in the area. The other three stations where situated in the study area at different elevations, measuring 5 min averages of wind speed at two levels, wind direction, air temperature and relative humidity. The AWS were maintained continuously during the season and the height of the instruments was frequently corrected after snow accumulation at the stations.

4.1 Results

The calibration of the radar system was done ten times at each survey. The resulting speed of electromagnetic waves is shown in table 1 and can be compared with the snow pit results. During the calculations, the radar calibration values were applied. The measured snow depth varied from zero to 2 m water equivalent. The distributed data was positioned using the GPS data and interpolated by minimum curvature with 50 m spacing. The result is plotted on the basis of 1996 topography data by the Norwegian Polar Institute. Typical features of the snow distribution can be seen in Figure 3 which

Table 1: Densities and wave speeds at the two surveys

	In situ Radar calibration		Control (snow pit data)	
Date	Electromagnetic wave speed m/ns	Bulk density kg/m³	Electromagnetic wave speed m/ns	Bulk density kg/m³
23. March 1999	0.212 ± 0.02	386 ± 12	0.214 ± 0.02	373 ± 19
19. April 1999	0.210 ± 0.02	398 ± 12	0.213 ± 0.03	382 ± 21

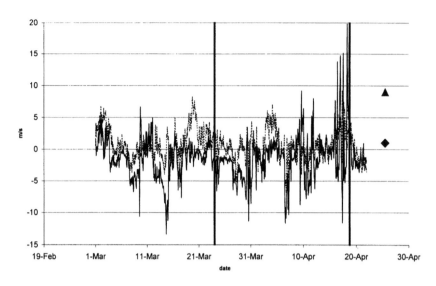

Figure 6. Wind speeds deviations from Station 1 at Station No.2 (...) and No.3 (-). Simulated values are shown for Station No.2 (♦) and No.3 (▲). Black vertical lines show the date of the snow surveys.

gives the result from the second survey: Exposed areas like ridges are nearly all free of snow, while steep leeward slopes accumulate large amounts of snow. The more gentle slopes in the center of the study area show a uniform snow distribution, which also can be found, on large areas further down the glacier.

Comparison of the two surveys allows an indication of the changes in snow cover distribution. Figure 8 shows a map with positive and negative changes from 23. March 1999 to 19. April 1999. Losses can mainly be observed in the exposed areas of the terrain like ridges and tops. Gains can be seen in lee-slopes and in areas with weak slopes. However, the pattern is much more complex than the expected erosion / accumulation pattern. Both losses and gains are observed in areas where the opposite would be expected from the map.

Wind roses from two of the four weather stations are shown in Figure 5. The domination of easterly winds at station No. 3 turns to a more southern direction at station No. 1. On the mountaintop, Dröntoppen, the station suffered from heavy icing during the season. Only data from ice-free occasions can be used. Therefore the results are not included in this study. No wind direction data was available from station No. 2. Figure 6 shows the deviations from wind speeds measured at station No. 1. In average station No. 3 shows lower wind speeds than station No. 1, while is No. 2 is higher. The two additional points show the model result for the corresponding area.

At friction velocities over 0.20 m/s drifting snow can be expected. This is the case during more than 35% of the time at station No. 2. Using the same threshold friction velocity as a filter value gives the two wind roses for drift events. The wind direction is turned to the east, which indicates the dominating easterly winds at drift occasions.

5 NUMERICAL MODEL

In a first modeling approach, the non-hydrostatic numerical mesoscale model MEMO (Flassak 1990, Moussiopoulos 1994, Grönås & Sandvik 1999) was used to generate a wind field for the study area. The model solves the momentum equation, the continuity equation and optionally several transport equations for scalars on a staggered grid. The applied model version includes the following dependent variables: The three wind components, pressure, potential temperature and turbulent kinetic energy. The vertical coordinate is normalized to the underlying terrain. One-way nesting is featured by MEMO. A nested model includes the effect of topography upstream of the integration area and reduces the influence of the lateral boundary conditions.

In the present study the high resolution study area (50 m x 50 m) was enclosed in three nesting levels up to 500 m. Topographical data, available in a 20 m x 20 m grid, was interpolated to the model grid. Vertically, the atmosphere was divided exponentially into 25 terrain following layers from 10 m to

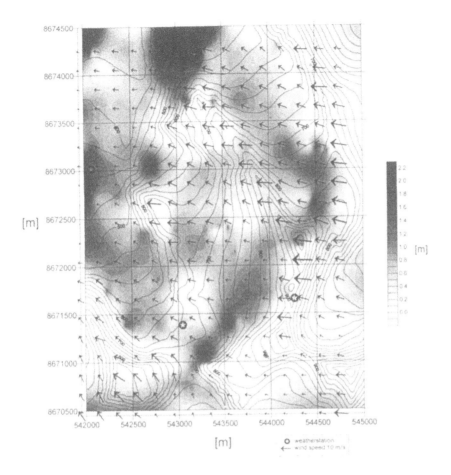

Figure 7. The simulated wind field in the study area. Arrows indicate the wind direction and strength.

4000 m. Only two surface types, snow covered land ($r_0 = 0.1$ m) and ocean ($r_0 = 0.001$ m), were used.

The wind field in the study area will be highly influenced by the atmospheric stability, wind speed and direction. The winter climatological wind direction at Spitsbergen is easterly. At this direction the upstream conditions are dominated by low-level inversions due to net radiative loss of energy from the surface. As a first approach it was decided to use a stable stratified atmosphere (constant potential temperature gradient equal to 0.7108 °C/100 m), and a uniform easterly wind of 10 m/s was imposed. The model computation was stopped when it reached nearly steady state after one hour.

5.1 Results

A mountainous terrain surrounds the study area at Drønbreen in all directions. Apparently the nearest mountains have the strongest influence on the wind field and snow distribution. But it was found that also the topography further upstream influences the model results. The wind field 10 m above the ground for the nested 50 m grid simulation is shown in Figure 7.

Comparison with the snow distribution map (Fig. 8) shows a good agreement of wind speed and snow distribution. High wind speeds correspond to erosion areas with little or no snow, while low wind speeds and strong negative gradients are located near accumulation areas with deep snow. Further experiments with different wind directions and atmospheric stabilities will be done to increase our knowledge of the wind field in the study area.

6 DISCUSSION

The chosen study area proved to be suitable for the objective to gain distributed snow depth data in a larger area of several square kilometers. The terrain is complex enough to produce a rapidly changing snow distribution in space and is still not too steep

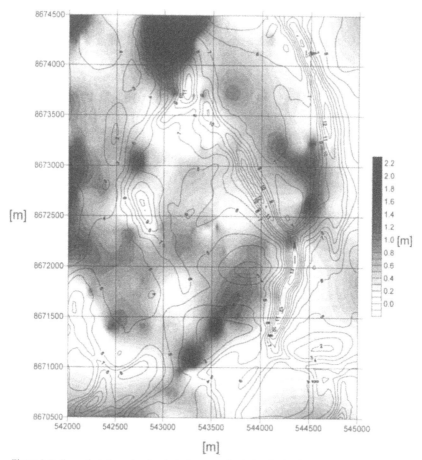

Figure 8. Isolines of wind speed and underlaying snow depth distribution in metres water equivalent.

for snow depth measurements and the numerical model.

The accuracy of the snow depth data is mainly dependent on the calibration of the radar system while the accuracy of the distribution is dependent on the positioning of the profiles. The calibration of the radar system shows only a small variability of snow properties. Using the average value of 0.22 m/ns yields an error of +/- 0.06 m water equivalent in the snow depth measurements. Profile positions collected by differential GPS allow an accuracy of +/- 2 m in x/y direction. The accuracy of the bare ground area positions is difficult to quantify, but considering the different locations of observation, the error should not be more then ten meters in x/y direction.

The achieved maps give a good impression of the snow distribution in the study area. Both areas of accumulation and erosion can easily be located. Strong gradients from erosion zones on the crests to areas of deep snow in the corresponding slopes indicate strong snow fluxes from the crests into the lee slopes. On the other hand, the uniform snow distri-

bution on the gentle slopes further down the glacier, indicate equilibrium of accumulation and erosion.

The losses and gains, shown in Figure 4, give a complex pattern of accumulation and erosion, which is not obvious from a first guess. This stresses the importance of high-resolution numerical snowdrift models, which can simulate such small-scale features. Some of the losses can be explained by evaporation from the transported snow. This can amount up to 40% of the suspended snow mass (Pomeroy 1997). A suitable routine, to quantify the evaporation losses, has to be considered in the development of high-resolution snowdrift models.

The collected weather data builds a good basis for comparison of the wind field with any model results. In addition, temperature and humidity data can be used later in evaporation modeling.

As shown in section 5 the idealized numerical simulations produced a realistic wind field. Details, like the changing wind direction from East at station No. 3 to South at station No. 1, are indicated by the model results. The simulated wind velocity at station

No. 3 is higher then the measurements while station No. 2 shows comparable velocities. It has to be kept in mind, that the model results are not based on a real but on an idealized weather situation. In addition the measurements are made at 2 m while the simulated wind is at 10 m. The wind model is the most important part of any snow drift model and the correspondence of the wind speeds with the snow distribution is a first promising results into the direction of a coupled wind / snowdrift model.

7 CONCLUSION

A data set of high resolution distributed snow depth data has been achieved. The data is of high quality and situated in an area ideal for modeling approaches. Additional weather and wind data from four stations allow a comparison of the natural and modeled wind fields. A first wind field simulation shows good agreement with both the collected wind data and snow depth data. The achieved data will be used in further investigations to test different distributed snowdrift models.

8 ACKNOWLEDGEMENTS

The topographical data was provided by the Norwegian Polar Institute. Supercomputing resources have been made available by the Norwegian research Council. Thanks to the field workers and technical unit at the University courses on Svalbard.

REFERENCES

Andrée, S.A. 1883. Om ytsnön I de arktiska trakterna, *Kgl. Vet. Akad. Förh. No. 9: 33 - 41*

Annan, A. P. & Cosway, S. W. & Sigurdsson, T. 1995. GPR for snowpack water content, *Fifth International Conference on Ground Penetrating Radar Vol 2(3),* Kitchener, Ontario, Canada

Budd, W. F. & Dingle, W. B. J. & Radok, U. 1966. The Byrd snow drift project: outline and basic results, in Ruben, M. J., ed: *Studies in Antarctic meteorology*. Washington, D.C., American Geophysical Union, (Antarctic research series, Vol. 9): 71-87

Dyunin, J.A. & Kotlyakov, V.M. 1980. Redistribution of snow in the mountains under the effect of heavy snow storms, *Cold Regions Science and Technology*, 3 (4): 287-294

Flassak, T. 1990. Ein nicht-hydrostatisches Modell zur Beschreibung der Dynamik der planetaren Grenzschicht, Ber. VIDZ, Reihe 15, *VDI-Verlag, Duesseldorf Germany*

Föhn, P. M. B. 1980. Snow transport over mountain crests, *Journal of Glaciology, 26, No. 94:* 469-480

Föhn, P.M.B. & Meister, R. 1983. Distribution of snow drifts on ridge slopes: Measurements and theoretical approximations, *Annals of Glaciology, 26:* 469-480

Gauer, P. 1998. Blowing and drifting snow in alpine terrain: numerical simulation and related field measurements, *Proceedings of the international symposium on snow and avalanches, Chamonix, Annals of Glaciology, 26,* in press

Grönås, S. & Dagrun Sandvik, A. 1999. Numerical Simulations of local winds over steep orography in the storm over north Norway on October 12, 1996, *Journal of geophysical research, 104,* Nr. D8: 1-12

Hagen & Liestol & Roland & Jörgensen 1993. Glacier Atlas of Svalbard and Jan Mayen, *Norwegian Polar Institute, medeleser,* No. 129,

Iversen, J. D. 1980. Drifting snow similitude transport rate and roughness modelling, *Journal of Glaciology, 26,* No. 94. 393-403

Liston, G.E. & Sturm, M. 1998. A snow-transport model for complex terrain, *Journal of glaciology, 44:* 498 - 516

Mellor, M. (1965): Blowing Snow, *Cold regions science and engineering part iii, section A3c,* pp 80

Moussiopoulos, N. 1994, The Zooming Model, Model structure and application, *EUROTRAC Int. Sci. Secr., Garmisch Partenkirchen, Germany,* 266

Norem, H. 1974. Utforming av veger i drivsnöområder (design of roads in drift snow areas), *Norges Tekniske Högskole, Trondheim*

Pomeroy, J. W. & Gray, W. M. 1990 Saltation of snow, *Water resources research, 26,* 1583-1594

Pomeroy, J.W. & Marsh, P. & Gray, D.M. 1997. Application of a distributed blowing snow model to the Arctic, *Hydrological processes, 11:* 1451-1464

Purves, R. S. & Barton, J. S. & Mackaness, W. A. & Sugden D. E. 1998. The development of a rule based spatial model of wind transported and deposition of snow, *Annals of Glaciology, 26:* 197 - 202

Sand, K. & Bruland, O. 1998. Application of Geopradar for snow cover surveying, *Nordic Hydrology, 2, 4-5:* 361 - 370

Sigurdsson T. 1998, The Electromagnetic Theory of GPR Technique, *Course Notes on Georadar in Geophysics and Geology,* Norfa, UNIS Longyearbyen, Svalbard

Tabler, R.D. 1988. *Snow fence handbook,* Tabler association, Laramie, Wyoming

Snow Engineering, Hjorth-Hansen, Holand, Løset & Norem (eds) © 2000 Balkema, Rotterdam, ISBN 90 5809 148 1

The snowdrift pattern around a small hill in the High Arctic

Christian Jaedicke
The University Courses on Svalbard, Norway

Thomas K. Thiis
Narvik Institute of Technology, The University Courses on Svalbard, Norway

Børre Bang
Narvik Institute of Technology, Norway

ABSTRACT: The snow distribution in complex terrain is important for the planning of infrastructure, so that large snow depths can be avoided in residential areas and in connection with roads and railroads. Numerical models can give valuable information about the snow drift pattern. Tests of model results against field data are necessary to validate the models. For this purpose, the snow distribution around a 8 m high hill is investigated. The hill is situated at 78° North at Spitsbergen in an area of low precipitation. The deposition is mainly a result of snow, which is redistributed by wind. Ground penetrating radar and a manual snow sonde are used for the snow depth measurements. The snow drifting and wind pattern in the terrain is simulated using a two-phase numerical flow model. The computed snow distribution corresponds well with the measurements. The results encourage application of numerical simulation to larger areas of complex terrain.

1 INTRODUCTION

Drifting snow can cause significant problems in open terrain covered by snow. Snowdrifts are usually developed behind crests, in surface depressions and around obstacles such as buildings and other structures. Mobility on traffic lanes and the accessibility of buildings can be very difficult if these effects are not taken into account during planning of new structures in snow drifting areas.

Drifting snow around single structures such as buildings and snow fences has been studied in wind tunnels (Haehnel & Lever 1995, Iversen 1980, Kikushi 1981) and in full-scale experiments (Tabler 1980, Haehnel & Lever 1994, Mellor 1965, Thiis & Gjessing, 1999). Recently, numerical two-phase flow models have been applied to model snowdrift around buildings (Bang et al. 1994, Thiis 1999) and in complex terrain (Gauer 1998). To predict the effects of drifting snow on buildings and structures in complex terrain, the effect of the terrain must be considered. To allow modeling of these effects, full-scale field data is necessary to validate the model results. In the present study the snow distribution around a small, well-defined hill of 8 m height was measured using ground penetrating radar (GPR). The snow distribution around the hill is assumed to have reached equilibrium with no further accumulation or erosion of the snowdrifts. The collected data is used to validate results from a numerical model, which has previously been used to determine the snow distribution around buildings (Thiis 1999).

There are three processes involved in transport of snow: creep, saltation and suspension. Creep corresponds to particles rolling along the surface without loosing contact with it. The contribution of creep to the overall mass transport is in this connection insignificant and is not considered further. The shear stress τ ($\tau = \rho_a u_*^2$) of the wind on the surface can be described by the friction velocity u_* and the density of air ρ_a. As the shear stress decreases, u_* reaches a threshold value, u_{*t}, where the snow particles stop moving. Above this value, the snow particles are carried along the surface in saltation. At a further increase of wind speed, particles are suspended in the turbulent eddies of the boundary layer and carried over long distances. A more detailed description of saltation is given by Pomeroy & Gray (1990) and of suspension by Budd et al. (1966). Generally, snow is eroded where the concentration of snow in the air is lower than the transport capacity of the air and is accumulated where it exceeds the transport capacity. The value of u_{*t} is dependent on the physical properties of the snow. Old hardened snow has a higher u_{*t} values than newly fallen light snow.

Ground penetrating radar (GPR) has been used to measure the snow depth around the hill. The method offers wide possibilities for underground investigations and has mainly been used in geology and glaciology (Siggurdsson, 1998). Studies of snow depth by GPR were done by Annan et al. (1995) and on Svalbard by Sand & Bruland (1999). The system sends out electromagnetic waves into the ground (snow pack) which are reflected by the interface between the snow and the underlying material (soil, rock, ice). As the GPR only measures the travelling time between sender and receiver, the system must be calibrated by finding the speed of electromagnetic waves in snow. From this the snow water equivalent can be calculated (Annan et al. 1995).

2 FIELD MEASUREMENTS

The studied hill is situated in inner Advent Valley on Spitsbergen at 78° 16' North. The location in the center of a larger valley causes the wind pattern to be fairly unidirectional. An automatic weather station (AWS) was mounted upwind of the hill, measuring wind speed at two levels as well as wind direction and air temperature at a five minute sampling rate. Radar measurements were done twice (03. March 1999 and 20. April 1999) during the season. The profiles were marked in north south direction through the center line of the hill with 5 m and 15 m spacing each about 300 m long (Fig. 1). The radar system had a center frequency of 900 MHz, which gives a resolution of 6 cm to 12 cm. The system was triggered every 0.5 m by an odometer wheel and was pulled by a snow mobile at about 15 km/h. The system was calibrated by manual snow depth measurements between the radar antennas. The detailed topography of the hill was mapped using land surveying equipment.

2.1 Results

The wind rose in Figure 2 shows the wind directions at drift events (u* > 0.20 m/s). During drift events the wind direction is mainly from northeast. Wind speeds reached a maximum in gusts of 16 m/s. The snow distribution at the 20. April 1999 is shown in Figure 3. The snow depth distribution shows minimal alteration from the measurements six weeks earlier, and indicates equilibrium of the snow pattern around the hill. Snow depth range from zero to 1.0 m water equivalent. On the southwest side of the hill, there is a large snow depth maximum (mark "a" in Fig. 3). This is the leeward snowdrift, caused by the decreased shear stress in the wake of the hill. Upwind the hill, another snow accumulation zone has developed (mark "b" in Fig. 3). This zone together with the increased snow depth on the southeast side of the hill, (mark "c" in Fig. 3), might be caused by an increased snow concentration due to the plowing effect of the hill. The blowing snow is plowed to the sides of the hill, and the snow concentration in the air is increasing above the carrying capacity of the air. On the top of the hill, all snow is eroded due to the increased shear stress in this area.

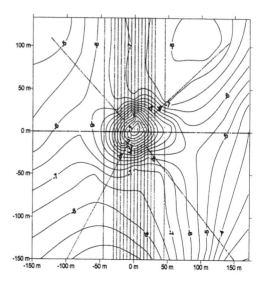

Figure 1. Position of the radar profiles around the hill

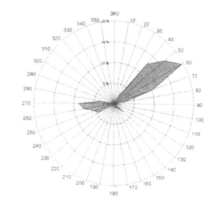

Figure 2. Wind directions upstream the hill at drift occasions [u* > 0.2 m/s]

3 NUMERICAL SIMULATIONS

3.1 Setup

A finite volume Computational Fluid Dynamics (CFD) solver is used to solve the wind flow and blowing snow concentration in the terrain. The time averaged Navier-Stokes equations are closed with the k-ε turbulence model and solved for a three dimensional grid. The grid size is 90 x 90 x 45 cells with grid refinement near the surface. The inlet wind profile follows the logarithmic expression

$$u(z) = \frac{u_*}{\kappa} \ln\left(\frac{z}{z_0}\right) \tag{1}$$

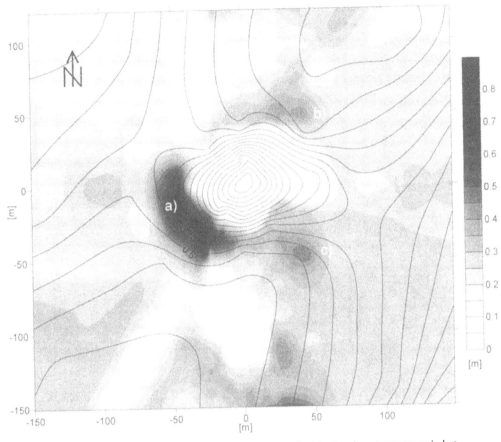

Figure 3. Snow distribution around the hill 20.04. 1999. Snow depth is given in metres water equivalent (shaded). Isolines show the topography of the hill (height equidistance is 0.5 meters).

where $u(z)$ is the wind speed in height z, z_0 is the roughness of the surface, set to 0.002 m, κ is the von Karman constant, equal to 0.4 and u_* is the friction velocity set to 0.6 m/s. This friction velocity is three times the threshold value and corresponds to a wind speed of 12.7 m/s at ten meter height. The wind direction is set to 45°. The snow phase is coupled to the wind with the drift flux approximation, which is also previously used for this purpose (Thiis, 1999; Flow Science, 1997; Bang et.al., 1994). The snow phase is here allowed to drift trough the air phase with the relative velocity

$$u_r = D_f f_1 f_2 \left(\frac{\rho_2 - \rho_1}{\rho_1} \right) \frac{1}{\rho} \nabla p \qquad (2)$$

Here ∇p is the pressure gradient, f is volume fraction, ρ is density and the subscripts 1 and 2 denote the two different fluids. The total mixture density is $\rho = f_1 \rho_1 + f_2 \rho_2$. D_f is a coefficient related to the friction between the snow phase and the air, found by balancing the drag force on a sphere with the buoyancy force.

$$D_f = \frac{2r_0^2}{9\nu} \qquad (3)$$

Here r_0 is the mean radius of the snow particles and ν is the kinematic viscosity of the air. The viscosity of the air at −20 °C is set to $1.63*10^{-5}$ N*s/m² and the air density is set to 1.38 kg/m³. The density of the snow particles is set to 400 kg/m³. To slow down the snow phase in areas of high snow concentration and/or low shear stress, an apparent viscosity is applied to the snow phase. The apparent viscosity is found from the threshold friction velocity of the steady state snowdrift profile, u_{*t} and the expression

$$\mu_{app} = \rho_{air} u_{*t}^2 \left(\frac{dz}{du} \right)_{surface} \qquad (4)$$

The threshold friction velocity is set to 0.2 m/s and the inverse vertical wind profile is found from equation 1.

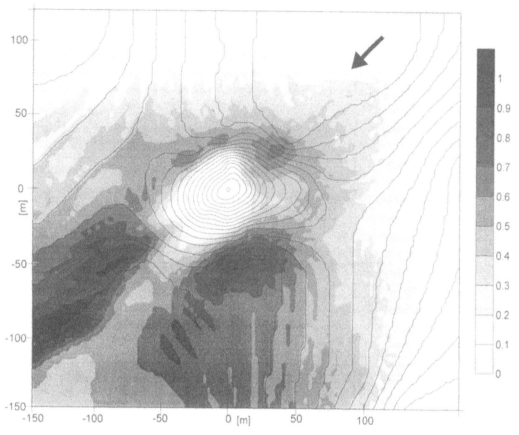

Figure 4. Simulated relative snow distribution around a hill, wind direction is indicated with arrow, height equidistance is 0.5 meters.

The method does not consider the building of a new snow surface, which will change the wind flow. The snow phase is slowed down, and blowing snow density increased where the surface shear stress is low. The increased blowing snow density will again slow down the wind speed. The presented result must therefore only be considered an initial snowdrift growth area.

3.2 Results

Figure 4 shows the computed normalized distance between the iso-surface of blowing snow density $\rho=1.51$ kg/m³ and the surface. This can be interpreted as the relative snow distribution in the area, where the darker areas corresponds to zones with snow accumulation. The three features appointed in the measurements in figure 3 are to some extent reproduced in the simulations. However, the downwind snowdrift is positioned too far downwind and stretched further downwind. The position of the upwind snowdrift, and the snowdrift positioned southeast of the hill is however fairly accurately determined. Figure 5 shows the relative wind velocity near the surface. The speedup at the top of the hill, and the low velocity wake region is clearly visible.

4 CONCLUDING DISCUSSION

The chosen hill proved to be ideal for model validation. Wind directions at drift events are primarily from one direction and the flow is not influenced from other terrain features. The resulting snow accumulation gives a clear pattern of erosion and accumulation areas around the hill. Snow is eroded on top of the hill, where the friction velocity is high and accumulated in the adjoining lee side with low friction velocities. The measurements can be trusted near the hill where the radar profiles cover the area tightly. At larger distances from the center of the hill and especially in the corners of Figure 3 the data coverage is lower and the quality of the interpolated snow depth map poorer.

The simulations give a good picture of the snow distribution around the hill. The measured snow dis-

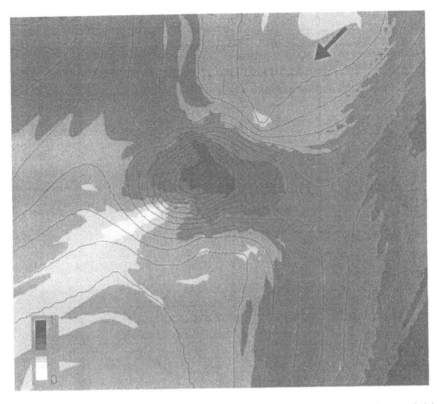

Figure 5. Normalized wind speed pattern around a hill, wind direction is indicated with arrow, height equidistance is 0.5 meters.

tribution is a result of a longer history with slightly varying wind direction. Such variation in the wind direction is not accounted for in the simulations and is probably increasing the area of accumulation compared to conditions of one distinct wind direction. To include this effect in the simulations, a deposition model, incorporation the development rate of the snowdrifts is necessary.

The achieved data of snow distribution around a small hill is valuable to validate results of snowdrift modeling. The applied model reproduced the measured snow accumulation pattern well and shows that it is possible to model relative snow distribution in terrain using a two-phase flow model.

REFERENCES

Annan, A.P. & Cosway, S.W. & Sigurdsson, T. 1995. GPR for snowpack water content, *Fifth International Conference on Ground Penetrating Radar Vol 2(3)*, Kitchener, Ontario, Canada

Bang, B. & Nielsen, A. & Sundsbö, P. A. & Wiik, T. 1994. Computer simulation of wind speed, wind pressure and snow accumulation around buildings (snow-sim), *Energy and buildings, Vol. 21*: 235-243

Budd, W. F. & Dingle, W. B. J. & Radok, U. 1966. The Byrd snow drift project: outline and basic results. In Ruben, M. J. (ed), *Studies in Antarctic meteorology.* Washington, D.C., American Geophysical Union, (Antarctic research series, Vol. 9): 71-87

Flow Science Inc., (1997), "Flow 3D User's Manual", *Flow Science Inc., Los Alamos.*

Gauer, P. 1998. Blowing and drifting snow in alpine terrain: numerical simulation and related field measurements, *Proceedings of the international symposium on snow and avalanches, Chamonix, Annals of Glaciology, 26*: 174 - 178

Haehnel, R.B. & Lever, J.H. 1994. Field measurements of snow drift, *Proceedings of ASCE / ISSW workshop on the physical modelling of wind transport of snow and sand*, Snowbird, UT: 1-14

Haehnel, R.B. & Lever, J.H. 1995. Scaling snowdrift development rate, *Hydrological processes, Vol. 9*: 935-946

Iversen, J. D. 1980. Drifting snow similitude transport rate and roughness modelling, *Journal of Glaciology, Vol. 26, No. 94*: 393-403

Kikushi, T. 1981. A wind tunnel study of the aerodynamic roughness associated with drifting snow, *Cold Regions Science and Technology, 5*: 107-118

Mellor, M. 1965. Blowing Snow, *Cold regions science and engineering part iii, section A3c*: 80

Pomeroy, J.W. & Gray, D.M. 1990. Saltation of snow, *Water resources research, Vol. 26*: 1583-1594

Sand, K. & Bruland, O. 1999. Application of Georadar for snow cover surveying, *Nordic Hydrology, 2, 4-5*: 361 - 370

Sigurdsson T. 1998. The Electromagnetic Theory of GPR Technique, *Course Notes on Georadar in Geophysics and Geology, Norfa*, UNIS Longyearbyen, Svalbard

Tabler, R. W. 1980. Self-similarity of wind profiles in blowing snow allows outdoor modeling, *Journal of Glaciology, Vol. 26, No. 94*: 421-432

Thiis, T. K. 1999. A comparison of simulations and full scale measurements of snowdrifts around buildings, *Wind and structures*, in press

Thiis, T. K. & Gjessing, Y. 1999. Large-Scale measurements of snowdrifts around flat roofed and single pitch roofed buildings, *Cold Reg. Science and Tech., 30 1-3*: 175-181

Snow Engineering, Hjorth-Hansen, Holand, Løset & Norem (eds) © 2000 Balkema, Rotterdam, ISBN 90 5809 148 1

Aerial photogrammetry of roof snow depths using three different aircraft

O. Joh
Hokkaido University, Sapporo, Japan

S. Sakurai
Hokkai Gakuen University, Sapporo, Japan

T. Shibata
Hokkaido Institute of Technology, Sapporo, Japan

ABSTRACT: The method of kite balloon aerial photogrammetry is proved to be applicable in surveying snow depths on roofs. First, features of three different aircraft (airplane, unmanned kite balloon and model helicopter) for mounting aerial cameras are compared from the viewpoint of surveying snow depths on roofs of actual buildings. Then, measurement errors of roof snow depths by the kite balloon aerial photogrammetry are investigated, and also the distributions of snow depths on two flat roofs are discussed. From the results, practical use of the kite balloon is clarified.

1 INTRODUCTION

Depth and weight of snow accumulated on roofs are usually different from those on the ground around the roofs because of moving by wind and/or melting by heat loss transmitted from inside of the building. A shape and size of roof effects remarkably on this difference and also the distribution of snow depths on the roof. The other hand, sufficient data on roof snow loads for structural design are not available because there is difficulty in measuring the intensity and distribution of snow load by climbing onto roofs. Especially on sloped roofs, snow sliding caused by the measurement work not only makes the long-term measurement almost impossible but also takes lives of people in some cases. It is important to develop an efficient and accurate method for surveying the roof snow loads. Here, the aerial photogrammetry is adopted in order to survey roof snow depths.

In a previous paper (Sakurai et al. 1992), the authors discussed the measurement accuracy of the aerial photogrammetry using mainly airplane for surveying roof snow depths and showed the probability of this method for snow survey. A chartered flight for aerial survey, however, is very expensive and it is inconvenient that users can not operate the airplane or camera by themselves. After that the authors has been developing an aerial photogrammetry using a kite balloon as a carrier of camera in order to solve these problems.

The present paper explains the performance of the kite balloon system developed by the authors and compares it with the cases using airplane and unmanned-radio control helicopter as the carrier. Furthermore the paper discusses the measurement

accuracy of the new system because the quality of the camera which can be mounted on the balloon has the superior limitation in weight, then the camera is different from that on the actual airplane, and some investigated results of snow depth distribution on flat roofs are shown.

2 PROCESS FOR MEASUREMENT OF ROOF SNOW DEPTHS

Roof snow depths can be measured by subtracting the elevation value of roof surface from that of roof snow surface. Figure 1 shows how to survey the elevation values of roof surface and those of roof snow surface by aerial photogrammetry. As the pass points to use the orientation, manholes on the ground near of the measured buildings are in general useful, however

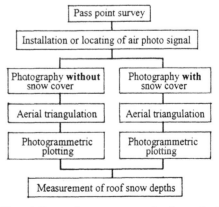

Figure 1. Process for measurement of roof snow depths

square plates of air photo signal printed a cross mark were selected, and they were set at the corners on the roofs (See figure 5 mentioned after) and at some places on the ground in this study.

When the roof surface deflects from the snow weight on the roof, the roof snow depths measured by this method has some errors caused by difference between vertical cross-shapes of the roof with and without snow. These errors, however, are negligible small usually for reinforced concrete structures, and can be corrected nearly by using deflections calculated with structural analysis for steel and timber structures.

3 KITE BALLOON AERIAL PHOTOGRAPHY SYSTEM

3.1 Composition of Instruments

The kite balloon aerial photography system is composed of a kite balloon itself, an instrument for aerial photography hanged below the balloon, a remote control transmission apparatus, a receiving antenna and a receiving apparatus for monitoring a camera view on the ground. These are shown in Figure 2 and Photos 1 and 2.

3.2 Property of Instruments

Property on each instrument used in this system is the following:
(1) Kite balloon (Photo 1)
 Type: vinyl chloride membrane with helium gas
 Dimension: length = 5 m, diameter = 2 m,
  Volume = 20 m³
 Weight: 85 N, Carrying capacity: 59 N
 Control rope: nylon, diameter = 3 mm, length =
  300 m, tensile strength = 1620 N/string
(2) Aerial photography instrument (Fig.2-a, Photo 2)
 TV-camera: focal length = 1/2 inch,
  color CCD with 380,000 elements
 Still camera: MKWE made in Hasselbrad Co.
 Direction: panning = 360-degree,
  tilting = downward to horizontal
 Altimeter: 0 to 300 m, resolution = 1 m
 Electric power: DC12V, 1.5-hour cont. operation
 Weight: MKWE = 24 kN, others = 28 kN
(3) Remote control transmission apparatus (Fig.2-b)
 Type: PCM 5ch, proportional system
 Frequency: 40.710 MHz
 Remote control: panning, tilting, shuttering &
  switching of power on/off
(4) Receiving antenna (Figure 2-c)
 Frequency: 473 MHz
 Element number: 5
(5) Receiving apparatus (Figure 2-d)
 Frequency: UHF 13ch, weak electric wave
 Altimeter: LED 3 figure indication
 Monitor TV: 10-in. screen with 8-mm VTR deck
 Battery pack: 14 Volt, 10.8 Ah

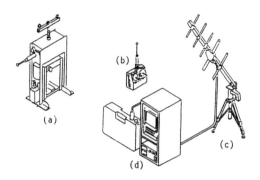

Figure 2 Composition of aerial photography system

Photo 1 Kite balloon and aerial photography instrument controlled by three ropes

Photo 2 Detail of aerial photography instruments consisted of camera and mounting frame

Performance of the aerial instruments, which were projected at the start of this study and confirmed in the test, are the following.

(1) Lift of kite balloon
Floating power is made from the difference between specific gravity of air and helium gas. Since the power is about 10 N per cube meter, the needed volume of balloon is estimated from the total weight of the

82

balloon itself and mounted equipment. An additional volume, however, is required in order to get a stable lift against effects of wind, descended air and temperature. The total weight of this system including three control ropes of 300 meters in length is 165 kN, consequently the volume used is 20 cube meters. The rope weight increases applied load to the balloon in proportion to flight altitude of the balloon. The flight altitude limit was about 150 meters in our winter measurement on condition that temperature was -5 Centigrade and wind velocity was 3 to 4 m/sec.

(2) Performance of camera direction control
The direction of camera can be faced to the ground always by using a suspension frame attached the camera. This frame is suspended from the balloon in order to keep the frame in vertical direction whenever the balloon inclines. Photographing in a stationary state of the balloon is easy, but doing in moving is a little difficult. The direction of camera can be turned also optionally within down angle for tilting and in whole directions for panning by a radio control motor set in the frame.

(3) Quality of camera
Using a lens without distortion, a large size film and an automatic exposure function improves the accuracy of pictures. The camera used in this study is MKWE Model made in Hasselbrad Corporation, the focal distance is 38.4 mm, the distortion of lens is 0.06 mm and the film size is 53 mm x 53 mm. The exposure is a manual type and is adjusted to weather condition just before lifting the balloon. It is good to focus the lens to infinite distance because the photographs are usually taken with over 50 m distance in height. Cross marks at every mesh points are recorded on the film for a figure analysis.

4   DISCUSSIONS

4.1   *Comparison of features of three different aircraft*

The features of three different aircraft for aerial photography were compared from the viewpoint of surveying roof snow depths. The aircraft are airplane, unmanned kite balloon and model helicopter.

(1) Airplane
The airplane discussed below is limited to business one, which a company of aerial photography has. The company entrusted carries out to fly the airplane mounting a cameraman and aerial cameras and to take photos at above objects. That is clients need not to have any technique nor work for aerial photography.

[Advantage]
a) The practice of photographing covers a exceedingly wide area with a short time.
b) No hindrance under the condition that a

photographing place is in a great distance, in a remote island or inconvenient on transportation.
c) No required condition to surroundings of an object if the object is visual from the sky.
d) The most suitable equipment for aerial photography is usable because the mounting weight is scarcely limited.
e) It is the strongest against wind condition.

[Disadvantage]
a) Necessity of a booking on a photographing day and a photographing time, difficulty with a urgent changing the time by weather condition at a photographing place.
b) The prohibition from taking a photograph under 300 meters in flight altitude according to the aviation law of Japan results the limitation of the measurement accuracy.
c) High expenses of photographing.

(2) Kite balloon
The kite balloon, which is treated in this paper, is an unmanned and small type. It is positioned with ropes from the ground. The visual monitoring system supports to control the direction of aerial camera. Users purchase the kite balloon photography instrument and operate it by themselves.

[Advantage]
a) The first investment in purchase of this system is necessary, but is cheaper than that of a model helicopter system. The running cost is economical because the main consumption rate is from gas.
b) Photographing from an optional flight altitude is possible within the ceiling ability of a balloon.
c) The operation is safe and relatively easy, and needs no skill.
d) The time and date for photographing are able to be set up optionally by user's mind.
e) This way is the best to do a stationary observation in the sky for long time.
f) Observation of snow cornice or flapped snow is practicable because of ability to take a photograph in lateral and slant directions.

[Disadvantage]
a) The maximum mounting weight and ascending altitude depend on the size of a balloon.
b) Becoming unable to take a photograph due to difficult positioning of a balloon in strong wind.
c) Some open space on the ground around an observed object and no obstacle of an electric wire etc. in the sky are required to operate the ropes joining a balloon.
d) Storage in a balloon hangar after each observation or recovery of the gas is necessary. Release of the gas runs up expenses.
e) Unsuitability for measuring a wide area or a large number of objects due to the bad mobility.

f) Many works to prepare and remove a balloon are needed before and after every photographing, respectively. The most number of manpowers for operation also is needed among the three aircraft.

(3) Model helicopter
The model helicopter discussed below is an unmanned-radio control type. The aerial camera and the visual monitoring equipment have the same property as ones used in the kite balloon system in this study. Users should operate the helicopter photography system by themselves in case of purchase, and they do not need to operate it in case of entrusting the photography. The used helicopter is shown in Photo 3 and the property of the helicopter is followings:

Type: TSK GSII improved version
Total length: 2150 mm (at rotor turning)
          1500 mm (body only)
Weight: 62 N, Carrying capacity: 39 N
Displacement of engine: 33 mm^3

[Advantage]
a) The high mobility makes measurement of a wide area and a large number of objects within a short time, next to the airplane.
b) Approximately no limits of an ascending flight altitude and no required condition to surrounding of an object.
c) In case of purchasing helicopter photography system, the time and date of photographing are able to be set up optionally by user's mind.
d) Observation of not only snow cornice or flapped snow but also an object halfway up a cliff or next to a tall building, is practicable because of ability to take a photograph in lateral and slant directions.

[Disadvantage]
a) From difficulty in operating the helicopter, users need to leave the operation to a technical specialist or to master it with a suitable time.
b) Limitation of the mounting weight depends on the size of a helicopter.

Photo 3   Model helicopter holding still and video cameras below the body

c) In case of purchasing the photography system with a helicopter of which specifications require a low vibration, the initial cost is expensive. In case of entrusting the photography, the running cost is expensive because a payment of each observation for a technical specialist and helicopter is so high.
d) Flying of a helicopter is affected sensitively by wind velocity and operation technique. Therefore, it needs to suppose a fall accident and to settle the matter.

The above mentioned features of the aerial photography with three different aircraft depend on some various conditions: geographical conditions of a site, surroundings of an object, total times of observations, duration of observation, existence of an operator, economical condition etc. The advantages of three different aircraft make up disadvantages of each other. Therefore, users have to select the best way among these systems according to the purpose of their investigations and their executive capacity. The authors selected the kite balloon aerial photography system in this study because of the following reasons: 1) their economical condition was low, 2) the number of observed sites was small and the small number of buildings was expected to be observed continuously, 3) they had many man power by students but no operator for a helicopter, and so on.

4.2   *Operational techniques of kite balloon system*

Operational techniques of the kite balloon and performance of the camera control system were investigated in field works.

(1) Charging and exhausting of gas
The balloon is stuffed with gas from gas cylinders through a pressure regulator. If the gas pressure in the balloon is too low, the balloon becomes unstable because it can not keep its shape, and if the pressure is too high, the floating power decreases because the specific gravity becomes high. The gas can be recovered by using a reclaimer and recycled at every observation, but this way is not so useful because the reclaimer is expensive, its movability is low and the floating power becomes lower every recycling.

(2) Control of camera view
The control of balloon positioning and camera directing is very important because the photographic targets that are observed roofs and air photo signals on the ground must be taken surely in the camera view. Altitude of the balloon can be read to one meter by an altimeter. The horizontal and vertical positions of balloon are controlled with three ropes hanged from the balloon usually. In strong wind four ropes are used, but the control is unable in wind velocity of more than four meters. Applying the monitoring method used by Koizumi (1986), the camera view can be monitored

on the ground through a small video camera attached beside the still camera. The lens of camera can be kept always toward the ground by using a hanging mechanism and also can be turned to any directions by a radio controlled motor. Consequently, operators can take easy photographs with the best camera view.

(3) Counterplans against halation

Halation is caused by the reflection of sunlight from the snow surface or roof surface. Our previous papers (Sakurai, 1992 & 1993) pointed out the necessity of investigation about the technical plan against the halation because the measurement accuracy of the aerial photography dropped remarkably and the portions occurring halation missed the snow depth data in case of hard halation.

The following ideas were checked up as the counterplans against halation: a) photographing on a thin cloudy day, b) photographing at early morning or late afternoon, c) photographing a few times with time lag, d) using a polaroid filter. The way of a) was effective on avoiding halation as the cloud interrupted direct sun light. However, it was not so easy to meet the good cloud condition for photography because the shade of surface was lost by cloud too heavy. The way of b) was in expectation that the camera did not receive the reflection of sunlight because the sun's ray had low angle. Even in this way the halation occurred at steep roofs or at edges of roofs and the hours possible for photography were very short in a day. The way of c) was in expectation that a few photographs could compensate the parts lost by halation to each other. That is the halation occurred at different positions within a roof because the orientation of the sun changed while photographs were taken at different time. This way needed many hours for photography because a large time lag was better. The way of d) could avoid halation for almost any conditions. Consequently, the way of using a polaroid filter showed the best performance in practical use.

4.3 Investigated results of snow depth on roofs

A measurement accuracy of this system and distribution of snow depth on roofs were discussed by investigation of accumulated snow on roofs of dugouts beside a baseball ground in Sapporo. Since the baseball ground was located in the Ishikari Plain and had no large building nor large hill around it, deviations of wind direction and velocity distribution by the local surroundings did not occur probably.

As shown in the location, Figure 3, two dugouts were similar in shape and located perpendicularly each other. The baseball ground in front of the dugouts was completely flat, however the ground surface behind the dugouts had a slope as the stands as shown in the section, Figure 4 and in Photo 4. The dugouts were 1.61 m high and had a rectangular and flat roof of 12.84 m in length and 2.34 m in width for wing A (2.31m in width for wing B), and their facades were open. They were a reinforced concrete structure and were not used nor heated in snowy season.

The snow condition covered on the roof at measurement is shown in Photo 5. Shortly after the

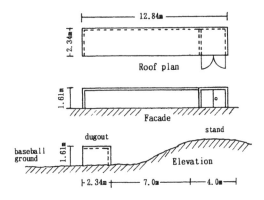

Figure 4   Configuration of dugout and stand

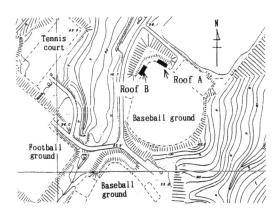

Figure 3   Location of observed baseball dugouts

Photo 4   Dugouts for players and bank for spectators
(Dugout A in near side, B in far side)

Table 1. Measurement errors of snow depths for flat roof by kite balloon

Assigned flight altitude	20 meters		50 meters		100 meters		150 meters	
Snow condition	Photograp. without snow cover	Photograp. with snow cover	Photograp. without snow cover	Photograp. with snow cover	Photograp. without snow cover	Photograp. with snow cover	Photograp. without snow cover	Photograp. with snow cover
Photography date	1994.11.30	1995.3.3	1994.11.30	1995.3.3	1994.11.30	1995.3.3	1994.11.30	1995.3.3
Actual flight altitude	19 m	18 m	49 m	46 m	89 m	97 m	138 m	142 m
Actual photo scale by flight	1/500	1/470	1/1290	1/1210	1/2340	1/2550	1/3630	1/3740
Measurement errors of roof snow depths.	n = 36 e =. -13 mm. σ = 23 mm		n = 36 e =. -53 mm σ = 27 mm		n = 36 e =. -49 mm σ = 44 mm		n = 36 e = 51 mm σ = 62 mm	

Notes: n = Number of measured points, e = Mean of measurement errors, σ = Standard deviation of measurement errors

Photo 5　Snow condition accumulated on dougout A

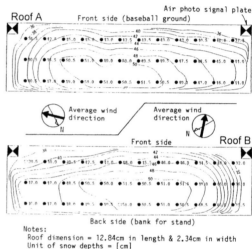

Notes:
Roof dimension = 12.84cm in length & 2.34cm in width
Unit of snow depths = [cm]

Figure 5　Distributions of roof snow depths

aerial photography of snow on the roofs was done by the kite balloon system, the snow depth at 36 points on each roof as shown in Figure 5 was measured directly with rulers by climbing onto the roofs (hereafter referred to this value as a measured value.) Using the aerial stereo-photographs taken in snowy season mentioned above and those taken in a season without snow on the roofs, elevation values at the 36 points were analyzed and the difference between these elevation values for two seasons was calculated as snow depths (hereafter referred to this value as an evaluated value.)

(1) Measurement accuracy
The measurement errors were defined as the difference of the evaluated values from the measured values. The mean values and the standard deviations of the errors were obtained from the investigated results at the 36 points of the wing A.

Table 1 shows the results of investigation on the effect of the flight altitude and on the measurement errors for the aerial photogrammetry. In practical application, the measurement errors can be considered as a normal distribution with the mean –13 mm, -53 mm, -49 mm and 51 mm at the assigned flight altitude of 20 m, 50m, 100m and 150m, respectively. The standard deviations at the assigned flight altitude of

these are 23 mm, 27 mm, 44 mm and 62 mm, respectively. The measurement results show that the mean values and the standard deviations increase according to increase of the flight altitude. This behavior is reasonable also theoretically. However, the result that the mean value at the assigned flight altitude of 50 m is larger than that at 100 m and 150 m is caused by a little halation. This means using the polaroid filter leaves a small room for improvement still. These analyses show that the aerial photogrammetry by kite balloon is accurate enough for practical use to survey roof snow depths. Users should choose the flight altitude properly within the necessary accuracy.

(2) Distribution of roof snow depths
Figure 5 shows the distributions of snow depths measured by climbed onto the roof A and roof B. The average values of 36 measured points on the roof A and roof B are 457 mm and 461 mm, respectively. The curved lines drew in the figure express the contour lines of the snow depths every twenty millimeters

within range of the depth over about 300 mm.

The distributions of snow depths on the roofs A and B are similar. The depths in the middle and back rows are about 1.1 times as deep as those in the front row and the depth at center part is deeper than the depths at both ends. The average wind direction of January and February in the year was the northwest at the Sapporo Meteorological Observatory. This direction corresponds to an axis along a long side of the roof A and to an axis along a short side of the roof B. The deepest snow positions appear at the backside on the roof A and also on the roof B, although such positions are not located on leeward of the average wind direction at the observatory. This means local wind blowing from the front side toward the back side is the strongest at each dugout because of a bank behind them, and the local wind effects larger on the roof snow distribution than the average wind direction in the plain. However, the effect of the average wind direction can be seen slightly in a behavior that the snow depth is a little deep on the leeward at the roof A.

5 CONCLUSIONS

The present paper is concerned with the application of aerial photogrammetry for surveying roof snow depths. The results are as follows:

1) The results of comparison among the features of aerial photography using an airplane, a kite balloon and a model helicopter show that the advantages of three different aircraft make up disadvantages of each other and that the features depend on some various conditions. Therefore, users should properly select the best system in consideration of geographical conditions of a site, surroundings of an object, total number or duration of observations, expenses for measurements and so on.
2) The kite balloon aerial photography system is enough for practical use to survey roof snow depths because the measurement accuracy is allowable. The cost performance is the best among the three systems. This system is suitable especially for the long term-continuous observation of objects concentrated in one area.
3) The following ideas were checked up as the counterplan against halation: a) photographing on a thin cloudy day, b) photographing at early morning or late afternoon, c) photographing a few times with time lag, d) using a polaroid filter. Consequently, the way of using a polaroid filter shows the best performance in practical use because this way can avoid occurring of halation for almost any sunlight conditions and does not restrict the photographing within limited time.

ACKNOWLEDGMENT

The authors gratefully acknowledge the economical support provided by the Japan Ministry of Education, Science and Culture (Grant-in-Aid for Developmental Scientific Research No.06555167) and the technical support by Mr K.Kodama who is a manager of Shin Technical Consultant Company.

REFERENCES

Koizumi,T.,Murai,S.,Koike,T. & Manabe,H. 1986. An Automated System and its Applications for Aerial Photography Using a Kite Balloon. *Journal of Photogrammetry and Remote Sensing* 25(3):12-23 (in Japanese)
Sakurai,S.,Joh,O. & Shibata,T. 1992. Survey of Roof Snow Depths by Aerial Photogrammetry. Special Report of Second International Conference on Snow Engineering: 93-104
Sakurai,S. & Joh,O. 1993. Fundamental Study on Applicability of Aerial Photogrammetry in Surveying Roof Snow Depths. *Journal of Structural and Construction Engineering, Architectural Institute of Japan* 450: 25-36 (in Japanese)

Snow Engineering, Hjorth-Hansen, Holand, Løset & Norem (eds) © 2000 Balkema, Rotterdam, ISBN 90 5809 148 1

Perspective of earthquake disaster mitigation in snow season – A case study of earthquake at Yuza Town, Yamagata prefecture, Japan

Tomohiro Kimura & Kiyomichi Aoyama
Research Institute for Hazards in Snowy Areas, Niigata University, Japan

ABSTRACT : Lying in the active earthquake belt, Japan is suffering from frequent earthquake disasters, which cause heavy toll of lives and properties. The damage potential of such disasters will obviously be high if it occurs in snow period.

Considering the above mentioned possibilities, study has been conducted on the earthquake event that was occurred in Yuza Town of Yamagata Prefecture, Japan. The earthquake was occurred on 26th February 1999 at 14:18 local time. The magnitude of the earthquake was recorded to be 5.4 respectively at Yuza town, Yamagata Prefecture and Kisakata Town, Akita Prefecture. This earthquake can be considered as weak although it did several severe damages like damage to roads, houses and compound wall, toppling of furniture and housing utilities, liquefaction effect and so on. The field survey was conducted on March 1999, after the disaster. Study on various parameters like damage condition, stove burning condition, personal injuries, people's concept on disaster prevention etc. were made by administering the pretested questionnaire. Meteorological data has also been collected and analyzed to relate with the disaster situation.

This paper deals with the analysis of disaster situation in Yuza Town and explains the importance of disaster mitigation and risk management for the earthquakes during snow season through the real case study, aging ratio and other related factors.

1. INTRODUCTION

Earthquake at Yuza Town is an example of the earthquake during snow season. The day of the earthquake was warmer than the other days. Authors collected meterological data and distribution of temperature and snow depth for various places. Again we analyzed the result of questionnare survey on disaster awareness and the extent of damages.Likewise we insighted various issues from the questionnare and suggested appropriate disaster prevention. Finally, we overviewed circumstances of aging ratio because authors noticed the increase of aging people those vulnerable to disaster [1].

2. OUTLINE OF EARTHQUAKE

The earthquake was occurred on 26th February 1999 at 14:18 local time. The magnitude of the earthquake was recorded to be 5.4 and 5 (weak as per J.M.A. scale) at Yuza town, Yamagata Prefecture and Kisakata Town, Akita Prefecture respectively. The depth of 11 Km and the epicenter was geographically located at 39.2N 139.8E.

The earthquake event at Yuza Town has broken 26 compound walls, formed severe cracks in 7 roads, partially damaged some houses with the induction of cracks on the walls and collapse of tiled roofs. The disaster day was warmer in comparison to previous years. So, many people did not use the heaters. According to the people who used the heater, few people could not extinguish it because of strong ground shaking. Corroded reinforcements in the building structures and various damages related with geo-technical aspects were observed in the area. Main entrance made of stone in Mikami shrine, one of the open public places in the area was collapsed. Plenty of open spaces like parks and parking areas can be found nearby disaster area which shows good possibility of evacuation. Various notice boards could be seen on the way to inform about the evacuation center which clarify the improvement on non-structural measures for regional planning on the disaster mitigation strategy with the point of view of aged and disabled people.

3. METEOROLOGICAL DATA BETWEEN VARIOUS PLACES

The day of earthquake had T_{max} of 7.8°C and T_{min} of 3.4°C. Usually, below 10°C, many people use stove, and danger of fire is higher, but, earthquake day was warmer than average. Epicenter Yuza Town was not covered with snow, but, near by area were snow covered and colder.(Fig. 1~4).

If we think on disaster mitigation we must know different temperature and snow depth, variation in various years.

4. RESULT OF QUESTIONNARE SURVEY

This earthquake has little relation with snow, but, earthquake was occurred in winter. Hence we must think on disaster mitigation for earthquake and snow hazard. The objective of investigation was to know the concept of public awareness on earthquake disaster mitigation and condition of cold. 284 residents in total answered correctly on questionnaire.From insite survey conducted in March 1999, many people used stove and feel fear of fire but this earthquake did not have fire hazard.

However, other issues gave importance to evacuation. Every municipal offices make regional disaster prevention strategy including evacuation center, but, in Yuza Town, several evacuation camps are outside parking areas. Region near Japan Sea has tendency of stronger wind and more precipitation than Pacific region. More than 80 % people know evacuation center because of many displayed maps of evacuation camp around the corner in order to protect themselves from Tsunami due to Niigata Earthquake which occurred on 16th June 1964.(Fig. 5). But the municipal government should preserve camp inside.

According to results of questionnaire, many pepople worry about earthquake during snow season and then, faer of this earthquake (Fig. 6 and 7).

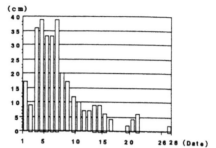

Fig.1 Snow depth near epicenter where is Yuza Town on Feb. (Resouce : Construction branch at Shohnai Office)

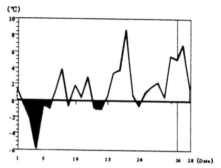

Fig.2 Temperature at 8 AM. near epicenter where is Yuza Town on (Resource : Construction branch at Shohnai Office)

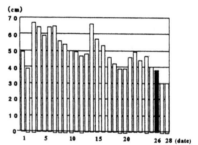

Fig.3 Snow depth at Shiroi-shinden near epicenter on Feb. (Resouce : Construction branch at Shohnai Office)

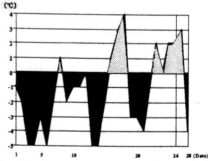

Fig.4 Temperature at 8 AM. where is Shiroi-shinden on Feb. (Resouce : Construction branch at Shohnai Office)

Fig.5 Perception of Evacuation Center (%)

Fig.6 Think to Occurr Earthquake during Snow Season (%)

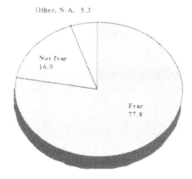

Fig.7 Fear of This Earthquake (%)

The extent of damage were crack in walls, toppling of furnitures and line busy telephone circuits and so on.(Fig. 8).

Therefore, municipal government should make disaster prevention policies adaptable snow and cold , especially, drifting snow.

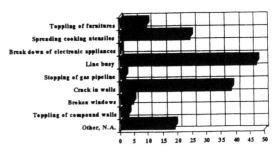

Fig.8 Extent of Damage Induced by Earthquake (Multiple Answer, %)

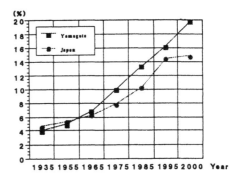

Fig.9 Trend of Aging Ratio in Yamagata Pref.
(Source : Yamagata Prefecture)

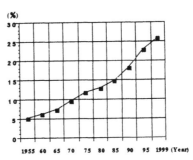

Fig.10 Trend of aging ratio at Yuza Town
(Resouce : Municipal office at Yuza Town)

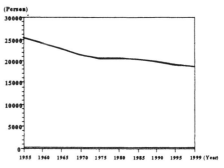

Fig.11 Trend of population at Yuza Town
(Resouce : Municipal office at Yuza Town)

91

5. CURRENT SITUATION OF AGING RATIO AT YUZA TOWN

Yuza Town is suffering from increase in aging ratio. Yamagata Prefecture is also suffered from increasing aging ratio. (Fig. 9~11).

According to the statistical data, we must make decision to disaster mitigation plan especially, rescue operation for old people primalrily [2]. Municipal governer advise them not to deploy furnitures in bed room, this attitude should be prevail in society.

6. CONCLUSION

These data show severe meteorological condition, blowing snow and aging ratio. According to the result of questionnaire survey, residents feel afraid of occurrence of earthquake during snow season and fear of strong ground motion but, most of the respondents know evacuation center. Still more, we must be alert of fire spreading because of tendency of strong wind even there is no arson in this earthquake. It is necessary to provide disaster mitigation education too.

7. REFERENCES:

1) T. Kimura and K. Aoyama : Earthquake induced disaster during snow period (Questionnaire survey on earthquake occurred in Niigata Prefecture during snowfall period). Proc. of International Conference on Applications of Statistics and Probability (ICASP · 8). vol. 1. A. A Balkema Pub. pp.535 − 543. 1999.

2) T. Kimura and K. Aoyama: Review on Risk Management for Earthquake Mitigation at Heavy Snow Areas, Tsuchi to Kiso, volume 47, No.1, pp.31-34, The Japanese Geotechnical Society, 1999. (in Japanese).

Snow Engineering, Hjorth-Hansen, Holand, Løset & Norem (eds) © 2000 Balkema, Rotterdam, ISBN 90 5809 148 1

Comparison of dry granulation and wet agglomeration of snow

Toshiichi Kobayashi
Nagaoka Institute of Snow and Ice Studies, NIED, Niigata, Japan

Yasuaki Nohguchi
Atmospheric and Hydrospheric Science Division, NIED, Japan

Kaoru Izumi
Research Institute for Hazards in Snowy Areas, Niigata University, Japan

ABSTRACT: In sufficiently developed wet snow avalanches, the snow sometimes changes into rounded lumps like snowballs. This process is called "Dry granulation". On the other hand, when large-scale slush flows occur, the snow sometimes changes into rounded lumps like snowballs immersed in water. This process is called "Wet agglomeration". To simulate such succeeded in formating snowballs on both dry granulation and wet agglomeration by the use of the same new snow. As a result, the particle size distribution of snowballs in dry granulation was wider than that in wet agglomeration. And the mean particle diameter of snowballs in wet agglomeration was larger than that in dry granulation. In addition, the number of granulated snowballs in dry granulation was equal to about the square of that in wet agglomeration.

1 INTRODUCTION

The making of snow particles is called granulation, and in particular the making of particles in air is called dry granulation. On the other hand, the making of particles in water is called wet agglomeration. The dry granulation is sometimes observed when wet snow avalanches occur. In addition, the study on a pneumatic conveying of wet snow have been carried out by the use of the dry granulation method. The wet agglomeration is sometimes observed when large-scale slush flows occur. To investigate the difference between dry granulation and wet agglomeration of snow, some tests were performed.

2 EXPERIMENTAL METHOD

The tests on tumbling granulation were carried out using a polyethylene bucket. As shown in Figure 1, dry granulation tests were performed by rotating or shaking a polyethylene bucket after the wet snow was thrown into the bucket. Wet agglomeration tests were performed in a similar manner after water was poured into the bucket to a level of 5 cm. The rotating speed of the bucket is about 100 rpm, and the granulation or agglomelation time is 3~5 minutes. The snow types tested are new snow, fine-grained snow and granular snow. After each test, the diameter and the weight of each snowball was measured. In addition, the hardness of each snowball was measured by the use of a load cell.

3 RESULTS

3.1 *New snow*

Figures 2 and 3 show snowballs which were formed in a dry granulation test and in a wet agglomeration test when about the same amount of wet new snow was used. The number of granulated snowballs in a dry granulation test was equal to about the square of that in a wet agglomeration test. Figure 4 shows distribution curves of the diameters of snowballs in dry granulation and wet agglomeration. The peak of the distribution of the diameters of snowballs in dry granulation is 1.5~2.0 cm, and that in wet agglomeration is 4.0~4.5 cm. The range of the distribution of the diameter of snowballs in dry granulation is wider than that in wet agglomeration. In the case of increasing the amount of wet snow thrown into the bucket, the number of the granulated snowballs decreased and the mean diameter of snowballs increased in a wet agglomeration test as shown in Figure 5.

3.2 *Fine-grained snow*

According to Kuriyama (1984), the van shear strength of fine-grained snow was 50 times as large as that of new snow. On the other hand, Kobayashi (1985) reported that the vane shear strength of fine-grained snow immersed in water for 3 minutes was 5 times as large as that of new snow. Then it is difficult to disaggregate the bond of fine-grained snow.

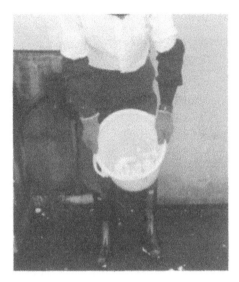

Figure 1. A tumbling granulation test using a polyethylene bucket.

Figure 2. Snowballs which were formed from new snow in a dry granulation test.

Figure 3. Snowballs which were formed from new snow in a wet agglomeration test.

Therefore, in dry granulation and wet agglomeration of fine-grained snow, the snow in the bucket was merely changed into a rounded lumps as shown in Figures 6 and 7.

3.3 Granular snow

In dry granulation and wet agglomeration of granular snow, it is easy to disaggregate the bond of granular snow by rotating or shaking a bucket. After a dry granulation test was carried out for 3~5 minutes, no snowballs were formed. But some loose lumps were formed as shown in Figure 8. On the other hand, after a wet agglomeration test was performed for 3~5 minutes, a skeleton-type snowball was formed as shown in Figure 9. As soon as the skeleton-type snowball was picked up from the bucket, the water that filled the openings of the snowball was drained away.

3.4 Hardness of snow and snowball

Figure 10 shows the relation between the ratio of increase of hardness from snow to snowballs Hb/Ho and the mixing ratio of snow to water. In this case, Hb/Ho means the ratio of the hardness of snowballs (Hb) to the hardness of original snow (Ho). Three symbols on the ordinate axis when the mixing ratio of snow to water is 100 % show the data in dry granulation, and other symbols show the data in wet agglomeration. The rate of increase of hardness from snow to snowballs in wet agglomeration was smaller than that in dry granulation except for the data from granular snow to the skeleton-type snowball when the mixing ratio of snow water was 36 % in wet agglomeration.

4 CONCLUSION

To investigate the difference between dry granulation and wet agglomeration of snow, small tests were performed by the tumbling granulation method. The results of these tests are as follows:

(1) In the case of new snow, the granulated snowball size in wet agglomelation was larger than that in dry granulation, and the number of snowballs in dry granulation was equal to about the square of that in wet agglomeration. The hardness of snowballs in wet agglomeration was smaller than that in dry granulation.

(2) In the case of fine-grained snow, no snowballs were formed, and the snow thrown into the bucket was merely changed into a rounded lump.

(3) In the case of granular snow, no snowballs were formed in dry granulation, but a skeleton-type snowball was formed in wet agglomeration.

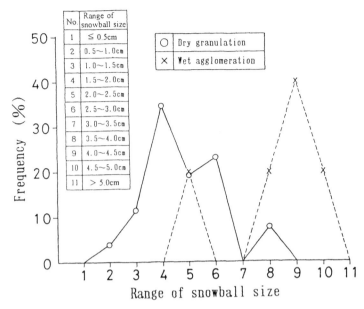

Figure 4. Distribution curves of the diameters of snowballs in dry granulation and wet agglomeration tests.

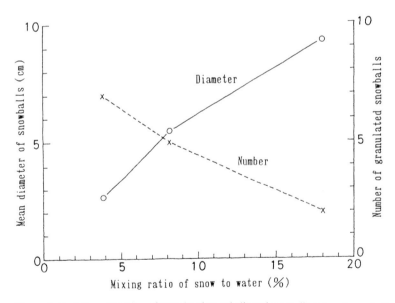

Figure 5. Variation of number of granulated snowballs and mean diameter of snowballs in wet agglomeration tests.

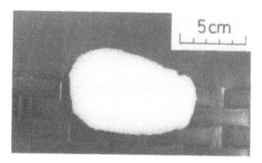

Figure 6. A rounded lump formed from fine-grained snow in a dry granulation test.

Figure 9. A skeleton-type snowball formed from granular snow in a wet agglomeration test.

Figure 7. A rounded lump formed from fine-grained snow in a wet agglomeration test.

Figure 8. Loose lumps formed from granular snow in dry granulation test.

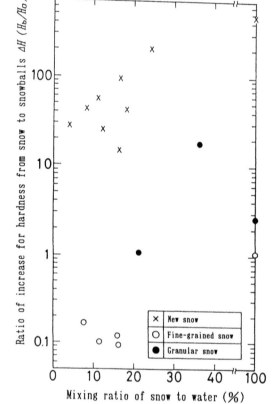

Figure 10. Relation between the ratio of increase for hardness from snow to snowballs and the mixing ration of snow to water.

REFERENCES

Kobayashi, T. 1985. Vane shear strength of snow immersed in water (1) - Relation between vane shear strength and immersion time -. *Journal of the Japanese Society of Snow and Ice*. 47:55-62.

Kobayashi, T. & Nohguchi, Y. 1998. Studies on tumbling granulation of wet snow (Granulation in water and that in air). *Proceedings of '98 Cold Region Technology Conference*. Vol.14, 1-5.

Kuriyama, H. 1984. Vane shear strength of snow (1) - Effect of vane angular velocity -. *Journal of the Japanese Society of Snow and Ice*, 46:101-108.

Nohguchi, Y., Kobayashi, T., Iwanami, K., Nishimura, K. & Sato, A. 1997. Granulation of snow. *Snow Engineering:Recent Advances*. 167-170.

Snow Engineering, Hjorth-Hansen, Holand, Løset & Norem (eds) © 2000 Balkema, Rotterdam, ISBN 90 5809 148 1

Hardening of snow surface – "natural/artificial"

Toshiichi Kobayashi
Nagaoka Institute of Snow and Ice Studies, NIED, Niigata, Japan

Yasoichi Endo
Tokamachi Experiment Station, FFPRI, Niigata, Japan

Yasuaki Nohguchi
Atmospheric and Hydrospheric Science Division, NIED, Japan

ABSTRACT: This paper describes two phenomena for the hardening of snow surface. One of them is natural hardening of snow surface by radiative cooling, which phenomenon is called "Shimi-watari" in Japanese. In Japan, people living in snowy regions have been enjoying such phenomenon for many years. Another is artificial hardening of snow surface. When snow festivals are held in early spring or in midsummer, snow surface was frozen by the use of sodium chloride (NaCl) preventing the snow surface from softening. This technique is the same as the arrangement of Alpine skiing racecourse by the use of "snow cement" or ammonium sulfate.

1 INTRODUCTION

Walking on a frozen surface of deep snow cover is called "Shimi-watari" in Japanese. The custom has been acted for many years in the Hokuriku and Chubu Districts of Japan. This phenomenon occurs in the early morning in early spring because the wet snow surface is frozen by radiative cooling. After that, the snow surface become so hard that people can walk around on the deep snow cover anywhere, i.e., on rice fields and brooks covered with snow, without wearing skis or snowshoes. Then some children go to school by taking a short cut to the school, and some people take a walk on the snow cover with their dogs. On the other hand, when a snow festival using the storage snow are held in early spring or in midsummer, sodium chloride (NaCl) is spread on the snow surface to prevent the snow surface from softening. But after 2 to 3 hours, the surface hardness of snow cover decreases. Therefore, it is necessary to spread sodium chloride again on the snow surface. In this paper, the observations and the experiments on these two types of hardening of snow surface are reported.

2 HARDENING OF SNOW SURFACE - NATURAL

"Shimi-watari" is a traditional Japanese custom. More than 10 years ago, this phenomenon sometimes occurred early in the morning from the end of February to the end of March. However, it only rarely occurs recently. On February 23rd of 1998,

the phenomenon occurred on the ground of our institute. Figure 1 shows a rough surface of the snow cover. In this case, the thickness of the freezing layer was 11 cm. The hardness of the snow surface was about 23 N/cm^2, which was hard enough to walk on the snow cover. Endo *et al.* (1995) observed from beginning to end of the phenomenon. They carried out snow pit study and measured density of snow (ρ), Ram hardness (R), snow temperature (Ts) and air temperature (Ta) about each snow layer as shown in Figure 2. In that case, the snow surface began to freeze by radiative cooling from 17:00 p.m. of previous day. Actually people could walk around from 5:00 to 8:00 a.m.

Figure 1. A rough surface of the snow cover when "Shimi-watari" occurred.

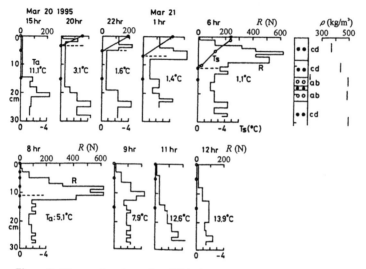

Figure 2. Observation date when "Shimi-watari" occurred (Endo *et. al*, 1995)

Figure 3. A snow pyramid in the Nagaoka Yukishika Matsuri Festival.

Figure 5. The trenches dug on the snow pyramid.

3 HARDENING OF SNOW SURFACE - ARTIFICIAL -

3.1 *Observations*

In Japan, many snow festivals are held in early spring or in midsummer. Snow structures are constructed in their festivals, i.e., a snow pyramid in the Nagaoka Yukishika Matsuri Festival (see Figure 3), a snow stage in the Ojiya Snow Festival (see Figure 4). In their events, sodium chloride (NaCl) is spread on the snow surface to prevent the snow surface from softening. In the 1998 Nagaoka Yukishika Matsuri Festival, children were enjoying climbing or sledding on the snow pyramid. However, on 11:00 a.m., after 1 hour from the opening, trenches were dug on the surface of the snow pyramid because of softening of the snow surface (see Figure 5), and it

Figure 4. A snow stage in the Ojiya Snow festival.

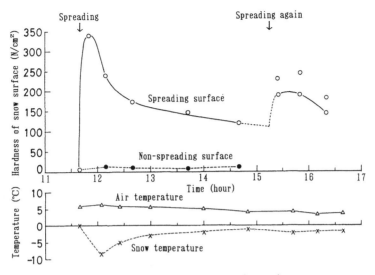

Figure 6. Variations of hardness of snow surface and temperatures.

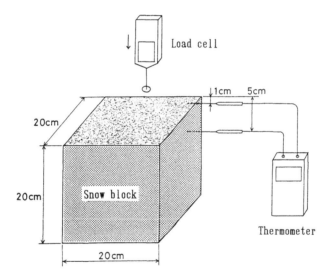

Figure 7. Experimental set up for spreading sodium chloride on snow surface.

was dangerous to climb or sled on the snow pyramid. Therefore, sodium chloride was spread on the surface of the pyramid at a rate of 500 g/m². Figure 6 shows variations of hardness of snow surface and temperatures. The hardness of snow surface had a maximum ten minutes after spreading, after that the hardness gradually decreased. The maximum hardness of snow surface was equivalent about 50 times to that before spreading. Three and a half hours after the first spreading of sodium chloride, sodium chloride was spread again at the same rate of 500 g/m².

The hardness of snow surface also had a maximum ten minutes after the second spreading, but the ratio of increase for hardness was 1.6 times. On the other hand, the children who were sledding on the snow pyramid said that the snow surface became hard to slide as compared with the condition before spreading sodium chloride. According to the standard for arrangement of Alpine skiing race course, the objective of hardness of snow surface was 70~100 N/cm² in the Sapporo Winter Olympic Games in 1972 and that was 100~120 N/cm² in the Morioka & Shizu-

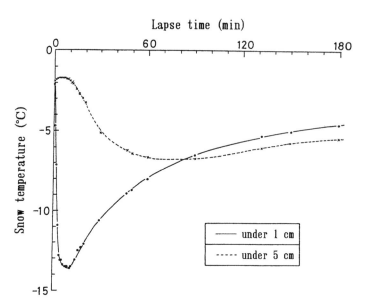

Figure 8. Time variation of snow temperature (room temperature: 0 °C).

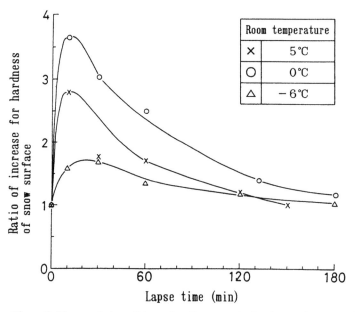

Figure 9. Time variation of the ratio of increase for hardness of snow surface.

kuishi World Cup Alpine Skiing Games in 1993 (Yaginuma *et al.* 1996). Then it is considered that the hardness of the snow pyramid was too hard to sled on it.

3.2 *Experiments*

In order to investigate the effects of hardening of snow surface by spreading sodium chloride, some experiments were carried out in a cold room kept at room-temperature of -6, 0 or 5 °C. As shown in Figure 7, after spreading sodium chloride at a rate of 300 g/m^2 on a snow block (20 x 20 x 20 cm), snow temperatures were measured by a thermometer located under 1 cm and 5 cm from the surface of the snow block. The hardness of the snow surface was measured by the use of a load cell. As a result, the

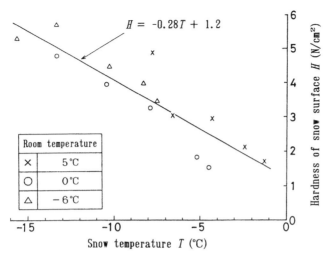

$$H = -0.28T + 1.2$$

Room temperature	
×	5 °C
○	0 °C
△	− 6 °C

Snow temperature T (°C)

Figure 10. Relation between the snow temperature and the hardness of snow surface.

snow temperature 1 cm under the surface had a minimum ten minutes after spreading. After that the temperature gradually increased, and the snow temperature under 5 cm kept the same value for ten minutes, after that the temperature gradually decreased at the room temperature of 0 °C as shown in Figure 8. At the room temperature of -6 or 5 °C, the results were about the same as at 0 °C. On the other hand, the hardness of the snow surface had a maximum ten minutes after spreading. After that the hardness gradually decreased as shown in Figure 9. The relation between the snow temperature and the hardness of snow surface were obtained as

$$H = -0.28T + 1.2 \qquad (1)$$

where H is the hardness of snow surface and T (°C) is the snow temperature (see Figure 10).

4 CONCLUSIONS

Observations and experiments for hardening of snow surface were carried out. The results of them are as follows:
(1) In the case of natural hardening of snow surface by radiative cooling, the hardness of snow surface of more than 20 N/cm^2 is needed to walk on the snow surface.
(2) The ratio of increase for hardness of the snow surface depends on the amount of sodium chloride for spreading in the range of less than 500 g/m^2.
(3) It is necessary that the amount of sodium chloride for spreading on the snow surface is decided according to the purposes.

REFERENCES

Endo, Y., Kominami, Y. & Niwano, S.1995. Observations of "Shimi-watari" - about walking on a snow surface -. *Preprint of the Japanese Society of Snow and Ice 1995 Meeting.*103.
Kobayashi, T., Mizuno, Y., Kamata, Y., Kawashima, K. & Nohguchi, Y. 1997. Hardening of snow surface by spreading sodium chloride (NaCl). *Proceedings of '97 Cold Region Technology Conference.* Vol.13. 41-45.
Kobayashi, T. 1998. Change in the hardness of snow surface for spreading sodium chloride (NaCl) on snow. *Preprint of the Japanese Society of Snow and Ice 1998 Meeting.* 122.
Yaginuma, R., Kaneta, T. & Kikuchi, M. 1996. Arrangement of Alpine racecourse (III) - The use of numerical value for making Alpine race course -. *Journal of the Japan Society of Ski Sciences.* Vol.6. 52-56.

Snow Engineering, Hjorth-Hansen, Holand, Løset & Norem (eds) © 2000 Balkema, Rotterdam, ISBN 90 5809 148 1

Design criteria for cylindrical masts exposed to snow creep forces

Jan Otto Larsen
Department of Geotechnical Engineering, Norwegian University of Science and Technology, Trondheim, Norway

ABSTRACT: Based on 25 years of measurements of snow creep forces on cylindrical masts, a relation between the pressure and snow parameters as snow depth, density and temperature, has been established. The long observation period has made it possible to find series of pressure values which have been used to develop an empirical model. In the model, the pressure at different depths in the snowpack can be calculated by factors which depend on the total snow depth. Long term observations on masts with different diameters, have also made it possible to estimate the influence of mast diameter on the pressure. Finally, an inclination dependent factor is taken into account.

1. INTRODUCTION

The project has been carried out to investigate the forces of snow creep applied to different types of fixed constructions on a snow covered slope with an inclination of about 25° The experiment site is on a slope consisting of rock slabs at 1200 m a.s.l. below the Grasdalen glacier, close to NGI's research station Fonnbu, Grasdalen, Western Norway (Larsen et al. 1989) and Larsen (1998). The aim of the project has been to develop design criteria for masts and other constructions exposed to snow creep on snow covered slopes.

2. OBSERVATION METHODS

The installations consist of two tubular steel masts of 6 m and 4,5 m height. The taller mast, erected in 1975, has been a part of the project for 25 years. The shorter mast was erected in 1983, and broke due to the high snow pressure during the winter 1988/89.

The constructions are fitted with opposing pairs of Geonor vibrating wire strain gauges, that are connected to data loggers situated in a steel tower near by. For more technical specifications refer to Hansen (1980) and Larsen et al. (1989).

In addition to measuring the strain in the constructions, snow investigations have been carried out to record the properties of the snow. The standard snow investigations have been done in accordance with the ICSI (1990) snow classification standards.

During the winter 1996/97 a temperature logger/sensor was placed on the ground at the foot of the construction, and another in a connector box above the snow surface at the steel tower. Other weather parameters were measured at NGI's field station at the valley floor 270 m below the test site.

Snow gliding has been observed at the test site, but this does seem to have any significance with regards to the pressures recorded as this have only occurred on special occasions in the late winter, and after the forces have exceeded peak values.

A cylindrical chair lift tower situated at similar elevation and aspect as the structures at NGI's test site a few kilometres to the south, has also been studied. The tower broke as a result of creep pressure during the winter of 1990. The design of the tower was based on preliminary observations at the NGI test site.

The fracture occurred because the lift manufacturer did not believe in the load estimates given. However, the loads were even higher than the estimates because of unusual snow conditions with 10 m snow depth and a long creep period due to early isothermal snowpack (Larsen, 1998). Back-calculation of the forces acting on the tower, have been used to supplement the data from NGI's test site.

3. SNOWPACK, WEATHER AND CREEP

The measurements seem to indicate that as long as the snowpack is cold (<0°C), the temperature of the snow is of less significance to the pressure than the increase in total water equivalent of the snowpack. The highest pressures occur normally in the transitional period when temperature in the snow approaches 0°C throughout the snowpack. We observe that the pressure acting on the structure in the end of the winter has fairly large daily fluctuations due to changes in the air temperature from night to day. In addition, warming of the steel by direct sun radiation, and subsequent melting around the mast reduce the pressure. Prolonged periods with cold temperatures and no sun after earlier warm periods in the spring, can however increase the pressure towards new peak values.

4. DESIGN CRITERIA

In development of design criteria for mast constructions, observations from late winter when the peak.

To find the pressure dependence related to the diameter of cylindrical masts, a comparison between the acting pressures on the 42 cm diameter tall mast and the 22 cm diameter short mast, has been done. As mentioned before, the shorter mast unfortunately broke during the winter 1989 because of overloading, and was not re-erected. The relatively short measurement period from 1983 to 1989 gave however some indication of the diameter influence. The smaller diameter mast was situated at the same site in a distance of 9 m from the tall mast (Larsen et. al. 1989).

Comparison of measurements for the masts exposed to similar snow conditions indicate a diameter dependency which can be expressed by factor f_β:

$$f_\beta = d^{0.63} + 0.42 \tag{1}$$

f_β = diameter pressure factor
d = tube diameter (m)

In figure 1 the relation between the "diameter pressure factor" and diameter of the mast is illustrated.

The density of the snow normally increases with snow depth. The maximum density in a seasonal snowpack will normally not exceed 600 kg/m³, and the snowpack will become increasingly homogeneous towards the late winter. To be able to calculate the design pressure at a specific level in the snowpack the density distribution has to be taken into account. A pressure index related to the snow depth is given in figure 4. This snow depth index is used in expression 2 below to calculate the snow pressure at different depth levels in the snow.

$$p(z) = K \cdot f_\beta \cdot z^e \tag{2}$$

p (z) = pressure at a depth z (kN/m)
K = the correlation factor shown in figure 5
z = depth (m)

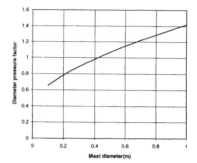

Figure 1. Diameter pressure factor f_β as a function of the mast diameter.

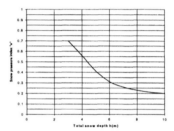

Figure 2. Snow pressure index e dependence on total snow depth h.

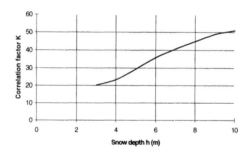

Figure 3. Correlation factor K dependency on total snow depth h.

Equation (2) gives the snow pressure against a mast at a certain level for a slope at 25° (the inclination at the test site). As pressure depends on the inclination φ, the inclination factor f_φ has to be taken into account:

$$f_\varphi = \sin\varphi / \sin 25° \tag{3}$$

The correlation factor K is dependent on the total snow depth h. Values for K which fit the results from our measurements are given in figure 3.

To find the total snow creep pressure against mast, the calculated pressure has to be integrated over the total snow depth:

$$R = \int p_{(z)} f_\varphi dz \qquad (4)$$

Figure 4 shows the exponential pressure distribution for a 42 cm mast at a 25°steep slope at different snow depth h.

5. DISCUSSION

Design criteria for snow creep pressure on cylindrical masts with diameters ranging from 0.3 m and 1.0 m can be estimated by the proposed expressions. There are, however, limitations to the data base, as it has relatively few measurements for snow heights above 5 m, and few measurements on masts with diameters exceeding 42 cm.

The proposed method gives an upper limit of the pressure measured in the field under environmental conditions typical for the Norwegian west coast climate. For other climate zones, the observed pressures may be different. It is reasonable to expect lower pressures in continental climates due to lower density and stiffness.

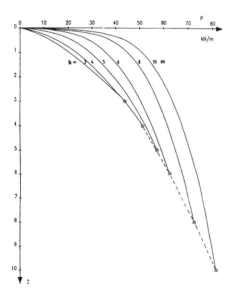

Figure 4. Snow pressure distribution for different snow depths. (In this case the mast has a diameter of 0.42 m, and is erected on a 25° steep slope).

6. CONCLUSION

The method of using the measured data from long time series for developing design criteria gives a good estimate of the expected pressure. The limitation lies in the fact that only masts with three different diameters, 22 cm, 42 cm and 100 cm have been tested. As this method is time consuming in practical use, a more user- friendly numerical model based on the present data should be developed in the future.

7. ACKNOWLEDGEMENT

The authors wish to tank the Norwegian Power Grid Company for funding of the project. Further thanks to the colleagues at NGI for support by fieldwork and discussions, and especially to Rolf Lauritzen for his contribution to the analyses.

REFERENCES

Bader, H., Haefeli, R., Bucher, E., Neherer, J., Eckel, O., Thams, C. and Niggli, P. 1939. *Der Schnee und seine Metamorphose, Beitrag zur Geologie der Schweiz.* Geologische Serie, Hydrologie, Lief. 3.

International Commission on Snow and Ice of the International Association of Scientific Hydrology, 1990 *The International Classification for Seasonal Snow on the Ground*

Hansen S. B., 1980. Resultater av målinger på rør vinteren 1978/1979(Results from measurements on tubes the winters 1978/1979). Report no.: 75420-4, Norwegian Geotechnical Institute, Oslo Norway.

Kristensen, K. 1998. *Snow CreepProject. Field measurements winter 1996/97.* Report No. 581300-7, Norwegian Geotechnical Institute, Oslo, Norway.

Larsen, J.O. 1982. *Snøens sigetrykk mot konstruksjoner i bratt terreng (Snow creep forces on constructions on steep slopes).* Report No. 75420/58110-1, Norwegian Geotechnical Institute, Oslo, Norway.

Larsen, J.O., McClung, D.M. and Hansen, S.B. 1985. *The temporal and special variation of snow pressure on structures.* Canadian Geotechnical Journal **22** (2), 166-171

Larsen, J.O., Laugesen, J. and Kristensen, K. 1989. *Snow-creep pressure on masts.* Annals of Glaciology, **13**, 154-156.

Larsen, J. O., 1991, *Snøens sigetrykk på konstruksjoner i bratt terreng.* (Snow creep forces on constructions at steep slopes) NGI publication no 581300 -3

Larsen, J. O., 1998. *Snow creep forces on masts.* Annals of Glaciology, **26**, 19 – 21.

McClung, D.M., Larsen, J.O. and Hansen, S.B. 1984. *Comparison of snow pressure measurements and theoretical predictions.* Canadian Geotechnical Journal, **21** (2), 250-258.

Snow Engineering, Hjorth-Hansen, Holand, Løset & Norem (eds) © 2000 Balkema, Rotterdam, ISBN 90 5809 148 1

Design criteria for avalanche supporting structures exposed to snow creep forces in maritime climate

Jan Otto Larsen
Department of Geotechnical Engineering, Norwegian University of Science and Technology, Trondheim, Norway

ABSTRACT: Since 1975 Norwegian Geotechnical Institute and Norwegian Power Grid Company have performed field investigation of snow creep forces on an avalanche retaining structure at the research site in Grasdalen, Stryn Mountains in Norway. A 15 m long and 3.1 m high supporting structure has been erected at a 25° steep slope and exposed to snow creep forces in a 2- 6 m deep snow-pack. Instrumentation of beams and supporters make us able to obtain the stresses in different elements in the structure. Snow glide has been controlled at the rock face above the construction. During the winter snow profiles have been made systematically where snow depth, densities and temperatures are measured. Snow type and moisture content are also observed. By relating the measured stresses and moments in the construction elements to the snow parameters it has been possible to find the snow pressure distribution and fluctuation with time. Based on long time observations we are able to introduce design criteria, which can be used for avalanche supporting structures on mountain slopes in maritime climate.

1 INTRODUCTION

Supporting structures are used in avalanche defense by anchoring the snow in the release areas. In the 60's and 70's Norway used the Swiss guidelines in design of supporting structures in avalanche defense. Damage to constructions because of overloading, made the designers suspicious. When Norwegian Geotechnical Institute, NGI, started active snow and avalanche research in the early 70's, one of the obvious questions was to test out these type of structures for snow in Norwegian climate which was supposed to be different from snow in the continental climate in Switzerland. In 1975 an instrumented steel/aluminum structure was erected at a 25° steep slope in NGI's research field at an elevation of 1150 m a.s.l. in the Stryn mountains at the west coast (Larsen, 1982). This structure was together with a tower for transmission lines and two poles with different diameters a joint venture project between Norwegian Power Grid Company and NGI in research of snow creep pressure on structures. In 1990 the project was enlarged by a wire net for a comparative testing of this type of supporting structure.

Based on the results from the project theoretical models has been tested for general application of the results to design of constructions (McClung et al. 1984, McClung and Larsen 1989). In the 90's the results from NGI's research field was compared with

results from Iceland where the climate is identical to parts of the Norwegian coast, and similarities in snow density and pressures were found (Johannesson et al. 1998).

The Swiss guidelines (EISLF, 1990) are still recommended as bases for design of supporting structures, but there has to be taken into consideration the higher creep pressure, which is observed in our research project. A theoretical approach based on linear viscous material behavior can be used for finding designing creep pressure (McClung and Larsen 1989).

2 MEASUREMENTS

Snow density and loading of supporting structures have been monitored by NGI at the research station from 1976 to 1990. Of primary interest with regard to the design of supporting structures is the density and snow depth when the snow pressure on the construction reaches its maximum. This has for the rigid structure been observed to happen at the end of the winter season when the snow pack reach 0° -C isothermal condition, which normally is at the end of April or the beginning of May. For the snow net this is also true, but the high-density snow later in May can give stresses in the cables which are even higher. Table 1 gives the snow depth, density and pressure (maximum and average) found in our research period 1976-1990.

As observed in Table 1, the average snow density in periods of maximum pressure is found to be 459 kg/m³ with a standard deviation of 48 kg/m³ in a 2 and 5 m deep snow-pack

Figure 1. shows negligible increase in average density with height for this data collection. The density at maximum pressure is relatively low in winters with thin snow cover (1979,1985), and relatively high for winters with deep snow (1976,1989,1990). In the last cases the pressure also last for a longer period of time after the isothermal 0° C snow-pack occur, and get a higher density at the end of the period of high pressures than when this situation appear first time. Also increase in density in a deep snow pack is a normal consequence of the settlement in the late winter.

A study of development of average density with time is given in Figure 2 for representative winters with deep snow cover. Snow densification is developing through winter, and maximum densities are in May with values measured above 500 kg/m³. When maximum pressure occur in late April or early May, the densities are between 450 kg/m³ and 500 kg/m³.

For comparison of the measured pressure with snow depth and density we have introduced the expression "Body force index " which is a product of density, snow depth and acceleration due to gravity (ρ g D). Figure 3 give the trend in development of maximum and average pressure as a function of Body force index. In Figure 3 the data are widely spread, but the tendency is increasing, and can by expressed to have a linear trend.

The maximum pressure is found to act at the middle height of the construction with an average value of 1.43 times the average pressure.

3 DESIGN OF SUPPORTING STRUCTURES

In our research field in Grasdalen we have no observations of snow glide, which in the Swiss research have been found to have a high influence on the snow pressure (Larsen et al. 1985). This is identical with the observations in Iceland (Johannesson et al. 1998, Johannesson and Margreth 1999). Design of supporting structures is therefore based on a "no glide situation". We recommend however to control the actual glide before design due to the importance of glide on the pressure (EISLF, 1990).

Design pressure for supporting structures can be calculated with bases in a theoretical approach for linear viscous material behavior by equation (1) (McClung and Larsen 1989):

$$\frac{\overline{\sigma}_x}{\overline{\rho} \cdot g \cdot D} = \left[\frac{2}{1-v}\left(\frac{L}{D}\right)\right]^{1/2} \sin\varphi + \frac{1}{2}\frac{v}{1-v}\cos\varphi \quad (1)$$

σ_x = Average pressure on the construction I (kPa)

$\overline{\rho}$ = Average snow density (kg/m³).

φ = Slope inclination

v = Poisson ratio

g = acceleration due to gravity (m/s²)

D = snow depth perpendicular to terrain

L/D = snow creep parameter dependent on slope inclination and viscous Poisson ratio given by:

$$\frac{L}{D} = \frac{1}{4}\left(\sin\varphi\right)^{1/2} + \frac{1}{4}\left(\frac{v^2}{1-v}\right) \quad (2)$$

As the snow is acting with shear forces on the surface of the structure when the snow is settling, we also have to introduce a force perpendicular to the slope (McClung and Larsen 1989):

$$\left(\frac{\overline{\tau}}{\overline{\rho}gD}\right) = \frac{1}{4}\left(\cos\varphi\right)^{1/2}\left[1 - \frac{3}{2}\left(\frac{v}{1-v}\right)\left(\sin\varphi\right)^{\frac{1}{2}}\right] \quad (3)$$

$\overline{\tau}$ = Average shear force on the construction (kPa).

From the earlier chapter of measurements we will find a density ρ = 500 kg/m³ as appropriate for design. The snow depth D can be estimated based on maximum with occurrence at yearly probability of 3×10^{-2}. Poisson ration for the maritime snow cover in Norway is for design purposes found to have reasonable fit with our measurements for v = 0.36 (McClung et al. 1985).

For design of the cross section members the maximum pressure has to be taken into account. Recommended design pressure is according to equation (4).

$$\sigma_m = 1.5\,\overline{\sigma}_x \quad (4)$$

At the edges of the structure there are higher pressures than the average found by equation (1) due to a bridge effect caused by viscous creep of the snow around the end of the structure. There are no measurements done in the Nordic countries to verify the size of this effect. In design we recommend to use the requirements in the Swiss guide lines (EISLF, 1990) for no slip conditions, and multiply the forces due to the edge effect by 1.5, according to equation (5).

$$\bar{\sigma}_r = (1 + 1.5 \cdot 1.2)\, \bar{\sigma}_x \qquad (5)$$

$\bar{\sigma}_r$: Pressure acting over a width of $1/3 \cdot H_{Const.}$ measured from the edges.

4 DISCUSSION

Snow pressure is dependent on snow density and snow depth, but the high variation in snow pressure can also be explained by the temperature effect. The highest pressures by equal "Body force indexes" are observed in the winters 1989 and 1990 (Table 1). Added to deep snow cover we observed 0° C iso-thermal snow pack extraordinary early as February. The relatively warm snow pack caused higher deformations and deformation rates than in the other winters where isothermal conditions occurred in late April (Larsen 1997), and is the main reason for these high pressures shown in Figure 3.

The Swiss guidelines recommend two load cases in design of supporting structures with different snow depth and densities (EISLF, 1990). In NGI's research project there is not found any significant difference in density between the case of maximum snow depth in the period of maximum load on the construction and the case with reduced settled snow depth. I therefore recommend to use a density of $\rho = 500 \text{ kg/m}^3$ in design for different load cases.

Snow pressure differences due to elevation and exposure is negligible in the Nordic region.

In Iceland the Swiss guidelines is recommended used with correction in the gliding factor N, which will give snow pressure close to the double of the Swiss values for "no glide" conditions.

5 CONCLUSION

Based on measurements in 15 years at our research station we have been able to develop a tool for calculation the snow creep pressure on supporting structures. Different from guide-lines used in other countries we make corrections for higher density, and have developed a theoretical tool for calculation the design forces. There is not taken into account the effect of glide because there were no observation of glide in the research area. Corrections

Table 1. Snow depth, density, maximum and average pressure in the end of the winters between 1976 and 1990.

Year	Date	Snow dept D (cm)	Density kg/m3	Aver pres kPa	Max pres. kPa	Body force kPa	Max/min pressure
1976	14.- Apr.	490	426	16	23	20.9	1.44
	18 - May	370	520	13	20	19.2	1.54
1979	27.- Apr.	250	395	8	11	9.9	1.38
1981	08.- May	410	430	13	20	17.6	1.54
1982	19.- May	230	388	6	9	8.9	1.50
1983	01.- Apr.	350	397	9	11	13.9	1.22
	03.- May	310	513	8	12	15.9	1.50
1984	13.- Apr.	390	433	10	14	16.9	1.40
	30.- Apr.	350	483	8	12	16.9	1.50
	05.- May	330	508	11	16	16.8	1.45
1985	16.- May	180	479	3	8	8.6	2.67
1987	11.- Apr.	310	400	8	10	12.4	1.25
	14.- May	250	448	6	9	11.2	1.50
1989	21.- March	430	387	15	19	16.6	1.27
	11- Apr.	490	488	15	18	23.9	1.20
	25.- Apr.	370	494	13	15	18.3	1.15
	19.-May	440	474	13	15	20.9	1.15
	26.- May	360	531	12	15	19.1	1.25
1990	10.- Apr.	480	456	15	21	21.9	1.40
	19.- Apr.	460	465	15	21	21.4	1.40
	08.- May	380	520	10	13	19.8	1.30
Average		363.3	458.8	10.8	14.9	16.7	1.43
Stand.dev		87.4	47.8	3.6	4.5	4.4	0.31

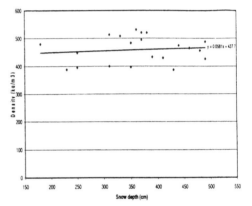

Figure 1. Density as a function of maximum winter snow depth in the period 1976 – 90.

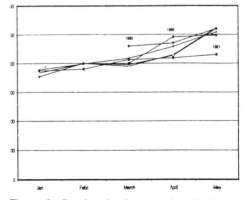

Figure 2. Density development through the winter (1976,81,83,84,89 and 90). Each point represents the month average based on one to three density measurements.

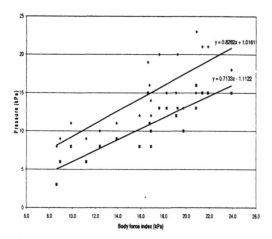

Figure 3. Maximum and Average pressure as a function of Body force index ρgD. ρ = density, g = acceleration due to gravity and D = snow depth measured perpendicular to the terrain.

have to be made if glide is observed in preliminary survey. Where constructions have a free end in the snow an local increase in load due to viscous creep around the edge has to be taken into account (EISLF, 1990)

6 ACKNOWLEDGEMENT

This work is based on research the author has been responsible for at Norwegian Geotechnical Institute, NGI, in the period 1975 – 1990. The section for Snow and Avalanche research at NGI has made valuable support to the project, and in particular thanks to K. Kristensen for his contribution in the field work. Norwegian Power Grid Company has funded the project, and I am grateful for their support.

7 REFERENCES

Bader, H. P., H. U. Gubler and B. Salm, 1989. Distribution of stresses and stain-rates in snowpacks. *Numerical Methods in geomechanics(Innsbruck 1988) Swoboda (ed.)*Balkema, Rotterdam, ISBN 906191 809 X.

Eidgenössishes Institut für Schnee- und Lawinenforschung, 1990. Richtlinien für der Lawinenverbau im Ambruchgebiet. Bern, Switzerland, 76 p.

Johannesson, T., J. O. Larsen and J. Hopf., 1998. Pilot project in Siglufjörður. Interpretation of observations from the winter 1996/97 and comparison with similar observations from other countries. *Veðurstofa Islands,* report VI-G98033-UR26, Reykjavik September 1998, 25 p.

Johannesson, T. and S. Margreth, 1999. Adaptation of the Swiss Guidelines for supporting structures for Icelandic conditions. *Veðurstofa Islands,* report VI-G99013-UR07, Reykjavik July 1999, 12 p.

Larsen, J. O. 1982. Snøens sigetrykk mot konstruksjoner i bratt terreng. (Snow creep pressure on constructions on steep slopes). *Norwegian Geotechnical Institute,* report 75420/58110-1, 11. June 1982, 40 p.

Larsen, J. O., D. M. McClung and S. B. Hansen, 1985. The temporal and spatial variation of snow pressure on structures. *Can. Geot. J.* Vol. 22(2) 166–171.

Larsen, J. O., 1997. Snow creep forces on masts. *Annals of Glaciology,* Vol. 26, 19-21.

McClung, D. M., J. O. Larsen and S. B. Hansen, 1984. Comparison of snow pressure measurements and theoretical predictions. *Can. Geot. J.* Vol 21(2) 250-258.

McClung, D. M. and J. O. Larsen, 1989. Effects of structure boundary conditions and snow-pack properties on snow-creep pressures. *Cold Regions Science and Technology,* 17, 33-47.

McClung, D. M. 1993. Comparison of analytical snow pressure models. *Can. Geot. J.,* Vol 30(6), 947-952.

Salm B., 1977. Snow forces. *Jour. of Glac.* Vol 19(81) 67-100.

Snow Engineering, Hjorth-Hansen, Holand, Løset & Norem (eds) © 2000 Balkema, Rotterdam, ISBN 90 5809 148 1

Combining snow drift and SNOWPACK models to estimate snow loading in avalanche slopes

Michael Lehning, Judith Doorschot, Norbert Raderschall & Perry Bartelt
Swiss Federal Institute for Snow and Avalanche Research, Davos Dorf, Switzerland

ABSTRACT: Snow drift is one of the major factors influencing avalanche activity in high Alpine terrain. This concerns catastrophic avalanches (February 1999 in the Alps) as well as regularly occurring 'touristic' avalanches. When discussing the role of snow drift in avalanche formation, four physically connected processes have to be taken into account: 1) the windfield over steep topography, 2) the preferential deposition of snow during snowfall events, 3) the possible redistribution of already deposited snow, and 4) the different developments of the snow cover at sites of ablation and accumulation. In this paper we present an integrated model approach for the estimation of snow loading in steep avalanche slopes. The assessment includes: 1) an atmospheric model analysis of the high resolution wind field over topography, 2) a novel formulation of wind transport in saltation and suspension that allows to treat preferential deposition and redistribution and 3) a distributed model assessment of the snow cover characteristics. The snow cover model is important to judge the erodability of the deposited snow and yields information on the status of the snowpack on the slope which is crucial for a final estimation of slope stability. Results are presented for a drift period in January 1999. A comparison with measurements shows a very good agreement. Over steep Alpine terrain (up to 50 degrees), the model is able to simulate snow distribution and snow cover status with a reasonable computational effort. A discussion is given of future developments towards an operational assessment of snow drift for the purpose of avalanche warning. The model is the first available tool to study the atmosphere-snow-vegetation interaction and the distribution of snow water equivalent in steep Alpine terrain.

1 INTRODUCTION

Snow transport by wind is one of the crucial factors influencing the seasonal build-up of the snow cover in Alpine terrain and the related avalanche activity. Because of its importance, it has been studied extensively over the last few decades, and a lot of progress has been made in modeling and understanding of snow drift. Despite these efforts, operational assessments of snow redistribution by the wind hardly exist due to the complexity of the physical processes associated with this phenomenon. The development of such models is of extreme interest for the improvement of avalanche forecasting, and deserves major attention in the studying of snow drift.

For snow drift, we usually distinguish between grains being transported in saltation or suspension mode. Saltation takes place in a thin layer directly above the surface, where the particle concentration is highest. In this transport mode the snow grains follow ballistic trajectories and return to the surface, possibly rebounding or ejecting other grains. For the mass flux in saltation, many investigators (Bagnold,

1941, Takeuchi, 1980, Pomeroy & Gray, 1990) have undertaken efforts to find empirical relations based on measurements and physical considerations. Only in the last decade numerical models have been developed (McEwan & Willets, 1991, Gauer, 1999, Shao & Li, 1999) which take into account the microscopic processes of aerodynamic entrainment, splash function, particle motion and particle-wind feedback. For neither approach, a model that explicitly takes into account the topographic effects has been developed so far. For obtaining the mass transport in suspension, a mass conservation equation for moisture needs to be solved in combination with the Navier Stokes equations for the turbulent air flow. This procedure is often simplified by assuming steady state and ignoring lateral effects, so that a one-dimensional diffusion equation remains (Kind, 1992, Liston & Sturm, 1998). However, we point out that for steep Alpine terrain, advective effects are very important and no such simplification is allowed.

A third term significantly contributing to the loading of avalanche slopes is preferential deposition of snow during snow fall events. Little is known about this tendency of precipitation to accumulate directly in a leeward slope, and this factor has seldom been included in simulations of snow distribution. The approach for modeling this term is however very similar as for the suspension layer, and its contribution can be taken into account as soon as a measurement of the precipitation rate is available.

The fourth process that needs to be considered is the development of the snow cover at sites of erosion and deposition. The snow type has a strong influence on the drift characteristics, since it affects model parameters like threshold shear velocity, effectiveness of rebound during saltation and the modification from saltation to suspension. In a windward slope, freshly fallen snow may be eroded until the old layer of snow with a higher threshold has been reached, thus providing an upper limit for the total mass transport. On the other hand, new snow that has been transported by the wind will be mechanically altered due to the high contact pressure during deposition.

In recent years, a number of numerical snow drift models has been developed that to some extent take topographic effects into consideration. The parameterizations of the involved processes differ greatly between these models, and therefore they also show various limitations for the situations they can be applied to and for the interpretation of their results. Whereas the model of Sundsbø (1998) for snow drift around obstacles is only two-dimensional, in the model of Green et al. (1999) advective terms are not taken into account, and it can only be used for gentle topographies. More advanced models have been developed by Naaim et al. (1998) and Uematsu (1993), however in neither of these models the preferential deposition of precipitation is included. The only 3-dimensional model which also calculates this factor is from Gauer (1999); the most important limitation of this model is its extensive computational time.

Fierz and Gauer (1998) pointed to the importance of combining snow drift and snow cover models and presented a first attempt to include snow drift effects in a SNOWPACK simulation. In this paper we present a new numerical model for the estimation of snow distribution in steep Alpine terrain due to blowing and drifting snow. An analysis of the high resolution wind field over topography is combined with newly developed formulations of drift and preferential deposition of snow. The inclusion of snow cover characteristics and development is accounted for by a coupling with the operational snow cover model SNOWPACK (Lehning et al., 1999).

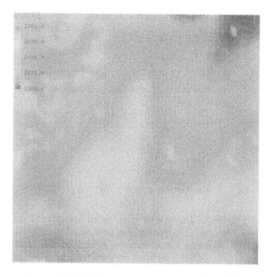

Figure 1. Windfield simulation for January 28 at 20:00 over Gaudergrat topography at the lowest grid level (Fig. 2). Isolines and shading show topography and the white arrows represent the wind. The flow is from NW. The speed-up over the ridge creates velocities up to 14 m s^{-1}. Vertical velocities at the steepest slopes (50°) reach 10 m s^{-1}.

First Nine Grid Levels over Gaudergrat Ridge

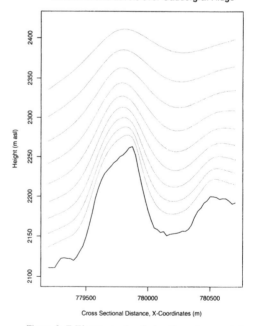

Figure 2. E-W cross sectional view through computational domain at y = 192250. The vertical grid stretching of the terrain following grid levels is illustrated.

114

2 NUMERICAL MODEL

2.1 Atmospheric model analysis

For obtaining the high resolution wind field over complex topography we use the mesoscale atmospheric model ARPS (Advanced Regional Prediction System) (Xue et al., 1995) in its SUBMESO version (Guilbaud et al., 1997). Our topography in the presented case is a mountain ridge with slope angles up to 50 degrees, a value which is typical for avalanche slopes. In ARPS, the non-hydrostatic compressible Navier-Stokes equations for a turbulent air flow are solved on a mesh, using the finite differences method. Novel is the use of a mesoscale atmospheric model with a resolution down to 25 m in the presented case. For turbulence parameterization in our drift model, we use the output of the turbulent kinetic energy (TKE) from ARPS.

2.2 Saltation layer

For calculating the snow mass that is transported in the saltation layer, we propose a physical model. One of its new features is the consideration of the influence of a sloping bottom. Furthermore, the effects of particle-wind feedback, particle trajectories and grain properties are included. Here we only sketch a rough outline of the model; a full description is in preparation for publication.

Saltation is regarded as a self-regulating process which develops to equilibrium due to the modification of the wind field by the transfer of particle momentum. Saltation starts when the shear stress at the surface is greater than the threshold shear stress for aerodynamic entrainment. The number of saltating particles increases until the maximum capacity that the air flow can carry is reached. Numerical simulations suggest that this steady state is reached within a time of 1-2 seconds. In principle there are three different ways for a snow particle to start a trajectory in saltation: aerodynamic entrainment, rebound from a previous impact, or ejection by the impact of an other grain. The last two mechanisms are referred to as the 'splash'-function, and the average number of particles entering saltation resulting from one impact is called the mean replacement capacity. In a saltation layer at equilibrium, there exists a statistical balance between a small probability of a grain not rebounding after impact and the also small probability of a new grain being dislodged with enough energy to start saltating (Anderson et al., 1991).

Within the saltation layer, the flying grains exert a stress on the air flow, the so-called grain borne shear stress, which is defined as the difference in horizontal momentum between up-going and down-going particles at a certain height. At steady state, the sum of the grain borne and fluid (or air borne) shear stresses equals the total shear stress above the saltation layer (McEwan & Willets, 1991). For an equilibrium to maintain, the mean replacement capacity has to be one, and this can only be the case if the air borne shear stress at the surface stays at impact threshold (see Owen, 1964). Furthermore, when the vertical coordinate z reaches above the saltation layer, the air borne shear stress has to approach the total shear stress. A profile which satisfies these conditions is given by (McEwan, 1993):

$$\tau_a(z) = \tau_s(1 - (1 - \sqrt{(u_{*t}^2 / u_*^2)}\exp(-z / h_s)), \quad (1)$$

where $\tau_a(z)$ is air borne shear stress, τ_s is the total shear stress, u_{*t} is the threshold shear velocity, u_* is the shear velocity, and h_s is the height of the saltation layer.

The main forces governing the particle trajectories are inertial force, drag force, and gravity. Thus, the equations of motion for a grain moving over a slope with slope angle α are given by:

$$x'' = -0.75\frac{\rho_a U_r}{\rho_p d}C_d(x' - U(z)) - g\sin(\alpha), \quad (2)$$

$$z'' = -0.75\frac{\rho_a U_r}{\rho_p d}C_d z' - g\cos(\alpha). \quad (3)$$

Here ρ_a and ρ_p are the respective densities of air and ice, d is the particle diameter, U_r is the particle speed relative to the air flow, and C_d is the drag coefficient, which is a function of the particle Reynolds number. The x and z-coordinates are aligned parallel and perpendicular to the surface, and x is in the direction of the mean air flow.

In our model, the following approximations and assumptions are made. Since we are primarily interested in the mass flux, only the stationary state of the saltation layer is considered. Furthermore, we only take into account the rebounding of grains, and neglect ejections that may result from an impact. The wind profile, modified by the saltating particles, is based on Equation (1). For the particle collision with the surface of a rebounding grain, we propose that the energy loss is a material constant, i.e. is only dependent on the type of snow:

$$v_{ej}^2 = r v_{imp}^2 \quad (4)$$

where v_{ej} and v_{imp} are the ejection and impact velocity, respectively, and $(1-r)$ is the percentage of energy that is lost in the collision. The ejection angle is taken to have a normal distribution around 30 degrees.

For obtaining the mass flux, a number of successive trajectories is calculated for a rebounding grain, until the process becomes stationary. From the average values of the horizontal momentum of up-going and down-going particles at the surface and the time between one impact and the next, the average grain borne shear stress caused by one particle trajectory is known, and this determines the maximum and equilibrium number of grains N that the air flow can carry. When the mean horizontal particle velocity is u_{mean} and m is the particle mass, the saltating mass flux Q (kg m^{-1}s^{-1}) is given by:

$$Q = Nmu_{mean}. \tag{5}$$

2.3 Suspension layer and preferential deposition

Particle transport in suspension from drift and snow fall are regarded in the same manner. The suspended particles are treated as passive tracers of the wind field, and no effects of turbulence damping or interaction between particles are considered.

We consider a small control volume, and apply conservation of moisture:

$$\frac{\partial c}{\partial t} + u\frac{\partial c}{\partial x} + v\frac{\partial c}{\partial y} + (w - w_s)\frac{\partial c}{\partial z} = 0, \tag{6}$$

where $\partial/\partial t$ stands for the partial derivative, c for the moisture concentration (kg m^{-3}), and w_s for the settling velocity. For the settling velocity, a parameterization is derived taking into account turbulence effects. In still air the settling velocity w_{s0} is governed by a balance between friction and gravity, resulting in (Stokes law):

$$w_{s0} = \frac{\rho_p d^2 g}{18\eta}. \tag{7}$$

Here is g the acceleration due to gravity and η the friction coefficient of air (viscosity). We make the assumption that the settling velocity in a turbulent air flow equals its value in still air minus a term to account for the turbulence:

$$w_s = w_{s0} - w_t. \tag{8}$$

For obtaining the term w_t, we turn to the definition of the turbulent kinetic energy, e:

$$e = 0.5(\overline{u'^2} + \overline{v'^2} + \overline{w'^2}), \tag{9}$$

where the dashed variables stand for the deviations from average of the respective velocities. When the turbulence is symmetric, this yields:

$$w_t = \sqrt{\overline{w'^2}} = \sqrt{\tfrac{2}{3}e}. \tag{10}$$

Thus, the total settling velocity w_s is given by:

$$w_s = \frac{\rho_p d^2 g}{18\eta} - \sqrt{\tfrac{2}{3}e} \tag{11}$$

The turbulent kinetic energy is obtained directly from the wind field simulations for every grid point, which allows for the computation of (6) and the suspended mass fluxes going in and out of every cell.

At the upper boundary of the domain, the boundary condition (BC) for the particle concentration is given by the precipitation rate. The lower BC is obtained using the saltation model. For finding the particle concentration at the height of the saltation layer, we consider the energy balance of the system. At saltation height, the wind velocity shows a deviation from the logarithmic profile $\Delta U(h_s)$, due to the energy that has been transferred to the saltating particles. Thus the particle concentration c_p is given by:

$$c_p(z = h_s) = \frac{\rho_a(\Delta U(z))^2}{u_{mean}^2}. \tag{12}$$

2.4 Snow cover characteristics: SNOWPACK

SNOWPACK is the finite-element based physical snow cover model of the Swiss Federal Institute for Snow and Avalanche Research (SLF), which is in operational use in connection with the Swiss network of approximately 50 high Alpine automatic snow and weather stations. It solves the instationary heat transfer and snow settlement equations and calculates phase changes and transport of water vapor and liquid water. Furthermore, it includes surface hoar formation and snow metamorphism (grain types). A more complete description of the model can be found in Lehning et al. (1999). For the present purpose, SNOWPACK has been coupled to our snow drift model for the assessment of the erodability of the snow cover and the development of the snow cover at different locations due to erosion and deposition of snow.

For determining the erodability of the snow cover we follow Schmidt (1980) and use the procedure as detailed in Lehning et al. (2000).

2.5 Numerical implementation

The purpose of the model is to allow a high resolution physical simulation of snow drift and snow cover with reasonable computational expense. The work of Gauer (1999) has shown that a full physical simulation of the drift (even without the snow cover) using a flow solver (CFX) leads to unacceptable long simulation times.

Therefore we adopt a different concept: For chosen time intervals (at present one hour), we model an average (stationary) wind field with ARPS using the time-averaged measurements to prescribe initial and boundary conditions. The missing meteorological parameters such as precipitation rates, temperature and are taken from measurements. From ARPS, the turbulent kinetic energy (TKE) and 3-dimensional wind velocity is then known at each grid point. The procedure continues as follows: 1) The snow cover is initialized and saltation is calculated for each surface grid point, if the threshold wind speed is exceeded. This yields the lower BC for the suspension model (Eq. 12). 2) The suspension model calculates for the given (stationary) wind and saltation conditions the concentration distribution for all grid points until a steady state is reached. 3) Erosion and deposition of snow are calculated and the snow cover status is updated by calculating energy- and mass balance, settling and the internal changes of the snowpack at every surface grid point. Then the model proceeds to the next global time step (stationary wind field).

At present, we use the ARPS computational mesh but solve the suspension equation (6) using a finite element (FE) technique: The ARPS grid cells define the finite elements.

The lower boundary of the computational domain is of particular importance. For determining whether snow drift is possible, the shear velocity at the surface is needed. Therefore, we take the TKE from the lowest grid point *above* the surface and use similarity theory (Stull, 1988):

$$u_*^2 = e / 5. \tag{13}$$

The lower BC for concentration is given by Equation (12) in case of saltation but is left variable in absence of saltation. Erosion will only take place in presence of saltation. By solving:

$$\frac{\partial r_{SA}}{\partial t} = \frac{\partial Q}{\partial x} + \frac{\partial Q}{\partial y}, \tag{14}$$

we get the erosion/deposition rate at the surface which is caused by saltation, r_{SA} (kg m^{-2}s^{-1}). The additional deposition or erosion due to suspension is problematic to determine because of the strong coupling of the lower BC to saltation (Eq. 12). We propose to calculate the suspension deposition/erosion rate, r_{SU} (kg m^{-2}s^{-1}), by:

$$r_{SU} = \begin{cases} c_p \, (w_s - f \, w) \,, & w_s - f \, w > 0 \,; \\ c_p \, (w_s - f \, w) \,, & w_s - f \, w < 0 \text{ and } Q > 0 \,; \\ 0, & \text{else} \,; \end{cases} \tag{15}$$

Here the particle concentration and the velocities are evaluated at the first grid level *above* the topography. The factor f is grid dependent and accounts for the fact that the vertical velocity has to go to zero as the ground is approached. Equation (15) states that suspension can only contribute to erosion in the presence of saltation.

3 RESULTS AND DISCUSSION

We present an analysis of snow drift from the first major drift event of the avalanche period 1999 and compare the model results to manual measurements. The simulated period starts on January 26 and ends on January 31. For this period, the snow height development has also been measured. The model has only recently been completed and the results may be regarded as preliminary.

3.1 Simulation Settings

We calculate snow drift for our experimental site Gaudergrat (Gauer, 1999) situated in a highly complex topography. The computational grid covers an area of 1.5 x 1.5 km. In Figure 1, an overview of the topography is presented together with the ARPS wind field at the first node above ground. The simulation is for January 29. The influence of topography on the wind field is clearly visible. A cross sectional representation of the topography and the vertical grid levels is found in Figure 2. The lowest grid cell has a vertical extent between 1 m at ridges and summits and 10 m at the flat boundaries. The horizontal resolution is 25 m and 30 vertical layers are calculated to a height of 5000 m a.s.l. The internal calculation time step for ARPS is 0.1 s and we calculate a stationary wind field for each hour between 26.01.1999 12:00 and 31.01.1999 12:00. The time step for calculating saltation and snow cover status is 1 hour. A calculation time step of 1 s for calculating suspension is found to be sufficient to ensure numerical stability. Typically after 10 to 20 iterations the concentration field will become stationary for a given wind field. The drift results are calculated with a factor f (Eq. 15) of 0.5. On a UNIX workstation (HP C3000), the simulation of the 120 hour period takes 105 minutes (excluding the wind field simulation which takes considerably more time).

3.2 Computational results

First, we present the time development of precipitation, wind speed and snow height at three selected grid points (Fig. 3). The first point is well in the luff of the Gaudergrat ridge, the second point is in the lee close to the ridge and the third point is in the lee at 75 m from the ridge. The (free-atmosphere) precipitation rate has been determined following Lehning et al. (1999) and is used as the upper BC. The presented wind speed is from the first node above the ground.

Upwind from the ridge (Fig. 3a) the snow depth changes are small. The old snowcover is eroded. During periods of high precipitation but low wind speeds (after hours 50 and 80), the snow depth

117

increases for some time but is eroded again as soon as the wind speed increases. The situation at the crest line is more complicated (Fig. 3b). The wind speeds are highest at the ridge. During the first half of the simulation period, the crest line shows similar characteristics as the point in the luff (Fig. 3a). Then, following hour 65 (with a change in wind direction, which is not shown), the model simulates the build-up of a cornice. Further down the slope (Fig. 3c), the wind speeds are considerably lower. During the whole period, snow accumulates at an almost constant rate. Of special interest is the fact that the settling of the snow cover can be seen during the periods with no deposition of snow (approximately at hour 110, for example).

The complete picture of snow distribution over the whole computational domain and at the end of the simulated period is presented in Figure 4. A picture with a realistic snow distribution results, where the prime accumulation areas are found around and in the lee of the ridges and summits. The simulation shows highly variable deposition patterns, which are are also found in the measurements (Fig. 5a).

3.3 Evaluation against measurements

We manually measured the snow distribution before and after the drift event for a small stripe over

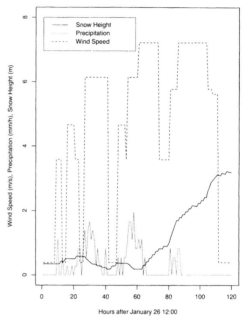

Figure 3b. As Figure 3a but for a point at the ridge.

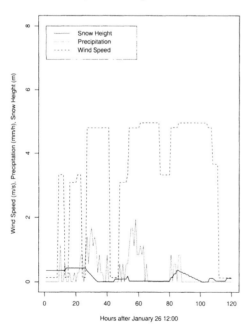

Figure 3a. Change of snow height, precipitation and wind speed during the simulation period at a point approximately 75 m in the luff of the Gaudergrat ridge.

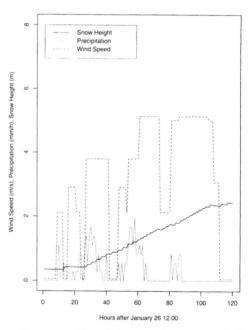

Figure 3c. As Figure 3a but for a point approximately 75 m in the lee of the ridge.

118

the Gaudergrat ridge. The setup and implementation of our measurement campaigns is described in Gauer (1999). The horizontal resolution of the measurements is between 2 and 8 m. Figure 5a gives the snow depth difference for the drift period as interpolated from the high resolution manual measurements. A very complex three-dimensional picture of snow distribution results. This distribution is to be compared against the model analysis for the same area in Figure 5b. Our horizontal grid resolution of 25 m is clearly insufficient to reproduce the very small scale variability as visible in Figure 5a. However, certain deposition patterns are reproduced by the model. In the luff of the ridge, we see a local deposition maximum close to point P0 (Figs. 5a,b) in the model as well as in the measurements. Closer to the ridge-top, at point P1, erosion is dominant. A local deposition maximum at point P2 is also reproduced by the model. The band of local deposition minima stretching from point P3a over P3b to P3c is less pronounced in the model than in the measurements and somewhat shifted. Other points with high deposition such as P3 are not as well reproduced in location and magnitude by the model. However, considering the model grid resolution, the overall agreement between the model and the high resolution measurements is satisfactory.

Finally, a cross-sectional, quantitative snow height comparison is made in Figure 6. The comparison confirms the findings from above: Even though the model suffers from insufficient lateral resolution and tends to deposit too much snow in the lee, the agreement is very satisfactory. The general pattern of luff erosion and lee deposition is correct, i.e. the model recognizes the erosion in the luff and a series of deposition maxima in the lee due to lee rotors (see also Fig. 5b). Even the approximate location of the maxima can be calculated with the resolution of 25 m. The overall amount of transported snow is also well reproduced.

4 CONCLUSIONS AND OUTLOOK

We have introduced a modeling system that combines wind field simulation with snow drift analysis and distributed snow cover simulation (the SNOWPACK results will be presented elsewhere). This physical model is the first complete system that is able to describe the topography-snow-drift-atmosphere interactions. The model will be a valuable tool in coming years in the disciplines of hydrology (distribution of snow water content), ecology and biology (snow-wind-vegetation-soil interactions), meteorology (momentum, energy and

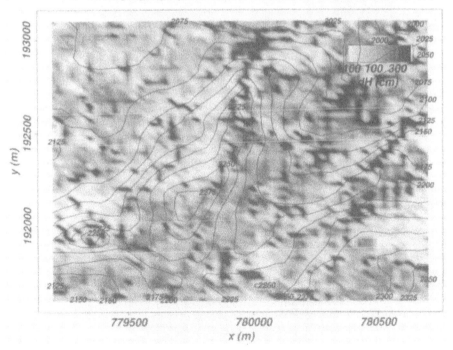

Figure 4. Overview of snow distribution at the end of simulated drift period for the entire computational domain

119

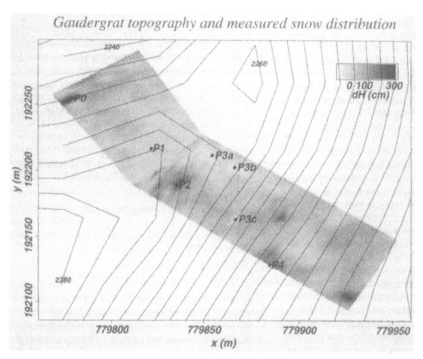

Figure 5a. Measured snow distribution across the Gaudergrat ridge for the simulation period. Note the three-dimensional small scale distribution pattern.

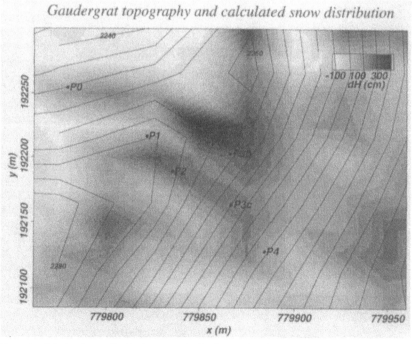

Figure 5b. Calculated snow distribution across the Gaudergrat ridge for the area of Figure 5a.

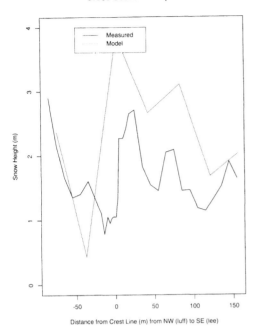

Cross Section Comparison

Figure 6. Comparison of the calculated total snow depth profile at the 7 grid points within the measured area (Fig. 5a) with the corresponding measured profile.

mass exchanges) and avalanche science. The first results as presented here are very encouraging. Development steps are directed towards a higher horizontal resolution, a two-way coupling between atmospheric model and drift model and a consolidation of the suspension and saltation models. Recently, we were able to measure remotely (laser scanner from airplane) the snow depth change during a drift period for a much larger area than up to now. We expect a major improvement with respect to our model evaluation efforts from these measurements.

This integrated model, which is the first to contain all major factors contributing to the snow loading of avalanche slopes, may also be considered a major step forward in the development of an operational assessment of snow drift for the purpose of avalanche forecasting. It will be used to develop parameterizations that can be applied to an even larger area.

ACKNOWLEDGEMENTS

We thank Walter Ammann, Tom Russi and Charles Fierz for their support. The work is funded by the Swiss National Science Foundation, the Swiss Federal Institute for Snow and Avalanche Research as part of the Swiss Federal Institute for Forest, Snow and Landscape Research and the Swiss Federal Institute of Technology.

Part of the simulations were made using the Advanced Regional Prediction (ARPS) developed by the Center for Analysis and Prediction of Storms (CAPS), University of Oklahoma. CAPS is supported by the National Science Foundation and the Federal Aviation Administration through combined grant ATM92-20009.

REFERENCES

Bagnold, R.A, 1941: The physics of blown sand and desert dunes, Methuen, London.

Fierz, C. Gauer, P.: 1998: Snow cover evolution in complex alpine terrain: measurements and modeling including snow drift effects, *Proceedings International Snow Science Workshop ISSW'98*, Sunriver, Oregon, U.S.A., 27 September – 1 October 1998. 284-289.

Gauer, P. 1999: Blowing and drifting snow in Alpine terrain: a physically based model and related field measurements. (Ph.D. thesis, ETH Zürich, Switzerland.).

Guilbaud, C., Chollet, J.P., Anquetin, S., 1997: Large eddy simulations of stratified atmospheric flows within a deep valley, in J.-P. Chollet et al. (eds.), Direct and Large-Eddy Simulation II, *Kluwer Academic Publishers*, 157-166.

Green, E.M., Liston, G.E. and Pielke Sr., R.A., 1999: Simulation of above treeline snowdrift formation using a numerical snow-transport model, *Cold Reg. Sci. Technol.*, **30**, 135-144.

Kind, R.J., 1992: One-dimensional aeolian suspension above beds of loose particles - a new concentration-profile equation, *Atmos. Environ.*, **26a**, no. 5, 927-931.

Lehning, M., Bartelt, P., Brown, R.L., Russi, T., Stöckli, U., Zimmerli, M., 1999: Snowpack model calculations for avalanche warning based upon a new network of weather and snow stations, *Cold Reg. Sci. Technol.*, **30**, 145-157.

Lehning, M., Doorschot, J., Bartelt, P., 2000: A Snow Drift Index Based on SNOWPACK Model Calculations, *Ann. Glac.*, **31**, *in press*.

Liston, G.E., Sturm, M., 1998: A snow-transport model for complex terrain, *J. Glaciol.*, **44**, 498-516.

McEwan, I.K. and Willets, B.B. 1991. Numerical model of the saltation cloud, *Acta Mech.* (Supplementum), **1**, 53-66.

McEwan, I.K., 1993, Bagnold's kink, a physical feature of a wind velocity profile modified by blown sand?, *Earth Surface Processes and Landforms*, **18**, 145-156.

Naaim, M., Naaim-Bouvet, F. and Martinez, H., 1998: Numerical simulation of drifting snow: erosion and deposition models, *Ann. Glaciol.*, **26**, 191-196.

Owen, P.R., 1964: Saltation of uniform grains in air, *J. Fluid Mech.*, **20**, part 2, 225-242.

Pomeroy, J.W. and Gray, D.M., 1990: Saltation of snow, *Water Resources Research*, **26**. 1583-1594.

Schmidt, R.A., 1980: Threshold wind-speeds and elastic impact in snow transport, *J. Glaciol.*, **26**, 453-467.

Shao, Y. and Li, A., 1999: Numerical modelling of saltation in the atmospheric surface layer, *Boundary-Layer Met.*, **91**, 199-225.

Stull, R.B. 1988. An introduction to boundary layer meteorology. Dordrecht, Kluwer Academic Publishers, 1st edition.

Sundsbø, P.-A., 1998: Numerical simulation of wind deflection fins to control snow accumulation in building steps, *J.*

Wind Eng. Ind. Aerodyn., **74-76**, 543-552.

Takeuchi, M., 1980: Vertical profile and horizontal increase of drift-snow transport, *J. Glaciol.,* **26**, no. 94, 481-492.

Uematsu, T., 1993: Numerical study on snow transport and drift formation, *Ann. Glaciol.,* **18**, 135-141.

Xue, M., Droegemeier, K.K., Wong, V., Shapiro, A., Brewster, K, 1995: ARPS Version 4.0 User's Guide. Available from Center for Analysis and Prediction of Storms, University of Oklahoma, Norman OK 73072, 380pp.

Snow Engineering, Hjorth-Hansen, Holand, Løset & Norem (eds) © 2000 Balkema, Rotterdam, ISBN 90 5809 148 1

Characterisation of snow structure in a cross-country race ski track

Sveinung Løset & Dag Anders Moldestad
Department of Structural Engineering, Norwegian University of Science and Technology, Trondheim, Norway

ABSTRACT: Characterisation of the sliding surface, i.e. the snow surface in the cross-country ski track, is necessary in order to understand ski base sliding friction and results from ski base sliding friction tests. It is therefore important to register essential snow parameters when performing accurate ski base sliding friction tests in-situ. The paper presents the mean snow grain diameter and the standard deviation of grain diameters calculated for 34 snow surface microscope images of ski tracks. The analysis of snow surface grains in ski tracks has shown that the typical grain size ranges from 0.08 to 2.59 mm.

1 INTRODUCTION

During measurement campaigns in ski tracks in Norway (1995-98), Hakuba/Japan (1996-98) and Sundance/USA (1999), the following snow and weather parameters have been registered: air temperature, relative humidity, net radiation, snow temperature, snow hardness, snow humidity, snow density, snow type, snow grain structure, electrolytic conductivity and ionic content of melted snow samples. The present paper focuses on snow type and grain structure. The major part of the paper is extracted from the dr. thesis of Moldestad (1999).

2 MEASUREMENT METHOD OF SNOW GRAIN STRUCTURE

A *LEICA MS5* stereomicroscope (max. 40× magnification) was used to observe the snow surface directly in-situ in the track. Microscope images of the snow surface were captured with a *Nikkon F70D* camera connected to the microscope through a phototube. The setup without camera connected to the phototube is shown in Fig. 1. At a later stage (indoor) the captured images were converted to digital form by a scanner, and analysed and characterised digitally using the general image processing program *Adobe Photoshop 5.0* (Adobe, 1998).

The in-situ measurement procedure has proved to be very efficient for ski tracks with snow surface grains larger than approximately 0.1-0.2 mm. For smaller snow surface grains it has been difficult to capture microscope images with sufficient quality.

The 2D snow grain surfaces in microscope images have been characterised using the following definitions:

j — snow grain surface number in a microscope image of a snow surface

d_j — length of the largest 2D diagonal for snow grain surface number j in a microscope image of a snow surface, mm

n_g — number of snow grains with distinguishable boundaries in a microscope image of a snow surface

$\bar{d}$ — mean of d_j for the n_g snow grains with distinguishable boundaries in a microscope image of a snow surface, mm

δ_d — standard deviation of d_j for the n_g snow grains with distinguishable boundaries in a microscope image of a snow surface, mm

$\bar{d}$ has also been classified according to the classification of grain size given in Colbeck et al. (1990), see Table 1.

Table 1. Classification of d according to the classification of grain size given in Colbeck et al. (1990).

Term	$\bar{d}$ (mm)
Very fine	< 0.2
Fine	0.2 - 0.5
Medium	0.5 - 1.0
Coarse	1.0 - 2.0
Very coarse	2.0 - 5.0
Extreme	> 5.0

Further, the snow type in the track has roughly been classified according to the following snow type categories:

1. Falling new snow
2. New snow
3. Combination of new snow and transformed snow, mainly new snow
4. Combination of new snow and transformed snow, mainly transformed snow.
5. Transformed snow
6. Artificial snow

The grain shape of the snow has furthermore been classified according to Colbeck et al. (1990), see Table 2.

3 RESULTS

We have measured snow parameters in 152 cross-country race ski tracks. The presented measurement results are based on measurement campaigns in Norway, 1995-98, Hakuba/Japan, 1996-98 and Sundance/USA, 1999. Some measurements have also been performed in the jumping ski track of the jumping hill in Granåsen, Trondheim, Norway, 1997-98. In addition an analysis of the electrolytic conductivity and ionic content of artificial Continental snow has been made, but is not reported in the present paper.

The mean snow grain diameter, $\overline{d}$, and the standard deviation of grain diameters, δ_d, have been calculated for 34 snow surface microscope images of ski tracks. These results are presented in Table 3 together with information of time, location, snow type, grain shape, snow humidity $W_{vol,\%}$, snow density ρ, snow hardness H and snow temperature T_s. Table 4 presents data on air temperature T_a, relative humidity Rh, cloudiness C and net radiation q_{net} at the time of snow microscope image registration.

A few examples of registered snow microscope images are shown in Figs. 2-4. The pictures highlight variations in grain sizes and snow types. Fig. 2 displays medium ($\overline{d}$ = 0.85 mm) snow grains of Snow type 5 while Fig. 3 shows a surface with very fine ($\overline{d}$ = 0.11 mm) snow grains of Snow type 3. Fig. 4 depicts a very wet ($W_{vol,\%}$ = 12.5 %) snow surface.

Fig. 5 presents a histogram of the grain size, i.e. $\overline{d}$, in 23 microscope images of ski tracks containing Snow type 5. The grain size in the microscope images has ranged from 0.32 to 1.88 mm with a mean value of 1.06 mm.

In the four analysed microscope images of Snow type 4, the grain size varied from 0.26 to 2.59 mm

Table 2. Classification of the grain shape of the snow according to Colbeck et al. (1990).

Basic classification	Graphic symbol
Precipitation particles	+
Decomposing and fragmented precipitation particles	/
Rounded grains (monocrystals)	●
Faceted crystals	□
Cup-shaped crystals and depth hoar	∧
Wet grains	O
Feathery crystals	∨
Ice masses	■
Surface deposits and crusts	∀

Fig. 1. Snow microscopy setup without camera connected to the phototube. The camera is used to take the setup picture.

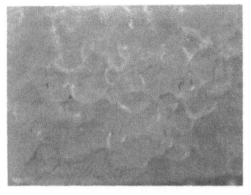

Fig. 2. Example of a snow surface with Snow type 5 and medium snow grains ($\overline{d}$ = 0.85 mm and δ_d = 0.38 mm): Image No. 32. The distance between the tick marks is 1 mm.

Fig. 3. Example of a snow surface with Snow type 3 and very fine snow grains ($\bar{d} = 0.11$ mm and $\delta_d = 0.03$ mm): Image No. 12. The distance between the tick marks is 1 mm.

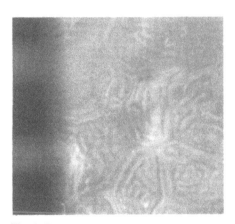

Fig. 4. Example of a very wet snow surface ($W_{vol.\%} = 12.5$ %, $\bar{d} = 0.86$ mm and $\delta_d = 0.29$ mm): Image No. 26. The distance between the tick marks is 1 mm.

with a median value of 0.41 mm. The highest value was obtained in a jumping ski track with below zero snow temperature after a night where a thin veil of new snow had settled on very transformed snow. When 10 ski jumpers had slid down the track, the new snow grains had either blown away or fallen between the old transformed grains. The old snow grains therefore dominated the microscope image of the surface, thus indicating that larger grain size than 1.88 mm also should be possible to register for Snow type 5.

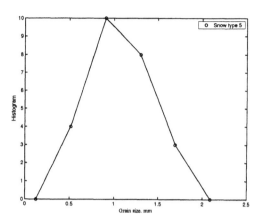

Fig. 5. Histogram of the grain size, i.e. $\bar{d}$, in 23 microscope images of ski tracks containing Snow type 5.

4 DISCUSSION

Throughout this text we have applied a simple classification system for snow types on snow surfaces. This system has been used quite extensively in practice by ski technicians due to its ease to use when no measurement equipment is accessible. It is for instance easy to interpret Snow type 5 in this system as snow where klister wax ought to be applied as kick wax, and Snow type 2 as snow where hard wax should be applied. The grain shape and grain size classification system presented by Colbeck et al. (1990) should be used in addition when a microscope or hand lens is accessible for inspection of the snow surface.

The maximum value of $\bar{d}$ for Snow type 4 in our measurements (2.59 mm) was higher than the maximum value registered for Snow type 5 (1.88 mm). This indicates that higher values of $\bar{d}$ also should be possible to obtain for Snow type 5. Schemenauer et al. (1981) have stated that individual snow crystals observed at the surface of the earth range in maximum dimension from approximately 0.05 mm to 5 mm. Wiesmann et al. (1998) found that grain sizes in snow covers ranged from less than 0.1 mm to 3 mm. In our few analysed microscope images of new snow types in ski tracks, the grain sizes ranged from 0.08 to 0.38 mm. Smaller grain sizes would probably also have been obtained if we had been able to register more snow microscope images of new snow types. In-situ microscope imaging of ski tracks is very difficult for new snow types with snow surface grains below 0.2 mm.

Table 3. Table of snow measurements for snow surface microscope images.

Image No.	Date (Time)	Location	Snow type	Grain shape	$d \pm \delta_d$ (mm)	$W_{vol.\%}$ (%)	ρ (g/cm^3)	H (Pa)	T_s (°C)
1	21.02.97 (12:54)	Granåsen cross-country	5	□	0.93±0.33 Medium	1 Moist	0.48	2.6×10^4 Medium	-6.2
2	23.02.97 (8:46)	Granåsen cross-country	5	O	0.74±0.33 Medium	1.7 Moist	0.57	2.2×10^4 Medium	-0.1
3	23.02.97 (11:40)	Granåsen cross-country	5	O	0.84±0.29 Medium	2.4 Moist	0.59	4.1×10^4 Medium	-0.1
4	23.02.97 (14:17)	Granåsen cross-country	5	O	0.72±0.20 Medium	6.8 Wet	0.8	2.6×10^4 Medium	-0.1
5	24.02.97 (8:52)	Granåsen cross-country	5	O	1.31±0.49 Coarse	1.2 Moist	0.52	6.6×10^4 Medium	-0.9
6	24.02.97 (11:45)	Granåsen cross-country	5	O	1.38±0.44 Coarse	1.5 Moist	0.41	2.2×10^4 Medium	-0.4
7	24.02.97 (13:49)	Granåsen cross-country	5	O	1.43±0.35 Coarse	1.8 Moist	0.44		-0.2
8	25.02.97 (8:36)	Granåsen cross-country	5	O	1.57±0.50 Coarse	0.8 Moist	0.55	1.9×10^4 Medium	-2.4
9	25.02.97 (11:38)	Granåsen cross-country	5	O	1.56±0.53 Coarse	1.4 Moist	0.49	8.3×10^3 Low	-1.7
10	26.02.97 (10:45)	Granåsen cross-country	5	□	1.89±0.52 Coarse	0.5 Moist	0.49	1.9×10^4 Medium	-7.7
11	27.02.97 (11:42)	Granåsen cross-country	3	/	0.08±0.02 Very fine	1.5 Moist	0.51	9.1×10^4 Medium	-3.5
12	28.02.97 (8:40)	Granåsen cross-country	3	/	0.11±0.03 Very fine	2.3 Moist	0.45	7.7×10^4 Medium	-2.1
13	28.02.97 (12:35)	Granåsen cross-country	4	O	0.46±0.34 Fine	3.4 Wet	0.57	1.3×10^5 High	-0.6
14	01.03.97 (8:47)	Granåsen cross-country	5	□	0.73±0.03 Medium	1.4 Moist	0.6	7.7×10^4 Medium	-3.9
15	01.03.97 (12:34)	Granåsen cross-country	5	O	1.49±0.64 Coarse	2.6 Moist	0.53	8.4×10^5 High	-1.7
16	02.03.97 (8:37)	Granåsen cross-country	5	O	1.03±0.39 Coarse			5.2×10^4 Medium	-0.1
17	16.12.97 (19:28)	Granåsen jumping hill	5	O	0.47±0.28 Fine			4.2×10^5 High	-0.9
18	27.12.97 (11:48)	Granåsen jumping hill	4	● (/)	0.26±0.09 Fine				-9.4
19	28.12.97 (10:15)	Granåsen jumping hill	4	□	2.59±0.71 Very coarse				-5.4
20	13.01.98 (14:49)	Granåsen jumping hill	5	O	1.32±0.44 Coarse			2.8×10^6 Very high	-0.8
21	25.03.98 (12:15)	Granåsen cross-country	2	/	0.21±0.07 Fine	1.3 Moist	0.6	7.7×10^4 Medium	-0.1
22	31.03.98 (13:49)	Granåsen jumping hill	5	O	1.26±0.36 Coarse	8.4 Very wet	0.6	4.3×10^3 Low	0.0
23	01.04.98 (13:44)	Meråker cross-country	5	O	1.24±0.48 Coarse	4 Wet	0.53	4.8×10^4 Medium	0.0
24	17.04.98 (12:35)	Lillehammer cross-country	2	O (/)	0.38±0.13 Fine	8.5 Very wet	0.69	2.7×10^4 Medium	0.0
25	20.04.98 (13:34)	Lillehammer cross-country	5	O	0.70±0.24 Medium	11.2 Very wet	0.74	2.7×10^4 Medium	0.0
26	21.04.98 (11:59)	Lillehammer cross-country	5	O	0.86±0.29 Medium	12.5 Very wet	0.61	1.1×10^4 Medium	0.0
27	07.11.98 (11:07)	Golå cross-country	2	/	0.12±0.07 Very fine	0.1 Dry	0.49	4.2×10^5 High	-6.8
28	28.02.99 (11:15)	Sundance cross-country	5	O	0.32±0.16 Fine			2.0×10^4 Medium	0.0
29	28.02.99 (12:46)	Sundance cross-country	5	O	0.40±0.16 Fine	3.8 Wet	0.72	<2×10^4	0.0
30	01.03.99 (11:30)	Sundance cross-country	4	O	0.49±0.22 Fine	4.5 Wet	0.79	1.2×10^4 Medium	0.0
31	02.03.99 (9:40)	Sundance cross-country	5	□	0.83±0.21 Medium			1.3×10^6 Very high	-3.3
32	02.03.99 (13:00)	Sundance cross-country	5	O	0.85±0.38 Medium	5.3 Wet	0.87	2.0×10^4 Medium	0.0

| 33 | 02.03.99 (13:43) | Sundance cross-country | 5 | O | 0.95 ± 0.34 Medium | 4.6 Wet | 0.89 | 2.2×10^4 Medium | 0.0 |
| 34 | 03.03.99 (10:55) | Sundance cross-country | 5 | O | 1.11 ± 0.54 Coarse | 0 Dry | 0.62 | 3.6×10^4 Medium | 0.0 |

Table 4. Table of weather measurements for the snow surface microscope images described in the snow measurements in Table 3.

Image No.	T_a (°C)	Rh (%)	C (octoparts)	q_{net} (W/m^2)
1	-0.1	58	3	243
2	1.6	89	8	0
3	2.6	87	8	131
4	6.7	64	8	103
5	-0.3	72	8	0
6	2.3	61	8	0
7	3.8	64	8	33
8	-2.3	61	8	0
9	4.1	57	8	250
10	-2.4	71	4	217
11	-0.3	79	0	74
12	3.6	60	7	57
13	5.2	60	8	152
14	3.6	60	2	0
15	5.2	64	4	119
16	5.7	80	8	0
17	1.3	96	0	-21
18	-5.1	57		
19	-3.2	72	7	
20	3.7	60	2	-33
21	4.7	43	4	415
22	4.3	80	5	415
23	0.0	74	8	312
24	3.1	83	7	164
25	5.1	57	8	205
26	5.8	53	0	753
27	-5.9	64	0	79
28	9.4	32	0	580
29	13.0	31	0	596
30	10.3	32	6	
31	1.6	48	0	
32	7.0	35	1	480
33	7.0	29	1	440
34	9.6	23	8	94

More microscope images of ski tracks need to be registered in order to find typical distributions of grain sizes for different snow types. The sizes of the snow surface grains are essential for the contact configuration between the ski base structure and the snow surface, and the design of optimum ski base structures for different snow conditions.

5 CONCLUSIONS

The mean snow grain diameter and the standard deviation of grain diameters have been calculated for 34 snow surface microscope images of ski tracks. The analysis of snow surface grains in ski tracks has shown that:

- The typical grain size has ranged from 0.26 to 2.59 mm for Snow types 4 and 5.
- The five analysed microscope images of Snow types 2 and 3 have ranged in typical grain size from 0.08 to 0.38 mm.
- In-situ microscope imaging of ski tracks is very difficult for new snow types with snow surface grains below 0.2 mm. Thus, snow microscope images of Snow type 1 have not been registered at present.

127

More microscope images of ski tracks need to be registered in order to find typical distributions of grain sizes for different snow types. The sizes of the snow surface grains are essential for the contact configuration between the ski base structure and the snow surface, and the design of optimum ski base structures for different snow conditions.

REFERENCES

Adobe (1998): Adobe Photoshop 5.0 - User guide. Adobe Systems Incorporated, San Jose, California, 390 p.

Colbeck, S. C., E. Akitaya, R. Armstrong, H. Gubler, J. Lafeuille, K. Lied, D. McClung and E. Morris (1990): International classification for seasonal snow on the ground. International Commission for Snow and Ice, World Data Center for Glaciology, University of Colorado, Boulder, Colorado, 23 p.

Moldestad, D.A. (1999): Some Aspects of Ski Base Sliding Friction and Ski Base Structure. Dr. thesis NTNU 1999:137, Norwegian University of Science and Technology,198 p.

Schemenauer, R. S., M. O. Berry and J. B. Maxwell (1981): Snowfall formation. In *Handbook of Snow. Principles, Processes, Management and Use* (D. M. Gray and D. H. Male, Ed.), Pergamon press, Toronto, pp. 129-152.

Wiesmann, A., C. Mätzler and T. Weise (1998): Radiometric and structural measurements of snow samples. Radio Science, Vol. 33, No. 2, pp. 273-289.

Snow Engineering, Hjorth-Hansen, Holand, Løset & Norem (eds) © 2000 Balkema, Rotterdam, ISBN 90 5809 148 1

Factors affecting ski base sliding friction on snow

Dag Anders Moldestad & Sveinung Løset
Department of Structural Engineering, Norwegian University of Science and Technology, Trondheim, Norway

ABSTRACT: The optimum ski base structure roughness varies under different snow conditions according to the generated frictional water film thickness under the ski and the roughness of the snow surface. Thick water films correspond to coarse ski base structures, while it is advantageous to use finer ski base structures and increase the water film thickness when the water film is thin. The friction under a ski is often dominated by solid friction in the front, surface tension in the middle and viscous drag at the rear. This indicates that the ski base structure ought to be different in the length direction of the ski. Thus, a fine structure that effectively induces and increases the water film production should be used at the front part of the forebody of the ski, at least for cold snow and snow with negligible free water content. As the development of water film possibly increases along the ski, structure with regularly higher roughness should be applied. The dry friction process is dominated and characterised electrically by accumulation of electrostatic charges in the ski base contact points. The frictional water film initiates discharge of potential differences between ski and snow due to the much higher electrical conductivity of water relative to snow. When the air gap volumes between the water film and the ski base structure, and the water film and the snow surface get small, the electrostatic pressures in the air gaps increase, and suction or drag may start occurring. The ski base structure topography and the snow surface topography is decisive for the electrical contact configuration between ski and snow.

1 INTRODUCTION

The low ski base sliding friction can be explained with lubrication by a thin water film created between ski and snow during skiing. Creation of water film under the ski is considered positive for minimum ski base sliding friction at cold snow temperatures. For snow close to or at the freezing point and containing a considerable amount of initial liquid content, water film creation may increase the ski base sliding friction due to suction or viscous drag.

Reynolds (1901) suggested that the water film was created by pressure melting, i.e. lowering of the freezing point due to high pressure under the ski. However, it can be shown that the pressure from a slider against snow or ice is insufficient to lower the freezing point significantly in most sliding situations. Mayr (1979) estimated that the freezing point was lowered by $0.00021°C$ for a 75-kg skier with a ski contact surface of 2700 cm^2.

Bowden and Hughes (1939) were the first to suggest the friction melting theory. They meant that frictional heating created the water film between slider and snow. The friction melting theory has been sup-

ported experimentally by e.g. Evans et al. (1976), Ambach and Mayr (1981), Warren et al. (1989) and Colbeck et al. (1997). Ambach and Mayr (1981) used a capacitor probe in the ski base to estimate water film thickness under the ski during skiing. They found increased film thickness as the snow temperature rose, thus implying the importance and effect of snow temperature, frictional heating and heat loss on water film generation.

Warren et al. (1989) installed thermocouples in an alpine ski base and measured a large thermal response from frictional heating. The temperature rise of the moving ski base increased with lower snow temperature due to more heat production caused by less water film development and higher sliding friction coefficient. It also increased with heavier loads. Long ski runs showed steady-state temperatures at the ski base that increased with snow temperature. Warren et al. (1989) also found that heat generation was more uniformly distributed when the snow was soft and could conform to the shape of the ski. Colbeck (1994) performed a similar experiment for skating skis. He found that the greatest temperature response at the ski base of a skating ski was just behind the foot, where the ski is heavily loaded. He also experienced increasing temperatures along the length of the ski. Further theoretical arguments for the friction melting theory have been provided by e.g. Oksanen (1980), Oksanen and Keinonen (1982) and Colbeck (1995b).

The influence of high-quality ski base structures on ski base sliding friction has been paid very little attention to in the literature. Slotfeldt-Ellingsen and Torgersen (1982) studied sliding properties of different polyethylene ski bases and the effect of different ski base grinding techniques. Their studies mainly considered manual grinding techniques and band grinding, but initial testing with stone grinding was also performed. Stone grinding of competition skis has been revolutionised since their study.

Mathia et al. (1989; 1992) reported the development of a 3D profilometer for systematic study of micro- and macro-topography of ski bases. Their research concentrated on non-dimensional parameterisation of ski bases and wear mechanisms associated with alpine skiing. The measurements of 60 areas on ski base surfaces for two manufacturing processes were reported to give skewness between -0.6 and 1.2 and kurtosis between 2.2 and 4.8 for wheel grinding, and skewness between -1.5 and 1 and kurtosis between 2.2 and 5 for belt grinding (Mathia et al., 1992). Skewness and kurtosis are defined as the normalised third and fourth central moments of the probability density function of the height distribution of an analysed surface. The present paper gives an introduction to the basis of ski base sliding friction, and the majority of the paper is extracted from the dr. thesis of Moldestad (1999).

2 BASICS OF SKI BASE SLIDING FRICTION

2.1 Tribology and ski base sliding friction

Tribology is defined as the science and technology of interacting surfaces in relative motion and of the practices related thereto (Hamrock, 1994). A tribological system consists of three parts:

- Upper surface i.e. moving ski with structure, pressure distribution, glide wax and base properties.
- Lubricant i.e. microscopic water film created by frictional melting and free water content in the snow.
- Lower surface i.e. snow.

By describing the three parts the frictional regime of the sliding friction coefficient, μ, can be found and determined. Fig. 1 shows the lubricant conditions for different frictional regimes. The left part of Fig. 1 shows the conditions when dry friction dominates.

Such conditions can exist under the ski when skiing on dry snow at cold snow temperatures, at least at the start of the forward contact line between ski and snow. The middle part of Fig. 1 indicates the conditions when both dry and wet friction are significant. The conditions when wet friction dominates are shown in the right part of Fig. 1. This is typically achieved when the snow contains a high free water content.

2.2 The slider (ski base)

The most essential ski parameters for ski base sliding friction are: the pressure distribution of the ski, the ski base material and quality, the glide wax or powder applied at the ski base, the ski base surface structure and ski speed.

The contact between the ski and the sliding surface is from a macroscopic point of view given by the pressure distribution of the ski and the snow hardness i.e. the bearing strength of the snow.

These parameters indicate the nominal contact between the ski and the sliding surface, i.e. the area on the ski where the contact between the ski and the sliding surface possibly can exist, when a given weight is applied on the ski. The real contact between the ski and the sliding surface is found from a microscopic point of view and is given by the ski base structure with applied ski wax or powder, the snow roughness and the orientation, size and hardness of the snow grains, the free water content in the snow and the development of frictional water film along the ski.

Research on measurement of mechanical properties of skis has been performed in Norway since 1969 (Stemsrud and Brun, 1976). The *Madshus Compuflex System* is a result of this research. This unique system measures e.g. the camber and kick zone of any ski along its entire length, thereby helping to identify the proper ski design for each type of *Madshus* ski. The ideal flex patterns for each *Madshus* model can therefore be developed and reproduced. Fig. 2 shows the pressure (or load) distribution for a typical dry snow skating ski (*Madshus 234 WC Supraflex Skate Dry*) measured by this system. The half skier weight has been applied 8 cm behind the balance point on the ski.

As can be seen from Fig. 2, the pressure distribution of a Nordic ski is characterised by two main peaks respectively on the forebody and afterbody of the ski. Ski preparation with glide wax or powder is therefore of course most important and effective in the areas under the two peaks, at least under hard packed snow conditions. The skating ski in Fig. 2 is specially designed for loose and uneven track conditions. The ski is said to perform better the softer the snow conditions are. This is natural due to the soft pressures in the tip and tail sections of the pressure distribution of the ski. Impact and compaction resistances against the ski are therefore minimised, and the ski does not bury down into the snow. Due to the influence of such resistances a stiffer ski may have less glide in soft, loose snow and under some soft wet snow conditions where suction occurs.

The material properties of the ski base do also affect the friction. High quality ski bases are normally made of sintered ultrahighmolecular-weight and high-density polyethylene (UHDPE). The UHDPE is modified with graphite (carbon black) and/or other additives. Some manufacturers use e.g. fluorocarbon additives in their top model ski bases. The graphite content of the ski base influences both the thermal and electrical conductivity.

The most important ski base material parameters are the thermal conductivity, the electrical conductivity, the graphite content, the molecular weight, the hydrophobicity (contact angle), glide wax and powder absorption, and hardness.

From the science of fluid-film lubrication, the sliding friction coefficient (μ) between two interacting surfaces in relative motion is known to vary with the film parameter, Λ. For skiing the non-dimensional film parameter, Λ, can be defined as:

$$\Lambda = \frac{h_{\text{min,wf}}}{\sqrt{R_{q,\text{sbs}}^2 + R_{q,\text{sn}}^2}} \qquad (1)$$

where:

$h_{\text{min,wf}}$ -   minimum water film thickness between a specific area on the ski and snow at a specific time during skiing, μm

$R_{q,\text{sbs}}$ -   root mean square roughness of the ski base structure surface in the specific area, μm

$R_{q,\text{sn}}$ -   root mean square roughness of the snow surface, μm

The characteristic dependence between μ and Λ is shown in Fig. 3 (based on a figure in Hamrock and Dowson, 1981). Under snow conditions where impact and compaction resistance contributions are negligible or small due to a vertical snow compaction distance that approaches zero, i.e. the snow surface has snow hardness above a certain threshold level, the friction situation in an area under the ski at a given time during skiing is possible to approach by exploiting the knowledge of Eq. (1) and Fig. 3. The dry and wet friction parts can then be treated by one tribological framework, instead of a theoretical division into two frictional parts (dry friction and wet friction).

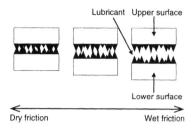

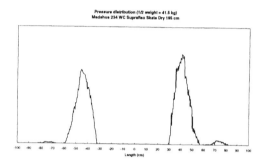

Fig. 1. Lubricant conditions showed for different friction regimes.

Fig. 2. Pressure (or load) distribution for a *Madshus 234 WC Supraflex Skate Dry* skating ski measured by the *Madshus Compuflex System*. Half skier weight (41.5 kg) has been applied 8 cm behind the balance point of the 195 cm long skating ski. The balance point is the zero point of the length axis (forebody to the left).

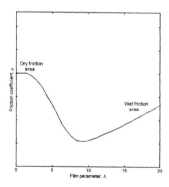

Fig. 3. Variation of friction coefficient, μ, with film parameter Λ (after figure in Hamrock and Dowson, 1981).

From Fig. 3 it can be deduced that the optimum glide, i.e. minimum sliding friction coefficient, arise when $\Lambda \approx 8\text{-}10$. A classic ski should therefore seek to have $\Lambda \approx 8\text{-}10$ in the sliding phase and $\Lambda \approx 0$ in the kick phase in order to have both optimum sliding and grip properties.

It is favourable to increase Λ when $\Lambda < 8$. Such frictional sliding conditions can possibly exist for instance under cold snow conditions where the free water content is small. Λ can then be increased by water film generation, i.e. increase of water film thickness $h_{min,wf}$, or decrease of $R_{q,sbs}$, i.e. use of a finer ski base structure with less roughness. A finer structure has little effect if $R_{q,sbs}$ is small compared to $R_{q,sn}$, i.e. snow surface roughness.

Oksanen and Keinonen (1982) showed that the dynamic sliding friction coefficient decreased with increased speed (v) at cold snow temperatures ($T_{sn} = -15°C$, $0.5 \leq v \leq 3$ m/s), thus indicating:

- The positive contribution of increased frictional melting, water film thickness $h_{min,wf}$ and film parameter Λ for better glide with increased speeds at cold snow temperatures.
- The movement from the left side of the minimum point of the graph in Fig. 3 towards the optimum point with increased speeds at cold snow temperatures.

It is favourable to decrease Λ when $\Lambda > 10$. Such wet friction sliding conditions can possibly exist for instance when the snow contains a considerable amount of free water content (Snow properties that are important for ski sliding friction is elaborated on in a separate paper by Løset and Moldestad, 2000). Λ can then be decreased by:
- Water film dilution, i.e. decrease of water film thickness $h_{min,wf}$.
- Increase of $R_{q,sbs}$, i.e. use of a coarser ski base structure with higher roughness.

The free water content of the snow or snow humidity which is possible to register in-situ by means of snow humidity measurement equipment can be viewed as:

- An indicator of initial water film thickness under the ski.
- A potential start point for estimation of generated water film thickness under the ski during skiing at different speeds.

Oksanen and Keinonen (1982) showed that the dynamic sliding friction coefficient increased with increased speed at near zero snow temperatures ($T_{sn} > -1°C$, $0.5 \leq v \leq 3$ m/s), thus indicating:

- The negative contribution of increased water film thickness $h_{min,wf}$ and film parameter Λ on the sliding properties of a flat surface when sliding with increased speeds at near zero snow temperatures.

- Movement towards the right side of the graph in Fig. 3, i.e. wet friction area, with increased speeds at near zero snow temperatures, at least for flat surfaces.

Fig. 3 indicates that a ski base structure should be coarser, i.e. have higher roughness, when the water film thickness under the ski is large, compared to situations when the water film thickness is small. Examples of fine and medium structures are shown in Figs. 4 and 5, respectively.

Moore (1975) stated that friction between a ski and ice quite probably is dominated by solid friction in the front, surface tension in the middle and viscous drag at the rear. This indicates that the ski base structure ought to be different in the length direction of the ski.

Thus, a fine structure that effectively induces and increases the water film production should be used at the front part of the forebody of the ski, at least for cold snow and snow with negligible free water content. As the development of water film possibly increases along the ski, structure with regularly higher roughness should be applied in order to stay in the optimum point of Fig. 3 all along the ski. The increased bottom temperatures found along the length of skating skis during skiing by Colbeck (1994), support the hypothesis of increased water film thickness along the ski. In spite of this it is very difficult at present to assume and predict water film thickness development along the ski under different snow, weather and skiing conditions. Another important issue is that the water film development should be a function of the weight of the skier and the ski length. Given equal ski length, heavier skiers should tend to have coarser structures.

3 THE EFFECT OF ELECTRICAL CHARGING AND ELECTROSTATIC PRESSURE ON SKI BASE SLIDING FRICTION

Petrenko (1994) showed in laboratory experiments that a 3 kV bias applied between ice and a slider can increase the friction by 2/3 for polyethylene and by 2/5 for metal sliders on ice. Prior to this he had found that electric fields developed during ice friction due to frictional self-electrification. Electric fields with a strength in the order of 2×10^6 V/m had been induced by friction at an ice/polyethylene interface at a temperature of $-31.5°C$ and sliding velocity of 8 m/s. Metal sliders (stainless-steel and aluminium foils) gliding on ice under the same conditions developed a potential difference of up to 1.6 kV between the ice and the sliders. These experimental results and basic electric field theory considerations showed that frictional self-electrification may play a significant role in forming ice and snow friction coefficients.

According to Petrenko (1994) the change in frictional force due to electrification can be explained by means of electrostatic pressures on the slider, that increase the normal pressures of the slider on the ice or snow surface. It is known from electric field theory that the electrostatic pressure, P_{el}, in a gap, L, filled with a dielectric material between which a potential difference, V, exists, is given by:

$$P_{el} = \frac{\varepsilon_0 \varepsilon E^2}{2} = \frac{\varepsilon_0 \varepsilon V^2}{2L^2} \qquad (2)$$

where:
V - potential difference in the dielectric gap, V
E - electric-field strength in the dielectric gap, V/m
L - thickness of the dielectric gap, m
ε_0 - dielectric permittivity of vacuum, 8.854×10^{-12} F/m
ε - relative dielectric permittivity for the material between which a potential difference exists, $\varepsilon = 1$ for vacuum

In a frictional ski base sliding situation at time, t, electrostatic pressures can be defined in:

• The air gaps: 1) Between the ski base and snow. 2) Between the ski base and the water film. 3) Between the water film and snow.
• The interfacial contacts: 1) Between the ski base and snow. 2) Between the ski base and the water film. 3) Between the water film and snow.
• The water film.

The relative dielectric permittivity values found in literature for ice, water, polyethylene, paraffin and fluorine are given in Table 1. Values for the electrical conductivity of snow and the electrolytic conductivity of melted snow are given in Table 2. It can be seen from Table 2 that the electrical conductivity of snow is several orders of magnitude less than the electrolytic conductivity of melted snow samples. Moldestad (1999) has shown that:

• The electrolytic conductivity increases and the grain size decreases when precipitation is introduced to the snow surface. The electrolytic conductivity can also increase with snow density.
• The electrolytic conductivity decreases as the snow goes through melt-freeze cycles and the grain size increases.
• The electrolytic conductivity of melted snow samples follows similar trends with respect to snow metamorphosis and snow density as the results referred by Kopp (1962) for the electrical conductivity of snow.

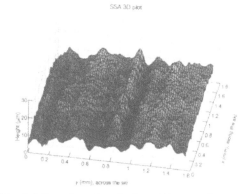

Fig. 4. 3-D surface plot of a fine ski base surface.

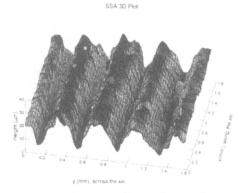

Fig. 5. 3-D surface plot of a medium ski base surface.

Table 1. Relative dielectric permittivity values (ε) found in literature for ice, water, polyethylene, paraffin and fluorine at different temperatures T.

Material/Fluid	ε	T (K)	Reference
Ice, static	100.8	263.2	Petrenko (1993)
Ice, high-frequency	3.2	263.2	Petrenko (1993)
Water, static	87.9	273.2	CRC (1997)
Water, high-frequency	5.7	273.2	CRC (1997)
Polyethylene	2.37	258.2	Petrenko (1994)
Paraffin	≈ 2		Van Pelt and Knol (1983)
Fluorine (F_2)	1.20	144.3	After equation in CRC (1997)

Table. 2. Electrical conductivity of snow and electrolytic conductivity of melted snow samples from ski tracks. Electrical and electrolytic conductivity are denoted by σ.

Material/Fluid	σ (μS/cm)	Reference
Snow (-30°C)	$10^{-6} - 10^{-3}$	Kopp (1962)
Snow (-10°C)	$10^{-5} - 10^{-2}$	"
Snow (-2°C)	$10^{-4} - 10^{-1}$	"
Melted snow	3.8 - 169.7	Moldestad (1999)
Melted snow (Trondheim)	11.8 - 61.7	"
Melted snow (Hakuba)	4.7 - 21.9	"
Melted artificial snow (Toblach)	169.7	"
Distilled water	0.5 - 1.7	"

In the measurement campaigns prior to the 1998 Olympics in Japan and prior to and during the 1997 World Championship in cross-country skiing in Trondheim, two to three times higher electrolytic conductivity values were registered for melted snow samples of new-fallen snow from ski tracks in Trondheim than in Hakuba, Japan.

Colbeck (1995a) has done field measurements of the potential difference between the ski base and the upper part of the ski during ski runs. During the start period of a typical ski run when according to Colbeck the ski was warming and adjusting to heat generated by friction, a sudden decrease to a negative voltage was found in his measurements. Once the ski temperature and rate of meltwater generation stabilised, the voltage began to climb toward positive values i.e. zero voltage and lower absolute potential difference according to Fig. 5 in Colbeck (1995a). This observation by Colbeck can be explained by using the knowledge of:

1. Higher electrical conductivity of water compared to ice or snow
2. From Point 1 it is intuitive that given approximately constant roughness for snow and ski base, $R_{q,sn}$ and $R_{q,sbs}$ in Eq. (1), a developed water film under the ski under dry conditions, i.e. a water film with larger water film thickness $h_{min,wf}$ and larger Λ in Eq. (1) compared to at the start time of the frictional situation and Λ still on the left side of the minimum point in the graph in Fig. 3 (dry friction area), at some time will introduce higher electrical conductivity between ski and snow compared to a situation where the water film thickness is lower or almost non-present i.e. equal to the tiny liquid layer present on the dry snow surface when the observed frictional situation i.e. ski run started.
3. The higher electrical conductivity explained in Point 2 gives larger discharge from the ski to the snow.
4. Larger discharge from the ski to the snow gives the lower absolute voltages observed by Colbeck.

A present interpretation of friction is suggested combining knowledge of tribology, electrical charging and electrostatic pressures (Moldestad, 1999):

- The dry friction process is dominated and characterised electrically by accumulation of electrostatic charges in the ski base contact points. The frictional water film initiates discharge of potential differences between ski and snow due to the much higher electrical conductivity of water relative to snow.

- When the air gap volumes between the water film and the ski base structure, and the water film and the snow surface get small, the electrostatic pressures in the air gaps increase, and suction or drag may start occurring. The wet friction process is characterised electrically by electrolytic behaviour.
- The ski base structure topography and the snow surface topography is decisive for the electrical contact configuration between ski and snow.
- The electrolytic conductivity of a melted snow sample may indicate the rate of ions introduced to the interface between snow and ski by frictional melting and thereby the rate and ease of discharge between ski and snow through the frictional water film during skiing. Given otherwise equal snow conditions and frictional situations, sliders on highly electrical conductive snow should discharge easier than sliders on snow with lower electrical conductivity. Larger frictional electrification should take place on snow with low electrical conductivity compared to snow with high electrical conductivity.

4 CONCLUSIONS

The ski base sliding friction is discussed and explained by use of tribology, and impact and compaction resistances. The major conclusions are as follows:

- The optimum ski base structure roughness varies under different snow conditions according to the generated frictional water film thickness under the ski and the roughness of the snow surface. Thick water films correspond to coarse ski base structures, while it is advantageous to use finer ski base structures and increase the water film thickness when the water film is thin.
- Possible increase of water film thickness along the ski implies an increase of the ski base structure roughness along the ski.
- Measurement of snow humidity can be viewed as an indicator of the initial water film thickness in the ski track and the frictional water film thickness that is possible to attain during skiing.
- Impact and compaction resistances are important when the snow hardness is below a certain limit, and when the water film thickness is low relative to the roughness of the ski base structure and the snow surface.
- The dry friction process is dominated and characterised electrically by accumulation of electrostatic charges in the ski base contact points.
- The frictional water film initiates discharge of potential differences between ski and snow due to the much higher electrical conductivity of water relative to snow.

- When the air gap volumes between the water film and the ski base structure, and the water film and the snow surface get small, the electrostatic pressures in the air gaps increase, and suction or drag may start occurring.
- The ski base structure topography and the snow surface topography is decisive for the electrical contact configuration between ski and snow.

REFERENCES

Ambach, W. and B. Mayr (1981): Ski gliding and water film. Cold Regions Science and Technology, Vol. 5, pp. 59-65.

Bowden, F. P. and T.P. Hughes (1939): The mechanism of sliding on ice and snow. Proceedings of the Royal Society of London, Series A172, pp. 280-298.

Colbeck, S. C. (1994): Bottom temperatures of skating skis on snow. Medicine and Science in Sports and Exercise, Vol. 26, No. 2, pp. 258-262.

Colbeck, S. C. (1995a): Electrical charging of skis gliding on snow. Medicine and Science in Sports and Exercise, Vol. 27, No. 1, pp. 136-141.

Colbeck, S. C. (1995b): Pressure melting and ice skating. American Journal of Physics, Vol. 63, No. 10, pp. 888-890.

Colbeck, S. C., L. Najarian and H. B. Smith (1997): Sliding temperatures of ice skates. American Journal of Physics, Vol. 65, No. 6, pp. 488-492.

CRC (1997): CRC handbook of chemistry and physics 78th edition 1997-1998, D. R. Lide (Editor-in-Chief), CRC Press, Boca Raton, New York, pp. 6-13, 6-139-6-187 and 13-12.

Evans, D. C. B., J. F. Nye and K. J. Cheeseman (1976): The kinetic friction of ice. Proceedings of the Royal Society of London, Series A347, pp. 493-512.

Hamrock, B.J. and D. Dowson (1981): Ball bearing lubrication - The elastohydrodynamics of elliptical contacts. Wiley-Interscience, New York, pp.121.

Hamrock, B. J. (1994): Fundamentals of fluid film lubrication. McGraw-Hill, New York, 690 p.

Kopp, M. (1962): Conductivité électrique de la neige, au courant continu. Z. Math. Phys., Vol. 13 (In French), pp. 431-441.

Løset, S. and D. A. Moldestad (2000): Characterisation of snow structure in a cross-country race ski track. Proceedings of the Fourth International Conference on Snow Engineering, Trondheim, Norway, June 19-22 2000, (in press).

Mathia, T. G., A. Midol, P. Lanteri and R. Longeray (1989): Topography, physico-chemistry and wear in sliding of ski soles in regard to rheology of snow. In Proceedings of Eurotrib 89, Elsevier, Amsterdam, pp. 253-260.

Mathia, T. G., H. Zahouani and A. Midol (1992): Topography, wear, and sliding functions of skis. International Journal of Machine Tools Manufacturing, Vol. 32, No. 1/2, pp. 263-266.

Mayr, B. (1979): Ein Beitrag zur Physik des Schigleitens: Elektronische Messung des Wasserfilms beim Gleitvorgang. Dissertation zur Erlangung des Doktorgrades an der Naturwissenschaftlichen Fakultät der Leopold-Franzens-Universität Innsbruck (in German), 152 p.

Moldestad, D.A. (1999): Some Aspects of Ski Base Sliding Friction and Ski Base Structure. Dr. thesis NTNU 1999:137, Norwegian University of Science and Technology,198 p.

Oksanen, P. (1980): Coefficient of friction between ice and some construction materials, plastics and coatings. Laboratory of Structural Engineering, Espoo, Report 7, 73 p.

Oksanen, P. and J. Keinonen (1982): The mechanism of friction of ice. Journal of Wear, Vol. 78, pp. 315-324.

Petrenko, V. F. (1993): Electric properties of ice. Monograph, USA CRREL Special Report 1993-20, 80 p.

Petrenko, V. F. (1994): The effect of static electric fields on ice friction. Journal of Applied Physics, Vol. 76, No. 2, pp. 1216-1219.

Reynolds, O. (1901): Papers on mechanical and physical Subjects, Vol. 2. Cambridge University Press, pp. 734-738.

Slotfeldt-Ellingsen, D. and L. Torgersen (1982): Gliegenskaper til skisåler i polyetylen. Report No. 820113-2 (in Norwegian), SI, Oslo, 64 p.

Spring, E. (1988): A method for testing the gliding quality of skis. Tribologia, Vol. 7, No. 1, pp. 9-14.

Stemsrud, F. and H. Brun (1976): Skiforskning ved Norges Landbrukshøgskole og Sentralinstitutt for Industriell forskning. Note from SI, Oslo, Norway (in Norwegian).

Van Pelt, T. P. and E. H. Knol (1983): Elektriciteitsleer 1, B. V. Uitgeverij Nijgh & Van Ditmar, The Hague, The Netherlands (in Dutch), 220 p.

Warren, G. C., S. C. Colbeck and F. E. Kennedy (1989): Thermal response of downhill skis. CRREL Report 89-23, 43 p.

Snow Engineering, Hjorth-Hansen, Holand, Løset & Norem (eds) © 2000 Balkema, Rotterdam, ISBN 90 5809 148 1

Daily observation of snowdrifts around a model cube

S.Oikawa
Institute of Technology, Shimizu Corporation, Tokyo, Japan

T.Tomabechi
Hokkaido Institute of Technology, Sapporo, Japan

ABSTRACT: An investigation of the snowdrifts around a model cube in a field observation was conducted in January and February 1998 in Sapporo, Japan. The cube, which was made from plywood sheets, 1.0 m on each edge, was set on the false floor. The observation duration of each run was one day. The results show that the snowdrifts with the largest depths were windward of the cube, followed by large snowdrifts on each side and leeward of the cube. The strong erosion areas were observed to be nearly surrounding the cube. It was found that when the wind speed exceeded about 4 m/s, the wind scoop, which was a combination of areas with a high snow deposit and those without, formed. This wind speed corresponded to the threshold friction velocity of 0.15-0.20 m/s, which agrees with those found in previous studies.

1. Introduction

In the vicinity of buildings the snowdrift phenomenon is one of the major problem in the snowy regions. In order to predict the snow accumulation patterns, a wind tunnel using fine particles to present the snow, has often been used. However, before the laboratory results can be accurately applied to the real world, it is necessary to know what features of the real snowdrift phenomenon are important, i.e. what the similarity criteria should be. There have been a number of observations of the snowdrifts around a building in the past, but few observations have been made clarify the weather conditions of the observation period. Most observations have been conducted under various weather conditions during a long period. The work described in this paper has been aimed at clarification of the snow accumulation patterns around a model cube in a short period i.e. the unit for one day, under the stable weather conditions.

2. Observation Site and Experiment Setup

The field study was conducted in Sapporo, Japan, in January and February, 1998. The observation site was located 10 km northwest from the center of the city. Residential dwellings, which are uniformly seven meters high, begin 200 meters to the northwest of the site, and continue 2 km towards the coast. The observation was conducted whenever the wind blew from the northwest. The model cube, which was made from plywood sheets with 1.0m on each edge, was set on false floor (0.5m H × 10m W × 10 m L). In the present observation period, there were no difference the height between the false floor height and windward snow accumulation height. To get the stable wind conditions and only newly fallen snow, one-day observations term were selected. Before each observation, the snowpack on the false floor from the previous day was removed. There were approximately 700 sampling points arranged in a rectangular-grid around the cube on the false floor. The snow depth of each point was measured with a snow depth gauge. Normalized snow accumulation ratio R is given as D_g/D_{up}, where D_g is the measure of snow accumulation depth at each point, D_{up} is measure of snow accumulation depth at the standard point of the snowfall box. Therefore, the larger the number R, the more the high snow accumulation depth. The one three-dimensional ultrasonic anemometer (Kaijo Corporation) were mounted at the model height. Data from the sonic anemometer was collected on a digital recorder at 1Hz sampling rate. A one-hour averaging time was calculated.

Figure 1 shows a photographic view of a model setup with no snow accumulation on the false floor. Figure 2 shows a photographic view of the cube on the false floor with snow cover. Figure 3 shows a schematic of the field site.

3. Results and Discussion

3.1 *Weather Condition*

The field study was conducted from January 4 to February 4, 1998. Atmospheric conditions are shown in Table 1. The observation time of each case is for 24 hours. Here, U is the day mean wind speed, and σ_u is the standard deviation of the fluctuating velocity, U_{max} is the day maximum value of the 1 hour average wind speed, θ is the wind direction measured from the plane perpendicular to the upwind face of the model, and σ_θ is the standard deviation of the fluctuating wind direction. A total of eleven runs were conducted and 7 runs were selected for their comparatively stable wind direction conditions. As there is a temperature within $-11 \sim +1\,°C$ in the observation period of all cases, this study is an observation conducted under powdery dry snow conditions. Figure 4 shows the vertical profile of the snowdrift mass flux above the surface. The snow flux increases with the increasing surface.

Figure 5 shows the relationship between 1 hour mean wind speed (U_h) and the friction velocity (u_*) at the 1m height. The friction velocity u_* was derived from the eddy correlation measurements. The friction velocity increased, the wind velocity increased. The friction velocity u_* is the index which is deeply concerned with the snow transport. In the dry uncompacted snow, the values of the threshold friction velocity (u_{*_t}) were observed by Iversen: 0.14 m/s, Tabler: 0.25 m/s and Kind: 0.15 m/s . Here, the value of $u_{*_t} = 0.15$ m/s is shown by the dotted line.

3.2 *Snow patterns around the cube*

Figure 6 show examples of the normalized snow-accumulation patterns on the false floor from 700 set data points. The areas of strong erosion were observed to be closely surrounding the cube. Further out from the position of the cube

Table 1   Summary of field experiment conditions (values are day average data)

Run	D_{up} (cm)	U (m/s)	σ_u (m/s)	U_{max} (m/s)	θ (deg)	σ_θ (deg)
SN09	20.0	1.7	0.84	4.3	-15	41
SN14	5.0	5.2	1.57	6.4	8	18
SN15 [*1]	5.0	4.6	---	7.5	23	--
SN19	10.0	3.4	1.10	5.1	-1	20
SN24	5.0	2.4	1.18	5.0	-11	38
SN36	20.0	3.1	1.10	5.4	1	28
SN39	7.0	4.1	1.26	5.5	7	17

*1: Flow data of SN15 from cup anemometer

Fig.1   A photographic view of a model setup with no snow accumulation on the false floor.

Fig. 2   A photographic view of cube with snow cover.

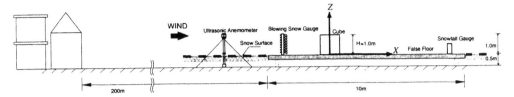

Fig. 3 A schematic of the field site.

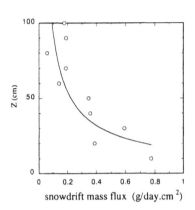

Fig. 4 Vertical profile of the snowdrift mass flux above the surface.

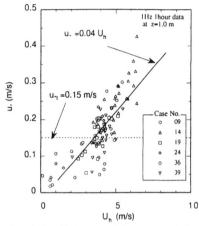

Fig. 5 Relationship between 1 hour mean wind speed (U_h) and the friction velocity (u_*) at the 1 m height.

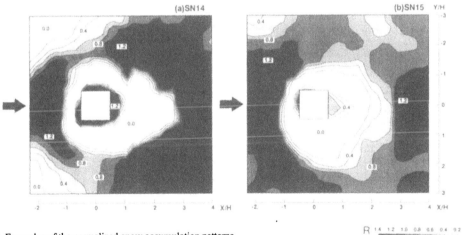

Fig. 6 Examples of the normalized snow accumulation patterns
(a) Case SN14 (U=5.2 m/s) (b) Case SN15 (U=4.6 m/s)

the large snowdrifting areas were observed windward of the cube and on each side and the leeward of the cube. That is to say, the wind scoop, which was a combination of areas with a deep snow deposit and those without, formed.

3.3 Relationship between snow depth of the leeward and wind speed

Figure 7 shows the relationship between the normalized snow depth (R) and the day average wind speed (U) at four positions (X/H = 1.5,

2.0, 3.0, 4.0) behind the cube along the model centerline (Y/H= 0). In the position of X/H=1.5 and 2.0 (Fig. 7a, 7b), R becomes also 0 when the day average wind speed 4 m/s is exceeded, and erosion occurs. However, at X/H=3.0 (Fig. 7c), R decreases a little when the wind speed increases. The region R=0 has not appeared. At X/H= 4.0 (Fig. 7d), even if the wind speed increases, remarkably erosion did not occur. It is shown that the strong erosion around the cube has been produced when the wind speed exceeded about 4 m/s.

In Fig. 8, the vertical section of R in the leeward on the model centerline is shown. The position of X/H= 0.5 shows the rear face of the cube. There was no strong erosion when the wind speed were less than 4 m/s (Fig. 8a). However, when it exceeded about 4 m/s, the strong erosion occurred X/H≒1.5～2.5 behind the cube (Fig. 8b) . After x/H= 2.5, snowdrift accumulation gradually occurred .

3.4 Relationship between snow depth of the windward and wind speed

The vertical section of R on windward of the model centerline is shown in Fig. 9. The position of X/H= -0.5 shows the front of the cube. Snowdrifts have also been formed both cases of below 4 m/s (Fig. 9a) and over 4 m/s (Fig. 9b) in windward direction a little but faraway from the cube (X/H ≒ -1.5). Note that the strong erosion which reaches R= 0 has been formed in case of over 4 m/s near the cube front (Fig. 9b). It is clear that wind scoop occurred when the wind speed exceeded 4 m/s from the result of Fig. 8 and Fig.9. The wind speed of 4 m/s corresponded to the threshold friction velocity of 0.15～0.20 m/s (Fig. 5) , which agrees with those found in previous studies.

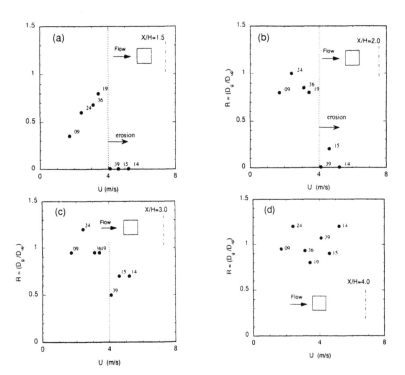

Fig. 7 Relationship between the normalized snow depth R and the day average wind speed (U)
behind the cube along the model centerline
(a) X/H = 1.5 (b) X/H = 2.0 (c) X/H = 3.0 (d) X/H =4.0

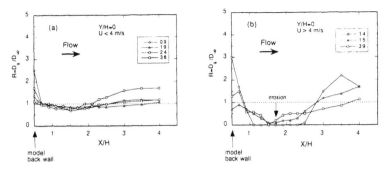

Fig. 8 Vertical section of R on the leeward of the model centerline
(a) U < 4.0 m/s (b) U >4.0 m/s

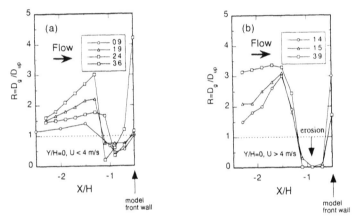

Fig. 9 Vertical section of R on windward of the model centerline
(a) U < 4.0 m/s (b) U >4.0 m/s

4. Summary

Daily observation of snowdrifts around a model cube were observed in a field study performed in Sapporo, Japan. It was found that the wind scoop which combined erosion and deposit patterns around a cube occurred, when the wind speed exceeded 4 m/s. This wind speed correspondents well to the threshold friction velocity (u_{*_t}) 0.15-0.20 m/s, which agrees with those found in previous studies.

References

Iversen,J.D.1979. *Drifting Snow Similitude*, J. Hydraul. Div., Am.Soc.Civ.Eng.,105(HY6) 737-753

Kind,R.J. 1976. *A Critical Examination of the Requirements for Model Simulation of Wind-Induced Erosion/Deposition Phenomena such as Snow Drifting*, Atmos. Env., 10, 219-227

Tabler,R.D. 1980. *Self-similarity of Wind Profiles in Blowing Snow Allows Outdoor Modeling* , J.Glaciology, 26, 421-434

Snow Engineering, Hjorth-Hansen, Holand, Løset & Norem (eds) © 2000 Balkema, Rotterdam, ISBN 90 5809 148 1

Estimation of equivalent density of snow accumulation for short period

S. Sakurai
Hokkai Gakuen University, Sapporo, Japan

O. Joh
Hokkaido University, Sapporo, Japan

T. Shibata
Hokkaido Institute of Technology, Sapporo, Japan

ABSTRACT: In the Third International Conference on Snow Engineering in 1996, we have proposed a practical method to estimate the daily ground snow weight during snow cover period, which is based on the data of daily precipitation and daily mean air temperature. In the present paper, this method is applied to estimate the equivalent snow densities for the short period at 126 observational points in Japan. As a result, it is found statistically that the equivalent snow densities for the period of 1,3,7 and 14 days can be considered as 2.1, 2.1, 2.3 and 2.6(kgf/m^2/cm) respectively, which values correspond to 95 % probability in the lognormal distribution from the data of 126 observational points.

1 INTRODUCTION

In the region where snow cover period is not so long, it is necessary for the structures to estimate design snow load properly during short periods like 7 days. There are many data of snow depths as meteorological observations in Japan, however, quite a few data about snow weights so far, especially in such regions. On the other hand, in recent years, huge membrane structures like a gymnasium have been constructed in both of light snow regions and heavy snow regions, which have some kind of active control systems to reduce roof snow load. In this case, it is necessary to determine design snow load during short period too, which period is needed for the repairs if the equipment of the system could not work. Therefore it has become so important to obtain a valuable information concerning snow loads during short periods.

In the present paper, we show the equivalent snow densities for the short periods such as 1,3,7 and 14 days, which are evaluated statistically by using the meteorological data of daily precipitations, daily mean air temperatures and daily snow depths at 126 observation points in Japan. And also, the prediction equation of equivalent snow densities throughout the period of winter in heavy snow regions, which we have proposed, are discussed concerning that validity.

2 OUTLINE OF PROPOSED METHOD

Because of the lack of sufficient database of snow weights in light snow regions and heavy snow regions, we have proposed a practical method to estimate daily ground snow weights during snow cover period as defined below [Kaneko, Joh & Shibata 1987; Sakurai, Joh & Shibata 1992] :

$$P_n = \sum_{i=1}^{n-1} P_i \qquad [T_i < 2\,°C] \qquad (1)$$

in which P_i (mm) is the daily precipitation; P_n (mm) is the accumulated value of P_i from the first day (i=1) of a period of continuous snow cover to the (n-1) day, which is regarded as the ground snow weight (kgf/m^2) of the nth day. Data of P_i should be accumulated when daily mean air temperature T_i (°C) is less than the air temperature limit ,+2°C [Tamura 1992], which is used to judge that the data of P_i means the value of snow fall or not. Needless to say, this accumulation should be ended if the data of snow depth H (cm) of nth day is zero. Kamiura and Uemura (1992) discussed the efficiency of their method using meteorological data in Nagaoka and Tokamachi, in which accumulated daily precipitation and melting coefficient were used.

Figures 1(a)~(c) show the examples of the comparison of the above mentioned method with the real observed data W (kgf/m^2) of snow weights at Sapporo (Hokkaido, 1979~'80), Kamabuchi (Yamagata Prefecture, 1980~'81) and Tokamachi (Niigata Prefecture 1980~'81) located in heavy snow regions, respectively. In these three cases, the snow cover periods are almost three months, five months and five months, respectively. As is obvious from these examples, we can understand that the proposed method can trace the increasing process of ground snow weights approxi- mately throughout the period to the peak value. Changes of weights about increasing snow accumulation for 7 days are shown in Figures 2(a)~(c), which are based on the daily data plotted in Figures 1(a)~ (c). As seen in Figures 2(a)~ (c), the estimated annual maximum values by the proposed method, which are 74 (kgf/m^2) at Sapporo, 130(kgf/m^2) at

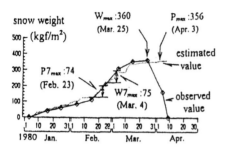

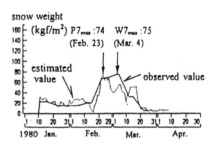

(a) Sapporo (1979-'80)

(a) Sapporo (1979-'80)

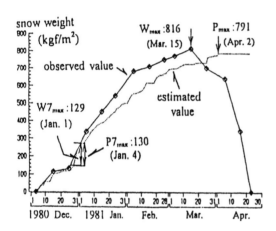

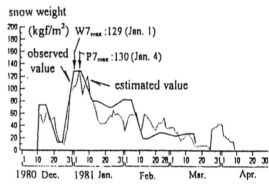

(b) Kamabuchi (1980-'81)

(b) Kamabuchi (1980-'81)

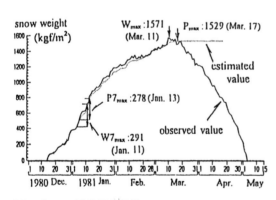

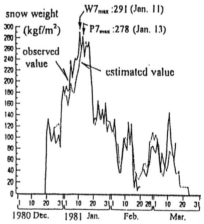

(c) Tokamachi (1980-'81)

Figure 1. Change of snow weights during a winter

(c) Tokamachi (1980-'81)

Figure 2. Change of snow weights about increasing snow accumulation for seven days

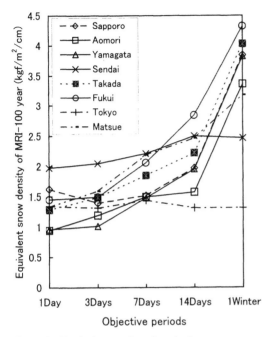

Figure 3. Typical examples of equivalent snow
densities for short periods and throughout
the winter at eight observation points

Kamabuchi and 278 (kgf/m²) at Tokamachi, are similar to
the observed values, which are 75(kgf/m²), 129(kgf/m²)
and 291(kgf/ m²), respectively.

3 EQUIVALENT SNOW DENSITIES

Fortunately, we have obtained the digital data of daily
precipitations, daily mean air temperatures and daily
snow depths during about 33 winters (1961/'62 ~1993
/'94) at 126 observation points, which are located widely
in Japan. In applying these data (named SDP DATA) to
Equation 1, the daily snow weights throughout the period
to the peak value could be estimate easily in each winter
at 126 points. Herein, as a basic statistic of this study,
we deal with the annual maximum values of increasing
snow weights and increasing snow depths for short
periods in the winter such as 1,3,7 and 14 days, and also
annual maximum values of snow weights and snow
depths throughout of the period of the winter. From these
basic statistic, mean recurrence interval(MRI) values on
snow weight, W_{MRI}(kgf/m²) and snow depth, H_{MRI} (cm),
are evaluated for using Type I probability distribution
function respectively. The topic chosen here, equivalent
snow density, D (kgf/m² /cm), is evaluated by dividing
W_{MRI} by H_{MRI}[Sakurai & Joh, 1998].One should note that
equivalent snow density is a statistic snow density
depending on MRI values, does not mean the density of
newly fallen snow for short period.

3.1 Typical examples of Equivalent densities at eight observation points

Figure 3 shows the typical examples of equivalent snow
densities for 4 kinds of short periods and throughout the
period of winter at eight observation points, which are
located from north area to south area in Japan such as 6
points of Sapporo, Aomori, Yamagata, Takada, Fukui and
Matsue in heavy snow regions, and 2 points of Sendai
and Tokyo in light snow regions. The values of equivalent snow densities are evaluated with respect to MRI-
100 year.

It is clear that those values throughout the period of
winter at 8 points separate into 2 groups roughly, that is,
the one including 6 points in heavy snow regions which
increases much more than that for 14 days, and other one
including 2 points in light snow regions. That is because
snow cover periods in heavy snow regions are usually
longer than 14 days. This result suggests that equivalent
snow densities throughout the period of winter in heavy
snow regions should be estimated in separate from those
in light snow regions. This will be discussed in section
3.3.

The equivalent snow densities for the periods less
than 14 days, especially 7 days show much smaller
scatter compared with those throughout the period of
winter, although they include the values both in heavy
snow regions and in light snow regions. In the case of 7
days, they take the value from 1.44 (kgf/m²/cm, Tokyo) to
2.21 (kgf/m²/cm, Sendai). The constructions, which have
some kinds of active control systems to reduce roof snow
loads, tend to be sensitive to the variation for snow loads
because their dead load is smaller than usual. Therefore,
it is practical to give the common value of equivalent
snow density for each short period to all area in Japan as
described next section.

3.2 Histograms of equivalent snow densities on 126 observation points

Figures.4 (a)~(d) show the histograms of equivalent
snow densities, D1, D3, D7 and D14, for the period of
1,3,7 and 14 days at 126 observation points respectively
Mean values (kgf/m²/cm) and Standard Deviations (S.D.,
kgf/m²/cm) of them are 1.40/0.387, 1.44/0.353, 1.60/
0.375, and 1.78/0.456 in order. And also coefficients of
variation (C.O.V.) are 0.276, 0.245, 0.234 and 0.256,
therefore it can be considered as 0.25 approximately. It is
important to give a probabilistic value as equivalent snow
density for design snow load. Herein, we propose the
values of 2.1, 2.1, 2.3 and 2.6 (kgf/m²/cm) as the equivalent snow densities for the period of 1,3,7 and 14
days, which correspond to 95 % probability in the
lognormal distribution. The above-mentioned studies are
based on the MRI (100 year), however, there are not big
differences in other MRI such as 10, 30 and 50 year.

3.3 Investigation for the relationship between equivalent snow densities and MRI values of snow depth

In Figure 5(a), equivalent snow densities, D3 and D7 at 126 observation points are plotted in related to H_{MRI} (100 year). Both of D3 and D7 (triangle and circle signs) are distributed in the range from 1.0 to 2.5 (kgf/m²/cm) approximately. It is obvious from the histograms in Figures. 4(b) and (c), too. The correlation coefficients between D3 and H_{MRI}, and D7 and H_{MRI}, take the values, –0.230 and 0.063 respectively. Therefore, in both of case, the correlation between equivalent snow densities and MRI values of snow depth can be neglected approximately. In the case of D1 and D14 in Figure 5(b), they show a similar tendency as to the above-mentioned D3 and D7.

On the other hand, we have reported a prediction equation for equivalent snow densities throughout the period of winter in heavy snow regions as follows [Joh & Sakurai 1993]:

$$D = 0.073\sqrt{H_{MRI}} + 2.37 \tag{2}$$

This square root equation has been adopted in AIJ Recommendations for Loads on Buildings 1993 using round number. It is based on the other data different from SDP DATA, which are surveyed snow weights (not based on daily precipitation) and snow depths at 12 observation points in heavy snow regions. We have concluded from these data that the values of D have a tendency to increase as the values of H_{MRI} increase. In Figures 5(a)

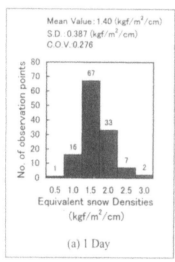

(a) 1 Day

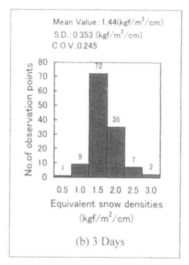

(b) 3 Days

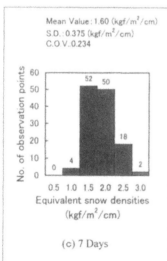

(c) 7 Days

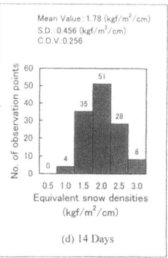

(d) 14 Days

Figure 4. Histograms of equivalent snow densities for short periods at 126 observation points

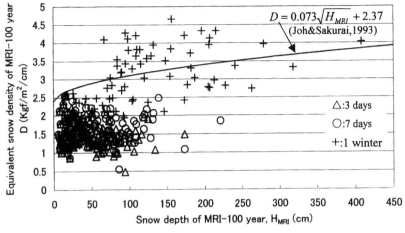

(a) Periods for 3 days, 7days and 1 winter

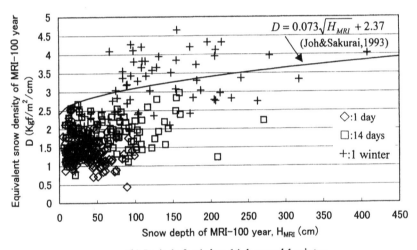

(b) Periods for 1 day, 14 days and 1 winter

Figure 5. Relationship between MRI values of snow depths
and equivalent snow densities for the objective
periods

and (b), Equation 2 is shown by solid line. Furthermore, in these Figures, the values of D, based on SDP DATA at 126 points, are shown by plus signs.

We can draw an interesting conclusion that Equation 2 shows the average of equivalent snow densities through- out the period of winter and the upper bound of those for the period of 1,3,7 and 14 days, approximately.

4 CONCLUSIONS

Considering the lack of the database of daily snow weights in Japan, which is necessary for estimating design snow loads of structures, we have proposed the simple and useful method to estimate them using the data

of daily precipitation and daily mean air temperature since 1987. In applying the digital data (SDP DATA) to this method at 126 observations, which are located in both light snow regions and heavy snow regions, it has become easily to investigate the snow weights statistically in detail. The results in this paper are as follows:

(1) The equivalent snow densities for the short period such as 1,3,7 and 14 days are estimated based on the SDP DATA. That value can be calculated by dividing the MRI values of snow weights by those of snow depths. From a practical point of view, we recommend the common values of 2.1, 2.1, 2.3 and 2.6 (kgf/m²/cm) as the equivalent snow densities for

the period of 1,3,7 and 14 days, respectively, which correspond to 95 % probability in the lognormal distribution. They are applicable to the sites located in both light snow regions and heavy snow regions.

(2) We have proposed the prediction equation for equiv--alent snow densities throughout the period of winter in heavy snow regions since 1993. It is introduced from the surveyed snow weights and depths at 12 observation points. Comparing this equation with the equivalent snow densities due to the method based on the daily precipitation etc., that shows the average of them throughout the period of winter and the upper bound of those for the period of 1,3,7 and 14 days, approximately. Therefore, this equation has the practical efficiency.

REFERENCES

Kaneko,H.,Joh,O.& Shibata,T.1987.Estimation of snow weight on the ground from daily precipitation and temperature data observed in Sapporo city. *Architectural Institute of Japan, Summaries of technical papers of annual meeting B(Structures I)*: 1411-1412 (in Japanese)

Sakurai,S.,Joh,O.& Shibata,T.1996.Estimation of ground snow weight based on daily precipitation and daily mean air temperature. *Third International Conference on Snow Engineering: Snow Engineering Recent Advances: 185-192*

Tamura,M.1990. Snow and Rainfall frequencies in Nagaoka. *Journal of the Japanese society of snow and ice*, Vol.52,No.4:251-257 (in Japanese)

Kamimura,S.& Uemura,T.1992.Estimation of daily snow mass on the ground using air temperature and precipitation data. *Second International Conference on Snow Engineering: CRREL Special Report 92-27*:157-167

Sakurai,S.& Joh,O.1998.Estimation of equivalent snow density of snow accumulation for nth days. *Architectural Institute of Japan, Summaries of technical papers of annual meeting B-1 (Structures I)* : 85- 86 (in Japanese)

Joh,O.& Sakurai,S.1993.Estimation of equivalent snow densities in heavy snow regions. *Journal of Snow Engineering, Japan Society for Snow Engineering* No.27: 112-114 (in Japanese)

AIJ Recommendations of loads on buildings 1996,Architectural Institute of Japan

APPENDIX

In AIJ Recommendations of loads on buildings 1996, classification of regions like light snow regions and heavy snow regions are not defined clearly. In this paper, as light snow regions, we image the regions such as H_{MRI} -100year is less than 50 (cm). However, further investigation is necessary about it.

Snow Engineering, Hjorth-Hansen, Holand, Løset & Norem (eds)© 2000 Balkema, Rotterdam, ISBN 90 5809 148 1

Snow and snowmelt models in Aomori Prefecture

M. Sasaki & T. Takeuchi
Hachinohe Institute of Technology, Japan

ABSTRACT: In this study, two models for snow and snowmelt were extended. Using those models, runoff forecast was carried out. The results were compared with the observations. The outflow estimated by using the snow model in winter is in good agreement with the observations, and the snowmelt runoff predicted in spring, April, May and June, by using the snowmelt model is also at the good accuracy. The snowmelt runoff begins to rise from the early April, and it reaches the peak in the early May.

1 INTRODUCTION

In Aomori Prefecture located in the northeast in Japan, the damage by the snowmelt runoff has been generated every year in spring. In May in 1997, the damage of 24 hundred million yen came out by the flood by the snowmelt in Tsugaru district in Aomori Prefecture. Then, it tried to carry out snowmelt runoff analysis of this district in this study.

Purposes of the present research are to establish a prediction model for snowmelt runoff in Aomori Prefecture at the good accuracy and to investigate and discuss the mechanism of the snowmelt runoff in the north district in Japan.

2 RIVER BASIN

There is Aseishi River in the mountain district in Aomori prefecture. In the Aseishi River, Aseishi River Dam has been constructed. In the dam, the inflow to the dam, a temperature and precipitation have been observed at every each time. Then, the river basin of Aseishi River Dam was chosen as the object river basin in the mountain district in Aomori Prefecture in the present study. Figure 1 shows the object river basin.

The basin area is 226 km². The top of Aseishi River Dam is in above sea level 201m, and the highest mountain in the river basin is about 1550m. Precipitation per a year is about 1600mm, however, the precipitation in the mountains in the basin area may exceed 2000 mm. The mean temperature of month in January and February is about −2 ℃, and the day mean temperature in January and February becomes −5 ℃ occasionally.

3 MODELS FOR SNOW AND SNOWMELT

There are get just the three kinds of data, temperature and the inflow to the dam and precipitation measured at Aseishi River Dam. However, the precipitation is precipitation from snow in cold day in winter. Other hand, the precipitation means only rain in warm day even in winter. Then, we should take out the effective rainfall from the precipitation data in winter. Besides, to calculate runoff in winter and spring, we have to estimate the effective rainfall originating from the snowmelt in the river basin covered with snow.

Now, we assume that snowfall, h_s, that corresponds water and snowmelt, R_m, are given as

$$h_s = afr \qquad (1)$$
$$R_m = bcT \qquad (2)$$

where r and T are precipitation (mm) and temperature (℃), and f is a coefficient which makes snow increase in the higher region, and a is a coefficient when rain becomes snow, and b is a coefficient when snow becomes the effective rainfall in river basin, and c is a coefficient when snow in the basin area melts per hour due to increasing temperature. The coefficients a and b are determined by temperature and the difference of elevation. Those three coefficients are given as

$$a = \begin{cases} 0 & T \geq T_m \\ 1 & T \leq T_i \\ 1 - \dfrac{1}{T_m - T_i}T & T_i \leq T \leq T_m \end{cases} \qquad (3)$$

$$b = \begin{cases} 1 & T \geq T_m \\ 0 & T \leq T_i \\ \dfrac{1}{T_m - T_i}T & T_i \leq T \leq T_m \end{cases} \qquad (4)$$

$$c = \frac{1}{12} \qquad (5)$$

where

$$T_i = 0 \quad {}^{\circ}C \qquad (6)$$

$$T_m = 2 \quad {}^{\circ}C. \qquad (7)$$

T_m is a limit temperature when the snow begins to melt in all of the river basin, and T_i is a limit temperature when all rain becomes a snow in all of the river basin. As shown in equations (6) and (7), the limit temperatures T_i and T_m are taken at 0 and 2 ℃ in this study. However, these limit temperatures reach another value, if the temperature is measured in another place. Thus, the snow depth corresponding to water and the effective rainfall at any time are given as

$$H_{s_t} = H_{s_{t-1}} + h_{s_t} - R_{m_t} \qquad (8)$$

$$R_t = R_{r_t} + R_{m_t} \qquad (9)$$

where

$$h_{s_t} = afr_t \qquad (1)$$

$$R_{r_t} = (1-a)r_t \qquad (10)$$

$$R_{m_t} = \begin{cases} bcT_t \\ H_{s_{n-1}} & bcT_t \geq H_{s_{t-1}} \end{cases} \qquad (11)$$

The subscripts t and t-1 denote time and the time one hour ago. Using equation (8), one can estimate the snow depth. Then, the amount of outflow from the river basin can be calculated by using equation (9) in winter and spring.

4 CALCULATION FOR SNOW DEPTH AND SNOWMELT

In this study, we use a tank model shown in Figure 2 to estimate runoff. As shown the figure, the model is composed of the tank of three steps.

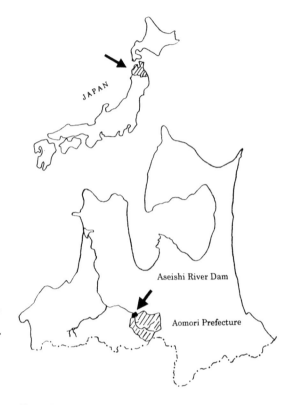

Figure 1. River basin of Aseishi River Dam.

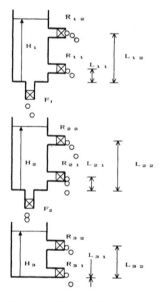

Figure 2. Tank model used in the present study.

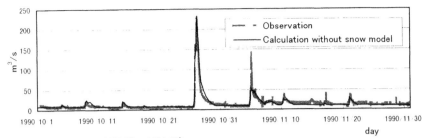

(a) Runoff in autumn (1990.10 ~ 1991.11)

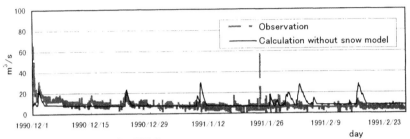

(b) Runoff in winter (1990.12 ~ 1991.2)

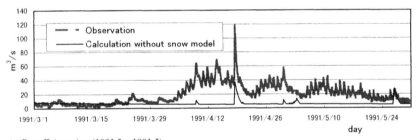

(c) Runoff in spring (1991.3 ~ 1991.5)

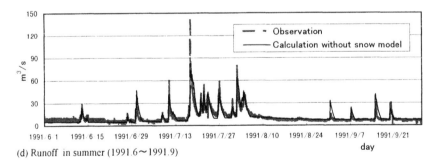

(d) Runoff in summer (1991.6 ~ 1991.9)

Figure 3. Calculation for runoff without snow model.

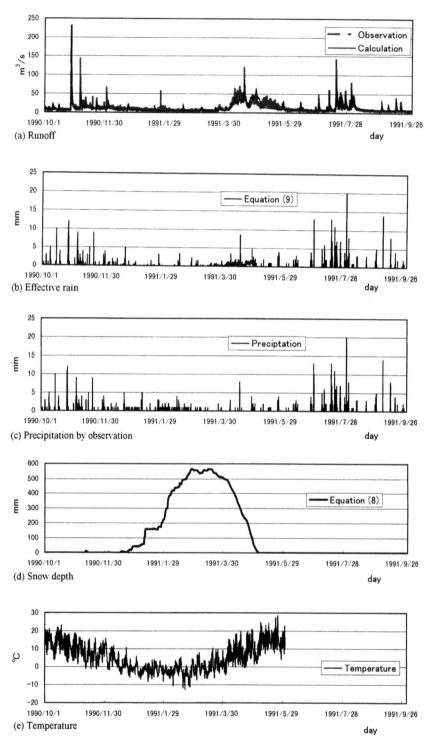

Figure 4. Runoff given by the present model in 1990 and 1991.

152

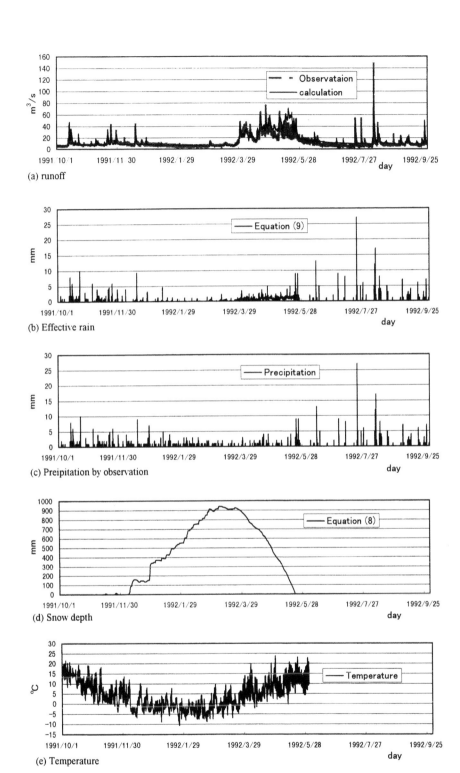

(a) runoff

(b) Effective rain

(c) Preipitation by observation

(d) Snow depth

(e) Temperature

Figure 5. Runoff given by present model in 1991 and 1992.

In the model, the outflow discharges from the upper tank, middle tank and lower tank show the runoff amount from the surface runoff, the middle runoff and the underground water runoff. The amount of evaporation is considered in the calculation. Evaporation volume of water is pulled from the water tank in the upper row. If the evaporation amount is larger than the water depth in the upper tank, the difference is pulled from the middle water tank. In the present study, the tank constants shown in Figure 2 are given as follows.

$$
\begin{array}{ll}
R_{11} = 0.6/24 & R_{12} = 0.9/24 \\
R_{21} = 0.013/24 & R_{22} = 0.014/24 \\
R_{31} = 0.007/24 & R_{32} = 0.007/24 \\
L_{11} = 10 \quad mm & L_{12} = 50 \quad mm \\
L_{21} = 10 \quad mm & L_{22} = 40 \quad mm \\
L_{11} = 0 \quad mm & L_{12} = 30 \quad mm \\
F_{1} = 0.4/24 & R_{22} = 0.05/24
\end{array} \tag{12}
$$

where R and F are constants for outflow hole and infiltration hole and L is the height of outflow hole. The depths of the tank are given as initial condition as follows.

$$
\begin{array}{ll}
H_{1} = 10 & mm \\
H_{2} = 80 & mm \\
H_{3} = 200 & mm
\end{array} \tag{13}
$$

Those depths are chosen as a value close to the residual depth after one year.

Figure 3 shows an example of the estimation for runoff without the snow model. In the figure, forecast runoff is shown from October in 1990 to September in 1991. From figure 3(a), it is understood that the calculation values are in good agreement with the observation discharges in summer and autumn. However, as shown figure 3(b), the calculation value in winter grows more than the observation value occasionally, and the agreement of both is not good. Other hand, the calculation discharge is larger than the observations in April and May as shown in Figure 3(c). The difference between the calculation value and the observation value in the winter means snow which falls and accumulates in the river basin, and the difference of both in the spring means snowmelt.

Figure 4 and 5 show the results of the calculation for the snow depth given by Equation (8) and runoff given by Equation (9) for a year from autumns in 1990 and 1991. Figures 4(e) and 5(e) are temperature observed at the Aseishi River Dam. In the calculation, the temperature is used in equations (3), (4) and (11) to determine coefficients a and b and snowmelt R_m. Figures 4(c) and 5(c) show precipitation that also observed at the Aseishi River Dam. The precipitation is divided into rain and snow depending on Equation (10). Figures 4(b) and 5(b) show the effective rainfall given by Equation (9).

As shown in Figure 4(a), the calculation runoff is corresponding well with the measurement value in winter and spring. This means that equations (10) and (11) can generate snow and snowmelt well. In figure 5(a), the calculation runoff is compared with the observation one as well as Figure 4(a). From the figure, the runoff given by Equation (9) is good agreement with the observation value even in the winter and the spring. The figures suggest that the snowmelt runoff predicted by using the present snowmelt model in April, May and June, is at the good accuracy.

Figures 4(d) and 5(d) show the snow depth given by Equation (8). From those figures, snow depth reaches the peak in the late February or the early March. The snow depth given by the present snow model is corresponding to the observation shown by Sasaki and others(1996).

5. CONCLUSION

In this study, a tank model was used in runoff analysis. Before the computation for the runoff, the precipitation was divided into snow and rainfall by using a distinction method by the temperature. Runoff forecast was carried out using the present snow and snowmelt models. The calculation was compared with the observation. Within the scope of the present study, following conclusions were derived.

(1) The runoff was estimated by using a snow model proposed in the present study in winter. The results are in good agreement with the observations.

(2) The snowmelt runoff predicted in spring, April, May and June, by using the snowmelt model given in the present study is at the good accuracy.

(3) The snowmelt runoff given by Equation (9) begins to rise from early April, and it reaches the peak in the early May.

REFERENCE

Sasaki M., Shuto N., Sawamoto M., Nagao M. and Kazama S. 1996. Density change of ground snow in Hakkoda, Proceedings of the 3rd ICSE:pp.609-612.

3 Structural engineering

Snow Engineering, Hjorth-Hansen, Holand, Løset & Norem (eds) © 2000 Balkema, Rotterdam, ISBN 90 5809 148 1

Global and European standardization, ISO and CEN

Kristoffer Apeland
Oslo School of Architecture, Norway

ABSTRACT: The paper discusses the somewhat artificial situation between ISO's global standardization work and CEN's regional standardization work. It is a setback for global standardization that the European Common Market has become a somewhat self-contained market in the field of standardization. The situation is exemplified by the fact that the new CEN's ENV 1991-2-3 on snow loads has not taken advantage of the revised ISO 4355 on snow loads on roofs.

It is proposed that in the field of environmental actions there should always be a governing general ISO Standard, which regional and National standards should comply with.

1 INTRODUCTION

The ISO 4355 "Bases for design of structures – Determination of snow loads on roofs", second edition, dated 1998-12-01 has been published recently [1]. The process has been very slow, considering that the revised ISO 4355 was adopted as a new ISO standard in 1995.

The new European PreStandard for snow loads: ENV 1991-2-3: "Actions on Structures – Snow Loads" has been evaluated and tried out for a few years [2]. A proposal for the final EN-version is expected to be presented for finalization in 2000.

It is a fact that the ENV document primarily is based on the old ISO 4355, although a few changes have been made on the basis of the new findings which have been incorporated in the revised ISO 4355. This somewhat astonishing decision of the Project Team of CEN, is in the author's opinion not based on scientific grounds. It is felt that the decision stems from the fact that the European Community has become a strong, almost self-contained market, see [3], for which ISO at present has been given low priority.

This is not restricted to the field of actions on structures, nor is it a new problem. As early as in 1991, the Vienna Agreement between ISO and CEN was set up by the highest directional authorities of CEN and ISO. The Vienna Agreement stated that all duplication of work shall be avoided.

The paper discusses the CEN/ISO question from a general and a more specialized point of view.

2 SELF-CONTAINED MARKETS AND GLOBAL STANDARDIZATION

International standardization has as one of its main objectives to minimize trade obstacles. Additionally, the practical importance of limitation of variants is indisputable. In particular, this is the case for product standards. There is general international consensus about these major objectives.

On the other hand, for large markets that may be said to be self-contained, the need for international standardization may not be as strong as in smaller markets. The United States used to be one such self-contained market which, in the building standards field, had been reluctant to invest great effort in international standardization. However, this attitude has been changing over the last decade. In the work on the revision of ISO 4355, and in TC 59, Building Construction, the United States has become an active participant.

Canada has been an active member of ISO for many years.

For the whole Pacific area, global activity has also increased. Japan has been involved in ISO work for a long time. It should also be mentioned that countries like Australia and New Zealand are very active in the field of building construction.

In Europe, the development has to a great extent gone in the opposite direction over the last decade. With the establishment of the European Common Market, a new and almost self-contained market came into existence, which has resulted in a considerable slowing down of global standardization in Europe.

In particular, this has been the case in the building and construction field. A joint effort, with the aim of developing a uniform code for Europe, the "Eurocode", was transferred to CEN in order to develop European Standards. All of the members of CEN, i.e., primarily EEC/EFTA countries, have taken part in this development during the last decade.

The ambitious aim of developing a complete set of CEN standards has been given top priority by the CEN member countries, so that during these years ISO work took on lower importance.

The aim of making the CEN standards into a complete set of documents has in a number of cases led to the establishment of new CEN work items, which were already under development in ISO committees. This problem went right up to the highest ISO/CEN directional level, and resulted in the Vienna Agreement, 1991, which stipulated that all duplication of work is to be avoided. Problems of this nature still arise, however, between CEN and ISO.

At the start of this intense CEN development, the hope was that these efforts would be of short duration. Although ISO was set up in Europe, and has benefited from a large European input and financing, there is a tendency nowadays for CEN to develop into a self-contained standardization organization. The author, therefore, feels that endeavours need to be made to concretize a Vienna Agreement Part 2, stating that after the first generation of CEN documents is finished, ISO should re-become priority number one, and that CEN should take part in activities in transforming the CEN documents into ISO standards.

3 BUILDING REGULATIONS, CODES AND STANDARDS

The field of national legal regulations and requirements, e.g. building regulations, has shifted from specific requirements in the national legal documents into a system of "reference to standards". This means that if a standard is deveioped on the

basis of minimum requirements, the law will thereby be respected, providing the products comply with the requirements of the standard.
This principle has been – and is – of great significance, in particular in connection with building regulations. It has also become very important, in order to minimize trade obstacles, to develop global standards for the field of regulations and for performance requirements.

The European Commission has established a framework of harmonizing directives. The directives present requirements, which to a great extent are based on performance of all kinds of products and also address the requirements to the various agents of the building sector. In particular, this is the case in connection with the building product directive. Requirements regarding health, environment and safety are becoming an important part of legal building regulations.

4 BUILDING CODES FOR ENVIRONMENTAL ACTIONS

Building codes for environmental actions on buildings, e.g. snow and wind loads, must take into consideration local situations, i.e. topography, local climate, etc..
It follows that environmental actions are less suited for standardization than are other kinds of actions which are not dependent of location.
Nevertheless, there has always been a considerable demand for standards which may guide the structural designer. This was the grounds for starting the work on the first ISO 4355 back in the early seventies.

When we started the work on the first ISO 4355, the Working Group had a long discussion about which phenomena that could or should be standardized.

Concerning snow load on the ground, the WG concluded that each country would have to establish maps and/or other information concerning the geographical distribution of ground snow load in that country.

From the "Scope and field of application" I quote:

"This International Standard specifies methods for the determination of snow load on roofs.

It is intended to serve as a basis for the development of national codes for the determination of snow load on roofs

National codes may supply statistical data of the ground snow load in the form of zone maps or tables.

The shape coefficients presented in this International Standard are prepared for design application, and may thus be directly adopted for use in national codes, unless justification for other values is available."

There was considerable discussion about the decision to standardize the shape coefficients. However, based on the fact that many countries had very little or no specifications on the distribution of snow on roofs, it was decided to make the shape coefficients normative. The shape coefficients were primarily based on USSR, Canada, USA-specifications supplied with some European and Japanese experiences.

5 REVISION OF ISO 4355

Revision of the old ISO 4355 was initiated in view of the fact that new research and snow surveys had shown that the shape coefficients should be revised. In particular this was the case for the shape coefficients for pitched roofs, curved roofs and multilevel roofs.

During the revisional work some additional new concepts were introduced, among which were the exposure coefficient, the thermal coefficient and the surface material coefficient.
In addition, the shape coefficients were presented in trigonometric functional form with the purpose of making them easier for programming, and at the same time avoid artificial discontinuities.

In my opinion the new ISO 4355 is a considerable improvement compared with the old one, and may serve as a basis for the development of national codes.

If one prefers to represent the shape coefficients in polygonal form, it is easy to make approximations to the ISO 4355 curves, see for example Figure 1 and Figure 2 for pitched roofs in which also 45 degrees has been introduced as a break in the curve.

Personally, I would have been much more satisfied if this approximation to the ISO 4355 curve had been proposed in the CEN document instead of the old ISO 4355 curve, which has been applied without further questioning.

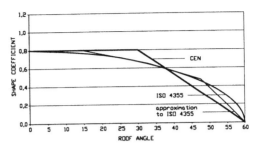

Figure 1 Pitched roof, windward side Variation with roof angle

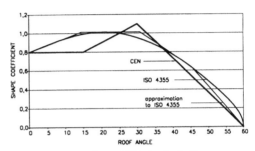

Figure 2 Pitched roofs, leeside Variation with roof angle

6 PROPOSAL FOR GLOBAL STANDARDIZATION

From the reasoning presented above, it should be clear that it is the author's opinion that in the field of environmental actions ISO should produce general global standards which will form a basis for the development of regional and/or National standards. The National standards should be compatible with the ISO standard. However, where it is found appropriate, the National standards may simplify the format, and even omit parts of the ISO standard.

It is felt that by this approach the quality of the standardization work will improve, and the process will become more efficient, thus bringing solutions at the time they are needed. In order to secure this process, publication of ISO standards must be speeded up considerably.

7 TRENDS IN GLOBAL STANDARDIZATION

In general, in the past, the procedure for the development of an ISO standard has been to develop an international consensus standard on the basis of an already existing national document or documents. Owing to the tremendous development in information communication technology over the last decades, however, it has proved more efficient now to develop international solutions for the problems in question right from the outset.

As a consequence, ISO has taken a leading role, in particular in global management systems, for which the ISO committees and working groups are instrumental in pre-normative research and development for the fields in question. The ISO 9000 series on quality management, and the ISO 14000 series on environmental management are the most distinguished examples of this new major field of standardization. For the building and construction sector, both of these series of generic standards are important.

It follows that ISO will again become the leading organization, and National and regional standardization bodies like CEN will adopt the ISO standards either in full text, or with amendments.

8 REFERENCES

ISO 4355, 1998. Bases for design of structures-Determination of snow loads on roofs.
CEN ENV 1991-2-3: 1995. Actions on structures- Snow loads
Apeland, K. 1999. Building on Standards
From product standards to global management standards. In ISO Bulletin October 1999.

Snow Engineering, Hjorth-Hansen, Holand, Løset & Norem (eds) © 2000 Balkema, Rotterdam, ISBN 90 5809 148 1

Statistics of ground snow loads in Italy

R. Del Corso & P. Formichi
Department of Structural Engineering, University of Pisa, Italy

ABSTRACT: The implementation of the Eurocode on actions led the European Commission DGIII/D-3 to sponsor a large prenormative research programme on snow loads on buildings. One of the main objectives of the research, started in December 1996 and concluded in June 1999, is the definition of a common statistical procedure to be adopted in all CEN countries, to get the characteristic snow loads values on the ground, from which derive a new European snow map.
This activity allowed to obtain a wide and updated snow database in 18 CEN countries and to perform new statistical calculations, taking into account the variation of the mean climatic conditions across Europe.
This paper gives information about the collection and the statistical elaboration of the Italian snow data.

1 INTRODUCTION

The implementation of the Eurocode on snow loads led the European Commission DGIII/D-3 to sponsor a large prenormative research programme aimed at the definition of a sound common scientific basis for the snow loading criteria within the CEN countries.

Eight Institutions belonging to six different European countries participated to this research programme.

The research is divided in two consecutive phases, each one divided in two tasks, according to the following scheme:

Phase I (Sanpaolesi et al. 1998)

Task Ia: "Development of models for the determination of snow loads on the ground"

Task Ib: "Development of models for exceptional snow loads"

Phase II (Sanpaolesi et al. 1999)

Task IIc: "Definition of criteria to be adopted for serviceability limit states verifications"

Task IId: "Analytical study for the definition of shape coefficients for the conversion of ground snow loads into roof snow loads".

One of the main objectives of the research is the definition of an European ground snow load map, derived with homogeneous criteria to cover 18 CEN countries. This study was conducted under the task Ia.

In order to achieve the definition of the map basic snow data were collected in all CEN countries, excluding Czech Republic.

The best statistical approach to study the variation of the snow loads on the ground, taking into account the different climatic conditions, was then investigated, and a common procedure was set up.

The characteristic ground snow load values obtained were employed as basis for the spatial interpolation analysis, performed using the most updated GIS techniques, from which the European snow map was derived.

The University of Pisa provided both Italian and Greek data and performed the relative statistical elaboration.

The present paper illustrates the main achieved results with regards to the analysis of the Italian data.

2 BASIC SNOW DATA

The scope of the collection of basic snow data is to get, from the snow time series - as a sample extracted from the population of the ground snow loads at the site considered – the representative values of such population, among which the characteristic value, with a given return period, as defined in the Eurocode 1 Basis of Design (CEN, 1994).

Since the mean return period for civil structures is generally fixed in 50 years, a minimum record period equal to 30 years was judged reasonable to allow an adequate statistical elaboration.

As a consequence of the above the following criteria, to identify the weather stations where collect information were established:

a) the quality and reliability of the meteorological records must be very high;
b) time series must show consecutive recorded years and possibly use the same years for all the weather stations analyzed;
c) due to the prenormative scope of the research it is desirable to focus the attention on low altitude stations, since in Italy the majority of cities and villages are usually located below 500 m a.s.l.;
d) the chosen stations must be homogeneously spread all over the country in order to allow a correct application of the interpolation formulas during the mapping phase;
e) the availability of snow loads rather than snow depths would be desirable, otherwise it is necessary to use an appropriate density model for the conversion of snow depths into snow loads.

In Italy there are two services recording meteorological data: the Italian Hydrographic Service (SIMI) and the Technical Institute of Italian Airforce (ITAV).

The research is based upon the study of the available data at SIMI, which collects meteorological records from 3000 weather stations, since 1910.

The meteorological service daily records the depth of the ground snow cover instead of the snow load intensity. It is therefore necessary to convert these values into snow loads via an appropriate density model.

The registered data relevant to the present research are:
- daily depth of the ground snow cover in cm;
- depth of the 24h snowfall in cm;
- 24h minimum temperature;
- 24h maximum temperature.

The measurements of the snow depths are performed using graduated poles.

The recorded data are published on the Hydrological Annual Reports as follows:
- daily snow depth measurements are available until 1940;
- from 1941 to 1970 are available the snow depth measurements taken at the 10th, 20th and 30th of each month;
- from 1971 on only one value of the snow depth measurements is available per month. It is also given the total number of snowy days for each month.

Due to the limited availability in the annual reports of daily snow data after 1940, these were collected directly at the archives of the Meteorological Service.

The present study, for Italy, is based upon the analysis of the snow data coming from 125 weather stations, 55 of which located below 500 m; 33 above

500 m and below 1.000 m; 26 above 1.000 and below 1.500 m and 11 above 1.500 m.

The location of the stations is shown in figure 1.

Particular attention was paid to the local, meteorological, geographical and environmental features, such as the presence of a valley or of a lake, which could have caused great influences in the determination of the characteristic snow load at the weather station, leading to discrepancies in the load value if compared with the surrounding ones. These stations were discarded from the database.

3 DENSITY MODEL

The density model used for the analysis has been adopted according to the observation's results obtained at about 15 experimental stations belonging to the Italian Research Council, where mean snow density of the cover has been directly measured since 1922 (Comitato Glaciologico Italiano, 1953).

Since the model has been elaborated for climatic stations located at high altitudes, it was distinguished its application for those places where the snow cover on the ground lasts more than 10 days, from other places where the snow cover usually melts within 10 days.

The governing parameter for the application of the density model is therefore the duration of the snow cover on the ground, which is directly linked to the structural modification of the snow layers, originating from different snow events, forming the cover. From a light fresh snow we get heavier snow with time due to the compression of the lower snow layers and, at the end of the period of presence of the snow cover on the ground, due to the melting process of the snow itself.

Figure 1 Location of the 125 Italian weather stations.

It is judged reasonable to neglect the above variation of density with time if the snow cover lasts on the ground less than 10 days.

The model adopted for different duration of the snow cover is illustrated below.

a) Snow duration (D) longer than 10 days:

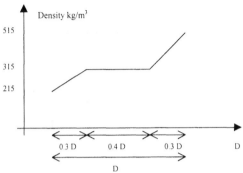

Figure 2. Snow density model for snow cover duration longer than 10 days

b) When snow duration is less than 10 days, a constant value for density is taken equal to $\rho = 250 \text{ kg/m}^3$.

The analysis of the time series of snow heights on the ground put in evidence that, for the Italian climate the model a) is suitable for stations located at altitudes greater than 500 m above the sea level.

The density model adopted was deeply checked during an experimental measurement campaign, which has taken place in the eastern Italian Alps and in the central Apennine, aimed at the definition of the roof shape coefficients (task IId of the research). During the winters 1997/98 1998/99 were measured the densities of sampled snow volumes extracted from the ground snow cover. The obtained results are fully in accordance with the model adopted for sites where snow cover lasts for a long time. In particular it was confirmed the increasing tendency of the density at the end of the period of presence of the snow cover on the ground, due to melting as well as the extreme values reached by the density itself at that time (Del Corso, Formichi, 1998).

4 STATISTICAL ELABORATION

The statistical elaboration was performed with reference to the annual maxima of the ground snow loads all cross CEN member states. This method is, in fact, the most widely used in the definition of climatic actions on buildings.

As first step they were analyzed the time series of the annual maxima drawing the graphs illustrated in the following figures. These show the variation with altitude of the mean values of annual maxima, of their standard deviation and of coefficient of variation, at each station.

It can be noticed that the mean and the standard deviation of the annual maxima increase with the altitude, while the coefficient of variation has a ten-

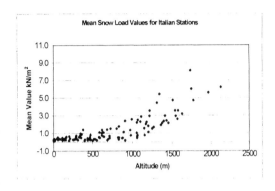

Figure 3. Mean values of annual maxima versus altitude for Italian weather stations

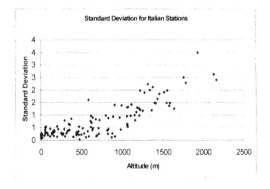

Figure 4. Standard deviation of annual maxima versus altitude for Italian weather stations

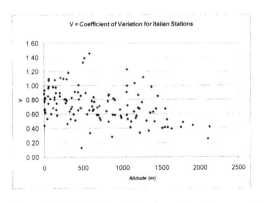

Figure 5. Coefficient of variation of annual maxima versus altitude for Italian weather stations

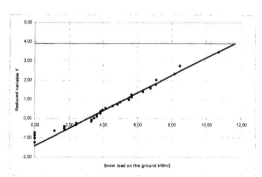

Gumbel Type I probability plot at Abetone - 50 years return period value = 11,65 kN/m² – Correlation coefficient = 0.992

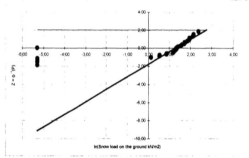

Lognormal probability plot at Abetone - 50 years return period value = 15,83 kN/m² - Correlation coefficient = 0.959

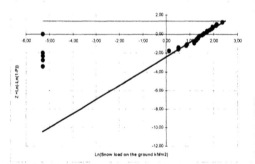

Weibull probability plot at Abetone - 50 years return period value = 12,05 kN/m² - Correlation coefficient = 0.980

Figure 6. Probability plots at Abetone 1340 m a.s.l. – Italy.

dency to decrease. It is also evident that the values of the coefficient of variation are widely scattered, from 0,13 up to 1,45.

These considerations suggest to follow different statistical approaches, in order to define the best fitting cumulative distribution function, describing the variation of the ground snow load at each weather station.

Climatological considerations have to be added to the above. At weather stations located at low altitudes and in the central and southern Italy, extremely irregular snowfalls are recorded and it is quite fre-

quent to have whole winter seasons without snow. To take duly account of the no-snowy winters it was applied the so called "mixed distribution approach".

Let P_{snow} be the probability to have non zero annual maxima within the time series at a given station, the probability of the characteristic ground snow load value s_k not being exceeded is given by:

$$F(X \leq s_k) = P_{snow} \, F_{snow}(X \leq s_k) + (1 - P_{snow})$$

where F_{snow} is the cumulative distribution function relative to the non zero values of the sampled annual maxima (Ellingwood et al. 1983).

This method (applied in all CEN countries where snow is intermittent and irregular) allows to take account of the zero yearly maxima in the ranking of the sampled data, without leading these zero values to influence the fitting of the non zero ones.

The next step is to perform the statistical elaboration of the ground snow load annual maxima, available at all the 125 Italian weather stations, by means of the most commonly used CDFs in these kind of analysis: Gumbel (Type I), Lognormal and the Weibull distributions.

The 50 years characteristic value (with a corresponding probability of being exceeded equal to 0,02) has been derived for each CDF.

As it is well known, in the Gumbel formulation, the probability for the random variable X not to exceed the value x is given by (Gumbel 1958):

$$F(X \leq x) = \exp[-\exp(-\alpha(x - u))]$$

where $\alpha > 0$ and $u > 0$ are two parameters.

By means of the Lognormal distribution function the same probability is given by:

$$F(X \leq x) = \Phi\left(\frac{\ln x - \mu_Y}{\sigma_Y}\right)$$

where $Y = \ln x$; Φ = standard normal distribution function.

Finally in the Weibull formulation the probability for the random variable X not to exceed the value x is given by:

$$F(X \leq x) = 1 - \exp\left[-\left(\frac{x}{k}\right)^\beta\right]$$

where $\beta > 1$ e $k > 0$ are two parameters.

The parameters of the three distributions have been calculated by the least square method and the moment's method. The latter has shown unacceptable sensibility to the presence of zero values in the sample. Therefore, the least square method has been used as the reference method for the determination of the parameters defining the interpolation line in the probability plot.

In the figure 6 are shown, as an example, the

probability plots obtained at Abetone (1340 m a.s.l.). It is shown the mixed distribution approach followed: only non zero values are interpolated and all the sampled data are ranked and plotted.

During the statistical analysis particular attention was paid to the so called "exceptional snow loads". This phenomenon occurs especially in southern Europe, where isolated very heavy snow falls have been observed, resulting in snow loads which are significantly larger than those that normally occur. Including these snowfalls with the more regular snow events for the lengths of records available, may significantly disturb the statistical processing of the sample, as can be seen from the figure 7, where it is shown the probability plot for the Gumbel Type I CDF at the climatic station of Vercelli, located in northern Italy at 135 m above the sea level.

In the figure line A represents the interpolation of all the non zero values excluded the maximum value of the sample; line B represents the interpolation of all the non zero values of the sample. It is evident the influence of the maximum value.

A wide investigation was made on the exceptional values, which were also encountered in Spain, Portugal, UK, France and Greece (Sanpaolesi et al., 1998; Sims et al. 2000).

The following definition has been suggested: *if the ratio of the largest load value to the characteristic load determined without the inclusion of that value is greater than 1,5 then the largest load value shall be treated as an exceptional value.*

In the statistical analysis for the calculation of the characteristic ground snow load, if an exceptional value, according to the above definition, is encountered in the sample, it is discarded from it. Exceptional values are then dealt with separately.

In Italy exceptional snow loads were recorded at eight weather stations.

Once calculated the characteristic value at the given station, according to the three CDFs, the criterion adopted for the assessment of the best fitting probability distribution function, has been based on the comparison of the coefficients of correlation shown by the sample in the three probability plots.

The results of the investigation are as follows:
- at 70 stations the best fitting CDF is the Gumbel (Type I);
- at 42 stations the best fitting CDF is the Weibull;
- at 13 stations the best fitting CDF is the Lognormal.

The variation of the ground snow loads is better represented by the Gumbel CDF at 56% of the stations; in 34% of cases the Weibull is the best fitting function while only in 10% of cases the Lognormal fits better.

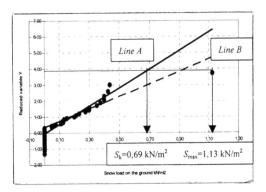

Figure 7: Extreme value plot in which a high value appears (Vercelli, 135m a.s.l.–Italy. S_k=0,69 kN/m², S_{max}=1,13 kN/m², S_{max}/S_k = 1,64)

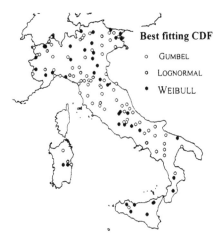

Figure 8. Weather stations where different CDFs fit better in Italy.

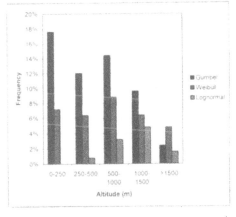

Figure 9. Variation of the frequency of the stations with best fitting CDFs versus altitude.

Similar conclusions were derived also in other CEN countries. These lead to assume, as reference statistical model for the calculation of the characteristic ground snow loads in Europe, the use of the mixed distribution approach with the Gumbel distribution function, evaluating its parameters by means of the least square method.

In figure 8 are shown the weather stations where the different CDFs fit better.

In figure 9 it is shown the variation of the frequency of stations with different best fitting CDFs with the altitude of the stations.

From figures 8 and 9 it is evident that the weather stations where Weibull and Lognormal fit better are mainly concentrated at medium-high altitudes and are located in the northern and central Italy and in mountainous areas. These CDFs appear, therefore, mainly adequate in sites where snowfalls are regular all over the winter.

Due to the prenormative scope of the present research, the indications obtained from weather stations located at low levels, where the Gumbel CDF results the best fitting function, are mainly relevant.

In addition it has to be noticed that the Weibull distribution function, although being the best fitting CDF in many stations, generally gives higher values than the Gumbel gives. Finally, the Lognormal CDF does not seems to be suitable for Italian climate.

5 CONCLUSIONS

The wide prenormative research activity in the field of snow loads on buildings, sponsored by the European Commission DGIII/D-3, made possible to review the current practice in the definition of ground snow loads all over the CEN area. In particular the study of a new European ground snow load map lead to an important improvement of the snow database, upon which calculate the characteristic snow load values.

Within this research programme in Italy were collected and analyzed a huge number of information coming from the archives of the Italian Meteorological Service, integrating the already available database with new daily records.

A wide investigation was performed to assess the best statistical procedure to be applied in all the different European climatic regions.

In Italy it has been observed that in general the Gumbel (Type I) probability distribution function is fairly the best fitting CDF for sites located at low altitudes (< 1000 m a.s.l.) and where snow is more intermittent and irregular.

The Weibull probability distribution function is the best fitting CDF for sites located at high altitudes (> 1500 m a.s.l.) and where snow accumulates and lasts for the whole winter season.

Finally the Lognormal probability distribution function does not seems to be adequate for temperate climates such as the Italian one.

The above considerations were also confirmed in other CEN countries, such as Greece, Spain and southern France.

REFERENCES

CEN, 1994. ENV 1991-1: 1994 Basis of Design and Actions on Structures, CEN Central Secretariat Brussels

CEN, 1995. ENV 1991-2-3: 1995 Actions on Structures – Snow Loads, CEN Central Secretariat Brussels

Comitato Glaciologico Italiano (Italian Committee for Glaciology), 1953. Relazioni su ricerche e studi promossi dall'ANIDEL (Reports on researches and studies promoted by ANIDEL). Rome.

Del Corso, R., Formichi, P. 1999 *Shape coefficients for conversion of ground snow loads to roof snow loads.* 16th BIBM International Congress. Venice May 25-28 1999.

Ellingwood, B., Redfield, R., 1983. Ground snow loads for structural design. Journal of Structural Engineering, ASCE, Vol. 109 n°4.

Gränzer, M., 2000 *New ground snow loads for an European standard.* Ibid.

Gumbel, E. 1958 *Statistics of Extremes,* N.Y./London, Columbia University Press.

Sims, P. et al., 2000 *European ground snow loads map: irregular and exceptional snow falls.* Ibid.

Sanpaolesi, L et al. 1998. *Scientific support activity in the field of structural stability of civil engineering works: Snow loads – Final Report Phase I,* Commission of the European Communities DGIII/D-3, Brussels

Sanpaolesi, L et al. 1999. *Scientific support activity in the field of structural stability of civil engineering works: Snow loads – Final Report Phase II,* Commission of the European Communities DGIII/D-3, Brussels

Sanpaolesi, L. et al., 1983. *Analisi statistica dei valori del carico di riferimento da neve al suolo in Italia (Statistical analysis of reference ground snow load in Italy).* Giornale del Genio Civile - Roma

Thom, H.C.S., 1966. *Distribution of maximum annual water equivalent of snow on the ground.* Monthly weather review (94): 265-271.

Snow Engineering, Hjorth-Hansen, Holand, Løset & Norem (eds) © 2000 Balkema, Rotterdam, ISBN 90 5809 148 1

Duration of the snow load on the ground in Italy

R. Del Corso & P. Formichi
Department of Structural Engineering, University of Pisa, Italy

ABSTRACT: The European Commission DGIII/D3 charged eight European Institutions to carry out a pre-normative research on snow loads, aimed to investigate such an action for the ultimate and the serviceability limit states verifications. In this context a large amount of snow data was collected, in many cases consisting of daily observations. In Italy, in particular, daily data are available for several years in 54 meteorological stations. These data are suitable for the analysis of the permanence of the snow layer on the ground, to establish a relationship between the snow load and its duration. The results can be used for the serviceability limit states verifications for structural elements made of materials with a creep behaviour. In this paper a probabilistic approach to the problem is presented and an application example for three meteorological stations placed in the Italian territory is developed.

1 INTRODUCTION

The European Commission DGIII/D3 promoted in 1997 a wide research programme for the study of the snow loads on structures, with the scope to harmonize the treatment of the snow data collected in the European territory in view of the implementation of the EN version of the Eurocode 1 "Actions on Structures". The study, carry out by eight research Institutions belonging to six different countries, was divided in two phases, ended respectively in March 1998 (Sanpaolesi et al. 1998) and in September 1999(Sanpaolesi. et al. 1999).

Among the tasks of the research programme there were the definition of an European characteristic ground snow load map and the study of some problems regarding the ultimate and the serviceability limit states verifications.

Daily ground snow data series for several years were collected at a notable number of meteorological stations. Such snow data are suitable for the determination of the duration of the snow load on the ground.

Since most of the structural codes define the snow loads on structures as the product of the characteristic value of the snow load on the ground for a conversion factor which is independent of time, it can be assumed that the duration of the snow load on the roofs is the same as that one on the ground.

In the present paper a statistical analysis procedure is presented and a correlation between ground snow load levels and their duration is established for three weather stations in the Italian territory, where long series of daily snow observations were available.

The results are useful for the serviceability verifications for materials with creep behaviour. In particular for wooden structures, very common in mountainous areas.

2 DAILY SNOW DATA

Daily observations of the snow on the ground usually consist in depth measurements of the snow cover. Rarely the water equivalent weight is measured, while only in a few cases snow pillows are used for the direct determination of the load.

In Italy, the main source of information on snow on the ground is the "Servizio Idrografico e Mareografico Italiano" (SIMI), which records the observations of the height of the snow cover in hundreds of weather stations. The data have been yearly published since 1910. A set of 125 stations has been selected, in which snow data were continuously collected for a period variable between 25 and 50 years. For 54 of them daily data are available.

It is a very large data base, which well represents the whole Italian territory, either for the location of the meteorological stations or for their altitude. The quality of the snow data has been carefully checked, disregarding those stations in which inconsistencies or errors were detected (also with regard to the other meteorological parameters) or large discrepancies with the results of the neighbouring stations were met.

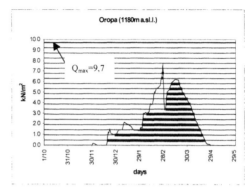

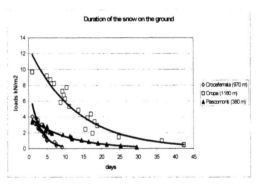

Figure 4 Ground snow load at Oropa, in 1973 (1180 m a.m.s.l.)

Figure 1. Localisation of the three meteorological stations

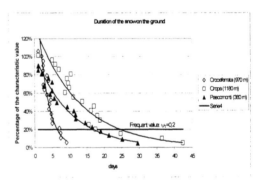

Figure 5 Correlation curves between ground snow load levels and their duration

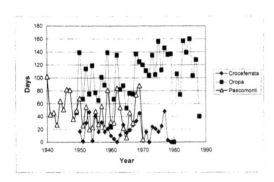

Figure 2. Total number of days with snow on the ground for each year

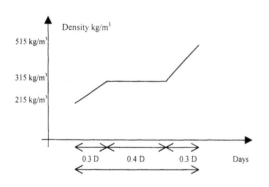

Figure 6 Correlation curves between ground snow load levels, referred to the characteristic value, and their duration

Three meteorological stations have been chosen to show an example of the determination of the snow load duration. All of them present significant periods of permanence of the snow on the ground. They are the stations of Oropa (1180 m a.m.s.l.), Pascomonti (380 m a.m.s.l.), Croceferrata (970 m. a.m.s.l.) (figure 1).

Figure 3. Snow density model for snow cover duration longer than 10 days

The selected stations are suitable to evaluate the phenomenon either at low or high altitude, in different climatic conditions and in different zones of the Italian territory.

For each year, the total number of days in which the presence of snow on the ground has been observed is represented in the graph shown in figure 2.

The snow depth measurements were converted into snow loads by applying an adequate density model, which, in Italy has been assumed to be a function of time of presence of the snow on the ground. The validity of the density function, shown in figure 3, has been widely verified for the Italian territory (Del Corso, R. & Formichi, P. 1999).

3 DATA ANALYSIS

Let us consider a building material with a time-dependent behaviour and a structural effect correlated with the duration of the snow load on a given roof (e.g. the deformation of a wooden truss). It is normal to assume such effect being correlated with the load intensity and with its duration.

In general it can be affirmed that the effect on the structure will disappear if the load causing it will be removed for a period of time as long as the loading one. On the contrary, if two or more loading periods are separated by very short unloaded phases, the effects caused by the loads should result in the sum of the single effects.

In the present analysis, for the sake of simplicity, it has been made reference only to a single loading phase, assuming the structure able to recover completely the relative effect before a further load's application.

As a consequence, it has been analysed only the maximum duration of the snow cover on the ground per each year.

The sampled data series can be usefully plotted on a graph, placing time on the abscissa and load levels on the ordinates. As an example, in the figure 4 it is represented the variation of the ground snow load at Oropa during the winter 1973-1974.

Let Q_{max} be the maximum load value registered during the N available years of observations. The threshold values Q_t are calculated by dividing the maximum value Q_{max} by a given number of levels m, which in the present investigation has been fixed equal to 20. The result is the set of the threshold values Q_t, with a constant step:

$$Q_t - Q_{(t-1)} = Q_{max} / m$$

Let consequently define as the random variable D, the yearly maximum duration of the ground snow cover relative to each load level. The daily ground snow data allow to get a sample of the r.v. D, as illustrated in figure 4.

Since for the highest threshold values the number of permanence periods is very low, reducing to one for $Q_t = Q_{max}$, it is not possible to determine a probability distribution function of the random variable for all the load levels. Therefore it has been retained, as representative value of the duration associated to each level, the mean value of the sample.

At each weather station are therefore available m pairs of load values and relative mean associated duration.

The correlation between load levels and mean duration periods, was determined by means of the least square method, as shown in figure 5.

A wider analysis of the daily snow data in CEN area could lead to the definition of a complete set of correlation curves for different climates and altitudes. Such information could be implemented into the structural codes. In fact, for materials with a time-dependent behaviour, the knowledge of the mean duration of a given snow load on the structure allows to define appropriate serviceability limit states verifications, which, in some cases, could govern the design.

At present, within the structural codes of CEN (European Committee for Standardization) countries, only the Swiss code SIA 160 (SIA, 1989) gives guidance about the snow load duration and the relevant limit states verifications.

In the Eurocode "Basis of Design" (CEN, 1994) and in the majority of codes on actions the ground snow load is defined on the basis of some reference values: the characteristic value, corresponding to the probability of 0,02 of not being exceeded during one year; the frequent value, defined such that the fractile of time during which it is exceeded is chose to be 0,05; the quasi-permanent value, is determined as the averaged value over the reference period of time.

The correlation curves, shown in figure 5, can be usefully be referred to the dimensionless ratio Q / Q_k, where Q_k is the 50-years characteristic ground snow load value for the weather station being studied, as it is shown in figure 6. On the ordinates axis it is indicated the ψ_1 coefficient, which represents the ratio between the frequent value and the characteristic value of the snow load on the ground. Such coefficient has not been calculated for each station but it has been set to 0,2, as recommended by the Eurocode "Basis of Design".

From figure 6 significant differences can be observed in the curves, due to the climatic conditions of the three stations.

At the Oropa, located in the Italian western Alps at the altitude of 1180 m a.s.m.l., severe and continuos snowfalls are observed. To the $\psi_1 Q_k$ value (with $\psi_1 = 0,2$) it corresponds a mean duration of 25 days, which undoubtedly represents a considerable loading period for a building material which shows creep.

Similar conclusions can be drawn from the analysis of the results obtained at Pascomonti, located in the southern Italian Alps at 380 m a.m.s.l. Notwithstanding a steeper correlation curve between duration and snow load, a duration of 17 days is expected when the load level is the 20% of the characteristic ground snow load.

Opposite considerations can be derived at Croceferrata, placed in southern Italy at 970 m a.m.s.l. This weather station is very close to the sea and the climate is very mild, therefore snowfalls are intermittent and the permanence periods of ground snow cover are usually short. As a consequence to the frequent value of the load it corresponds a 6 days permanence period, which is too short to produce a significant effects in materials with time-dependent behaviour.

4 CONCLUSIONS

Within the recent European research project on snow loads on buildings, sponsored by DGIII/D-3, it has been possible to collect in all CEN countries a huge database of daily ground snow data.

This data allowed, among other investigations, to study the ground snow load duration for three Italian weather stations.

The performed analysis put in evidence quite different correlations between the ground snow load and its duration at the three stations.

In areas where the climate is mild, the snow load duration is very short and has not any influence on the materials with time-dependent behaviour. On the contrary, where more severe climatic conditions occur much longer duration of significant snow load values are observed.

It is then worthwhile to extend the statistical procedure illustrated in the present paper to a greater number of stations, to define more precisely the duration of the snow load on the ground.

Since this problem is important for wooden structures, the results of a wider investigation could be conveniently used in the structural codes and, in particular, in the Eurocode 1 "Actions on Structures".

REFERENCES

CEN, 1994. ENV 1991-1: 1994 Basis of Design and Actions on Structures, CEN Central Secretariat Brussels

CEN, 1995. ENV 1991-2-3: 1995 Actions on Structures – Snow Loads, CEN Central Secretariat Brussels

Del Corso, R., Formichi, P. 1999 *Shape coefficients for conversion of ground snow loads to roof snow loads.* 16th BIBM International Congress. Venice May 25-28 1999.

International Standard Organisation, 1998, ISO/FDIS 4355 Bases for design of structures – Determination of snow loads on roofs. Geneve.

Sanpaolesi, L et al. 1998. *Scientific support activity in the field of structural stability of civil engineering works: Snow loads – Final Report Phase I,* Commission of the European Communities DGIII/D-3, Brussels

Sanpaolesi, L et al. 1999. *Scientific support activity in the field of structural stability of civil engineering works: Snow loads – Final Report Phase II,* Commission of the European Communities DGIII/D-3, Brussels

Sanpaolesi, L. et al., 1983. *Analisi statistica dei valori del carico di riferimento da neve al suolo in Italia (Statistical analysis of reference ground snow load in Italy).* Giornale del Genio Civile – Roma

SIA Norm 160, 1989 *Actions on Structures,* English translation. Zurich.

Snow Engineering, Hjorth-Hansen, Holand, Løset & Norem (eds) © 2000 Balkema, Rotterdam, ISBN 90 5809 148 1

Wind tunnel investigation of snow loads on buildings

M. Dufresne de Virel, P. Delpech & C. Sacré
Centre Scientifique et Technique du Bâtiment, Nantes, France

ABSTRACT: In connection with roof snow load assessment, experimental evaluations of roof shape coefficients were undertaken in a large climatic wind tunnel. The physical modeling of snowstorms was made thanks to the capabilities of the facility which make use of artificial snow and enables the simulation of a realistic air flow around the building models. Various roof shapes were investigated with controlled and reproducible snowstorm parameters. A snow loads database of basic shapes was set-up and compared with on site evaluations of natural snow loads on real buildings. The wind tunnel experimental approach and in nature observation appeared to be complementary to provide a sound scientific basis for the revision and harmonization of European building codes.

1 INTRODUCTION

In the frame work of the European Snow Loads research project, which aims at improving the evaluation of snow loads on roofs, a revised version of the European map of ground snow loads was first worked-out. It was then undertaken to refine the conversion process which enables the evaluation of roof snow loads from ground snow load measurements (Sanpaolesi 2000).

In European national building codes, the conversion is made by multiplying the ground snow loads by an empirical roof shape coefficient which takes into account all parameters influencing the snow loading. The revision work to perform should enable the setting-up of a new snow load model based on a physical approach. Two tasks were undertaken : in nature measurements of snow loads on real buildings correlated to climatic data and wind tunnel modeling of the snow loading on various basic buildings.

In nature observation is probably the only way to account for all physical parameters which actually interfere with the snow loading. But, due to the great diversity of roof geometry as well as climatic situations, only statistical series of measurements are valuable to reach a quantitative assessment of snow loads on roof.

On the other hand, the use of a climatic wind tunnel allows to investigate a large number of roof shapes in controlled and reproducible snow fall and wind conditions. Wind tunnel experiments enable the simulation of single snow events but they cannot easily simulate multiple snow falls as observed in nature. Consequently, one considers that there is no proper wind tunnel equivalent to the statistical ground snow loads used as a reference measurement in building codes.

It was appropriate to dedicate the wind tunnel simulation to the investigation of aerodynamic interactions of roof shapes with the snowstorm 2-phase flow. Hence, the aim of the experiments undertaken in the Jules Verne climatic test chamber was to identify the effects of wind on roof snow coverage (unbalanced roof loads).

Typical and simple roof shape tests were first dedicated to the parameter adjustment and calibration of the wind tunnel experiments. Then, a snow load data collection for reference building configurations (various typical roof shapes) was performed.

2 WIND TUNNEL EXPERIMENTAL SIMULATION OF SNOW LOADS ON BUILDING

2.1 *The Jules Verne Climatic Wind Tunnel*

The Jules Verne Climatic Wind Tunnel (JVCWT) run by Aerodynamic and Climatic Environment Department of Centre Scientifique et Technique du Bâtiment (CSTB) aims at developing the physical knowledge and understanding of the influence of wind and other climatic parameters on buildings, structure elements, industrial pieces of equipment and vehicles (Gandemer et al. 1996).

One of the strength of the facility is the possibility to simulate and fully control the combination of climatic parameters to perform tests on a full, or moderately reduced, scale basis which are often the only relevant experimental approaches.

The test chamber of the JV wind tunnel thermal unit is 25 m long, 10 m wide and 7 m high. The appropriate thermodynamic conditions (air temperature and humidity) are controlled by two heat exchangers (cold and warm) across the guided return of the wind tunnel.

The thermal unit equipment enables the control of the air temperature from −25 °C to +50 °C and relative humidity from 30 % to 100 %. According to the section of the adjustable nozzle (cross section : 18 to 30 m²), the maximum air flow speed can be set from 90 km/h to 140 km/h in the test chamber.

Three snow guns can be fitted into the test section nozzle. They are used to generate a freezing water spray which produces the snow mantle. Deposits of about 10 to 15 cm/h can be obtained on the 200 m² floor of the test chamber.

Experimental parameters as air/water ratio and pressure in the guns are controlled and enable the adjustment of the snow quality from dry to wet.

2.2 The wind tunnel artificial snow

Initially a complete characterization of the snow quality was achieved in the wind tunnel. The measurements made aimed to know the liquid water content of the snow and its evolution with respect to the air/water ratio of the spray or the distance from the snow guns (Boisson-Kouznetzoff et al. 1996). In some cases the snow quality, is adjustable.

The artificial snow particles do not look like fresh natural starry snow flakes. They actually could be compared to slightly aged and melted natural snow. They could also be compared to natural flakes, which would have been blown and eroded by wind (Figure 1).

2.3 The experimental procedure

The chosen model scale of 1/10 suited the test section width. This rather large scale presented a good compromise with regard to blockage effects and drifting similitude with full scale observations.

The turbulence intensity was adjusted, by fitting roughness upstream the model location, to be similar to the turbulence rate over an open field terrain.

The test temperature of −10 °C was set to ensure the production of rather dry snow (liquid water content < 2% in volume). Some tests were achieved at a lower temperature (-15 °C) in order to quantify the influence of temperature on the snow drifting mechanisms in the wind tunnel. The only visible effect of the temperature change was the increase of the bulk snow density. This change does not interfere with the snow repartition on the building model.

A little air flow is required in the test chamber in

order to maintain the hygro-thermal conditions and to enhance the heat exchange coefficient of the droplets. Two different experimental situations were simulated in the wind tunnel. A low wind speed experiment (< 1 m/s), with deflection of the air flow, was suitable to reproduce the uniform snow loading of the models. The snowstorm situation which aimed at reproducing the unbalanced loads on roofs was simulated with a wind speed of 4 m/s.

Intermediate wind speed situations revealed the sensitivity of the snow accretion phenomenon to the wind speed setting in the wind tunnel (i.e. the combined effect of additional air flow induced by the snow guns and the impingement effect of the droplets increased the loading on the windward side of the models).

Snow load simulations with higher wind velocities would require a particular adjustment phase. Since the snow is produced in the wind tunnel by the freezing of water droplets and since the freezing process requires a finite period of time (~ 2 or 3 s.), it is quite obvious that, if the distance between the snow guns and the model is unchanged, the increasing of the wind speed would generate a solid ice layer on the model instead of a snow mantle.

Consequently, higher wind speed snow testing are actually possible by changing the location of the snow guns in the wind tunnel in order to adjust the time of fly of the water droplets before they reach the model.

The experiment duration was set empirically to 1 hour in order to reach a "steady" snow loading situation (i.e. the snow load pattern does not change any more). Snow loads on model roofs were measured along the main axis of each roof by drawing their depth profiles on a piece of cardboard. These continuous measurements were post-processed thanks to digitalization which allowed to identify the snow load characteristic parameters.

At the end of experimental snowstorms, the snow loads on the ground was measured to provide a reference value to compare with the snow layer on the roofs. For the low wind speed experiments (< 1 m/s) the thickness of the snow layer on the ground was regular enough to be measured around the model. Measurements 1m upwind and 1m downwind the model were averaged to assess the snow depth on the ground : H_{ref} = 15.5 cm. To assess the ground snow load at the end of windy experiments (4 m/s), additional testings were performed without model and lead to 10 cm < H_{ref} < 12 cm at the model location.

3 PHYSICAL MODELLING OF SNOW LOADING AT REDUCE SCALE

3.1 Similitude requirements for reduced scale experiments

Snow accumulation is dominated by global and local

172

aerodynamic effects (boundary layer turbulence, vertical wind velocity gradient, separation and reattachment zones,...). To simulate wind induced snow drifting mechanisms at reduced scale it is necessary to reproduce the mean and turbulent wind flow and achieve the similarity of particles trajectories. Similarity conditions involve mean and turbulent flow characteristics, local flow behaviour, properties of the snow phase.

Let us consider the snow properties summarised in Table 1. Although the property discrepancies, the artificial snow storm characteristics could be related to natural snow fall. In particular, the saltation length seemed smaller than the model reference dimensions as specified by the similitude requirements.

The particle trajectories are similar if the ratio of gravitational forces to aerodynamic forces as well as the ratio of inertial forces to gravitational forces is conserved for both model and prototype.

Exact reduced scale 2-phase flow similarity with full scale experiment is not possible to achieve. The similarity of gravity and inertial effects is ensured if the densimetric particulate Froude number is conserved, but gravitational to fluid force ratio conservation leads to contradictory results.

Comparison with full scale in nature observation is probably the more reliable and easiest way to solve the problem. Snow load data with wind, recorded in the Italian Alpes on a particular building, were used as natural calibration data (Lozza 1999). Several wind tunnel tests were performed with a 1/10 scaled model of the reference building. They showed that the windward/leeward snow load ratio was conserved if the wind tunnel air flow speed was higher than the on site wind velocity : i.e. 2.7 m/s natural wind corresponds to about 4 m/s wind tunnel air flow.

The evaluation of the equivalent prototype snowstorm duration with respect to the model wind velocity and experiment duration is another difficulty. Due to a rather high snow particle density in the wind tunnel air flow, the calculation of the equivalent accumulation rate, by using known similitude relationships (Isyumov et al. 1997), did not lead to realistic values. It actually seemed more relevant to deduce the prototype/model equivalent snow event duration by comparing the results of prototype observations and model experiments.

The wind tunnel accumulation rate of snow on the ground was, for a single snow event with wind (4 m/s) of 1 hour : 12 cm. The equivalent prototype snow event would have induced a 120cm snow fall. From field observations (Lozza 1999) the accumulation rate due to heavy snow falls in nature is close to 5 cm/h. This means that an estimate of the prototype snow event duration, with respect to the wind tunnel modelling, is equal to 120/5 = 24 h. Actually, it is more realistic to consider that a wind tunnel experiment correspond to a period of several days in nature.

3.2 Influence of model size and test duration

As stated above, the model scale and the duration of wind tunnel experiments are probably the most important sources of discrepancy with in nature roof snow load situation. To deal with that difficulty, a basic parametric experimental study was undertaken to collect quantitative information about the influences of both model size and experiment duration.

3.2.1 Influence of model size or height

Three models with gable roofs (slope of 40°) were used for this test (Fig. 2). According to the modeling scale (1/10), the first model is a 2 floor building, the second is a ground floor building. The third model is similar to the first one with double surfaces (the main dimensions were multiplied by 1.41).

A single snow gun was used for the tests, the wind velocity was 4 m/s and the experiment duration was set to half an hour.

As shown in Table 2, the dimensionless snow depth factors H* (average snow height on the roof divided by average snow height on the ground) are conserved in all cases.

One can verify that the snow load pattern on the model is reproducible at two different height. Although the snow gun is localized in the nozzle of the wind tunnel, there is no significant vertical gradient of particle density in the air flow which would interfere with the snow loading mechanisms.

Similarly, the moderate increase of the model scale did not introduce a significant snow load discrepancy. This result tends to show that the snow loading pattern is little sensitive to the model scale.

3.2.2 Influence of test duration

During longer snowstorm experiments, with the "small high" model, the snow cover was measured every 15 minutes. The results show that the snow cover increases quite regularly with time (Table 3). Let us notice that the windward/leeward average snow cover ratio decreased at the beginning of the test and was getting constant after one hour (Fig. 3). This means that, after a 1 h snow storm, the unbalanced loading of the roof seemed maximum. Consequently the test duration was set empirically to one hour.

4 SNOW LOADS ON BASIC BUILDING ROOFS

4.1 General presentation of the results

For each case the snow load profile was recorded and processed to provide the characteristic load parameters. The average snow height H_{ave} was calculated by dividing the snow profile surface by the roof length L_{roof}. The maximum snow height H_{max} and the location of that maximum i.e. the distance from the windward roof edge, eaves or ridge L_{max} were measured. The dimensionless maximum location, $L*_{max}$ was calculated by dividing distance L_{max} by the roof length L_{roof} as indicated in Figure 4.

When appropriate, the experimental results are commented with regard to European Snow Loads Standards (CEN 1994).

The dimensionless snow depth factors $H*$, defined earlier is occasionally associated to the dimensionless snow load factors $w*$ calculated by dividing the actual roof snow load by the ground snow load if variations of snow density are noticeable.

Once again, since the dimensionless depth factors $H*$ or load factors $w*$ cannot be directly compared with building code shape coefficients, the unbalanced snow loads due to wind effects were estimated by calculating load ratios between the different parts of roofs (windward/leeward, high/low ...).

4.2 Flat roofs

4.2.1 Single flat roofs

Snowstorm tests (4 m/s wind) have been carried out with single flat roof models. Two models of 1.5 x 2 x 0.5 m and 1.5 x 3 x 0.5 m were used.

One can notice that the wind tunnel resulting snow loads are unbalanced (Fig 5 and Table 4). The length of the roof increased the snow load on the leeward part. These snow loads are higher than those calculated by Eurocode which specifies a roof shape coefficient of 0.4 on half of the roof (CEN 1994).

4.2.2 Two-level flat roofs

Investigations have been made for various two-level flat roof configurations, i.e. different building length a, step length b, step height c and wind directions as indicated in Figure 6 and Table 5.

Figures 7-15 show cross sections of building models and the snow cover at the end of the tests (wind comes from left).

The resulting snow loads cannot be compared directly to snow load profiles indicated in Eurocode. Maximum snow load near the abrupt change of level indicated in the Eurocode (roof shape coefficient = 2.1) is in agreement with most of the wind tunnel measurements. In few cases (Figures 10. 11 and 12) the wind tunnel experiments lead to more severe loads than those indicated by the Eurocode.

4.3 Gable roofs

The models are 2 floor buildings. Their ground surface is 1 x 1.2 m (10 x 12 m full scale). The roof slopes are 20° and 40° with eaves of 7.5 cm. Roof surface roughness due to the tiles is modeled by thin plywood plates of about 3 mm.

Table 8 presents the average snow load factors calculated by taking into account the density of the snow layer measured locally during the experiments.

Table 1. Properties of snow particles.

Property	Natural snow	Artificial snow
Diameter	0.5 to 5 mm	0.15 to 0.3 mm
Particle density	50 to 700 kg/m^3	910 kg/m^3
Fluid density	1.22 kg/m^3	1.34 kg/m^3
Snow cover density	100 to 600 kg/m^3	315 to 370 kg/m^3

Table 2. Snow depth factors measured at the end of the snowstorm (1 snow gun, ½ h, 4 m/s).

$H*$	Small high	Small low	Large high
Windward	0.78	0.81	0.88
Leeward	0.69	0.75	0.76

Table 3. Temporal evolution of the windward and leeward average snow height (2 snow guns, -10 °C, 4 m/s).

Duration. (mn)	15	30	45	60	75	90	105	120
Windward (cm)	1.6	3.2	4.5	5.8	7.3	8.2	9.7	11.6
Leeward (cm)	2.0	4.0	6.1	9.0	11.2	13.4	16	18.5

Table 4. Snow profile characteristics for flat roofs.

L_{roof} (cm)	$L*_{max}$	$H*_{ave}$	$H*_{max}$
200	0.91	0.41	0.75
300	0.75	0.56	0.91

Table 5. Flat roof model dimensions.

Shape	N°1	N°2	N°3	N°4
a (m)	2	2	3	2
b (m)	1	0.5	0.5	1
c (m)	0.25	0.25	0.25	0.5

Table 6. Results for two-level flat roofs, wind 4 m/s.

Name	wind	$L*_{max}$	$H*_{ave}$	$H*_{max}$	Ave. (Hi/Lo)
N°1P2	0°	1.00	0.73	2.08	0.66
N°1P3		0.94	0.48	0.83	
N°1P2	180°	0.91	0.26	0.42	0.24
N°1P3		0.32	1.08	1.41	
N°1P6	90°	0.83	0.37	0.75	0.95
N°1P2		0.95	0.35	0.58	
N°2P2	0°	1.00	1.16	2.32	0.47
N°2P3		0.92	0.55	0.75	
N°2P2	180°	0.90	0.23	0.33	0.18
N°2P3		0.63	1.30	1.66	
N°3P2	0°	1.00	0.93	2.16	0.40
N°3P3		0.98	0.37	0.58	
N°3P2	180°	0.22	0.15	0.17	0.22
N°3P3		0.34	0.69	1.00	
N°4P2	0°	1.00	1.83	4.15	0.26
N°4P3		0.91	0.47	0.91	
N°4P2	180°	0.12	0.17	0.25	0.27
N°4P3		0.84	0.64	0.83	

One can notice that the load factors are similar to the depth factors in low wind speed conditions. This indicates the uniformity of the snow density on the roof and on the ground in steady climatic situation.

In case of 4 m/s wind, the snow density on the windward side (384 kg/m^3) is higher than the density

Figure 1. Artificial snow particles produced in the wind tunnel.

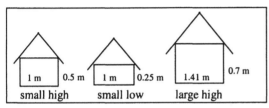

small high   small low   large high

Figure 2. Cross section of the three models.

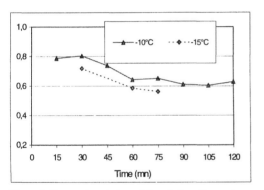

Figure 3. Temporal evolution of the windward/leeward snow cover cross section ratio.

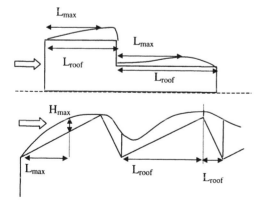

Figure 4. Definition of snow load characteristic parameters for flat and pitched roofs.

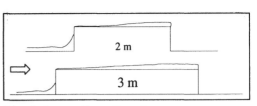

Figure 5. Snow profiles on flat roofs.

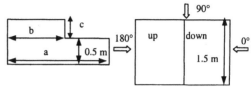

Figure 6. General shape of two level flat roof models and incident wind directions.

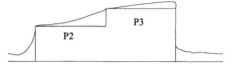

Figure 7. Roof N°1, wind 0°.

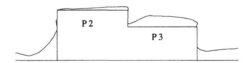

Figure 8. Roof N°1, wind 180°.

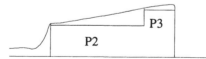

Figure 9. Roof N°1, wind 90°.

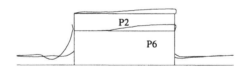

Figure 10. Roof N°2, wind 0°.

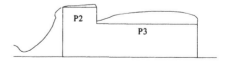

Figure 11. Roof N°2, wind 180°.

Figure 12. Roof N°3, wind 0°.

Figure 13. Roof N°3, wind 180°.

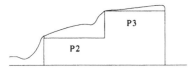

Figure 14. Roof N°4, wind 0°

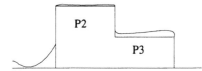

Figure 15. Roof N°4, wind 180°.

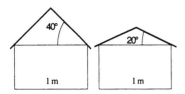

Figure 16. Double pitch models.

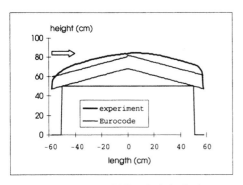

Figure 17. Cross section of 20° roof. wind < 1 m/s.

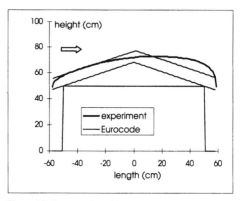

Figure 18. Cross section of 20° roof. wind 4 m/s.

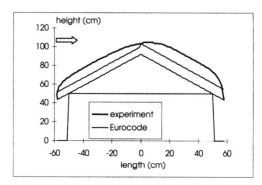

Figure 19. Cross section of 40° roof. wind <1 m/s.

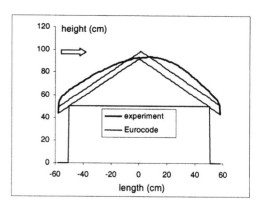

Figure 20. Cross section 40° roof. wind 4 m/s.

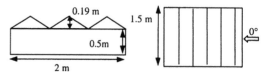

Figure 21. Symmetrical roof model with slope of 30°.

176

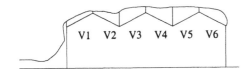

Figure 22. Cross section of symmetrical roof.

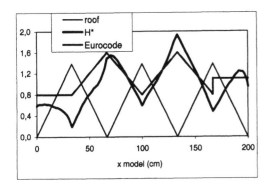

Figure 23. Snow load repartition for symmetrical roof. wind 0°.

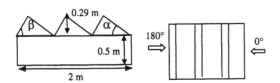

Figure 24. Non-symmetrical roof model with pitch angles of 60° and 30°.

Figure 25. Non-symmetrical roof. wind 0°.

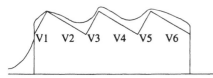

Figure 26. Non-symmetrical roof. wind 180°.

on the ground (360 kg/m³). In the same wind condition. the snow density on the leeward side (329 kg/m³) is lower than the density on the ground. This is probably due to the way the snow is packed on the roof sides by eddies and local high or low speed airflows. On the gable roofs. the difference of load factors between windward and leeward side is actually lower than the difference of depth factors meas-

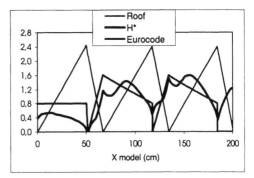

Figure 27. Snow load repartition for non-symmetrical roof wind 0°.

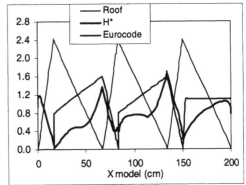

Figure 28. Snow load repartition for non-symmetrical roof. wind 180°.

Table 7. Average and maximum snow depth factor on roof.

Slope,Wind, Side		L^*_{max}	H^*_{ave}	H^*_{max}	Wind./lee.
20°	<1m/s wind.	0.52	0.89	1.25	0.82
20°	<1m/s lee.	0.61	1.08	1.47	
20°	4m/s wind.	0.37	0.70	0.83	0.57
20°	4m/s lee.	0.77	1.22	1.66	
40°	<1m/s wind.	0.41	0.88	1.23	0.78
40°	<1m/s lee.	0.56	1.13	1.61	
40°	4m/s wind.	0.19	1.00	1.33	0.81
40°	4m/s lee.	0.54	1.24	1.66	

Table 8. Average snow loads factor on roof.

Angle	Wind	w* windward	w* leeward	w/l
20°	< 1m/s	0.89	1.09	0.82
20°	4 m/s	0.74	1.11	0.67
40°	< 1m/s	0.88	1.13	0.78
40°	4 m/s	1.06	1.13	0.94

Table 9. Eurocode roof shape coefficient for gable roofs.

Angle	w* windward	w* leeward	w/l
20°	0.8	0.9	0.89
40°	0.53	0.73	0.73

177

Table 10. Results for multi-pitch symmetrical roof, wind 0°.

Profil	L_{roof} (cm)	L^*_{max}	H^*_{ave}	H^*_{max}
V1	33	0.23	0.53	0.62
V2	33	1.00	0.79	1.49
V3	33	0.10	1.19	1.54
V4	33	1.00	1.21	1.92
V5	33	0.00	1.24	1.92
V6	33	0.86	1.01	1.18

Table 11. Results for multi-pitch non symmetrical roof.

Profil	wind	L_{roof} (cm)	L^*_{max}	H^*_{ave}	H^*_{max}
V1	0°	50	0.31	0.42	0.53
V2		17	0.98	0.42	1.16
V3		50	0.46	1.13	1.44
V4		17	0.98	0.89	1.51
V5		50	0.45	1.28	1.60
V6		17	0.90	0.93	1.23
V1	180°	17	0.07	0.72	1.18
V2		50	1.00	0.53	1.36
V3		17	0.00	0.68	1.36
V4		50	1.00	0.85	1.70
V5		17	0.00	0.97	2.03
V6		50	0.88	0.79	1.04

ured at the same locations. The snow density variations tend to compensate the apparent unbalanced windward/leeward snow loads.

The experimental unbalanced loadings are more severe than what the Eurocode indicates (Table 9) for 20° roof. The opposite conclusion can be made for 40° roof.

4.4 *Multipitch roofs*

Three different configurations of two valley multipitch roofs were tested (symmetrical and non-symmetrical roof, Figs 21, 24). Snow profiles on each part of the roof are noted V1 to V6 starting from windward edge (Figs 22, 25, 26). Results are summarised in Tables 10 and 11. Figures 23, 27 and 28 enable the comparison of experimental snow profiles wirth Eurocode indications.

One can notice a light increase of snow load in leeward valley.

Except for the edges of roof with slope of 60°. where snow load was not expected the Eurocode provides conservative values of snow load.

5 CONCLUSION

The objectives of this work was the investigation of unbalanced snow loads on roofs of various shapes in order to provide updated background data to improve European building codes.

The wind tunnel modeling of roof snow load benefited from the possibilities of the Jules Verne

climatic wind tunnel which enables the production of artificial snow in a large test section. The main advantages of this approach was first to work with a rather large model scale (1/10) and second to observe specific snow packing which cannot be reproduced with less realistic modeling particles.

Several tests were designed and performed as a cross check of initial results. The modeling parameters were partly validated thanks to comparisons with in nature observations of snow loads on traditional gable roofs. For the time being. let us emphasize that the lack of full scale comprehensive and reliable results is an obstacle to further developments of the wind tunnel modeling.

In most cases the wind tunnel results were in good agreement with the Eurocode prescriptions. However. in some cases. the snow loads assessments seem under-evaluated in the codes. Those discrepancies need to be addressed by focusing the analysis on aerodynamic particularities of roofs. Alternatively. the influence of a single snow event should be evaluated quantitatively with respect to series of snow falls during the winter season.

REFERENCES

Boisson-Kouznetzoff S. & Palier P. 1996. Modeling and experimental verification of snow production within a closed space. *Eurotherm. 2nd European Thermal-Sciences and 14th UIT National Heat Transfer Conference. Rome. May 1996.*

CEN. European Prestandard. Eurocode 1 : Basis of design and actions on structures. Part 2-3 : Snow Loads. 1994.

Gandemer. J. Palier. P. & Boisson-Kouznetzoff. S. Snow simulation within the closed space of the Jules Verne Climatic Wind Tunnel. In M. Izumi. T. Nakamura & R. Sack (eds). *Snow Engineering : Recent Advances; Proc. 3rd intern. Symp.. Sendai. 26-31 May 1996.* Rotterdam: Balkema.

Isyumov. N. & Mikitiuk. M. 1997. Wind tunnel model studies of roof snow loads resulting from multiple snowstorms. In M. Izumi. T. Nakamura & R. Sack (eds). *Snow Engineering : Recent Advances; Proc. 3rd intern. Symp.. Sendai. 26-31 May 1996.* Rotterdam: Balkema.

Lozza. S. Elaboration of four Italian stations with instrumented roofs. Snow Loads Research project. Task D : internal report. ISMES Spa. April. 1999.

Sanpaolesi. L. Del Corso. R. & Formichi. P. 2000. European Snow Loads Research Project : Recent results. In Proc. 4th Intern. Snow Engineering Conf.. Trondheim. Norway. 2000.

Snow Engineering, Hjorth-Hansen, Holand, Løset & Norem (eds) © 2000 Balkema, Rotterdam, ISBN 90 5809 148 1

New ground snow loads for a European standard

M.Gränzer
Landesgewerbeamt Baden-Württemberg, Landesstelle für Bautechnik, Stuttgart, Germany

ABSTRACT:

In Europe the different existing national standards determining snow loads for buildings will definitively be replaced within the coming years by a general Eurocode. To supply the European Committee for Standardization (CEN) with the information required for setting up a general European code eight research institutions from different countries worked together on a pre-normative research project.

The investigation had to start with existing records of snow data observed at many stations in the 18 member states of CEN. In some countries only the snow depth is recorded, in others weights are available. Applying statistics of extremes a characteristic snow load could be determined for every station of observation.

Areas showing a similar correlation between the characteristic snow load on the ground and the altitude of the site above sea level are forming one region in which the snow load can be given by the same type of altitude function. Within one region deviations from the general trend due to topographical and meteorological influences are taken into account by subdividing this region into zones characterised by excessive or reduced amounts of snow in relation to the average of this region. Finally, regions and zones are represented in maps largely independent form the administrative borders within Europe.

1 INTRODUCTION

The construction of roofs, especially of light-weight constructions may predominantly depend on the snow load to be taken into account for designing. Snow load values are given in national standards. In Europe this means each of the 18 member states of the European Organization for Standardization (CEN) has established different regulations. In Germany which shares borders with nine other European countries the following test was made: The snow load, given by the current local code on either side of the border lines has been compared. At some places the snow loads were quite similar but sometimes big discrepancies appeared which could not be explained neither by inevitable inaccuracies nor by a different climate.

The project team setting up the first provisional European code on snow loads ENV 1991-2-3 (CEN 1995) had to face the fact that the national codes had been worked out completely independently from one another and were based on very different background information.

The European Commission aspiring to adjust the level of safety within the EU decided to fund an international research group comprising

Building Research Establishment (BRE), (UK)
CSTB, Centre de Recherche de Nantes, (F)
École Polytechnique Fédérale de Lausanne, (CH)
ISMES, Structural Engineering Department, (I)
Joint Research Centre, ISIS, (EU)
SINTEF, Civil and Environmental Engineering, (N)
University of Leipzig, Inst. of Concrete Design, (D)
University of Pisa, Dept of Structural Engineering (I)

One of the main tasks was to establish a common and reliable procedure allowing to fix snow loads for buildings in all member states of CEN.

2 SNOW LOAD ON THE GROUND

In most of the building codes the snow load on a roof is given in a format like

$$s_{k,roof} = s_{k,ground} \cdot f(C_e, C_t, \mu, ...) \tag{1}$$

where $s_{k,roof}$ = characteristic snow load on a roof; $s_{k,ground}$ = characteristic snow load on the ground; f(...) = function of coefficients, representing different influences like exposure, temperature or shape of the roof.

The present paper will give a general view on the steps performed by the research group in fixing the snow load on the ground. More detailed information and examples could be found in the Final Report (Sanpaolesi et al. 1998). Furthermore, special aspects will be presented by some of the participating research institutions during this conference.

3 DATABASE

3.1 *Geographical range*

Snow load data had to be collected in 18 CEN member states: Austria, Belgium, Denmark, Eire, Finland, France, Germany, Greece, Iceland, Italy, Luxembourg, Netherlands, Norway, Portugal, Spain, Sweden, Switzerland and UK. In the first phase of the research it was not yet possible to incorporate the countries of Eastern Europe.

It is not surprising that in a continent reaching from northern Norway (70° N) with strong winters down to the mild climate in Crete (35° N) the registration of snow is well organized only in countries where snow may cause significant difficulties in everyday life.

3.2 *Recording of snow*

In northern and alpine countries the amount of accumulated snow is important not only for the safety of roofs but also for traffic conditions, skiing resorts, avalanche risks, or hydroelectric power stations. These various interests have led to a continuous recording of the snow, which gives a good basis for a statistical analysis. It was more difficult to obtain enough measurements of the snow cover in the southern parts of Europe where snow occurs rather seldom and lasts only for some hours.

Unfortunately measuring snow doesn't belong to the standardized measurements imposed internationally by Geneva-based World Meteorological Organization (WMO). Some countries only keep records of the snow depth, others are weighing the snow cover at certain time intervals.

The participants in the research group had decided to base the statistical analysis on yearly observed snow load maxima, a procedure often used in elaborating national codes. Weights could be introduced directly into the statistical analysis, however, measured snow depths had to be converted into loads applying a mean density of the snow cover according to national experience. Some countries used simply a fixed density for all their data. Other countries introduced a density function increasing

with the total depth of the snow or they applied different values for different regions of the territory. Where available, existing statistics on the variation of the mean density of snow with date were used to set up a function, giving higher densities at the end of the winter. Inevitably, the different procedures of converting depths into loads imply uncertainties, but without recurring to depth measurements the database would have been too small.

All data had to be checked before introducing them into the statistical processing. Often it was necessary to make a choice, either to use measurements of as much stations as possible or to exclude stations with questionable data risking to end up with a very coarse-meshed network of stations.

3.3 *Density of measuring points and record length*

At the end of a laborious collection of data and after a scrupulous check the calculations were based on about 2600 stations covering a good part of Europe. The average distance from station to station ranges from about 20 km to 70 km in different countries. Even if the points of recording are located far from one another, satisfying results may be obtained in areas with homogeneous topography and uniform climatic conditions (e.g. Sweden or Finland), whereas regions like the Alps or the Norwegian mountains presenting big variations on small scale ask for much more detailed data.

At some stations the recordings covered only a few years, but most often snow was measured for a period of 30 to 50 years. The longest record of 101 years was found in Potsdam near the German capital Berlin. As mentioned in the next paragraph, snow fall is treated as a steady state process, allowing utilization of snow records with gaps but only if the station has not been moved in the meantime. Long records representing many winters will, of course, assure most significant statistical results. 20 years may be considered as a minimum.

4 STATISTICAL MODEL

In an early stage of the research the idea was to develop a model in order to derive the amount of snow from meteorological data (precipitation, humidity of the air, temperature, wind, radiation etc.). Since a complete set of these data normally is not available, this idea was abandoned in favor of a simpler and more statistical approach which finally was successful.

The daily measured amount of snow on the ground is considered to be an independent random variate, ignoring any inter-correlation in time and space. Since at present any general trend indicating a continuous reduction of snow (e.g. due to the global warming) or, on the other hand, indicating an increase (e.g. due to a multiplication of extreme meteorological phenomena) is not clear either, the snow accumulated on the ground is taken as resulting from a steady state process in time.

4.1 Procedure of observation

The easiest way of recording snow is done by simply measuring the depth of the snow cover in a strictly vertical direction at a protected place not affected by snow drift.

In some of the countries the water equivalent of the snow cover was registered directly. This is done by using a cylinder to cut a certain area out of the snow cover and weighing the melted snow. Afterwards a correction is made if the height of the cylindrical snow sample is less than the depth measured in unaffected snow.

Usually snow is registered every day or at least every three days. After converting the snow depths into loads (or water equivalents) each station disposes of a great number of short-term (daily) snow load values at the end of the winter.

4.2 Annual maximum snow loads

The statistical analysis is not based on records of all short-term values but only on their annual maximum. At a given station every winter a new maximum value can be sorted out. These top values may be taken as a sample of all imaginable and randomly distributed winter maxima at this station (statistical basic population).

In some parts of Europe where snow fall occurs very seldom many winters don't bring any snow at all. But sometimes several snow events with complete melting in between are observed during the same winter. Sticking strictly to the reference period of one year could mean a loss of information. In these cases, the statistical analysis was no longer based on annual maxima but on the maxima of single snow events taking into account an average number of snow events per year (Sims et al. 2000).

4.3 Distribution function

Due to the fact that all these values are maximum values out of random short-term measurements the distribution may be expected to fit to one of the well-known extreme value distributions. In fact, quite often data were in good accordance with the cumulated distribution function (CDF) of an extreme value distribution FISHER-TIPPET type I (Gumbel 1958) but at some stations a lognormal or a WEIBULL distribution gave slightly better correlation factors (Sukhov 2000). The final decision in favor of the extreme value distribution type I (cf. equation 2) was taken in order to avoid inconsistencies caused by switching from one type of distribution to another within the same region. Furthermore the lognormal distribution most often results in an unacceptable high design load.

$$F(s) = \exp\left\{-\exp\left[-\alpha(s-u)\right]\right\} \qquad (2)$$

where
F(s)   is the type I cumulative distribution function,
s   is the snow load on the ground (kN/m²)
u   is a position parameter (kN/m²),
α   is the dispersion parameter (kN/m², α > 0)

The parameters u and α are found by means of the least square method. In a GUMBEL scaled plot the distribution is represented by a straight line. From such a graph it is easy to read the snow load which will have a given probability of not being exceeded in any one year.

At some stations the utmost point of the measured annual maxima (or the event maxima) is much higher than expected by the general trend of the distribution. These exceptional snow falls often are due to extraordinary weather conditions. If the single value had a strong impact on the parameters of the distribution it was omitted in the least square fitting. The framework of the European safety system allows to treat these extremely rare snow loads as an accidental load with reduced coefficients of safety (Sims et al. 2000).

4.4 Characteristic snow load

According to the safety regulations given in the European preliminary code ENV 1991-1 'Basis of Design' variable loads will be represented by a characteristic value. Paragraph 4.2(8) of this standard defines the characteristic value as the load which has a probability of 98 % of not being exceeded within any one year. This corresponds to a mean return interval (MRI) of 50 years.

Once the best fitting Gumbel distribution is found for a given station of observation, the characteristic snow load on the ground can be read from the plot or it may easily be calculated.

4.5 Problems: Snowless winters, small database

This paper can only give an overview of the general procedure applied during the research project. Some problems encountered in practical work had to be solved, for example, the treatment of stations reporting a lot of snowless winters. Including the zero values into the record of all maxima leads to an erroneous characteristic snow load. In fact, we have a mixed distribution consisting of a certain percentage of snowless winters and an extreme value distribution for the non-zero values. In very mild climates the few remaining winters with snow may not allow to derive a distribution. Using snow events

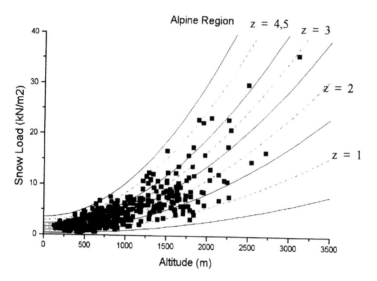

Figure 1. Scatter-plot of all stations in the Alpine Region
(graph due to JRC, Ispra)

zone 1 $s = 0.651 [1 + (A / 728)^2]$
zone 2 $s = 1.293 [1 + (A / 728)^2]$
zone 3 $s = 1.935 [1 + (A / 728)^2]$
zone 4.5 $s = 2.898 [1 + (A / 728)^2]$

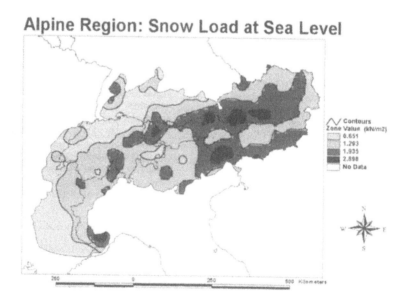

Figure 2. Example of a snow load map: Alpine Region
The given snow loads are valid for a theoretical altitude A = 0 m a.s.l.;
snow load for an altitude A see altitude functions given in Fig. 1

instead of years sometimes made it possible to still determine a characteristic snow load.

These problems will be dealt with more in detail by other authors (e.g. Sims et al. 2000).

5 ALTITUDE FUNCTION

According to a general opinion the amount of snow increases with the altitude of the site. This is true for large parts of Europe. Therefore many national codes do not present the snow load itself with all its variation with altitude, but they are mapping only a starting value and give a function which allows to calculate the snow load for an actual altitude. In order to find an appropriate type of altitude function for one region, the characteristic snow load has been plotted against the altitude for all stations. The correlation of this scatter-plot was very different from region to region. For the Alpine Region this plot shows clearly an increasing trend with altitude (Fig. 1).

In this case an applicable relation is given by a simple quadratic function. The parameters of the function can be optimised by the least square method. Further analysis in other parts of Europe revealed that the increase with altitude is not a general rule. In some regions the increase is only linear but in other areas the correlation between the snow load and the altitude of the place is scarcely distinguishable. This is the case in most areas of Norway and in Iceland. In those countries the snow load may be much more influenced by other circumstances as for example the distance to the sea than by altitude.

Finally, depending on the region in Europe, three very simple types of altitude functions are used:

$$S_{k.ground} = a \left[1 + \left(\frac{A}{b} \right)^2 \right] \quad \text{quadratic} \quad (3a)$$

$$S_{k.ground} = a + \frac{A}{b} \quad \text{linear} \quad (3b)$$

$$S_{k.ground} = a \quad \text{constant} \quad (3c)$$

where A = Altitude of the site; a and b are parameters allowing to adjust the function to a scatter-plot of stations. The different altitude functions obviously represent different climatic regions. In case of equation (3c) the snow load doesn't depend from the altitude, it has to be presented in maps directly.

6 REGIONALIZATION

Consequently, the research team had to divide Europe into different climatic regions. In each region a common relation between the characteristic snow load and the altitude applies. Forming coherent regions was simple in some parts of Europe (e.g. UK incl. Eire). In other parts it was only after several trials that the limits of the regions could be specified. In the end the following ten regions were identified:

Climatic Region	Type of altitude function
Alpine Region	quadratic
Central East (mainly Germany)	quadratic
Greece	quadratic
Iberian Peninsula	quadratic
Mediterranean Region	quadratic
Central West (mainly France)	linear
Sweden and Finland	linear
UK and Eire	linear
Iceland	constant
Norway	constant

The limitations of the regions often deviate from political borders. For instance the altitude function found for the South-East of France ("Provence") corresponds well to that one valid in most parts of Italy. Therefore, the Mediterranean Region was extended to France. The Alpine Region roughly limited by the contour at 500 m altitude includes not only the whole territory of Austria and Switzerland but also Southern Germany and the alpine regions of France and Italy.

7 ZONING

7.1 Procedure

The altitude function represents the average increase of the snow load with altitude for one climatic region. Some stations at the upper part of the scatter-plot (Fig. 1) obviously belong to especially snowy areas, whereas stations below the curve have less snow than the average of all stations situated at the same level of altitude. In order to take into account the possibility of deviations, the altitude function was split into a fan of similar curves covering all the stations in the scatter-plot. The curves divide the stations into different zones. The zone with the lowest number comprises the stations with the least snow. Mathematically, the splitting is done by varying the parameter a in the equations (3a...c) in a way that a is a linear function of the zone number z.

As a result every single station is located between two of these curves and may easily be classified into the corresponding zone. With regard to an easy handling only three or four zones have been chosen.

The characteristic snow load for all stations belonging to <u>one zone</u> will be given by the central altitude function situated between the upper and the lower limiting function. Inevitably some stations will be distant from the centre line towards the unsafe side (stations plotted above this centre line). This shortcoming is considered to be due to modelling and, consequently, will be covered by the safety factors.

All this sounds quite complicated. However, in practice it is very easy to handle. If an engineer needs the characteristic snow load on the ground for designing a roof construction he will simply have to sort out the region and the zone his building site belongs to by using a snow load map. Each zone is linked to an altitude function. Putting in the altitude of the site he immediately gets the characteristic snow load on the ground.

For those who had to elaborate the necessary maps defining the snow load zones geographically, the problem was much more difficult to solve.

7.2 *Maps*

The last step in setting up a European regulation for snow loads on the ground is the preparation of maps.

If the scatter-plot of the region contains the zone limiting curves, it is possible to specify the correct zone for each station. Since the fan of altitude functions varies unambiguously with the zone number z, it is possible to calculate the precise zone number (with decimals) which would give the exact snow load as determined by the previous statistical analysis for every station of this region.

Transferring these results onto a map of the region means that the target zone number is known at the location of all stations. Applying methods of spatial interpolation a continuous surface of the zone number may be calculated. At the location of the station the surface should give the correct zone value. Contour lines on this surface at the limiting zone numbers (e.g. z = 1.5 and z = 2.5 for zone 2) automatically represent the limits of a zone on the map. Using this method for every region a map can be traced automatically, showing the extension of the different zones. This computer work was done by the Joint Research Centre in Ispra (Italy).

Modifying the zones in the scatter-plot will have an impact on the complexity of the maps. The last version of the maps published recently (Sanpaolesi et al. 1999) are the results of several trials and improvements. Special aspects like interpolation near to the limits of regions or effects of smoothing procedures cannot be explained here.

7.3 *Verifications*

The maps have been tested at the locations of the stations. The characteristic snow loads found by using the map should equal those resulting from the statistical analysis of the recorded data. In general, the agreement is very satisfactory.

Another check was made along the border lines of contiguous regions. The correspondence between the snow load values on either side of the limits of regions is much better than those found so far in comparing the results of current national codes along political borders.

After all, characteristic snow loads read from the maps may even be better than those resulting from the statistical analysis for one single place. By the procedure of mapping disturbing local effects are partly compensated by the influence of neighboring stations. Sometimes, a big difference detected at a certain station led to recheck the data revealing single errors in the records.

8 RESULTS

Different climatic regions could be distinguished and delimited in Europe, characterised by a similar increase of the snow load with altitude. Maps has been traced for each region showing different snow load zones. The maps and the corresponding altitude functions can be used in a draft for a European snow load code.

The maps are based on snow measurements and after an appropriate choice of the zoning they may be plotted almost automatically. Nevertheless, the result has to be considered as a rough draft. Modifications and improvements are certainly possible in a continuous process of consultation with meteorological experts or experienced local observers.

9 CONCLUSIONS

Until now the different countries in Europe used heterogeneous ways to assess the snow load. Developing a common procedure applicable for all countries as it was done in this research has to be considered as a great step forward. The common procedure avoids shortcomings and allows for easy comparisons across national border lines.

This pre-normative research project has been completely sponsored by the European Commission. Apart from the obligation to establish European standards the Commission also had a second motivation for funding: by encouraging the cooperation of research institutions of different European countries problems of language and communication were to be tested. The fast communication by electronic mail and also the good cooperation of the research institutions were very helpful in respect of the tight schedule.

REFERENCES

CEN 1995. ENV 1991-2-3 Basis of design and actions on structures. Part 2.3: Snow loads. CEN Central Secretariat, Brussels.

del Corso R. et al. 1995. New European Code for Snow Loads; Background document. University of Pisa, Italy.

Gumbel E. 1958 (Reprint 1967). Statistics of Extremes. N.Y./London: Columbia University Press.

Sanpaolesi L. et al. 1998. Scientific support activity in the field of structural stability of civil engineering works-Snow Loads. Final Report Phase I. University of Pisa, Italy.

Sanpaolesi L. et al. 1999. Annex B to the Final Report Phase II: European ground snow loads map. Pisa, Italy.

Sukhov D. 2000. Statistical Investigation on Ground Snow Load in Germany. Proceedings of the 4th International Conference on Snow Engineering, Trondheim.

Sims et al. 2000. European Ground Snow Load Map: Irregular and Exceptional Snow Falls. Proceedings of the 4th International Conference on Snow Engineering, Trondheim.

Snow Engineering, Hjorth-Hansen, Holand, Løset & Norem (eds) © 2000 Balkema, Rotterdam, ISBN 90 5809 148 1

Snow load serviceability design with focus on the Nordic countries

Ivar Holand (1924 – 2000)
SINTEF, Civil and Environmental Engineering, Trondheim, Norway

Bernt J. Leira
Department of Marine Structures, Norwegian University of Science and Technology, Trondheim, Norway

ABSTRACT:
Codified design criteria relevant for the combined action of snow and other loads are reviewed with focus on the Nordic Countries. Formulations of the serviceability criteria as implemented in European and national codes are discussed.

To illustrate the derivation of the *characteristic, frequent* and *quasi-permanent* combination factors, which e.g. are required in the Eurocodes and ISO standards, two Norwegian data sets are applied. These data correspond to daily measurements for a continental snow climate. Sample distribution functions are applied directly, but fitted distributions are also considered. Furthermore, the quasi-permanent combination factor is estimated by a more elaborate procedure where also the time-scale of snow load fluctuations is represented in the model. The particular load combination of snow and wind is applied as an example.

1 INTRODUCTION

For the two basic reference documents, i.e. *CEN-ENV 1991-1 Basis of Design*, Ref./1/, and *ISO/FDIS 2394: 1998 General Principles*, Ref./2/, three different load combinations are applied. These are the Characteristic, the Frequent and the Quasi-permanent Combination. These are introduced in order to cover different types of consequences in relation to exceeding a given serviceability criterion. There seems to be some minor differences between the two documents as to distinction between the three combinations.

Procedures for derivation of the combination factors are mainly different for the Characteristic versus the Frequent and Quasi-permanent Combinations. Accordingly, procedures for derivation of the different factors are organized in two main categories:

- The Characteristic Combination which deal with the combination factor ψ_0. Three different procedures for derivation of this combination factor are proposed in the reference documents. For the most refined approaches, the time variation of the load processes enters the calculations.
- The Frequent and Quasi-permanent Combinations are investigated by consideration of cumulative probability levels for the short-duration maxima (i.e. daily measurements)

Derivation of these combination factors are illustrated for two particular Norwegian snow measurement stations. The combination factors implemented in national codes within the Nordic Region are first reviewed. The basic reference documents, Refs. /1,2/ and other relevant European codes are also discussed.

2 SNOW LOAD COMBINATIONS IN RELATION TO TIME VARIATION

In order to address the combination of snow loads with other loads, the variation of snow loads as a function of time needs to be considered. Equivalently, the duration of snow loads above a set of given threshold values is of relevance. The behaviour of the snow process is generally significantly different for a continental climate and a maritime climate. For the former, the snow cover is accumulated more or less continuously during the winter season. For the latter, the snow cover may in some cases last only for a few hours or a few days after each snow-fall.

The problem of load duration is not discussed in the ENV for snow loads, Ref./1/. The general problem of duration of snow loads is, however, described in Basis of Design, Ref/3/, Section 5. From Section 5.3 the following is quoted:

"Great differences in the representative values of ground snow loads are therefore expected as a function of the particular climatic regions. These obvious conclusions have not yet been checked through relevant research work valid for the whole European territory. Nor are normative indications available, except for the method presented by the Swiss standard SIA 160."

and further:

"In order to obtain such a relation, it is necessary to have at one's disposal daily measurements of snow depth over several years and in several places. For each meteorological station, the stochastic process is analysed and, for any fixed load level, the mean duration is assessed. It is then possible to plot the results on a graph with duration as abscissa and load as ordinate. Such studies have been performed for Switzerland alone."

3 OTHER EUROPEAN AND INTERNATIONAL STANDARDS

ENV 1991-1 Basis of design and actions on structures, Ref./2/ describes the principles and requirements for safety, serviceability and durability of structures. The standard includes the following definitions (see 1.5.3.14 and Annex A, A.4 (3)), that are particularly relevant for combination of loads:

- The characteristic value F_k chosen so as to correspond to a prescribed probability of not being exceeded on the unfavourable side during a reference period (for snow loads the probability is 0.02 and the reference period one year)
- The combination value $\psi_0 F_k$ is determined in such a way that the probability of combined action effect values being exceeded is approximately the same as when a single variable action only is present
- The frequent value of a variable action $\psi_1 F_k$ corresponds to the value which is exceeded either 5 % of the time or 300 times a year; the highest value should be chosen
- The quasi-permanent value $\psi_2 F_k$ corresponds to the time average or to the value with a probability of being exceeded of 50 %

The following boxed (not mandatory) values are specified for snow loads on buildings:

- Combination value $\psi_0 = 0.6$
- Frequent value $\psi_1 = 0.2$
- Quasi-permanent value $\psi_2 = 0.0$

ENV 1998-1-1, Design provisions for earthquake resistance of structures, Ref./4/, requires that variable actions are considered with their quasi-permanent values in seismic design situations. Under certain restrictions quasi-permanent loads shall also be considered in the mass evaluation.

ISO 2394, Ref./5/ is under revision, and an ISO/FDIS has been sent out for voting in 1997-98. The document contains an informative Annex F "Combination of Actions and Estimation of Action Values". In this annex the sections F3 "Estimation of combination values", F4 "Estimation of frequent values", and F5 "Estimation of quasi-permanent values" are of particular interest.

The application of infrequent, frequent and quasi-permanent values in design for serviceability is related to the time-dependent properties of the structural materials. The application may be illustrated by reference to concrete structures.

ENV 1992-1-1: Design of concrete structures, Ref./6/ requires for instance:

- For a quasi-permanent combination of loads a maximum design crack width of about 0.3 mm is specified, and the calculated sag of a beam, slab, or cantilever should not exceed the span divided by 250.
- For a frequent combination of loads all parts of the tendons or ducts shall lie at least 25 mm within the concrete in compression

Such requirements govern the choice of suitable definitions of representative values.

The ENV 1995-1-1 for timber structures, Ref/7/, also applies combination value, frequent value and quasi-permanent value and refers to ENV 1991-1 for the specification of these values. However, for serviceability limit states only the frequent combination is required to be used, whereas characteristic, frequent, and quasi-permanent combinations are specified in ENV 1991-1.

In addition, load-duration classes are defined as in Table 1 in accordance with traditions for timber structures. It appears that the classes defined in Table 1 are similar, but not identical to those defined above.

The load-duration classes are used for specifying a reduction of material strengths in the ultimate limit states and modified stiffness moduli for serviceability limit states. Modified stiffness moduli are determined by dividing the instantaneous values of the modulus for each member by the appropriate value of $(1 + k_{def})$. These values are specified for three different service classes depending on moisture (class 1 indoor, dry, class 3 outdoor, moist) and type of wood-based material as exemplified in an extract from Table 2 below.

Table 1. ENV 1995-1-1 Load duration classes
for timber structures (Table 3.1.6 in
original document)

Load-duration class	Order of accumulated duration of characteristic load	Examples of loading
Permanent	More than 10 years	Self weight
Long-term	6 months - 10 years	Storage
Medium-term	1 week - 6 months	Imposed load
Short-term	Less than 1 week	Snow[1] and wind
Instantaneous		Accidental load

[1]In areas which have a heavy snow load for a prolonged period of time, part of the load shall be regarded as medium-term

Table 2 ENV 1995-1-1.Values of k_{def}
for solid timber and glued extract from laminated timber.(Table 4.1 in original document)

Load duration class	Service class		
	1	2	3
Permanent	0,60	0,80	2,0
Long-term	0,50	0,50	1,50
Medium-term	0,25	0,25	0,75
Short-term	0,00	0,00	0,30

Table 3. Combination factor for snow load as a function of load level normalized 50-year value. Characteristic snow period is 15 days.

Weibull Shape Factor	Normalized Snow Load effect		
	0.25	0.50	1.00
3.0	0.30	0.40	0.55
2.0	0.15	0.30	0.50
1.0	0.00	0.15	0.40

Table 4. Combination factor for snow load. Characteristic snow period is 90 days

Weibull shape factor	Normalized Snow Load effect		
	0.25	0.50	1.00
3.0	0.30	0.50	0.65
1.0	0.20	0.35	0.50

The table illustrates the influence on deflections of classifying snow loads as medium-term or short-term loads. The modulus is reduced only when calculating deflections from, in our case, snow loads, and the reduction is particularly important for light roofs.

4 DENMARK

Daily water equivalents have been studied for a ten-year period 1971-80 at one station: Store Jyndevad, Ref /8/. DS 410, Addendum Snow loads 1988, Ref /9/, specifies a snow load on ground 1.0 MPa for the whole country and a combination factor $\psi_0 = 0.5$ for snow loads. The return period for snow load has not been specified in the standard, but a reference to /8/ implies a return period of 50 years.

5 FINLAND

Finland has for many years had an extensive recording programme for snow loads. The values presently used are based on the winter seasons from 1946/47 to 1995/96, and snow recordings each 1st and 16th in the winter months. A physical model has been used for interpolating daily values between the recordings, Ref /10/. A reliable database for analysis of load duration at various locations is thus available.

In load regulations in Finland, Ref /12/, snow loads have been reduced only in combination with wind loads. In that case a resulting combination with 50 years return period has been obtained by applying one load with 25 years return period combined with the magnitude of the other based on a 2 year return period. (However, instead of 25 years return period for wind, the values for 50 years return period have been used for simplicity).

Values for two years return period for snow are about equal to the mean value of annual maxima, which in southern Finland is approximately equal to 0.5 times the 50 years characteristic value, and in northern Finland about 0.6 to 0.7 of the characteristic value. In the regulations only one value 0.5 has been used, as a simplification.

Deflections are analyzed for an annual maximum equal to 0.5 times the characteristic value with 50 years return period (approximately equal to a two years return period). The total duration of snow load is assumed to be 6 months, with a maximum value after 5 months, and a duration curve determined by a linear increase and decrease.

In a National Application Document for ENV 1991-1, $\psi_0 = 0.7$ has been used for the whole country, Ref /11/.

6 NORWAY

Statistical parameters for duration of snow loads have not been studied.

Representative values for snow loads on buildings are in a new standard specified as follows:

- combination value ψ_0 = 0.7
- frequent value ψ_1 = 0.4
- quasi-permanent value ψ_2 = 0.2

Regulations concerning loadbearing structures in the petroleum activities, Ref /14/, specify reduced values of environmental loads directly, and not as a product of a characteristic load and a ψ-factor. Thus, in load combinations for temporary conditions during construction, installation and operation, reduced values of environmental loads (here mainly loads due to wind, waves and current) may be applied under specified conditions. Examples of duration and return periods applied for evaluating the reduced loads are:

• duration until three days with adequate weather forecast
• duration in excess of three days with a one year return period for the time of the year in question

7 SWEDEN

Statistical parameters for duration of snow loads have been studied in Ref /17/. The basic data are registrations of snow depths by SMHI (Sveriges meteorologiska och hydrologiska institut) the 15th and 30th each month, and maximum snow depth each month. The duration found is thus duration of a certain snow depth and not of a water equivalent.

The observation periods are

• 40 years for 114 stations
• 30 years for 49 stations
• 66 years for 10 stations

The reference includes a number of duration curves. The duration for a certain snow depth d is defined as the accumulated time T_d (of a total considered time T_0) where the snow depth is equal to or larger than d.

The relative duration is defined as

$$\Pi_d = T_d / T_0$$

The continuous duration for a certain snow depth d is in an analogous manner defined as the largest continuous time T_{dc} during the winter where the snow depth is equal to or larger than d. In the analyses, the snow depth has been assumed to be unchanged in the 15 days period without observations and to change discontinuously at the end of the period. The resulting duration diagrams are thus discontinuous. It has also been assumed that snow during a winter is continuous without zero periods in-between.

The relations between ψ_0 and Π_d have been taken as follows, Ref/16/ (ψ_0 is denoted ψ in the reference)

$$\Pi_d = 1 \qquad \psi_0 = 1.0$$
$$\Pi_d = 0.1 \qquad \psi_0 = 0.5$$
$$\Pi_d < 0.1 \qquad \psi_0 = 0.25$$

A linear interpolation is used between these values.

Load reduction factors are specified from $\psi_0 = 0.6$ for the snow load 1.0 MPa to $\psi_0 = 0.8$ for snow load 4.0 kPa Ref/17/. These values are based on studies of duration of snow loads Ref/15/ also considering the results in Ref /16/ presented above.

8 CALCULATION OF ψ_0 FOR NORWAY BASED ON SIMPLIFIED METHOD

The simplified method for derivation of ψ_0 is described in Refs/2,5,19/. Essentially, the Coefficient of Variation (CoV) of the annual extreme snow loads are required. In addition, the "characteristic number of snow periods" during each winter season is required. For an increasing number of snow periods (i.e. "repetitions of snow load" per winter) the value of ψ_0 will be decreasing. Application of a lower bound for the number of repetitions will accordingly give an upper bound for the combination factor.

45 stations are available for investigation. They are located on altitudes between 0 m and 687 m above sea level. The values of the CoV vary between 0.2 and 0.6 and show weak dependence with respect to altitude, see Figure 1. The stations can be subdivided into 2 groups: one with a minimum CoV = 0.3 (altitude < 300m), and the second with a minimum CoV = 0.2 (altitude > 300m).

The number of events per winter, r, can be chosen equal to 1 in order to be on the safe side. A Gumbel distribution has been applied as the basis for computation of the combination factor. The following values are then obtained for the two groups of stations:

CoV	r	ψ_0
0,3	1	0.6
0,2	1	0.68

Accordingly, a ψ_0 value equal to 0.6 is obtained for the low altitude stations, while a ψ_0 value equal to 0.7 is obtained for high altitudes. By application of a Weibull distribution instead of the Gumbel, these values are increased to 0.75 and 0.83.

9 EXAMPLE OF CALCULATION OF COMBINATION FACTOR ψ_0 FOR A PARTICULAR NORWEGIAN STATION

A more refined assessment of the combination factor related to wind and snow loads is performed by application of time-process models for the two loads. The basis for the analysis is the Borghes-Castanheta model as outlined in Refs /3,5/.

As an illustration of analysis related to the characteristic combination, the measurements from Blindern are applied as an example. The two load processes are normalized by their 20-year return period values. The cumulative distribution function for the snow load is calculated by employing the fitted Weibull function. The shape coefficient is assumed to be constant, and the two-week maxima are then obtained by taking the square-root of the distribution of monthly maxima.

For the wind load, the Weibull exponent of the distribution of annual wind-speed maxima is taken as 2.0, which implies that the Weibull exponent of the wind pressure becomes 1.0 (corresponding to an exponential distribution).

The basic time period for the wind load is taken to be 3 days. To establish the distribution function corresponding to a period of 15 days, the basic distribution is raised to the power of 5. The distribution function corresponding to a reference period of 1 year is subsequently computed by raising the 15 day distribution to a power of 12 (corresponding to a 6 month snow season).

The 0.95 fractile for the annual distribution of the combined load effect is now identified as 1.20. The expression identifying the proper value of the combination factor then becomes $(1. + \psi_0*0.5) = 1.20$, which results in $\psi_0 = 0.40$. Similar calculations have been performed for a range of values of the wind load contribution factor from 0.25 to 1.0.

The results are shown in Table 3 An upper bound of the combination factor is identified as 0.50 for the present analysis.

Similar calculations are performed for a range of Weibull shape factors for the snow load. The value of the combination factor generally increases for an increasing value of the shape factor.

Similar calculations have been performed with a characteristic snow period equal to 90 days. The results are presented in Table 4 and indicate that the combination factor increases, which complies with the results for the simplified analysis method. The upper value of the combination factor (i.e. 0.65) is also comparable to that obtained by the simplified method above.

10 CALCULATION OF COMBINATION FACTORS ψ_1 AND ψ_2 FOR TWO NORWEGIAN STATIONS

Examples of data samples for derivation of the combination factors ψ_1 and ψ_2 are given in Figures 2 and 3. These are based on daily measurements. The characteristic values corresponding to 50% and 95% fractiles can be identified directly from the present data samples. However, for stations with extremely short record lengths (i.e. just a few years) the sample values may be somewhat un-

reliable. The data samples correspond to "total time normalization", which is referred to as Model 2 in Refs/18,19/.

For all the samples, a Gumbel probability paper is employed to provide a uniform frame of reference. Fitted distributions represented by straight lines based on regression analysis are also given in each case. The values of the regression coeffients are given in the figures, and these are quite high for both stations (roughly 0.92). However, it is noted that the upper parts of the sample distributions are fitted less well than the lower part. A relevant issue in this context is obviously the stability of the upper part of the sample distribution for increasing record lengths.

The choice of probability paper giving the best fit to the sample will generally vary from one measurement station to the next. Other types of distributions than the Gumbel, such as Normal or Weibull, may give better fits in a number of cases. It must be noted that the relative ranking of the regression coefficients for the different types of distribution will generally also be different if the Model 1 approach is adopted (corresponding to a normalization of probabilities by the number of days with snow rather than the total number of days as for Model 2).

The combination factors obtained for the two stations are obtained as $\psi_1 = 0.46$,$\psi_2 = 0.10$ for Blindern, and as $\psi_1 = 0.59$, $\psi_2 = 0.17$ for Susendal. The reported values of correspond to the average value of the snow load rather than the 50% fractiles.

11 CONCLUSIONS

Existing international and national Nordic codes have been reviewed with respect to implementation of load combination factors relevant for snow load serviceability criteria has been performed. Derivation of the combination factors has been illustrated in relation to data from Norwegian measurements.

Typical values of the combination factor (corresponding to ψ_0) as implemented in the Scandinavian countries ranges from 0.5 to 0.7. This is largely in compliance with values obtained by application of statistical procedures. Values of the combination factors ψ_1 and ψ_2 are not generally implemented in the codes. However, in Ref./13/ values of 0.4 and 0.2 are given for these factors. Processing of data samples for the two Norwegian stations yields somewhat higher values than 0.4 for ψ_1 and somewhat lower values for ψ_2. Results from some of the measurement stations in Finland give values of ψ_1 as high as 0.67, Ref/19/.

Obviously, more detailed investigations are required in order to draw firm conclusions in relation to codified implementation. However, the proce-

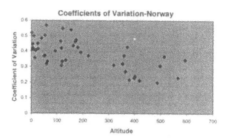

Figure 1. Coefficient of Variation of annual
snow load maxima versus altitude for 45
Norwegian measurement stations.

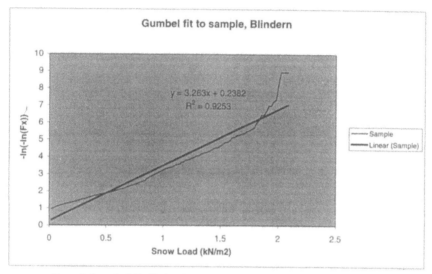

Figure 2. Snow load distribution function. Gumbel distribution fitted to sample for Blindern.

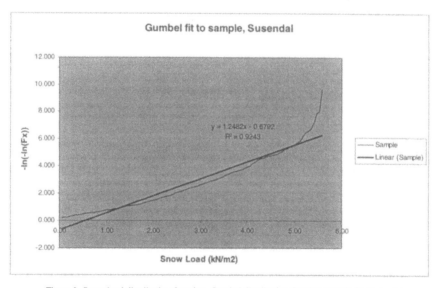

Figure 3. Snow load distribution function. Gumbel distribution fitted to sample for Susendal.

dures outlined can be applied for particular design and construction projects if high levels of accuracy are required. Analysis of relevant measured snow and possibly wind data is then required.

12 ACKNOWLEDGEMENT

Mr. Arve Bjørset is acknowledged for his contribution to processing of snow data at the initial stage of analysis.

13 REFERENCES

/1/ ENV 1991 Part 2-3 Actions on structures - Snow loads

/2/ ENV 1991-1 EUROCODE 1: Basis of Design and Actions on Structures. Part 1 Basis of Design 1994

/3/ New European Code for Snow Loads. Background Document. Proceedings of Department of Structural Engineering, University of Pisa, N° 264, 1995.

/4/ ENV 1991-8 EUROCODE 8 - Design provisions for earthquake resistance of structures - Part 1-1: General rules

/5/ ISO/FDIS 2394:1998(E) General principles on reliability for structures 1997

/6/ ENV 1992-1-1 Eurocode 2: Design of concrete structures. Part 1-1 General Rules and Rules for Buildings

/7/ ENV1995-1-1 Design of Timber Structures Part 1-1: General rules and rules for buildings

/8/ Allerup, P. et al: Snow Pack Maximum Water Equivalent in Denmark. Danish Meteorological Institute Climatological Papers No. 14, Copenhagen 1985

/9/ DS 410. Loads for the Design of Structures 3. Edition 1982 (Translation Edition 1983) Addendum Snow loads June 1988 (only available in Danish).

/10/ Kuusisto, E.: Snow accumulation and snow melt in Finland. Publication of the Water Research Institute 55. National Board of Waters, Finland, Helsinki 1984.

/11/ Nylund, A.: Note 1997-06-16 (in Swedish).

/12/ Finland's Code of Practice for Buildings and structures. Safety of Structures and Loads. Regulations 1983 (in Swedish).

/13/ NS3490. Design of structures. Requirements to reliability. First edition June 1999.

/14/ Regulations concerning loadbearing structures in the petroleum activities. Norwegian Petroleum Directorate 1992

/15/ Forsler et al.: Snow cover depth and water equivalent. (In Swedish) LTH Report 24, 1971.

/16/ Statens Byggnorm avd 2A, :5222 Statens Planverk (in Swedish)

/17/ Åkerlund, S: Snow and wind loads (In Swedish) Boverket 1994.

/18/ Leira,B.J; Sukhov,D. and Brettle, M.(2000):"European snow loading: serviceability load criteria", to be presented at Snow Engineering, Trondheim.

/19/ CEC, 1999. Scientific Support Activity in the Field of Structural Stability of Civil Engineering Works, Snow Loads. Comission of the European Communities DGIII-D3. Final Report, Phase II. University of Pisa. Department of structural engineering. Pisa.

Snow Engineering, Hjorth-Hansen, Holand, Løset & Norem (eds) © 2000 Balkema, Rotterdam, ISBN 90 5809 148 1

Estimation of snow load on a large-scale inclined roof of Tajima Dome

I. Kurahashi & A. Honda
Nagasaki R&D Center, Mitsubishi Heavy Industries Company Limited, Japan

T. Tomabechi
Hokkaido Institute of Technology, Sapporo, Japan

M. Fukihara
Kobe Shipyard and Machinery Works, Mitsubishi Heavy Industries Company Limited, Kobe, Japan

ABSTRACT: Tajima Dome is a newly opened dome structure, which looks like a mountain hut. The northern half of its roof is covered with steel plate. The southern half is movable and made of glass fiber cloth. At the stage of structural designing, snow load was considered to be dominant compared to other loads. The conventional way to estimate snow load is based on the accumulation of snow depth through winter season. However, it is not unusual to observe that snow on an inclined roof sometimes slips off to the ground. The authors took this phenomenon into account in estimating the snow load. This paper describes field observation of snow slipping in model roofs at the construction site, simple physical modeling of snow slip phenomenon and statistical analysis of snow load from meteorological data near the site.

1 INTRODUCTION

Tajima Dome is a newly opened dome structure in northern Hyogo Prefecture, Japan. The area is famous as a ski resort. The snow there is rather wet and estimating snow load was an important item for structural designing. The dome, which looks like a mountain hut as in Figure 1, is 60 meters high and 200 meters long.

In the view of the surface material, the roof can be classified into two parts. The northern half of its roof has multi-layered triangular-shape and is covered with Galvalume fluorine steel plate. The southern half is movable with large span and the material is 4-fluride-ethylene-resin-coating glass fiber cloth. The slope is also 30 degrees for both parts.

At the stage of structural design, snow load was considered to be dominant compared to other loads. The conventional way of estimating snow load in Japan is based on the accumulation of the depth of snow through winter season (AIJ 1993). However, it is not unusual to observe that snow on an inclined roof sometimes slips off to the ground, which suggests decrease of snow load. Therefore, the authors tried to estimate the snow load considering the 'snow slip' effect.

2 FIELD OBSERVATION OF 'SNOW SLIP'

'Snow slip' seemed to be affected by many factors, such as temperature, roof material and depth of snow. Therefore, field observation using a model roof was conducted during winter season from January 1996 to March 1998.

2.1 *Model and test arrangements*

Figure 2 shows the model for the roof made of glass fiber cloth. The model was set at the construction site and is 9 meters wide and 6 meters long. The whole model was supported by five load cells in order to monitor the snow load acting on the model.

For the steel roof, the model is 1 meter wide and 5 meters long. It was also installed at the site as shown in Figure 3. The surface consists of many layers. The detail of this test is described in Kurahashi et al. (1997).

In both of the two models, the slope was fixed to 30 degrees as in the prototype roof.

The atmospheric temperature, roof temperature, wind speed and wind direction was measured and recorded in personal computers. 'Snow slip' phenomenon were monitored using a video camera. The snow depth on the model roof and on the ground was also evaluated.

2.2 *Condition for 'snow slip'*

Figure 4 shows a typical 'snow slip' phenomena observed on the cloth roof. The slip is accompanied by the decrease of the snow load.

From field observation, the snow on the cloth roof model never slipped when the temperature was rather low. Once the temperature became high enough, the snow began to slip gradually and finally dropped off to the ground. The temperature was assumed to be the most dominant parameter for 'snow slip'.

Figure 5 show the time series of atmospheric temperature and snow load before and after 'snow slip.' Around 7:30 AM, the temperature began to in-

Figure 1. Bird's view of Tajima Dome

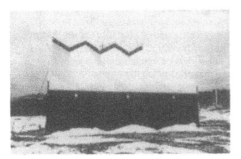

Figure 5. 'Snow slip' on glass fiber roof model

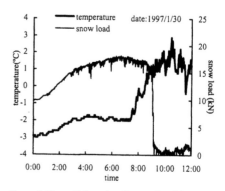

Figure 2. Roof model made of glass fiber cloth

Figure 6. Snow on steel roof

Figure 3. Roof model covered with steel plate

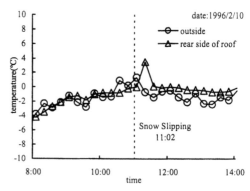

Figure 7. 'Snow slip' on steel roof model

crease and the snow on the roof began to decrease gradually. At 9:00 AM, almost all of the snow on the roof slipped to the ground in a short period and the snow load decreased to almost zero.

Figure 6 shows the snow accumulated on the steel roof model. In this type of model, 'snow slip' occurred less frequently compared to cloth roof.

Figure 7 shows the temperature both for the atmosphere and the steel roof. 'Snow slip' occurred at 11:02 AM, just after the roof temperature exceeded +1 °C.

Figure 4. 'Snow slip' on glass fiber cloth roof

3 PHYSICAL MODELLING OF 'SNOW SLIP'

'Snow slip' is understood to occur when the friction force between the snow and the roof decreases. This friction force seems to be affected by temperature, pressure and other factors.

From the field observation at the site, the critical condition for 'snow slip' was studied.

3.1 Snow slip on roofs made of glass fiber cloth

Figure 8 shows the snow depth on the roof and the atmospheric temperature, which was measured when snow slip occurred. The temperature was assumed to be the most dominant factor for snow slip.

As the snow depth increases, snow slip seems to occur at lower atmospheric temperature. However, the number of data is not enough to determine the effect of the snow depth. Except high snow depth condition, the critical condition for 'snow slip' was simplified as follows:

$$T > -1°C$$
where T is the atmospheric temperature.

3.2 Snow slip on roofs covered with steel plate

Figure 9 shows the critical condition in case of steel roof.

The temperature when 'snow slip' occurs seems to be higher than that of cloth roofs. The critical condition for steel roof was simplified as follows:

$$T > 2 °C$$

4 STATISTICAL ANALYSIS OF SNOW DEPTH

To determine snow load for structural design, maximum snow depth must be examined. Using long-term meteorological data measured at a weather station near the construction site, the incremental snow depth was estimated considering the effect of 'snow slip.' Conventional definition of incremental snow depth was modified by accumulating snow depth only when the temperature is below the critical value for 'snow slip.'

4.1 Annual maximum incremental snow depth

Meteorological data was available at a weather station 'Muraoka', which is located not far from the construction site.

The annual maximum of the modified incremental snow depth was plotted using Gumbel's distribution, and the value for 100-year return period was obtained by least square method as in Figures 10-11.

The design value at the station 'Muraoka' was shown in Table 1.

Table 1. Estimated snow depth at Muraoka

Material of roof	cloth	steel
Critical temperature	-1 °C	+2 °C
Annual maximum snow depth	88 cm	120 cm

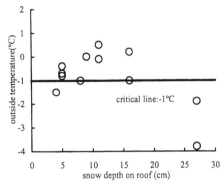

Figure 8. Critical condition for snow slip on roofs made of glass fiber cloth

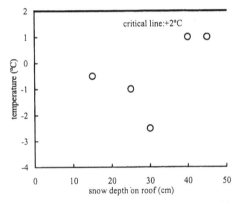

Figure 9. Critical condition for snow slip on steel roof model

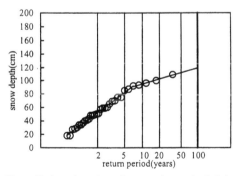

Figure 10. Annual maximum incremental snow depth (below +2 °C)

4.2 Correlation of snow depth between the meteorological station and the site

From the measured data of snow depth on the site ground, the correlation of snow depth between the weather station and the site was examined (Kurahashi et al. 1998b). The correlation is plotted in Figure 12. The snow depth at site was estimated to be 1.2 times of that at the meteorological station.

197

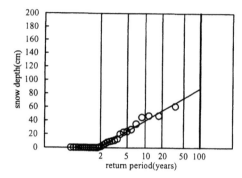

Figure 11. Annual maximum incremental snow depth (below -1 °C)

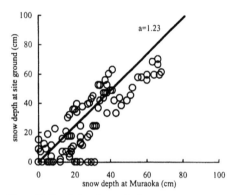

Figure 12. Correlation of snow depth

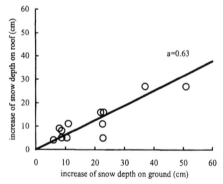

Figure 13. Snow depth on the ground and roof

5 EVALUATING SNOW LOAD

5.1 Snow depth on the ground and roof

During field observation of snow slip at the site, the snow depth on the model roof was found to be smaller than on the ground even before snow slip was observed. The ratio of snow depth was about 0.6 to 1 as in Figure 13.

The main reason of this difference was assumed to be the wind. Snow accumulated on the roof is

sometimes blown off during strong wind. AIJ (1993) suggests 'blown off factor' of 0.9 in this case. The value is dependent on roof shape.

In deciding the design load, this value was employed on the safe side. Further study would explain this difference to obtain more fair estimation of snow load.

5.2 Estimation of snow load

When the maximum snow depth on roof is estimated, design snow load can be obtained multiplying it by snow density. The authors assumed snow density considering the low temperature below the critical value for snow slip (Kurahashi et al. 1998a).

The consideration of snow slip in estimating snow load can be applied when the roof is inclined and the atmospheric temperature during snow season is high enough for snow slip. The critical condition must be decided from the roof material, slope and other factors.

6 CONCLUDING REMARKS

From the study of estimation of snow load considering 'snow slip', following conclusions were reached.

1 Snow slip phenomenon was examined in the field observation with roof models.
2 Condition for snow slip was obtained in a simplified form from observed data.
3 Maximum snow depth was estimated using meteorological data in a weather station.
4 Reduced design snow load was decided for Tajima Dome.

The above result was used in the structural design process and Tajima Dome was completed in 1998.

Because of low snow load compared to conventional estimation, the structural members supporting the roof are rather thin as a dome structure in a snowy area. This helps introducing more sunshine through the cloth roof as in Figure 14.

Figure 14. Inside View of completed Tajima Dome

Figure 15. Snow slip occurred in Tajima Dome

Since the opening of Tajima Dome, the roof has already experienced winter season twice. The snow on its roof has been reported to slip as expected thus far, as shown in Figure 15.

7 ACKNOWLEDGEMENT

The authors would like to pay special thanks to Prof. K. Ishii of Yokohama National University and Prof. H. Mihashi of Tohoku University for their advises on this study.

REFERENCES

Architectural Institute of Japan. 1993. Recommendations for loads on buildings. Tokyo: Maruzen.
Kurahashi, I., Tomabechi, T., Nagata, K., Fukihara, M., Tanabe, S. & Honda, A. 1997. A study on snow load for metal roof of Tajima Dome (in Japanese), Summaries of Technical Papers of Annual Meeting, Architectural Institute of Japan: 119-120.
Kurahashi, I., Tomabechi, T., Nagata, K., Fukihara, M., Tanabe, S. & Honda, A. 1998a. Examination of the snow load of large-scale building built to many snow area (in Japanese), Journal of Structural Engineering, Architectural Institute of Japan 44B: 95-100.
Kurahashi, I., Tomabechi, T., Nagata, K., Fukihara, M., Tanabe, S. & Honda, A. 1998b. Result of snow-load drop technique (in Japanese), Summaries of Technical Papers of Annual Meeting, Architectural Institute of Japan: 95-96.

Snow Engineering, Hjorth-Hansen, Holand, Løset & Norem (eds) © 2000 Balkema, Rotterdam, ISBN 90 5809 148 1

Application of mathematical planning to snow load for structural design

S. Kurita
Tohoku University, Sendai, Japan

H. Katukura & M. Izumi
Izumi Research Institute, Shimizu Corporation, Tokyo, Japan

ABSTRACT: A group including authors has formed to study and develop Mathematical Planning. It has proposed to build a network called Virtual Research Institute and made the prototype. It believes this method is applicable to almost all rational planning. Snow load for structural design should be determined rationally. In the paper, the concept of mathematical planning is explained and the application of the method to design snow load is described.

1. INTRODUCTION

It is self-evident that planning be rational. Rationality consists of mathematical and logical processes. Japanese structural design of buildings has relied on specification-based-design. Recently performance-based-design method is also accepted by the government. The new method requires more rationality than the old one. Otherwise, structural designers cannot escape from litigious society. Mathematical planning may give a solution to the new circumstance.

2. CONCEPT OF MATHEMATICAL PLANNING

2.1 Gap, problem and planning

When people find out gaps between a desirable state and the real state in daily trifle things or in important global matters, they recognize the gaps as problems. Then, they analyze the problems and think of the methods and steps to decrease the gaps. This act of thinking is the act of planning. The contrived methods and steps are called plans. A plan is selected out of the plans through decision-making process, and it is fruited to a real project via steps called ' design '.

2.2 Mathematical planning

Many data, information and knowledge are essential in the process of getting the plan, and the process can be called ' mathematical planning' only when it is ruled by the common language of science; mathematics and logic. There are some opinions like " planning is inherently rational, therefore it is always mathematical" ,or , on the contrary, "planning is an act of abduction, and it differs from mathematics where deduction and induction are basic concepts". Authors believe that planning is an act of repetition of analyses and syntheses and a plan is obtained through decision-making process, and that mathematical planning must exist. A group was formed to study the context and practical methods and several papers are already published .

2.3 Virtual research institute

For the real application of the mathematical planning the group presented the concept of VRI (Virtual Research Institute) ; a network where specialists and institutions of various field are connected, and, under a request of a sponsor who has a problem to be solved, VRI searches for information or knowledge useful for the planning in the connectors, and , when finds out it , pays for it and use it for analyses and syntheses. Information and knowledge are so arranged that almost all data and information are hide behind except those really available to the process of the planning. The technology to make knowledge into securities or knowledge-securitization may be an important study field in the near future. The prototype of the network has been made. The speed of computation may improved by study effect.

2.4 Analogy between FEM and mathematical planning

Mathematical planning may be used as an all round player like FEM (Finite Element Method) in applied mechanics. In FEM, even the deformation-nodal coordinate relation is an approximation, small-size elements can easily make a given form of a structure, and minimize the influence of the single element to the whole structure. Rigidity of elements and behavior of the whole structure is controlled by Energy Principles, that assures the reliable approximation method.

Planning is for the people, and the principle that rules the mathematical planning may the human essence. As for the human essence, there are various features, and different selections of features may construct different types of mathematical planning. Our group selected the 'selfish gene' as the principle. In a plan, the human selfishness is controlled under the allowable minimum level, and a fruit is obtained as the result of compromise. The plan that assures the highest allowable level may be appreciated.

In the fundamental theory of applied mechanics, constitutive equation plays important role which connects the mechanics to material-behavior. When it is expanded from linear relation to generalized non-linear viscoelastic relation including memory functions, the available range of the applied mechanics remarkably spreads, but the whole system loses the simplicity and clearness. In the mathematical planning, relation like the constitutive equation may exist in the relation between the plan and society. Expanded relation may increase the complexity in the plan. The mathematical planning has the structure shown in figure 1.

2.5 *Future estimation*

A plan is evaluated through out the durable period.(Design working life is more convenient terms in theoretical discussion.) In planning stage, the planner should assess the future conditions, that makes the accuracy of estimation very low and increase the risk. To the estimation of future events, the following three methods are often applied.

1. Assume an unchangeable circumstance. Most of stochastic processes and methods like System Dynamics belong to this category.
2. Assume tendency of circumstance-change toward the future on the bases of past experiences and empirical laws. Changes like ripples are usually neglected.
3. Assume a possible range of future states by a set of experts opinions.

Accuracy of estimation concerning future events depends on the events. For example, population-construction in an advanced country in the future may predict with a high accuracy, but the future economical activity of the same country is hard to estimate as it suddenly changes with an artificial and small trigger. Even very hard, one cannot neglect the future state of economy in planning.

3. DESIGN SNOW LOAD

3.1 *Design load*

Design load or load for structural design is a model of an actual load that is used to analyze the behavior of the model of real structure in target. Design load should be balanced to the model structure in both accuracy and simplicity , enable the structural designer to analyze the behavior of the model structure , and make him estimate the real behavior of real structure under real loading. Usually, the risk of damage & collapse and cost to be paid for the safety are important factors to determine the size of design load.

3.2 *Snow load applied to buildings*

Out of actual loads applied to houses and buildings, snow load has comparatively good characteristics. Prediction of heavy snow fall is possible, the acceleration is small and force direction is vertical except snow slide and avalanches, size of load is visible, and the premonitory symptom usually appears in the structure prior its collapse. There are some unfavorable properties such as long term load and viscoelastic character of the snow. Wind effects also make the model of snow load be complicated. As a whole, the structural designer can rationally determine the balanced point of this risk and return problem, because human-life protection is not very tight in comparison with such as seismic load. In winter, Japan is separated into two regions at the high backbone mountains; heavy snow area faced to the Japan Sea and dry area of the Pacific Ocean side. Even the dry area, it snows tentatively when a low pressure vortex moves from the South-west to north east along Japanese islands. Snow disaster takes place in both heavy snow and dry areas.

3.3 *Problems of design snow load in Japan*

Japan is located between the largest ocean and largest continent, and situated at plates-boundary. Very high population density on narrow available but unstable areas keeps high disaster potential. So, there are many laws, regulations, codes, recommendations and guide-lines concerning design loads including snow load, which have been revised to meet up-to-date needs. A new problem occurred from outside of Japan. Foreign designers who intend to work in Japan need to read through a piles of documents written in Japanese and they believe this is a kind of barriers to protect Japanese market of building design from foreigners. To eliminate the suspicion, Japanese Government decides to accept two kinds of design methods; specification based design and performance based design. Unfortunately there is no experience of the latter type of design methods used in high potential disaster regions, and there is such a possibility that buildings are designed by lawyers instead of engineers. MP(mathematical planning) method may save the situations, as it is possible to show the logical planning steps to public.

4 APPLICATION OF MATHEMATICAL PLANNING TO DESIGN SNOW LOAD

4.1 Laws and guide-lines

Laws and regulations provide the people the minimum requirement for the social safety. Funds for raising the level of the safety should effectively used, and design loads are to be decided under the comparison of the building risk with other social risks. Very rare and powerful load like seismic force to buildings might be neglected in the code, when the improvement of public health could save much more lives than those lost in earthquake disaster.
Guide-line and recommendation have no legal power, and usually provide bigger values of design load than those described in the law. But even guide-lines do not answer to the question like; Why this values of risk and return is rational ?

In the paper, a process on VRI concerning the determination of design snow load is explained. It is the designer's work to make his own process, except when he selects a ready made process.

4.2 An example of process on VRI

Authors have mentioned knowledge-securitization. In the future, an international 'knowledge-rating-organization' may be established. In VRI, connectors are selected for the time being. When a structural designer would like to use VRI to get design snow load in his performance-based-design, he first opens the free general information page on the net, where items related to design snow load are presented. In free information page, he can get national and local laws and regulations, when he input the site address. Repeating clicking, he can come into rich information which is not free. Data and information which he needs may be:

1 Basic Snow Load
In Japan, basic data are records of piled snow depth on the ground observed by the Meteorological Agency. This means, average density or the piled snow is essential when the depth is transformed into weight. But the density is usually not recorded. The agency measures the quantity of rain-fall also, but relation between rain quantity and snow depth is not clarified yet. The designer should give a look to density information. As for the depth, three types of information such as normal value related to the return period, modification factor for the locality (hillside, valley etc.) and site condition are necessary. In some sites other data like lateral snow pressure and setting pressure may also be required.
2 Snow Load on Roofs
Additional input data are size and shape of roof including chimneys and snow stoppers, roof-materials, and heat loss from the inside of the building(this can be neglected). The circumstance conditions at present and in future are also important. Expectable output information is main wind direction and average velocity, rough shapes of piled snow on the roof for different depth of snow on the ground. Lateral and setting effects are also shown, if requested.
3 Building Life Cycle Cost
Construction companies and the engineers have the data. Decreasing of strength of a building caused by aging effects are not necessarily provided well.
4 Environmental Problem
The building under designing cannot give a big trouble to its environment. The limitation of the trouble is some times shown in regulations and laws. Troubles caused by snow are such as roof snow slid down to other's or public sites, wasted snow on the road etc.. In urban area, snow should be kept on the roof, if the site is not wide enough.
5 Social and Economical Conditions in Future
Social change has made the removal of roof snow and reduction of the weight impossible. Economical conditions may be one of the most important factors , when risk and return problem is discussed. But no high accuracy can be expected in this field. From experts' opinions, designer may have some image of

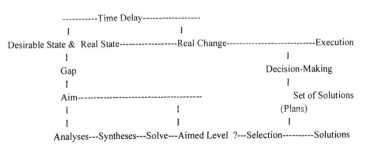

Figure 1. Structure of Mathematical Planning

the range where future economy may situate.

6 Insurance

Performance -based-design method only can be realized with the insurance system. Bigger design or construction companies have advantage to smaller ones. Well-use of the insurance system is recommendable to both owners and designers of buildings.

7 Legal Responsibility

The knowledge of legal responsibility may be essential in the new design method.

How to make documents may also be the output of VRI in the near future.

8 Risk and Return

The final output in this case is design snow load. Before getting the output, many comparisons are made for the time dependent risk values and correspondent returns. The value, strength, reparation costs of a building are functions of time and assumed risk. Present mathematical tool to solve this problem is the Monte Carlo Method. It is time consuming but can be applied to complicated problems. For the stable future conditions, VRI can give a solution.

5. CONCLUSION

Mathematical planning is a method that can be applied to various types of planning. Both the theory and practice is still insufficient, but this could be widely developed when it is opened to public.

REFERENCES

Izumi, M.,Kurita, S.,Katukura,H.&Nakai,S.1995. Mathematical planning. In Y.C.Loo (ed.); proc. 5thEASEC.,vol.1,913-918,25-27 July 1995. Gold Coast, Australia

Izumi, M. 1996. Mathematical science and mathematical planning. Bulletin of Tohoku Univ. of Art and Design, vol. 3, 52-63,23 July 1996, Yamagata, Japan

Snow Engineering, Hjorth-Hansen, Holand, Løset & Norem (eds) © 2000 Balkema, Rotterdam, ISBN 90 5809 148 1

European snow loading: Serviceability load criteria

B.J. Leira
Department of Marine Structures, Norwegian University of Science and Technology, Trondheim, Norway

D. Soukhov
Institut für Massivbau und Baustofftechnologie, University of Leipzig, Germany

M. Brettle
Construction Division, Building Research Establishment Limited, Garston, Watford, UK

ABSTRACT: Design checks which are relevant for serviceability criteria involving snow in combination with other loads are addressed. According to the relevant Eurocodes and ISO standards, three different load combination factors are required. These are typically referred to the characteristic, the frequent and the quasi-permanent combinations. Procedures for derivation of these factors are different for the *characteristic* versus the *frequent* and *quasi-permanent* combinations, all are considered in this paper. Derivation of corresponding combination factors is performed.

1 INTRODUCTION

The present study is concerned with procedures for derivation of load combination factors in relation to the serviceability limit state (SLS). For the two basic documents CEN (1994) and ISO (1998), three different basic combinations are proposed. These are the characteristic combination, the frequent combination and the quasi-permanent combination. These combinations are introduced in order to cover different types of consequences in relation to exceeding a given serviceability criterion.

In CEN (1994) the combinations considered for verification correspond to the following cases:

- Combination values for irreversible serviceability limit states.
- Frequent values for reversible serviceability limit states.
- Quasi-permanent values for reversible serviceability limit states and for calculation of long term effects.

In ISO (1998) these combinations are classified as follows:

- Characteristic combination when exceeding a limit state causes serious permanent damage.
- Frequent combination when exceeding a limit state causes local damage, large deformations or vibrations which are temporary.
- Quasi-permanent combination when long term effects are determinative.

There are also differences between the CEN (1994) and ISO (1998) regarding the structure of the combination format. This applies in particular to the characteristic combination. It is implicitly understood that the characteristic value of each load effect is to be applied in the load combination. This characteristic value is generally to be taken as that corresponding to a return period of 50 years, i.e. with a probability of exceedance of 0.02 when referring to the cumulative distribution of the annual maxima. The values of the partial coefficients γ_G and γ_Q are typically equal to 1.0 for the serviceability limit state.

Procedures for derivation of the combination factors are mainly different for ψ_0 versus ψ_1 and ψ_2. Accordingly, procedures for derivation of the different combination factors are organized in two main categories:

- The characteristic combinations which deals with the combination factor ψ_0. Three different procedures for derivation of this combination factor are described. One of the methods is applied extensively to snow data from a number of European climatic regions.
- The frequent and quasi-permanent combinations are investigated by consideration of the cumulative probability function for the short-duration maxima (e.g. one-day, one-week).

These procedures and corresponding results are summarized in the present paper.

2 DESIGN CHECK FOR SLS IN THE EUROCODE SYSTEM

2.1 *Combinations of actions*

The combination of actions to be considered for serviceability limit states depends on the nature of the effect of actions being checked, e.g. irreversible, reversible or long term. The main combinations are:

Characteristic (rare) combination:

$$\sum_{j\geq 1} G_{kj} + P_k + Q_{k1} + \sum_{i>1} \psi_{0i}Q_{ki} \tag{1}$$

Frequent combination:

$$\sum_{j\geq 1} G_{kj} + P_k + \psi_{11}Q_{k1} + \sum_{i>1} \psi_{2i}Q_{ki} \tag{2}$$

Quasi-permanent combination:

$$\sum_{j\geq 1} G_{kj} + P_k + \sum_{i\geq 1} \psi_{2i}Q_{ki} \tag{3}$$

where:

G_{kj} is the characteristic value of permanent action j;

P_k is the characteristic value of a prestressing action;

Q_{k1} is the characteristic value of the dominant variable action;

Q_{ki} is the characteristic value of the variable action i;

$\psi_0 Q_k$ is the combination value of the variable action;

$\psi_1 Q_k$ is the frequent value of the variable action;

$\psi_2 Q_k$ is the quasi-permanent value of the variable action.

ψ generally designates the combination and/or reduction coefficients. In these design checks, variable actions are defined as either dominant or non-dominant. The dominant action produces the most severe effect on the structure among those present.

2.2 *Combination factor ψ_0*

The combination value takes into account a reduced probability of simultaneous occurrence of the most unfavourable values of several independent actions. The combination value is determined in such a way that the probability of combined action effects being exceeded is approximately the same as when a single variable action only is present.

CEN (1994) establishes procedures of calculation of the combination factor ψ_0 for the ultimate limit states (ULS). This factor is also to be applied for the serviceability limit states, but with a different partial load factor. Accordingly, the resulting combination value $\gamma_Q \times \psi_0 \times Q_k$ contains the partial safety factor γ_Q

which is equal to 1.5 for the ULS and 1.0 for the SLS.

The combination factor should be calculated for all pairs of combined loads (for example for snow the most important combination is snow with wind). The actions are normally considered as stochastic process (Borghes-Castanheta, pulse-process etc). This requires substantial computation efforts due to numerical integration of probability functions or use of First Order Reliability Methods (FORM).

The CEN (1994) and ISO (1998) also propose to use simplified procedures which consider only the load itself (for example snow load). In this case the combination factor can be defined as:

$\psi_0 = Q_{non\,dom} / Q_{dom}$

where $Q_{non\,dom}$ = the design value of the snow load when the last one is the non-dominant load; Q_{dom} = the design value of the snow load when the last one is the dominant load.

The probability of exceeding the design value for the variable action when it is dominant is expressed as:

$\Phi(-0,7\times\beta) = 1 - 0.996 = 0.004$

where $\beta = 3.8$ is a reliability index and $\Phi(.)$ is the standardized Gaussian distribution function.

The probability of exceeding the design value of variable action when it is non-dominant (the combination value of variable action) is correspondingly equal to:

$\Phi(-0.4\times0.7\times\beta) = 1 - 0.856 = 0.144$

The combination factor ψ_0 can be defined either by means of Turkstra's rule or by means of Design Value Method.

2.3 *Factor for frequent value ψ_1*

The frequent value is determined such that:

– the total time, within a chosen period of time, during which it is exceeded for a specified part, or
– the frequency with which it is exceeded,
is limited to a given value.

The part of the chosen period of time or the frequency should be chosen with due regard to the type of construction works considered and the purpose of the calculations. Unless other values are specified the part may be chosen to be 0.05 or the frequency to be 300 per year for ordinary buildings.

For snow the duration of load above a given load level should be used as only criterion. For simplification it is also suggested that only the total (as opposite to continuous) duration during a year is considered. The purpose is to establish "duration-over-the-threshold" curves (curves showing the total time related to one year, during which the load is above a specified threshold). The frequent value is determined such that fractile of time during which it is exceeded would be equal to 0.05 (a basic case). Additionally, fractiles ranging from 0.01 to 0.10 can be relevant.

2.4 Factor for quasi-permanent value ψ_2

The quasi-permanent value is so determined that the total time, within a chosen period of time, during which it is exceeded is a considerable part of the chosen period of time. The part of chosen period of time may be set as a value of 0.5. The quasi-permanent value may also be determined as the value averaged over the chosen period of time.

3 PROCEDURES FOR DERIVATION OF THE COMBINATION FACTOR ψ_0

3.1 Introduction

The combination factor ψ_0 is to be applied on the snow load effect when the dominating load effect is due to some other external load, such as wind. Accordingly, a derivation of this combination factor strictly requires a refined modeling of both the snow and wind load. This also includes modeling of their variation with time. However, procedures based on such a refined modeling are typically time consuming with respect to collection of input data, numerical algorithms and computation time. As a consequence, simplified procedures are described in the CEN (1994) and ISO (1998) documents for derivation of this combination factor.

In the present investigation, three different categories of methods have been employed:

– Simplified methods based on the ratio of two design values of snow load: i.e. when it is dominant and when it is non-dominant.
– Modeling of the time variation for the two load effects by means of step-wise constant values. The characteristic time intervals for the two load effects are generally different from each other (referred to as the Borghes-Castanheta model).
– Modeling of both load effects as stochastic time depended processes with continuously changing intensity. This model is here referred to as up-crossing rate analysis.

Results obtained by application of these three methods are presented below.

3.2 Derivation of combination factor based on simplified methods

3.2.1 Turkstra's rule

According to ISO (1998), Annex F the combination factor based on Turkstra's rule can be found as:

$$\psi_0 = \frac{F_{Q_{max}}^{-1}\left\{\Phi(-0.4\alpha_s\beta)^r\right\}}{F_{Q_{max}}^{-1}\left\{\Phi(-\alpha_s\beta)\right\}} \tag{4}$$

where:
α_s- sensitivity factor for actions (equals -0.7 according to ISO (1998) and CEN (1994))
β - reliability index, equals 3.8 for design life of 50 years according to CEN (1994)
Φ - Gaussian normalized distribution
r - the number of an independent load repetitions during the reference time
Q_{max} the maximum value of action Q during the reference time
F_{Qmax} probability distribution function of Q_{max}

3.2.2 Design value method

According to ISO (1998), Annex F and CEN (1994), Annex A the combination factor based on the design value method may be found as:

$$\psi_0 = \frac{F_{Q_{max}}^{-1}\left\{\Phi(0.4\beta_c)^r\right\}}{F_{Q_{max}}^{-1}\left\{\Phi(\beta_c)^r\right\}} \tag{5}$$

where $\beta_c = -\Phi^{-1}[\Phi(\alpha_s\beta)/r]$ is the modified reliability index.

The specific expressions resulting from Turkstra's rule and the design value method are established for three different types of distributions:
1 Extreme value distribution type I for maxima (Gumbel),
2 Extreme value distribution type III for minima (Weibull),
3 Log-normal distribution.
Only the results for case 1 are reported here. Further details can be found in the report CEC (1999).

3.2.3 Explicit expressions for extreme value distribution, type I for maxima (Gumbel)

Applying the Gumbel distribution for Turkstra's rule (Equation 4) and using the moment method for determination of the distribution parameters the combination factor can be found as follows:

$$\psi_0 = \frac{1-0.78V\left\{0.577+\ln\left[-\ln(\Phi(-0.4\alpha_s\beta))\right]+\ln r\right\}}{1-0.78V\left\{0.577+\ln\left[-\ln(\Phi(-\alpha_s\beta))\right]\right\}}$$

where V = the coefficient of variation of the variable action (e.g. snow).

Applying the Gumbel distribution for the design value method (Equation 5) and using the moment method for determination of the distribution parameters the combination factor can be found as follows:

$$\psi_0 = \frac{1 - 0.78V\{0.577 + \ln[-\ln(\Phi(0.4\beta_c))] + \ln r\}}{1 - 0.78V\{0.577 + \ln[-\ln(\Phi(\beta_c))] + \ln r\}}$$

3.2.4 Analysis of statistical data from 10 different climatic regions in Europe

The design value method give results for the combinations factor ψ_0 which are slightly conservative in comparison with Turkstra's rule. Furthermore, the extreme value distribution Type I (Gumbel) is found to give the best fit to measured data for most climatic regions. Therefore it is decided to calculate combination factor by means of the design value method, using the extreme value distribution Type I (Gumbel).

In different parts of Europe the climatic conditions are different. The combination factor should accordingly also be different for these geographical parts. To calculate ψ_0 factors, the coefficient of variation of the annual maximum snow load (designated by CoV) and the number of load repetitions (designated by r in the following) are used. The last quantity can be obtained using the snow data collected during the work "European snow loads research project" (see reports CEC 1998, CEC 1999). The values of the ψ_0 factors will generally also vary with altitude inside each region.

The information about values of CoV and r can be found in CEC 1999. The resulting combination factors are summarised in Table 1.

3.3 Derivation of combination factor based on Borghes–Casthaneta model

3.3.1 Basic model assumption

The specific load combination related to snow in conjunction with wind is considered. In order to apply the Borghes-Castanheta model, the length of the characteristic time scale for the snow loads must be a multiple of the scale for the wind load. Setting the latter e.g. equal to 3 days and the former equal to 15 days, a factor of 5 is obtained. Furthermore, the extreme dynamic wind load is assumed to act constantly throughout the characteristic wind interval.

For simplicity, the dynamic wind component is represented by a single gust factor. This represents an approximation on the conservative side. However, for the purpose of load combinations it is believed to be sufficiently accurate.

The basic time varying load (or load effect) to be analysed can then be expressed as:

$$S(t) = a_1 \times Q_{snow}(t) + a_2 \times [Q_{wind\,(static)}(t) + Q_{wind\,(dyn)}(t)] \tag{6}$$

where a_1 and a_2 are fixed constants, the ratio of which determines the relative scaling of the snow and wind loads.

In addition to the intrinsic time scales, cumulative distribution functions of the loads are also required in order to perform a load combination analysis. For the static wind load, a Weibull distribution is frequently employed. The extreme dynamic wind load referred to stationary wind conditions is represented by a single gust factor.

For the snow load, it is generally found that Gaussian, Gumbel, Weibull or exponential distributions can be employed for daily snow loads. In the present example, a Weibull model is used.

The Borghes-Castanheta model is based on a simplified time variation of the load processes. The properties of the following simplified expression is investigated:

$$S = a_1 \times Q_{snow} + a_2 \times \max_{n=nref}[Q_{wind\,(static)} + Q_{wind\,(dyn)}] \tag{7}$$

where $n = n_{ref}$ is the number of repetitions of characteristic wind load "time intervals" within each characteristic snow load "time interval" (e.g. n is equal to 5 if the snow interval is two weeks and the wind interval is 3 days, as mentioned above).

3.3.2 Methodology

Given the present simplifications, numerical integration is performed to obtain the relevant probability functions. The reason is that closed form expressions can not be obtained. The computational procedure can be described in terms of the following four steps:

1. Establish cumulative distribution functions (CDF) for snow and wind loading for the basic reference periods.
2. Compute the cumulative distribution function for the maximum value corresponding to $n = n_{ref}$ repetitions of the compound variable from step 1.
3. Compute the distribution function for the total sum of the two main terms assuming independence between snow and wind loads. This involves calculation of a convolution integral.
4. Compute the distribution function for the maximum value of the sum S obtained in step 3, corresponding to a given number of repetitions. The number of repetitions corresponds to the chosen reference period for evaluation of the combined load effect.

In step 4, the reference period is chosen as one year. Two different characteristic intervals for the snow load are considered : one is 15 days, and the other is 90 days. The number of repetitions of the "snow interval" during one year is then equal to 12 for the first case an 2 for the second case if the snow season is set to 6 months (which is a representative duration for a continental climate).

The basic scheme for derivation of the combination factor itself (i.e. ψ_0) can subsequently be formulated as:

1 Calculate normalized values for the snow and wind load effects for the 50 year return period.
2 Compute the value of the combined load effect corresponding to the 50 year return period based on the CDF obtained from step 4 above.
3 This value of the combined (and reduced) load effect is divided by the sum of the normalized values from point (1) and the resulting combination factor is obtained.

For example: if the wind load effect is normalized to 1.0 and the snow load effect to 0.5, the sum of these values becomes 1.5. If the reduced value of the combined load effect from step 4 is equal to 1.2, then the combination factor will be found from the expression: $1 + 0.5 \times \psi_0 = 1.2$. Hence, the resulting value of ψ_0 becomes 0.4.

Steps (1), (2) and (3) are accordingly repeated for each new set of values for the normalized load effects. The value of the combination factor will also vary as a function of the ratio between the normalized wind and snow loads. For code checking purposes, a representative high value or an upper bound should generally be employed.

3.3.3 Numerical examples

The combination factor has been computed for a range of values of the normalized load effects. Cases where the wind load effect is dominating while the snow load effect is secondary has been addressed. This implies that the derived combination factor applies to the snow load.

The CDF for the snow load is calculated by employing the fitted Weibull function. The shape coefficient for the characteristic interval is varied from 1.0 to 3.0 in order to cover a range of relevant cases.

For convenience, the normalized 50 year wind load effect value is set to 1.0. The normalized snow load effect is varied at levels 0.25, 0.50, 0.75 and 1.0 of the wind effect. The first case hence represents negligible snow loads, while for the last case the snow load is of equal magnitude to the wind load.

Results for the case with a characteristic snow period of 15 days and a characteristic wind period of 3 days are given in Table 2 . The number of wind load repetitions for each basic snow period is accordingly equal to 5. The number of repetitions of the com-

Table 1: ψ_0 values for different regions in CEN members area

	Region	ψ_0	Altitude/Area
1	Alpine	0.65	above 1000 m
		0.5	below 1000 m
2	UK and Eire	0.4	
3	Iberian	0.5	above 500 m
		0.4	below 500 m
4	Mediterranean	0.5	
5	Central East	0.55	above 500 m
		0.4	below 500 m
6	Central West	0.4	
7	Greece	0.5	
8	Norway	0.7	above 300 m
		0.6	below 300 m
9	Finland - Sweden	0.65	above 250 m
		0.6	below 250 m
10	Iceland	0.6	

Table 2. Combination factor for snow load as a function of normalized 50 year value, if the 50 year wind load value is equal to 1.0 (Basic snow interval of 15 days).

Weibull shape factor	Normalized snow load effect			
	0.25	0.50	0,75	1.00
3.0	0.30	0.40	0.50	0.55
2.0	0.15	0.30	0.40	0.50
1.0	0.00	0.15	0.30	0.40

Table 3. Combination factor for snow load as a function of normalized 50 year value, if the 50 year wind load value is equal to 1.0 (Basic snow interval of 90 days).

Weibull shape factor	Normalized snow load effect			
	0.25	0.50	0.75	1.00
3.0	0.30	0.50	0.60	0.65
1.0	0.08	0.35	0.45	0.50

Table 4. Combination factor calculated by means of upcrossing rate.

Dominating load effect	Combination factor for non-dominating load effect
Snow	0.55
Snow and wind are of the equal magnitude	0.63
Wind	0.37

Table 5: ψ_1 and ψ_2 values for different regions of Europe

	Region		ψ_1	ψ_2
1	Alpine:	above 1000 m	0.45	0.10
		below 1000 m	0.30	0.00
2	UK and Eire		0.05	0.00
3	Iberian:	above 500 m	-	-
		below 500 m	0.00	0.00
4	Mediterranean		0.10	0.00
5	Central East:	above 500 m	0.40	0.10
		below 500 m	0.20	0.00
6	Central West		0.05	0.00
7	Greece		0.00	0.00
8	Norway:	above 300 m	0.50	0.20
		below 300 m	0.40	0.20
9	Finland - Sweden		0.50	0.20
10	Iceland:	South-west	0.20	0.00
		North-east	0.40	0.10

bined load effect per year becomes 12. The Weibull shape factor for the snow load is varied at the levels 1.0, 2.0 and 3.0.

Corresponding results for a case with the characteristic snow period equal to 90 days (i.e. 3 months) are given in Table 3. The number of repetitions of the wind load effect per characteristic snow period accordingly becomes 30. The number of repetitions of the combined load effect per year becomes 2. Two values of the Weibull shape factor are considered for this case: 3.0 and 1.0.

3.3.4 *Conclusions*
Considering the examples given above the following conclusions can be made:

1 The combination factor increases with increasing values of the Weibull shape factor (which implies a decreasing value of the Coefficient of Variation).
2 The combination factor increases with increasing values of the normalized snow load relative to the normalized wind load.
3 The combination factor increases with increasing length of the characteristic snow time scale (relative to the characteristic wind time scale of 3 days).

In relation to the last point, it is necessary to note that increasing length of the snow time scale implies a reduced number of repetitions of the combined load effect per year.

3.4 *Derivation of combination factor by upcrossing rate analysis*

3.4.1 *Basic model assumptions*
The following basic model assumptions are employed:
− The wind and snow load effect processes are both non-stationary (slowly varying) processes which are mutually independent.
− A long term period can be considered as a sequence of short term conditions (duration in the range of 1 to 3 hours) for which the wind load process can be considered stationary, and during which the snow load is constant.
− The wind velocity process is assumed to be Gaussian for each short term stationary condition.
− Weather systems with a time separation of three to four days are considered independent from each other.
Furthermore, since high levels of the combined and single processes are considered, the extreme value distributions are determined by the upcrossing rates for given levels.

3.4.2 *Methodology*
The extreme value distribution for the combined process is estimated by application of the upcrossing rate for the wind process, conditioned on a given level of the snow load. Subsequently, integration with respect to the probability distribution function of the snow process is performed.

The extreme value distributions for the individual processes are obtained by computation of the average upcrossing rate (for the total duration considered and for each given extreme value level). Details of the analysis are given in Naess & Leira (2000).

3.4.3 *Numerical example*
As example the climatic station Blindern, near Oslo is considered. A Rayleigh distribution is employed for the hourly mean wind speed. For the snow load, the monthly maxima are modelled by a Weibull distribution with scale parameter 685 and shape parameter 1. The distribution of daily maxima is subsequently back calculated from these parameters. For the calculations, a return period of 20 years is employed.

Resulting values of the combination factor are shown in Table 4.

Obviously, similar calculations need to be undertaken for other snow load distributions (and for Weibull with other shape factors) to generalise these results.

3.5 *Comparison of the results obtained by different methods and conclusions*

From the results above (Tables 1-4), the following conclusions can be made :

1 The combination factors for snow load obtained from the Borghes-Castanheta and upcrossing rate models coincide very well for continental climate if snow and wind are of equal magnitude and shape factor is equal to 3.0 (value of 0.65 from Table 3 versus value of 0.63 from Table 4). For other cases values from Borghes-Castanheta model are smaller.
2 Values of 0.6-0.7 from the simplified method (see Table 1) are obtained for continental climates. They coincide well with values of Borghes-Castanheta model (Table 3) for shape factor 3.0. These are equal to 0.6 if the snow load effect is 75% of the wind effect, and to 0.65 if the snow and wind effects are of equal magnitude.
3 Values of 0.6-0.7 from the simplified method coincide also well with the upcrossing rate model (value of 0.63 from Table 4) if snow and wind load effects are of equal magnitude.

Finally, the upper bound value for combination factor of snow load should be of the order of 0.7. This value is obtained from simplified method and from Borghes-Castanheta model for the highest value of the shape factor. This is also in agreement with the highest value from the upcrossing rate method. This value would apply to continental climate where the snow load during the winter season is mostly non-zero. For maritime climates where the snow loading is intermittent during the winter the reduced value of 0.5 - 0.6 may be used.

4 DERIVATION OF THE FACTORS ψ_1 AND ψ_2

4.1 *Factor for frequent value ψ_1*

For derivation of this factor, the duration of snow load above given load levels must be considered. The "duration-over-the-threshold" curves can be established either by statistical model 1 or by statistical model 2 as discussed below, see e.g. ISO(1998), Brettle et al (2000).

4.1.1 *Model 1*
The following data should be available:
- the record period of N years for daily snow depth (or load),
- the total number of days k per N years when snow covers the ground,
- the number of days n per year when snow cover is possible due to climatological conditions (if no data from meteorological sources are available then n can be set as 365 days - the total number of days per year).

4.1.1.1 Method of order statistics
All k values of snow depth (load) should be ranked in order and then cumulative distribution function(CDF) from k points should be obtained, i.e. P_k (x), probability of not exceedance of each snow value x, is calculated.

The probability distribution function for all days $P_{nN}(x)$ (including the non-snow days) can be then obtained through the function $P_k(x)$ as:

$$P_{nN}(x) = q + p \times P_k(x) \qquad (8)$$

where $p = k / nN$ is the probability of snow presence, $q = 1 - p$ is the probability that no snow is present.

To obtain the fractile of probability t of the distribution $P_{nN}(x)$ it is not necessary to compute this by Equation 8 but simply calculate a fractile of already existed function $P_k(x)$ with probability g which can be defined as:

$$g = (t - q) / p \qquad (9)$$

For example for $t = 0.95$, $p = 0.8$ and $q = 0.2$ the probability g will be equal to:

$$g = (0.95 - 0.2) / 0.8 = 0.9375 \qquad (10)$$

Thus fractile of this probability can be calculated using the distribution $P_k(x)$.

4.1.1.2 Bins method
The maximum value of snow depth (load) which occurs during the period of observation should be subdivided into m bins and probability that snow exceeds each of m levels is calculated. The difference with the method of order statistics is only that CDF based not on k observed values but consists of m steps (according the number of bins) and represents the probability of non exeedance of each of the m levels. Equations 8-10 are valid in this case.

4.1.2 *Model 2*
According to this method the CDF is obtained by consideration of all days including zero-days.

4.1.2.1 Method of order statistics
All nN values of snow depth (load) should be ranked in order and the CDF should be obtained. However, only k values are non-zero . The $(nN - k)$ values are equal to zero. Therefore the CDF begins with the value $(1 - k / nN)$ at point $x = 0$. The t-fractile is calculated directly from obtained CDF (i.e. $P(x)$, probability of not exceedance). The probability of exceedance is:

$$Q(x) = 1 - P(x) \qquad (11)$$

It is possible that the value $(1 - k / nN)$ will be greater than 0.95 (especially for maritime and mixed climate). Therefore the use of different probability plots to estimate these values was investigated. From these investigations it was concluded that the use of the exponential distribution gave results in which confidence could be placed.

4.1.2.2 Bins method
This method described above in section 4.1.1.2 is applied here but based on all nN values. The probability of non-exceedance of the first level (which is equal to maximum of snow load/depth divided by number m) is the probability that no snow is present and equal to $(1 - k / nN)$.

4.1.3 *Calculation of factor ψ_1 for frequent value*
The frequent value can be obtained directly as 0.95-fractile (by Model 2) or with the help of Equations 9-10 (by Model 1). The other fractiles from 0.9 to 0.99 can be also considered. The ratio of this fractile to the characteristic value of snow load is the coefficient ψ_1 for calculation of frequent value of snow load.

4.2 *Factor for quasi-permanent value* ψ_2

4.2.1 *Model 1*

Again the both method of order statistics and bins method can be used.

Using the same CDF as in section 4.1.1 and Equation 9, the 0.5-fractile of the snow load (depth) can be calculated based on the transformed probability:

$$P_{0.5} = (0.5 - q) / p \qquad (12)$$

This fractile is the quasi-permanent value of the snow load. The ratio of this fractile to the characteristic value of snow load is the coefficient ψ_2 for calculation of quasi-permanent value of snow load.

4.2.2 *Model 2*

The procedure is the same as in section 4.2.1. The CDF obtained in section 4.1.2 is used and the 0.5-fractile of non-exceedance of a given snow load (depth) can be directly calculated from it.

4.2.3 *Value of snow load averaged over the chosen period of time*

If all days in a year are used as a possible period of snow layer on the ground ($n = 365$ days) then 0.5-fractile is automatically equal to zero if snow lays less than 180 days per year (6 months). In this case another procedure for calculation of ψ_2 can be used. Quasi-permanent value is determined as value averaged over the chosen period of time. This period of time is normally chosen as one year (365 days). The ratio of this averaged value to the characteristic value of the snow load is the coefficient ψ_2 for calculation of the quasi-permanent value of the snow load.

This procedure is especially preferable in regions with heavy snow (e.g. mountains) where it gives more realistic results than the 0.5-fractile method.

4.3 *Summary of results for* ψ_1 *and* ψ_2

The procedures described in section 4 were applied to approximately 60 climatic stations across Europe. The stations were chosen to try to represent all climatic regions in the CEN area. The results are given in Table 5.

5 CONCLUSIONS AND SUMMARY

Three different procedures of varying complexity for derivation of the combination factor ψ_0 are outlined. These are:

1 A simplified procedure (CEN / ISO approach).
2 Borghes-Castanheta model.
3 An upcrossing rate method.

For procedures (2) and (3) the specific load combination of wind and snow is considered. The following conclusions can be made:

- Three procedures give comparable results. However, only the latter two models reflect that the combination factor increases as a function of the ratio between the snow load versus the wind load.
- The simplified method typically gives an upper bound for the combination factor.
- If wind data are not available then the simplified method can be used for rough calculations with quite a good accuracy and the representative values for various climatic regions can be obtained. The results obtained are found in Table 1.

Two different models for derivation of the ψ_1 and ψ_2 factors are proposed. For model 1, the CDF is based only on days with non-zero snow values. For model 2 all days are used to obtain the CDF. After computation of the CDF, different fractiles can be defined. For regions with heavy snow loading the factor ψ_2 can also be obtained through the value averaged over the chosen period of time. The calculations were undertaken for a number of stations in Europe, and the results are summarized in Table 5.

REFERENCES

Brettle et al. "Snow loading in the UK and Eire - Combination of actions", paper submitted for publication in the Structural Engineer during 2000.

CEC, 1998. Scientific Support Activity in the Field of Structural Stability of Civil Engineering Works, Snow Loads. Commission of the European Communities DGIII-D3. *Final report, Phase I. University of Pisa. Department of structural engineering.* Pisa.

CEC, 1999. Scientific Support Activity in the Field of Structural Stability of Civil Engineering Works, Snow Loads. Commission of the European Communities DGIII-D3. *Final report, Phase II. University of Pisa. Department of structural engineering.* Pisa.

CEN, 1994. ENV 1991-1: 1994 Basis of Design and Actions on Structures, Part 1: Basis of Design. CEN European Committee for Standardisation. Brussels.

ISO, 1998. ISO/FDIS 2394: 1998 General Principles on Reliability for Structures. ISO International Standard Organisation. Switzerland.

Næss, A. & Leira, B.J. 2000. Combination of Wind and Snow Loading. *5th International Conference on Snow Engineering.* Trondheim. Norway.

Snow Engineering, Hjorth-Hansen, Holand, Løset & Norem (eds)© 2000 Balkema, Rotterdam, ISBN 90 5809 148 1

Roof design in regions of snow and cold

I. Mackinlay, R. Flood & A. Heidrich
Ian Mackinlay Architecture Incorporated, San Francisco, Calif., USA

ABSTRACT: This paper discusses those factors the designer and builder of roofing systems must consider when working in areas where snow falls and temperatures below freezing are encountered. It analyzes the snow distribution on such roofs and the various factors that influence their design and construction. It considers conventional wood frame as well as concrete and steel construction, with normal insulation, superinsulation and ventilated (cold) roofs, with examples of each. It discusses the impact of vapor retarders, roof slope, snow arresters, and roof surface materials on roof design. It reviews roof geometry and projections through roofs on overall performance. The advantages and disadvantages of various schemes are contrasted.

1 INTRODUCTION

There are places in the world where it does not freeze and thus snow almost never falls. There are other places where it is quite cold in winter, but very little snow is found due to lack of moisture. Snow and freezing weather is very rare near sea level in the tropics, but even there snow and cold are present at higher elevations. In North America, for example, over eighty percent of the land area has snow, ice or freezing conditions in some months of the year that require the special attention of the architect or engineer engaged in construction projects. If these factors are disregarded in the design, grave consequences will result. At the least, the function of the project will be impaired; at the most the lives of the occupants will be in jeopardy. The design of buildings in regions of snow and cold is a complex subject because freezing conditions and the build up of snow and ice is an ever-changing phenomenon hard to simplify. Snow and ice can assume the aspect of rock or water. They can be adhesive or slippery. They can change characteristics in very short periods of time, moving from benign to pernicious, from beautiful to menacing, from tranquil to terrifying. This has made the hazards of building in areas of snow and cold difficult to quantify. Building codes have often left risk assessment up to local building officials, with uneven results. Many codes consider snow as simply applying vertical loads to roofs, and with the overly optimistic assessment that steep roofs will reliably

shed snow, will permit loads to be reduced as the pitch of the roof increases. Such a simplistic approach is not safe. There is no substitute for a comprehensive design approach which takes into such diverse factors as temperature, precipitation, structural geometry, roof materials, and projections, etc. In this paper we will attempt to briefly discuss the factors which influence roof design in regions of snow and cold. It will discuss in more detail one system of reducing problems on sloping roofs. (Figure 1).

II. THE INFLUENCE OF COLD ON ROOF DESIGN

Heated buildings dissipate internal temperature through their roofs during the winter months. If the interior is at living temperature, 20°C (68°F), the difference between inside and outside might be 40°C (72°F) or more and this differential becomes greater as outside temperature drops. If the area ratio between the roof and the walls is great, as for example, in a single story residential complex, most internal heat is radiated through the roof. Many current energy codes in the USA require twice the insulation resistance in the roof as in the walls of buildings. The rational for the excess insulation at the roof is energy conservation, but as will be seen below, the design requirements are much more complex than just adding insulation to reduce the heating bill.

Figure 1. Sugar Bowl Cabin, Norden, CA with standing seam metal roof. June 1999

Figure 2. Deep snow on a slippery metal roof. Sugar Bowl Cabin, February 1998

Figure 3. Vent housing at ridge. Sugar Bowl Cabin. Nov. 98

It has recently become popular in cold regions to require more and more heat resistance in roofs. Some designers advocate super-insulated roofs, where the roof insulation might exceed the wall insulation by three or even four times. This raises the first cost of the building and often cannot be justified by energy savings. In cold country, the problem with this approach is that the thickness or thermal capacity of the heat resistant material is not the only factor in effective thermal roof function. Often cracks between insulating batts allow warm, moist air to reach the outer roof surface, effectively circumventing the insulation regardless of its thickness.

The retardation of moisture is as important as thermal resistance. Warm air retains more moisture than cold air. It is desirable to keep residential building at approximately 50% RH during winter months. If the buildings are constructed without a proper vapor retarder on the warm side of the insulation, condensation may occur either inside the insulation (reducing its effectiveness) or on the under side of the roof surface. This condensation may appear as roof leaks, even though the roof surface itself is perfectly dry. The migration of moisture through the roof may actually reduce the thermal

properties of the roof as the insulation increases. The higher the internal moisture (vapor pressure) and the colder the outside temperature, the better the vapor retarder must be. Vapor pressure is a powerful force and moist air can find its way through the smallest crack.

Even in cold locations where there is very little precipitation, effective control of the moisture inside a building is vital to the success of roof design. Great internal damage can occur to buildings whose interior air is warm and moist if that air reaches the inside face of the exterior skin of the building. For these reasons the cold must be understood and be considered in the design of buildings.

III. THE INFLUENCE OF SNOW ON ROOF DESIGN

Snow adds load to roofs where it falls on them. These loads are often more like dead loads than live loads. On low sloped (flat) roofs these loads are often thought to be some fraction of ground snow load, but this can be a dangerous assumption as will be discussed below. Most designers assume that steeply sloping (4 in 12 or greater) roofs will shed the snow and many codes allow a reduction in load for such roof. But, the author's experiences have shown (Figure 2) that even slippery sloped roofs do not reliably shed snow. The snow can be held on the roof by projections such as vent pipes, skylights or even standing metal ribs. (Figure 1) Perhaps the most serious deterrent to sloping roofs sliding their snow blanket is the ice dams that form at roof edges and behind roof projections. Building heat or solar radiation can cause the snow on the roof to melt. The melt water runs down the roof slope under the insulating blanket of snow and refreezes when it comes in contact with cold outside air. The ice formed adheres to the roof surface and prevents the snow from sliding as is discussed below. These ice dams can greatly increase roof loads and can cause very serious leaks.

Complex roof geometry greatly increases the destructive influence of snow on roofs. Vents and chimneys and other projections through the roof surface should be at roof ridges, (Figure 3) not at eaves. Valleys that concentrate snow and ice can greatly increase roof loads. Roofs that cascade snow and ice from one level to another can cause serious damage to the lower roof. Snow, which falls from higher roofs to lower roofs, can rebound and break windows and walls.

Heated gutters at roof edges can reduce both ice damming and icicle formation, if properly designed, but if improperly constructed, they can be swept away.

215

Attention must be given to downspouts or they will freeze. They must either be interior or, if exterior, heat traced so as to convey melt water into warm areas.

Snow on roofs can be beautiful but it adds considerable complexity to their design. The architect or engineer must allow for these special factors in the snow country.

IV. LOW SLOPE (FLAT) ROOFS

The design of low slope (flat) roofs in regions of snow and cold have their own unique parameters to integrate into the same building geometry that a designer would normally contemplate for warm weather climes. A flat roof in snow regions usually lends itself to wind stripping (ie. some snow is blown away). Unless the building is relatively small and located in a very sheltered "hole" in the terrain, wind will play a role in removing some of the snow from the roof. Flat roof snow is usually considered to be about 70% of ground snow. This snow can be higher when the site is sheltered and may equal or even in excess of the ground snow. (ASCE 7-98)

During initial building planning, the designer should visit the site during cold and snowy conditions. Besides being able to experience the fixed geographical conditions, the designer can ascertain the unique microclimate of the site. Snow depth and storm wind direction is by far the most important climatic aspect to determine. The site reconnaissance is extremely important because surrounding terrain features can radically alter both storm wind direction and speed from available regional weather data. Furthermore, the designer can get a fairly good idea of how nearby similarly roofed buildings are performing. For example, logic will dictate that given a uniform snow distribution, the short dimension of a flat roof in line with the storm wind will allow the wind to blow the snow off the roof more readily (less length of roof snow). Conversely, a microclimate wind change of ninety degrees from the regional wind direction will put the long direction of the roof in line with the storm winds. Hence, more length of roof snow is subject to stripping and more snowdrift volume (per unit width) will be generated. The designer must be constantly aware of the storm wind direction when site planning the flat roof building. He must be aware of the snow distribution, the transport of the snow by the wind and the obstructions to snow transport.

On a low slope (flat) roof, there are almost always vertical roof projections. In many buildings these projections are mechanical equipment, duct-

work and piping, with related electrical conduit. Architecturally, the most common obstructions are perimeter parapet walls, stair and elevator penthouses. All of these elements impede the smooth flow of storm winds across the flat roof. On the windward side of these projecting elements snow will drift against them. This will change the normal uniform snow loading criteria to a surcharge loading situation in the immediate vicinity. This surcharge drift load can be quite high if the projecting element is tall and the lower roof lengthy. (ASCE 7-98)

Roof beams parallel with the drift surcharge load can easily be overloaded and fail if not sized to accommodate this additional drift weight. Conversely, when a flat roof steps down to a lower roof on the leeward side, a drift load will form at this step. This drift concentrated load must also be accommodated in the structural design in a similar fashion as for roof projections.

At the roof edges, the wind can gradually cantilever the snow beyond the roof edge support, causing a snow cornice. Snow cornices can grow quite large and can impose a high eccentric load on parapets and cantilevered roof structures. These cornices may pose a danger to anyone or anything that may be on the ground below when they break loose and fall. Sidewall projections, such as cantilevered surface mounted light fixtures, are highly susceptible to snow cornice damage.

If the project under design has light design snow loads, care must be taken to investigate the local history of freak storms. It is not uncommon to have the 50 year mean recurrence interval loading be substantially exceeded by freak storms within an average adult lifetime. When 1.5m (5 feet) of 25% density snow falls over a two day freak storm, the resulting 3.81 kN/m² (78 lb/ft²) is almost four times more than an original roof design load of 0.96 kN/m² (20 lb/ft²). High loads often result from rain on snow. This overload condition can potentially collapse these lightly designed structures.

Flat or low slope roofs should always be designed to slope toward roof drains and away from the perimeter curb or parapet. The drains should be located well inboard of the perimeter walls so that building heat will keep the melt water from refreezing. The rain water pipes and the drain bottoms should be insulated to prevent condensation on the pipe's outer surface and reduce the sound of dripping water. A moderate depth snow blanket on a low slope (flat) roof will usually insulate the drain and melt water from contact with the cold outside air preventing ice formation, although ice dams can form at drains when the drain grates are exposed to the freezing air. Overflow scuppers in flat roof parapet walls serve a very useful purpose. If these scuppers are dripping or show icicle formation, then the building occupant knows that the internal drains are blocked (either by ice dams or leaves and debris) and must be unclogged. As an additional safety factor, as long as the perimeter curb or parapet height does not exceed the weight of water height equivalent to the snow load, the bathtub formed will not cause collapse of the roof structure.

In summary, low slope (flat) roofs in snow country can have distinct advantages over steep sloped roofs. The snow is retained on the roof and (with the exception of snow cornices), does not fall to the ground. Ice dams are minimized and snow cornices can be relatively easily removed (as compared to eave ice dams) as the cornices are snow and not ice. Wind stripping can keep the roof snow depth to manageable levels, especially if favorable

building to wind orientation is utilized in site planning. Melt water can be internally drained.

V SLOPING ROOFS

The design of sloped roofs in snowy regions has its own special set of problems. As with flat roofs, site orientation with respect to sun and storm winds can either exacerbate or mitigate roof snow accumulation. For example, a gable pitched roof with the ridge running east/west with a south storm wind will probably have the south facing portion of the sloped roof wind stripped of most of its snow cover and some of this blown onto the north portion of the roof. This drift surcharge load on the north side can become quite substantial. This north drift can easily build up over time as it is in shade from the southern sun, and solar melting will be minimal. Con-

Figure 4. Cold roof snow avalanche. Sugar Bowl Cabin Nov. 98. R. Flood photo

217

Figure 5. Small icicle formation, cold roof mitigation of ice dams. Sugar Bowl Cabin. Feb. 99

would allow wind stripping to occur uniformly on each portion of the sloping gable roof in a manner somewhat similar to flat roofs. Drifting at projections,and roof steps needs to also be considered as discussed above.

Sloped roofs in snow country are best kept to very simple forms. Complex pitched shapes, dormers, valleys and multilevel steps all create a myriad of snow catchers and consequential unbalanced loading conditions and promote ice dams. Besides the loading problems, these shapes also create extremely difficult to solve flashing and waterproofing situations. Icing and ice dams can easily grow at these complex geometric intersections, further complicating the loading and waterproofing problems. Complex roof forms lead directly to roof leaks in snow country.

A sloped roof in snow regions should always have at least one gable end where entering the building will not be hazardous due to snow or ice avalanching off the roof. Too many projects are designed with no consideration of where shedding roof snow will land. Many an automobile has been buried or severely damaged due to shedding snow and ice. People suffer the same fate with resulting severe injury or death.

The designer must decide "Do I keep the snow on the roof or do I let it shed off?" Roofing surfaces such as composition shingles and wood or concrete shakes tend to hold the snow on the roof. The granules and lapping layers of composition shingles allow the snow/ice to grip the sloping roof surface; similarly with shakes. The snow on these high coefficient of friction roofing materials tends to creep down toward the eave. If not restrained, the snow/ice will curl and eventually break loose and come crashing down. On the other hand, smooth metal roofing is very slippery with a melt water interface layer below the sloped roof snow. The snow will build up until the frictional resistance is overcome and then it will avalanche off almost instantly. Standing seams in metal roofs often freeze into snow blankets on the roof and then, when the snow slips, act as rails to accelerate the avalanche. (Figure 4)

The ground impact zones for both the creep and sliding roof snow must be carefully allocated and designed to prevent damage or injury. Snow arresters (snow guards, fences) can be detailed to retain snow on the roof. Case Study Two, Mackinlay, Flood (1996) is an example of a steel bar as snow arrester at the eave. Concurrent with this arrester is an upslope, snow insulated and heat cable traced gutter to intercept the melt water before it is exposed to freezing air, thus mitigating ice dam and icicle formation at the eave. The engineering for

versely, some mitigating drift effect will be achieved if the storm wind comes from the north. The north roof snow would be blown onto the south portion of the gable roof where more melting would occur. Alternately, a change in the ridge orientation where the ridge was in line with the storm winds the loads imposed on snow arresters is explained by Tobiasson, Buska and Greaterex (1996).

Warm sloping roofs (ie. sloping roofs with little or no attic or rafter ventilation) will melt the roof snow faster and more continually than a "cold roof" for the same installed roof insulation. This continual melt water generation must be constantly intercepted and drained away before it is exposed to the freezing air at the eaves. If not, the melt water will refreeze and form ice dams and icicles. Icicles may be exquisitely beautiful but other than that, neither ice dams nor icicles have any redeeming social value. Ice dams cause a concentrated unbalanced loading condition at eaves. Ice dams cause melt water to pond and back up behind the dam. This in turn can easily lead to leaks if the roof substrate is not adequately waterproofed. Icicles and ice dams are deadly when they fall and destructive to the roof when they tear away. Ice dams usually take shingles, flashings, and other materials along when they fall. Shoveling off roof snow to mitigate ice dams is usually a case of wasted time and money with a hefty roof repair cost thrown in for good measure. Chipping away at ice dams with shovels, picks, etc. more often than not damages the roofing. Shoveling from the eave partially up the roof is a waste of time and energy as an ice dam will form just down slope of where the shoveling stopped (melt water hits the freezing air at that point). Only if the entire roof slope from eave to ridge is cleared of snow will the ice dam/icicle formation cease.

Cold sloping roofs (ie. sloping roofs where a great deal of natural ventilation is provided at the attic or rafter space) work well to mitigate ice dam formation. (Figure 5) With a properly designed cold roof, the large outside air ventilation air flow below the roof surface washes away the building heat loss migrating into the cold roof cavity. When this "washing away" effect is substantial, little or no interior heat warms the underside of the roof, and therefor the sloping roof snow cover is not melted by building heat. No melt water, no ice dam/icicle formation.

A cold sloping roof and its ice dam mitigation performance will be discussed in section VI.

VI ANALYSIS OF A COLD ROOF DURING THE WINTER OF 1998/1999

The journal article "Attic Ventilation Guidelines to Minimize Icings at Eaves" (Tobiasson, Buska, Greatorex, 1998) presents a case study of ice damming and formation of icicles at four identically built buildings in upstate New York. Outside temperatures and attic temperatures were measured. The conclusion of this paper is that icing problems at eaves occur only when outside temperatures are below -5.6°C (22°F), and attic temperatures are above -1.1°C (30°F) at the same time. Tobiasson (1998) refers to this situation as "Icing Envelope". Later mechanical ventilation in the attic space was introduced in 56 buildings with former icing problems. This ventilation was designed to keep the attic temperature under the critical -1.1°C (30°F). Maintaining these low temperatures, icing problems were avoided or significantly reduced. The article reports that "there were no reports of problematic ice dams or icings on any of the modified buildings". Tobiasson, Buska, Greatorex, (1998).

This test building is at Sugar Bowl, Norden, CA (near Donner Summit) in the Northern Sierrra mountains. (Figure 6) This cabin is the now constructed Third Case Study in the article "The Impact of Ice Dams on Buildings in Snow Country" Mackinlay, Flood (1996).

The cabin is located at 2146m (7040ft) elevation on the west side of the crest of the range. The area experiences heavy snowfalls at an average of 6m (204 inches) ground snow per winter. (Figure 7) On an average of 96 days per year the minimum temperature drops below freezing. During the monitored time period, November 1998 – April 1999 and November 1999 to January 2000, the lowest temperature measured was –19°C (-2°F). The time with the critical temperature of -5.6°C (22°F) or less for the formation of icing expands over 766 hours. This represents 17% of these six months. The Sugar Bowl Cabin is covered with a cold roof. A closed cell polyurethane insulation with a U-value of 0.026 (R38) is built in.

The ventilated air space between the insulation level and the roof shell is 152 m (6 inches) wide. The air intake is at the eaves, the air outlet through the metal ridge cap to the vent shafts, which extend above the snow cover on the roof. (Figure 3) This cold roof air space is equipped with sensors in various locations, which read the temperatures every 15 minutes 24 hours a day. Two of these sensors are located within the ridge cap of the roof, four in the air space at the eaves of the roof, and one sensor reads the temperature of the South facing metal roofing. Additional sensors take temperature readings of the outside ambient temperature, interior upstairs temperature and interior downstairs temperature.

The data obtained shows temperature readings in a typical cold weather situations (Figures 8+9). December 5 and 6, 1998 are two days within a cold weather period with temperatures mostly under -5.6°C (22°F), in which Sugar Bowl roofs are prone to icing. Ridge and metal roof temperatures are very close. This indicates snow cover on the roof.

Sensor readings show ridge temperatures mostly between -6°C and –4°C (25°F and 28°F). No icing develops on the roof at these cold roof air space temperatures.

Figure 10 shows temperature readings for November 1998 to April 1999, Figure 11 provides the same information for November 1999 to January 2000. The principles of a cold roof to work are perceived in a lower temperature at the eaves and a generally higher temperature at the ridge. Cold air enters at the eave and heats up as it passes the warm inner roof. As warm air is lighter than cold air, it rises to the ridge where it is vented. For that reason this analysis uses the highest temperature at the ridge as worst case scenario to determine the time period when icing at eaves occurs.

Ridge temperature readings exceed -1.1°C (30°F) in 51 readings which equals a total time of 12.75 hours in the winter 98/99. The longest consecutive time period of ridge temperatures above -1.1°C (30°F) is 1 hr 45 min or three cumulative hours within a seven hour time period. The times with high ridge temperatures usually range between one hour and 1½ hours per day. Icicle and ice dam formation at this cabin possibly occurred at 1.7% of the time with temperatures below -5.6°C (22°F) or at a total of 0.07% during the whole 6 months winter period. Climate conditions increase the formation of icing in Sugar Bowl. Unlike upstate New York where temperatures remain under freezing for extended time periods and snow melting is only caused by interior heat loss, temperatures in Sugar Bowl rise above freezing most days of the winter. Snow melts during the day and the melt water runs down the slope under the snow blanket to the eaves and there refreezes at night. Figure 12 shows a temperature chart for a typical day of this thaw and freeze cycle, which naturally creates more ice dams than conditions in New York.

The conclusion that can be drawn from Figures 10, 11 & 12 is that the naturally ventilated space in the cold roof performs the same way as the mechanically ventilated attic in the Tobiasson study. The time periods with temperature readings within the icing envelope when exterior temperatures are below –5.6°C (22°F) and ridge temperatures are above –1.1°C (30°F) are very rare and of short duration. Figure 5 underlines the result of the graph. Some icing occurs, which is more likely due to the freeze-thaw cycle than to ventilation problems.

The previously discussed condition of low exterior temperatures combined with snow cover on the roof is one typical situation of the roof.

Figure 13 shows a different typical situation. December 22 and 23, 1999 are in a warmer time period. Metal roof temperatures peak in the early afternoon. This is an indication that the roof is not covered with snow. The sensor, which takes these temperature

Figure 6. Cold roof analysis test building. Sugar Bowl Cabin. May 98

Figure 7. Deep ground and roof snow. Sugar Bowl Cabin. Feb. 99

readings is located on the roof itself on the South facing side, close to the ridge. Even on these shortest days of the year with no snow on the roof, solar radiation heats the metal roof significantly during the four to six hours it is exposed to the sun, independent of the air temperature.

The condition change from a period of temperatures below freezing with snow cover on the roof to a period of higher temperatures with the impact of solar radiation on the temperatures readings for the South facing metal roof can be observed in Figure 14. At temperatures below freezing, the ribs of the metal roof hold the snow on the roof. November 12–14, 1998 is a time period of continuous air space temperatures above or around the freezing point. This situation is caused by high outside temperatures and increased heat transfer through the roof based on raised interior temperatures due to occu-

pancy change. After a time period with enough warm air to warm up the metal roof ribs, the snow pack slides off the slippery metal roof in the afternoon of 13 November 1998. (Figure 4) From that time on the metal roof temperature exceeds the outside air temperatures due to solar radiation as shown in Figure 13.

VII CONCLUSION

The snow and cold greatly increases the complexity of roof design. Not only are roof loads increased in ways that are not often self evident, but such problems as ice dams and icicles can cause hazards that are difficult to visualize during periods of warm weather. The roof designer not only needs to understand the general principles of snow country design, but be aware of the special conditions that apply to the site under consid-

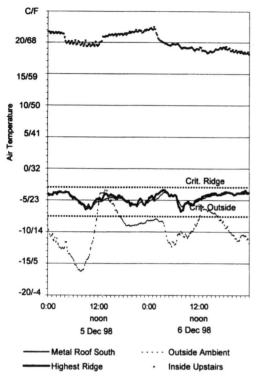

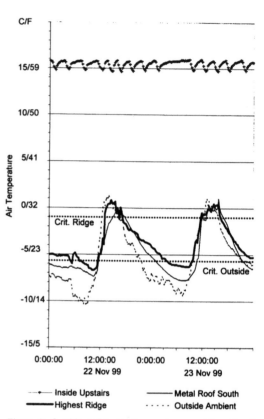

Figure 8: Outside and inside temperatures, metal roof and ridge temperature readings for 5+6 December 1998 as typical days in a cold weather period.

Figure 9: Outside and inside temperatures, metal roof and ridge temperature readings for 22+23 November 1999 as typical days in a cold weather period.

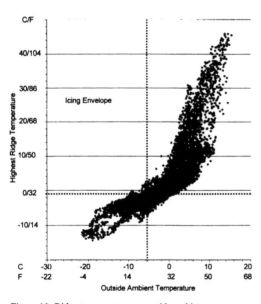

Figure 10: Ridge temperature vs. outside ambient temperature for November 1998 – April 1999, 4 temperature readings per hour, 24 hours per day, after Tobiasson, (1998)

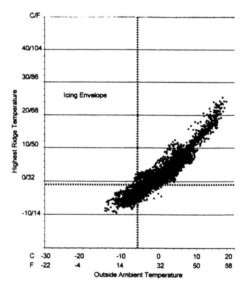

Figure 11: Ridge temperature vs. outside ambient temperature for November 1999 – January 2000, 4 temperature readings per day, 24 hours per day, after Tobiasson (1998)

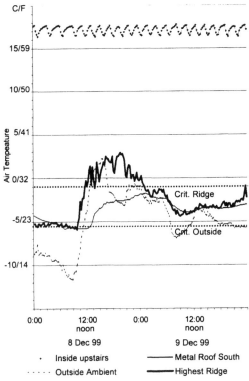

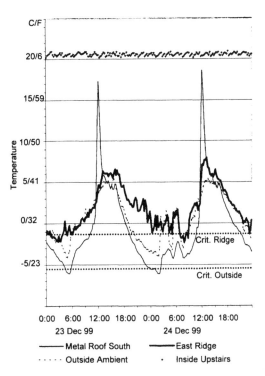

Figure 12: Outside and inside temperature, metal roof and ridge temperature readings for 8+9 December 1999 as typical days in a freeze-thaw cycle

Figure 13: Outside, inside, metal roof, ridge temperature readings 22+23 Dec 99, typical warm period days, shows the impact of solar radiation on roof temperatures

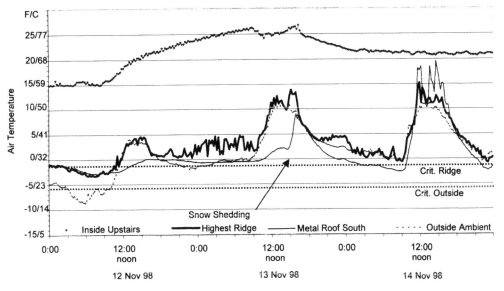

Figure 14: Outside, inside, metal roof and ridge temp readings for 12-14 Nov 98 as documentation for change in temperature readings connected to snow shedding on 13 November 1998.

Figure 15. There is deep snow and cold weather at Sugar Bowl in mid winter. Although this is a steep metal surfaced roof a great deal of snow and some ice is retained in late February 1998.

eration. Special design techniques, such as cold roofs and vapor retarders, can greatly reduce the impact of snow and cold on buildings. Cold weather can produce many varied effects on buildings. There is no substitute for experience when dealing with the problems of snow and cold.

REFERENCES

Mackinlay, Ian & Flood, Richard S. 1997. The Impact of Ice Dams on Buildings in Snow Country. Masanori Izumi (eds.), *Proceedings of the Third International Conference on Snow Engineering*, Sendai, 26-31 May 1996. Rotterdam: Balkema: 489-496

Tobiasson, Wayne & Buska, Wayne & Greatorex, Alan 1996. Snow Guards for Metal Roofs. *ASCE 8th International Conference on Cold Regions Engineering*, August 1996

Tobiasson, Wayne & Buska, Wayne & Greatorex, Alan 1998. Attic Ventilation Guidelines to Minimize Icings at Eaves. *Interface - Journal of the Roof Consultants Institute*. January 1998 (Vol. XVI, No.1): 17-24.

ASCE 7-98, American Society of Civil Engineers, *Minimum Design Loads for Buildings and Other Structures*, Reston, VA 1998.

Mackinlay, Ian & Flood, Richard S. 1999. Roof Slopes, Ice Dams and Cold Roofs. *Interface – Journal of the Roof Consultants Institute*, January 1999 (Volume XVII, No 1) 5-9

Photographs are by Ian Mackinlay except as noted.

"One never can know enough about snow", George Leigh Mallory, 1923.

Snow Engineering, Hjorth-Hansen, Holand, Løset & Norem (eds) © 2000 Balkema, Rotterdam, ISBN 90 5809 148 1

Long term stochastic modeling for combination of snow and wind load effects

A. Naess & B. J. Leira
Department of Marine Structures, Norwegian University of Science and Technology, Trondheim, Norway

ABSTRACT: The paper presents a study of the load effect combination problem for snow and wind actions on building structures. The goal is to assess simple combination rules that are implemented in existing international design codes. Combination rules for the serviceability limit state is considered. Two different approaches are employed. One simplified method based on the Ferry Borges-Castanheta model, which is complemented by a more exact method that is based on the long term average level upcrossing rate of the combined process. This allows a fairly accurate estimate of the extreme value distribution of the combined process thereby providing a means of testing the accuracy of the simpler approach.

1 INTRODUCTION

The focus of this paper is an investigation of the combination of characteristic values of snow and wind load effects for the purpose of design of a civil engineering structure. The work is carried out under some simplifying assumptions that makes it possible to use an accurate method for calculating the characteristic values of the combined load effect. This allows a study of the accuracy of combination factors employed in EUROCODE 1 (1994) and the ISO standard (ISO/FDIS 2394 1998) for checking against the serviceability limit states. In the former of these documents, the boxed value for the pertinent combination factor is 0.6. As part of the assessment, the same calculations are carried out by employing the simplified approach offered by Ferry Borges-Castanheta's load model.

The load effect combination format to be investigated in this paper can be expressed as

$$X(t) = W(t) + S(t) \tag{1}$$

where $W(t)$ = the wind load effect, $S(t)$ = the snow load effect. Both load effects are considered as stochastic processes, and they will be considered statistically independent. This is clearly not true in general, but the required information that would allow a rational treatment of possible dependence between the two load effects is not available for the data used in this study. In any case, it is believed that the conclusions of the present study is not influenced to a large extent by the assumption of statistical independence, unless the correlation is strong. The principal aim is to find the long term extreme value distribution of the combined load effect $X(t)$, which will allow us to calculate the relevant characteristic values.

2 THE EXTREME VALUE DISTRIBUTION OF THE COMBINED LOAD EFFECT

Let $X(t)$ be as above. Let $M_X(T) = \max(X(t); 0 \leq t \leq T)$ denote the extreme value of $X(t)$ over the time interval $(0, T)$. If $\bar{\nu}_X^+(x; t)$ denotes the (nonstationary) average rate of independent level upcrossing events of the process $X(t)$ at time t, then the long-term extreme value distribution of $X(t)$ can be expressed as

$$F_{M_X(T)}(x) = Prob\{M_X(T) \leq x\}$$

$$= \exp\left\{-\int_0^T \bar{\nu}_X^+(x; t)\, dt\right\} \tag{2}$$

A general discussion of extreme value theory is given by Castillo (1988), Gumbel (1958) and Leadbetter, Lindgren, and Rootzen (1983).

To be able to exploit equation (1), it is necessary to make the following observations. A long-term period can be considered as a sequence of short-term conditions during which the wind load process can be considered stationary and the snow load constant. This is certainly an acceptable assumption if the short-term condition is stipulated to last for one hour or three hours, say. Each short-term condition for the wind load is determined by the mean wind speed $\bar{U}$ and its direction θ, provided that a specific class of wind spectra is used and appropriate parameters determined. Let $f(\bar{U}, \theta)$ denote the long-term probability density function (PDF) of $\bar{U}$ and θ.

Actually, $\bar{U}$ and θ varies with time and can be considered as stochastic processes $\bar{U}(t)$ and $\theta(t)$. Let $Q(t) = (\bar{U}(t), \theta(t), S(t))$ denote the stochastic vector process of the slowly-varying quantities of this problem. In order to incorporate the statistical variability of $Q(t)$ into equation (2), it is necessary to impose a particular condition on the process $Q(t)$. This concerns the possibility of deriving estimates of statistical parameters from time averages, i.e. ergodicity (in some specific sense), cf. Naess (1984) or Schall, Faber, and Rackwitz (1991). Hence, the statistics of $Q(t)$ is generally determined by implicitly imposing an ergodicity assumption. The consequence of this is that the following relation holds

$$\lim_{T \to \infty} \frac{1}{T} \int_0^T \tilde{\nu}_X^+(x; t)\, dt = \int_q \tilde{\nu}_X^+(x|q) f_Q(q)\, dq \quad (3)$$

where $\tilde{\nu}_X^+(x|q)$ = the average rate of independent upcrossings of $X(t)$ during a short-term condition characterized by $Q(t) = q$. $f_Q(q)$ = the long-term PDF of Q. Statistical independence of $W(t)$ and $S(t)$ implies that $f_Q(q) = f(\bar{U}, \theta) \cdot f_S(s)$, where $f_S(s)$ = PDF of the snow load effect $S(t)$.

For statistical analysis of meteorological parameters, it is generally assumed that the weather systems with a time separation of three to four days can be considered independent. In keeping with this, we shall assume that upcrossing events of the combined process $X(t)$ separated by three to four days are independent. The reason for implementing this filter is to cope with the situation of large excursions of the snow load effect process combined with small but rapid excursions of the wind load effect process. In this case the statistical dependence is effectively determined by the underlying snow load effect process.

Letting $\nu_X^+(x|q)$ denote the average upcrossing

rate of $X(t)$ during a short-term condition characterized by $Q(t) = q$, then the following relation will ensure compliance with the assumption about statistical independence for a three to four day separation for the situation described above.

$$\tilde{\nu}_X^+(x|q) \to \min[\nu_X^+(x|q), 0.33 \cdot 10^{-5}] \quad (4)$$

when $s \to x \to \infty$, that is when the snow load effect is the dominating contribution to the combined load effect. Equation (4) expresses the effect of the three to four day filter in the sense that $0.33 \cdot 10^{-5}$ equals one upcrossing per 3.5 days.

Similarly, it is expected that in the situation of a dominating wind load effect, it is the statistical dependence properties of the crossing events of the wind loading process that are most relevant for the extreme value distribution of the combined load effect. This can be expressed as follows.

$$\tilde{\nu}_X^+(x|q) \to \nu_X^+(x|q) \quad \text{when} \quad s \ll x \to \infty \quad (5)$$

For the calculation of high fractile values of the extreme value distribution of the combined load effect, the following simple approximation will be adopted here.

$$\tilde{\nu}_X^+(x|q) = \min[\nu_X^+(x|q), 0.33 \cdot 10^{-5}] \quad (6)$$

Equation (6) is assumed to cover all load effect combinations where the wind load effect is not the dominating contribution. In such cases, the approximation implied by equation (5) should also be checked.

Let $\nu_W^+(w|\bar{U}, \theta)$ = the average upcrossing rate of $W(t)$ during a short-term wind condition determined by $\bar{U}$ and θ. Then, clearly

$$\nu_X^+(x|\bar{U}, \theta, s) = \nu_W^+(x - s|\bar{U}, \theta) \quad (7)$$

By introducing the following variant of a long-term crossing rate of the wind load effect process

$$\bar{\tilde{\nu}}_W^+(w) = \int_\theta \int_{\bar{U}} \tilde{\nu}_W^+(w|\bar{U}, \theta) f(\bar{U}, \theta)\, d\bar{U} d\theta \quad (8)$$

where $\tilde{\nu}_W^+(w|\bar{U}, \theta) = \min[\nu_W^+(w|\bar{U}, \theta), 0.33 \cdot 10^{-5}]$, then equation (3) assumes the following form

$$\lim_{T \to \infty} \frac{1}{T} \int_0^T \tilde{\nu}_X^+(x; t)\, dt = \int_s \bar{\tilde{\nu}}_W^+(x - s) f_S(s)\, ds$$

$$= E_S[\bar{\tilde{\nu}}_W^+(x - S)] \quad (9)$$

where $E_S[\cdot]$ denotes the expectation operator with respect to S.

The expression for the long-term extreme value distribution can then be written as

$$F_{M_X(T)}(x) = \exp\left\{-E_S[\bar{\nu}_W^+(x-S)]\,T\right\} \qquad (10)$$

which is a formula that is suitable for calculation purposes.

3 THE SNOW LOAD EFFECT PROCESS

On the basis of snow depth observations for the location Blindern over a period of 42 years, the following fitted distribution of monthly maxima of the snow load S (N/m^2) on the ground is adopted.

$$F_S^m(s) = 1 - \exp\left\{-\left(\frac{s}{s_0}\right)^{1.3}\right\} \qquad (11)$$

where the parameter $s_0 = 685$ (N/m^2).

Denoting the distribution of daily snow load values by $F_S^d(s)$, the monthly maximum distribution will be obtained as $F_S^m(s) = (F_S^d(s))^{30r}$, where r is a reduction factor to account for the fact that the daily snow load values are not statistically independent. Here we shall assume that $30r = 10$. Hence

$$F_S^d(s) = \left(1 - \exp\left\{-\left(\frac{s}{s_0}\right)^{1.3}\right\}\right)^{0.1} \qquad (12)$$

The distribution of yearly snow load maxima $F_S^y(s)$ is obtained by assuming an effective snow load season of 4 months (December - March) for Blindern. This gives the relation

$$F_S^y(s) = (F_S^m(s))^4 = \left(1 - \exp\left\{-\left(\frac{s}{s_0}\right)^{1.3}\right\}\right)^4 \qquad (13)$$

4 THE WIND LOAD EFFECT PROCESS

Let $U(t)$ denote the wind velocity process. The average level upcrossing rate $\nu_U^+(u|\bar{U},\theta)$ for $U(t)$ is commonly calculated by assuming that the wind velocity components are stationary and Gaussian in a short-term condition. This leads to the result

$$\nu_U^+(u|\bar{U},\theta) = \nu^+ \exp\left\{-\frac{(u-\bar{U})^2}{2\,\sigma^2} + \phi(u,\bar{U},\theta)\right\} \quad (14)$$

Here $\nu^+ = \nu^+(\bar{U},\theta)$ = parameter to be interpreted as an average upcrossing rate of a suitable ref-

erence level. $\sigma^2 = \sigma^2(\bar{U},\theta) = \int_0^\infty G_U(f|\bar{U},\theta)\,df$, where $G_U(f|\bar{U},\theta)$ = the (one-sided) spectral density at frequency f (in Hertz) of the wind speed process during a short-term condition specified by $\bar{U}$ and θ. $\phi(u,\bar{U},\theta)$ = function that, among other things, incorporates the coupling between wind components. This function will generally have an insignificant influence on the extreme value distribution, and it is therefore neglected.

The dependence of σ and ν^+ on the mean wind direction θ is not important here, and it will be neglected in the subsequent analysis. This should not be interpreted to mean that the wind direction is not important for a specific load combination problem. However, here we study load effects with specified relative magnitudes. And in this context, wind direction as such has minor significance. It is further assumed that there is a turbulence level given by the relation $\sigma/\bar{U} = 0.3$, which could represent a situation found in an urban area.

The wind load effect process $W(t)$ is assumed to be expressed as a quasi-static effect, that is, $W(t) = c\,U(t)^2$ for some constant c. For simplicity of calculation, the following simplification is adopted $W(t) = c\,[\bar{U}^2 + 2\,\bar{U}\,V(t)]$, where the zero-mean process $V(t) = U(t) - \bar{U}$ has been introduced.

Taking into account the assumptions above, we may now approximate the average level upcrossing rate of the process $V(t)$ as follows.

$$\nu_V^+(v|\bar{U}) \approx \nu^+ \exp\left\{-\frac{v^2}{0.2\,\bar{U}^2}\right\}$$

$$= \nu^+ \exp\left\{-5\left(\frac{v}{\bar{U}}\right)^2\right\} \qquad (15)$$

To see what values the parameter ν^+ will assume, we shall have to assume a class of wind velocity spectra. Here, we shall adopt the expression $(G_U(f) = G_V(f))$

$$\frac{fG_V(f)}{\kappa\bar{U}^2} = \frac{4\,\vartheta^2}{(1+\vartheta^2)^{4/3}} \qquad (16)$$

where the roughness/drag coefficient $\kappa = 0.015$, which matches the turbulence level assumed, the reduced frequency $\vartheta = Lf/\bar{U}$, and $L = 1200$ m = a length scale. It can be shown that $\sigma^2 = 6\,\kappa\bar{U}^2 = (0.3\,\bar{U})^2$, and $\dot{\sigma}^2 = \int_0^{f_c}(2\pi f)^2\,G_V(f)\,df \approx (0.126\,\bar{U}^{4/3}\,f_c^{2/3})^2$, where an effective cut-off frequency f_c has been introduced. This is done to take into account the fact that a building structure will be insensitive to frequencies above a certain threshold, which depends on the geometry and dimensions of the structure. Here we shall assume that $f_c \approx 0.5$ Hz. This gives a parameter value of $\nu^+ = \dot{\sigma}/(2\pi\sigma) \approx 0.1$ Hz.

Taking account of the relation $W(t) = c[\bar{U}^2 + 2\bar{U}V(t)]$, we find that

$$\nu_W^+(w|\bar{U}) \approx 0.1 \cdot \exp\left\{-1.25\left(\frac{w - c\bar{U}^2}{c\bar{U}^2}\right)^2\right\} \quad (17)$$

Finally, an expression for the long-term distribution of the mean wind speed is required. Adopting the hourly mean wind speed as the reference, the following Rayleigh distribution is obtained as a reasonable, representative distribution for the geographic location considered, viz. Blindern, Oslo.

$$F_{\bar{U}}(x) = 1 - \exp\{-0.022\,x^2\} \quad (18)$$

The long-term average level upcrossing rate of the wind load effect process $W(t)$, cf. equation(5), can then be written as

$$\bar{\nu}_W^+(w) = 4.4 \cdot 10^{-3}$$

$$\cdot \int_0^\infty x \exp\left\{-1.25\left(\frac{w - cx^2}{cx^2}\right)^2 - 0.022x^2\right\}dx \quad (19)$$

It is then obtained that the long-term extreme value distribution $F_{M_W(T)}(\cdot)$ of the wind load effect process will be given by the formula (Poisson approximation)

$$F_{M_W(T)}(w) = \exp\left\{-\bar{\nu}_W^+(w)\,T\right\} \quad (20)$$

5 THE FERRY BORGES - CASTANHETA COMBINATION PROCEDURE

5.1 Basic Modelling Assumptions

In order to apply the Ferry Borges-Castanheta (FBC) model, some degree of simplification of the load models is required (Melchers 1999). In particular, the length of the characteristic time scale for the snow loads must be a multiple of the scale for the wind load. Setting the latter e.g. equal to 3 days and the former equal to one month, a factor of 10 is obtained. Furthermore, the extreme dynamic wind load during the characteristic 3-day time interval is assumed to act constantly throughout the period.

The dynamic wind component is for simplicity

represented by a single gust factor, see Figure 1. This represents an approximation on the conservative side. However, for the purpose of load combinations it is believed to be sufficiently accurate.

The basic time varying load (or load effect) to be analysed can then be expressed as:

$$\tilde{X}(t) = \tilde{S}(t) + GF\,\tilde{W}(t) \quad (21)$$

where the tilde-sign signifies an FBC process, that is a process constant over time intervals specified for each process. In particular, $\tilde{W}(t)$ denotes the slowly varying wind load effect based on the mean wind speed, which is also referred to as the static wind load effect. The constant GF denotes a suitable gust factor. The corresponding design format ascertains that the resulting design load effect is properly selected in relation to the statistical properties of the combined load effect $X(t)$.

In addition to the intrinsic time scales, cumulative distribution functions of the loads are also required in order to perform a load combination analysis. For the static wind load, a Weibull distribution is employed as for the upcrossing-rate analysis.

For the snow load, it is generally found that Gaussian, Gumbel, Weibull or exponential models can be employed for daily snow-loads. In the present example, a Weibull model consistent with equation (21) is employed. This is also in compliance with the model applied in the upcrossing analysis.

The FBC model is based on a simplified time variation of the load processes. The properties of the following simplified expression is investigated:

$$\hat{\tilde{X}}(t) = \tilde{S}(t) + GF\,\hat{\tilde{W}}(t) \quad (22)$$

Here $\hat{\tilde{W}}(t)$ denotes the maximum of $\tilde{W}(t)$ over $nref$ time intervals, where $nref$ is the number of repetitions of characteristic wind load time intervals within each characteristic snow load time interval (e.g. $nref = 5$ if the snow interval is two weeks and the wind interval is 3 days, as discussed above). A normalization of this expression is subsequently performed such that a target ratio between the two load effects is achieved. The procedure is illustrated in Figure 1 for $nref = 2$.

5.2 Analysis Methodology

Given the present simplifications, numerical integration is performed since closed form expressions can not be obtained. The computational procedure can be described in terms of the following four steps:

Step 1: Establish cumulative distribution functions for the snow and wind loading, that is, $\tilde{S}$ and $GF\tilde{W}$.

Step 2: Compute the cumulative distribution function for the maximum value corresponding to n_{ref} repetitions of the wind loading variable from Step 1, that is, $GF\hat{\tilde{W}}$

Step 3: Compute the distribution function of $\tilde{X}(t)$, assuming independence between snow and wind loads. This involves calculation of a convolution integral, which is performed numerically.

Step 4: Compute the distribution function for the maximum value of the FBC process $\hat{\tilde{X}}(t)$ obtained in Step 3, corresponding to a given number of repetitions. The number of repetitions corresponds to the chosen reference period for evaluation of the combined load effect.

In step 4, the reference period is taken to correspond to one year. The number of repetitions is then equal to 4 if the characteristic time scale for snow is equal to one month and the length of the snow season is 4 months.

The basic scheme for derivation of the combination factor (i.e. ψ_0) can subsequently be formulated as:

1. Select values of the normalized wind and snow load effects with return periods of 20 years. Compute the sum of these normalized values.

2. Compute the level for the combined load effect which has a return period of 20 years (assuming that the two component processes are statistically independent).

3. Identify a combination factor ψ_0 which results in the value of the combined load effect computed in Item 2.

The two load processes are normalized by their 20-year return period values. The Weibull exponent of the distribution of annual wind-speed maxima is taken as 2.0, which implies that the Weibull exponent of the wind *pressure* becomes 1.0 (corresponding to an exponential distribution). The basic time period for the wind load is taken to be 3 days. To establish the distribution function corresponding to a period of one month, the basic distribution is exponentiated to the power of 10. The distribution function corresponding to a reference period of 1 year is subsequently computed by exponentiation the 1 month distribution to a power

of 4 (corresponding to a 4 month snow season).

An example of a probability density function and the corresponding distribution function is shown in Figure (2)

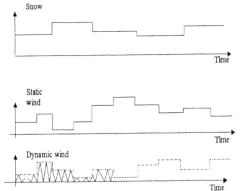

Figure 1: The Ferry Borges - Castanheta model of the wind and snow load effect processes

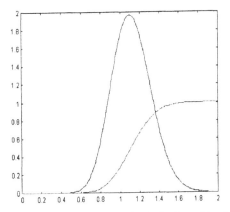

Figure 2: The PDF and CDF of combined wind and snow load

6 NUMERICAL RESULTS

The 20 year characteristic value of the snow load, s^{20y} is obtained as the 95 percentile of the extreme value distribution $F_S^y(s)$, that is, $F_S^y(s^{20y}) = 0.95$. It is obtained from equation (13) with $s_0 = 685$ that $s^{20y} = 2125$ for Blindern. This is transferred to a 20 year characteristic snow load effect, which will be written as $s^{20y} = 2.0\,l_0$, where l_0 denotes some suitable reference load effect value.

Case 1

First, a situation where the wind load effect is approximately half the snow load effect is studied. This is achieved e.g. by setting the constant $c =$

0.6, which gives a 20 year characteristic value w^{20y} of the long-term wind load effect given by $w^{20y} = 1.0\,l_0$.

As a check on the combined characteristic value, we may use the obvious relation

$$M_X(T) \leq \tilde{M}_X(T) = M_W(T) + M_S(T) \qquad (23)$$

which leads to the inequality

$$F_{M_X(T=1y)}(x) \geq F_{\tilde{M}_X(T=1y)}(x)$$

$$= \int_0^\infty F_{M_W(T=1y)}(x-s) \cdot f_S^y(s)\,ds \qquad (24)$$

Hence it follows that $x^{20y} \leq \tilde{x}^{20y}$, where x^{20y} (resp. $\tilde{x}^{20y}$) denotes the 95% fractile of $F_{M_X(T=1y)}(x)$ (resp. $F_{\tilde{M}_X(T=1y)}(x)$). It is obtained that $\tilde{x}^{20y} = 3.0\,l_0$.

Now, calculating x^{20y} using equation (10) gives the result $x^{20y} = 2.6\,l_0$. Observing that $w^{20y} + s^{20y} = 3.1\,l_0$, it follows that our prediction for the combination factor in this case is $\psi_0 = 0.55$, $(s^{20y} + \psi_0 w^{20y} = x^{20y})$.

For the FBC model, the combination factor is found to be $\psi_0 = 0.50$, which is slightly lower than that found from the upcrossing analysis above.

Case 2

The second case to be considered is a situation with approximately equal load effects. To achieve this, the parameter s_0 is adjusted to $s_0 = 342$. This leads to the value $s^{20y} = 1.0\,l_0$. Also, $\tilde{x}^{20y} = 1.9\,l_0$. It is found that in this case $x^{20y} = 1.7\,l_0$ when equation (10) is used. Since $w^{20y} + s^{20y} = 2.1\,l_0$, a combination factor $\psi_0 = 0.63$ is obtained $(s^{20y} + \psi_0 w^{20y} = x^{20y})$.

The FBC model now gives $\psi_0 = 0.58$, which is again somewhat lower than obtained by the up-crossing analysis.

Case 3

In the third case, the snow load effect is assumed to be half the wind load effect. This is achieved by setting $s_0 = 171$, giving $s^{20y} = 0.5\,l_0$. Also, $\tilde{x}^{20y} = 1.4\,l_0$. Equation (10) leads to $x^{20y} = 1.2\,l_0$. Now, $w^{20y} + s^{20y} = 1.5\,l_0$, giving a combination factor $\psi_0 = 0.38$ $(w^{20y} + \psi_0 s^{20y} = x^{20y})$.

As mentioned in Section 2, it is for the present case relevant to check the result obtained by adopting the approximation implicit in equation (5). This gives $x^{20y} = 1.3\,l_0$, leading to a combination factor $\psi_0 = 0.60$. The 'true' value should then be expected to lie between 0.4 and 0.6 for the present case.

For this example, the FBC model gives $\psi_0 = 0.40$, which is in agreement with the lower value obtained by the upcrossing analysis.

7 CONCLUDING REMARKS

Two procedures for calculating the long-term extreme value distribution of combined wind and snow load actions using an upcrossing analysis have been presented. For the snow load situation at Blindern, Oslo, the two methods give quite consistent results. But it is noted that the calculated load combination factor ψ_0 exhibits some variability over the chosen load effect combination scenario. Over the range from situations with dominating snow load effect to approximately equal load effects gives $\psi_0 \approx 0.5 - 0.6$, while situations with dominating wind load effect leads to a more unsettled conclusion with values in the interval 0.4 to 0.6.

The FBC model gives values for the combination factor that are largely consistent with those found by the upcrossing analyses.

By performing parameter variations with respect to the Weibull shape factor by means of the Ferry Borges-Castanheta model, it is observed that the combination factor increases for increasing shape factor values. A combination factor of 0.65 is obtained for a shape factor of 3.0 in the case of comparable snow and wind load effects. The boxed value of the combination factor in EUROCODE 1 (1994) is 0.6.

REFERENCES

Castillo, E. (1988). *Extreme Value Theory in Engineering*. San Diego: Academic Press Inc.

EUROCODE 1 (1994). *Basis of Design and Actions on Structures. Part 1: Basis of Design* (European Prestandard ENV 1991-1). Brussels, Belgium: European Committee for Standardization.

Gumbel, E. (1958). *Statistics of Extremes*. New York: Columbia University Press.

ISO/FDIS 2394 (1998). *General Principles on Reliability for Structures* (International Standard, Final Draft). Genève, Switzerland: International Organization for Standardization.

Leadbetter, R. M., G. Lindgren, and H. Rootzen (1983). *Extremes and Related Properties of Random Sequences and Processes*. New York: Springer-Verlag.

Melchers, R. E. (1999). *Structural Reliability Analysis and Prediction* (2nd Edition). New York: John Wiley & Sons, Inc.

Naess, A. (1984). On the long-term statistics of extremes. *Applied Ocean Research* 6(4), 227–228.

Schall, G., M. Faber, and R. Rackwitz (1991). The Ergodicity Assumption for Sea States in the Reliability Assessment of Offshore Structures. *Journal of Offshore Mechanics and Arctic Engineering* 113(3), 241–246.

Snow Engineering, Hjorth-Hansen, Holand, Løset & Norem (eds)© 2000 Balkema, Rotterdam, ISBN 90 5809 148 1

Code provisions for gable roof drift loads

M.J.O'Rourke

Rensselaer Polytechnic Institute, Troy, N.Y., USA

ABSTRACT: The 1998 version of the ASCE 7 Load Standard introduced new provision for unbalanced or drift loads on low sloped gable roofs. During the adoption process, issues were raised in relation to the need for such provisions, the magnitude of the proscribed loads, as well as the appropriate functional form for the relationships. Herein prior and current provisions for unbalanced loads on gable roofed structures in the ASCE 7 Standard are reviewed. Where available, technical justification for the new provisions will be provided, and the issues raised during the adoption process will be addressed. Finally potential future enhancements will be suggested.

1 INTRODUCTION

In the 1995 version, as well as prior versions of the ASCE 7 Load Standard, unbalanced or drift loads were not required for gable roofs with slopes less than 15^0. Loss experience from a March 1993 blizzard in the U. S. seemed to suggest that drift loads were in fact forming on gable roofs with slopes less than 15^0. This lead to a study of drift loads on low sloped gable roofs (O'Rourke and Auren; 1997a, 1997b) which in turn formed the basis for the new provisions eventually adopted in the ASCE 7-98 Load Standard (ASCE 1998).

A number of issues were raised during the consensus adoption process for the new gable roof drift load provisions. Some claimed that the new provisions were unnecessary since they address a load condition which, in that individuals opinion, was an infrequent cause of structural distress or collapse. Others thought that the proscribed drift loads seemed unrealistically large, particularly for sites with comparatively large design ground snow loads. Finally Wayne Tobiasson (1999) former chair of the ASCE 7 Snow Committee, challenged the form of the drift load relationship and proposed an alternate. Specifically, in the ASCE 7-98 provisions the drift surcharge load is an increasing function of the roofs aspect ratio, L/W (i.e., length over eave-to-ridge distance) and a linear function of the design ground snow load. Tobiasson suggested that the drift load be independent of the L/W ratio, and its increase with ground load be something less than linear.

This paper will present a brief overview of the unbalanced or drift load provisions in both the 1995 and 1998 versions of ASCE 7. To the extent that it is available, technical justification for the new provisions will be provided. Each of the three issues raised during the ASCE 7-98 adoption process will be addressed. Finally arguments supporting revision of the current linear relation between gable roof surcharge loading and ground snow load are presented.

2 PRIOR U.S. PROVISIONS

Required design snow loads for gable roof structures in the ASCE 7-95 Load Standard (ASCE 1996) are sketched in Figure 1. As can be seen, two separate load cases are considered. The first is the balanced load case, which specifies a uniform load of intensity p_s on both sides of the ridge line. The sloped roof uniform load p_s is the flat roof snow load p_f times a slope factor C_s

$$p_s = C_s \, p_f \qquad (1)$$

The flat roof load is related to the ground load p_g by

$$p_f = 0.7 \, C_e \, C_t \, I \, p_g \qquad (2)$$

where C_e is an exposure factor, C_t is a thermal factor, and I is an importance factor. The exposure factor varies from 0.7 to 1.3 as a function of terrain topology and roof exposure to wind. For a partially exposed roof in something other than large city

233

centers or a shoreline, $C_e = 1.0$. The thermal factor varies from 1.0 to 1.2 and is a function of the roof thermal environment. For a heated structure with something other than a cold ventilated roof, $C_t = 1.0$. The slope factor is related to the roof slope, slipperiness of the roof surface, and the thermal factor. For a warm roof with shingles and a slope less than 30^0, $C_s = 1.0$. The importance factor is 1.0 for typical occupancies. Finally p_g in the design ground snow load with a return period of 50 years. Hence for what one could argue are "typical" conditions (i.e., $C_e = C_t = C_s = I = 1.0$), the balanced load is 70 percent of the ground snow load.

The second load case is an unbalanced load of intensity $1.3\ p_s/C_e$ on the leeward side only. That is, the unbalanced load is not a function of the exposure factor

$$1.3\ p_s/C_e = 0.91\ C_t\ C_s\ I\ p_g \qquad (3)$$

and has an intensity of 91% of the ground snow load for a "typical" roof with $C_t = C_s = I = 1.0$. In ASCE 7-93 and prior versions, the unbalanced load multiplier was 1.5 as opposed to the 1.3 value used in ASCE 7-95. Using the multiple of 1.5, the intensity of the unbalanced load is only slightly larger (5% larger) than the ground load, again for the "typical" roof. Hence the unbalanced provision in either ASCE 7-95 or its precessors envision at least for a "typical roof" load that is not greatly different than the ground snow load. That is, the unbalanced condition , at least in terms of load for a typical roof, could be due simply to snow sliding off of one side of a gable roof when something close to the ground snow load exists on the other side.

The unbalanced load case was required only for gable roofs with slopes between 15^0 and 70^0. Note that the unbalanced load is not a function of the size of the snow source. That is, for an asymmetric gable roof the load on the leeward side is taken as the same irrespective of whether the windward side was the larger or smaller.

3 CURRENT U.S. PROVISIONS

The current ASCE 7-98 provisions for gable roof drift loads are sketched in Figure 2. As in Figure 1, there are two separate load cases. The first is a balanced load p_s. Depending upon roof geometry, one of three distribution for the unbalanced load is used. For most residential construction, that is, an eave to ridge distance W of 6.1m (20 ft) or less, the load is, in essence, the same as that in ASCE 7-93. That is, the windward side is unloaded while the multiple used for the leeward side is 1.5.

For "non-residential" construction, i.e., W greater than 6.1m (20 ft), one of two unbalanced distributions is proscribed depending on the storage space

available (i.e., roof slope) on the leeward side. The drift formation process envisioned by these provisions is sketched in Figure 3. The load distribution at three separate times is shown. Thus T_o is the initial condition without redistribution by wind, that is, symmetric loading about the ridge line. At time T_1, some of the snow on the windward side has been transported across the ridgeline and has settled in the areodynamic shade region near the ridge. After additional snow has been transported (i.e., at time T_2), the drift extends further from the ridgeline. During this process, that is as snow is added at the downwind end of the drift, the top surface of the drift surcharge remains nominally flat. Hence for a given percentage of the windward roof snow being relocated to a leeward drift, the lateral extent of the leeward surcharge is a decreasing function of the roof slope. That is, if the roof slope is small (i.e., small downwind snow storage area) the drift may well extend almost all the way to the eave. Conversely if the roof slope is large (i.e., large downwind snow storage area) the drift would likely be located quite close to the ridge line. Empirical evidence in O'Rourke and Auren (1997) support this model. Specifically, the average roof slope for case histories with a maximum leeside depth close to the eave was about 10^0, while the average slope for case histories with the largest depth at midspan or close to the ridgeline was about 20^0.

Specifically in ASCE 7-98 for moderate to large roof slopes (i.e., roof slopes between $275\ \beta\ p_f\ /\ \gamma\ W$ and 70^0) the uniformly distributed drift load on the leeward side varies between $1.5\ p_s/C_e$ and $1.8\ p_s\ /\ C_e$ as a function of the gable roof drift parameter β. That is, for a typical roof the leeward side load is 1.05 to 1.26 times the ground snow load. For these moderate to large roof slopes, one expects the maximum surcharge load to occur somewhere near the midway point between the ridge and eave (i.e., similar to the T_1 distribution in Figure 3). To simplify the design process, the equivalent uniform load is proscribed.

The gable roof drift parameter characterizes, in a sense, the percentage of windward roof snow which eventually forms the leeward drift. The percentage is $\beta/2$. It is taken to be a function of the roof aspect ratio L/W

$$\beta = \begin{cases} 0.5 & L/W \leq 1 \\ 0.33 + 0.167 L/W & 1 < L/W \leq 4 \\ 1.0 & L/W > 4 \end{cases} \qquad (4)$$

where L is the roof plan dimension parallel to the ridgeline and W is the ridge to eave distance. The parameter β is intended to quantify the likelihood of wind induced drifting across the ridge line. Specifically for small values of L/W one expects significant drifts for a wide range of wind directions. Since strong wind from any compass direction is possible and the direction tends to vary from storm to storm, gable roof drifts would occur more frequently if the L/W ratio is big.

For non-residential type construction (i.e., W>20 ft (6.1m), with smaller roof slopes, ASCE 7-98 proscribes a trapezoidal load. It consists of a balanced portion 1.2 p_f/C_e plus a triangular surcharge, with intensity β p_f/C_e at the eave. For a structure of ordinary importance (i.e. I = 1.0) the balance portion, 1.2 p_f/C_e corresponds to 0.84 p_g for a typical heated structure (i.e., C_t = 1.0) and about 1.0 p_g for an unheated structure (i.e., C_t = 1.2). That is, on the leeward side, one does not expect a significant amount of roof snow removal due to wind. Furthermore, if there is little or no roof load reduction due to thermal effects, one expects that the balanced load on the leeward side would be something close to the ground snow load.

For these smaller roof slopes, one expects that the maximum surcharge load would be located somewhat close to the eave, that is having a distribution similar to T_2 in Figure 3. For simplicity a triangular surcharge is proscribed.

As indicated above, the roof slope angle which differentiates between a small (peak drift load near eave) roof slope and a moderate (peak load somewhere between the ridge and eave) roof slope is 275 βp_f/γW. It corresponds to a leeward drift which extends exactly to the midpoint between the ridge and the eave. The model used to evaluate this angle is shown in Figure 4. Prior to wind redistribution, the balanced load on both sides of the ridge line is 1.2 p_f with a depth of 1.2 p_f/γ where γ is the snow density. The area of snow transported from the windward side is

$$A_w = \frac{\beta}{2}(1.2\frac{p_f}{\gamma})W \tag{5}$$

while the area of a triangular surcharge drift on the leeward side is

$$A_d = \frac{1}{2}(\frac{W}{2})(\frac{W}{2}tan\theta) \tag{6}$$

Using the small angle approximation (i.e. θ ≃ tan θ) for angles in radians and converting from radians to degrees,

$$\theta = 275\ \beta p_f/\gamma W \tag{7}$$

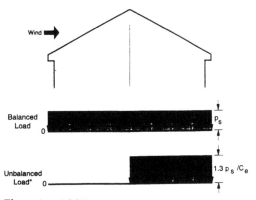

Figure 1. ASCE 7-95 Provisions

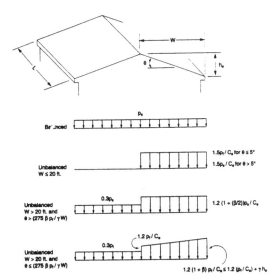

Figure 2. ASCE 7-98 Provisions

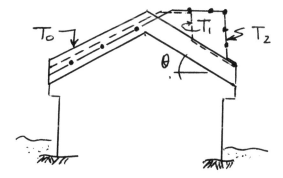

Figure 3. Gable Roof Drift Formation

235

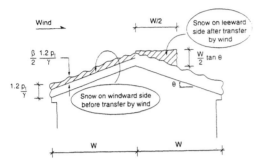

Figure 4. Gable roof drift midway between ridge and eave

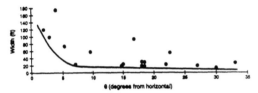

Figure 5. Scattergram from O'Rourke and Auren (1997a)

Based upon a suggestion from Tobiasson (1999), unbalanced or drift loads are not required for roof slopes (in degrees) less than 0.5 + 70/W where W has units of feet. This is intended to exempt low slope roofs, such as membrane roofs on which significant unbalanced loads have not been observed. There is both theoretical and empirical evidence to support this limit. Considering the drift formation process shown in Figure 3, substantial drifts can occur only when there is adequate space, that is when the elevation difference between the ridge and eave is fairly large. In other words, as noted by O'Rourke and Auren (1997), "overload and collapses are possible only when the product θ W (with θ in radians) is reasonably large".

Empirical evidence for the lower limit of 0.5 + 70/W is presented in Figure 5 which is a scatter diagram of the eave to ridge distance W (in feet) and roof slope θ (in degrees) for the structural collapse case histories analyzed in O'Rourke and Auren(1997a, 1997b). The ASCE 7-98 line, as proposed by Tobiasson lies a bit below the solid line in Figure 5.

4 ISSUES

As noted previously, there were a couple of issues raised during the consensus adoption process for the new gable roof drift load provisions in ASCE 7-98. Each of these issues will be discussed in this section.

4.1 *Importance*

The need for drift load provisions for low sloped gable roofs was questioned by one individual since, in his opinion, they were an infrequent cause of structural collapse and/or distress. In that individuals experience, structural distress due to snow was most commonly related to triangular drift loads at a roof step. While that type of load has lead to a significant number of snow load collapses in the past, it's the authors opinion that roof step design relations which relate the load to the size of the snow source area are, over time, effectively mitigating many of those losses.

In order to gauge the relative importance of gable roof drift loads, the author reviewed his forensic consulting practice files. It consists of about two dozen separate partial or complete collapses over the past three or four years. Knowing winter weather information at the various sites particularity the direction of strong wind during and after significant snowfalls (typically from a forensic meteorological report), the location of observed roof snow drifts, and/or the location of the initial structural distress; the primary snow loading mechanism could be determined. Out of a total of 23 separate cases, 7 were clearly gable roof drifting, 7 were roof step or parapet drifting, 6 were combinations such as roof step and gable roof drifting, while the final 3 were nominally balanced loading with some eave icing and/or rain-on-snow. Hence it appears, from this sample of recent loss experience, that gable roof drifting is in fact an important snow loading mechanism.

4.2 *Load Magnitude*

Another issue raised, in relation to the ASCE 7-98 provisions, was the overall magnitude of the load particularly for sites with comparatively large ground loads. This question is best answered by a comparison of observed gable roof drift loads with those predicted by the 1998 provisions.

Table 1 lists the pertinent roof geometry (width, length and roof slope), the ground snow load (both the observed value as well as the 50 year ground snow load from ASCE 7-98), and finally the observed and predicted roof load. For two of the case histories (F and G) there were no physical measurements for the "observed" roof snow load. However knowing the snow load used in the design, S, as well as the corresponding design dead load D (which includes any mechanical or collateral loads) the observed roof load $(P_r)_{obs}$ was estimated

$$(P_r)_{obs} = 1.67 (S + D) - D \qquad (8)$$

Case History	W (m)	L (m)	θ (deg)	$(P_g)_{obs}$ (kPa)	$(P_g)_{50}$ (kPa)	$(P_r)_{obs}$ (kPa)	$(P_r)_{pre}$ (kPa)
4	31	76	4.8	1.3	.48	3.0	1.5
19*	23	277	4.8	1.3	.48	2.6	2.7
21*	8	22	**	.86	1.2	1.9	2.4
26	37	125	1.9	.43	.72	1.5	0.9
27*	7	93	18	1.5	.96	2.6	3.1
28	7	38	7.1	.96	1.9	1.7	3.2
A	27	61	3.6	1.2	1.2	1.7	1.4
C*	31	69	4.8	1.4	2.4	1.8	3.3
D	12	46	4.8	1.4	1.4	2.1	2.3
E	27	84	2.4	1.5	1.7	2.6	2.6
F	31	152	2.4	1.4	1.4	2.0***	2.4
G*	15	76	4.8	1.4	1.4	1.8***	2.6

* $C_t > 1.0$

**   assume $(\theta < 275\ \beta P_f/\gamma W)$

***   Based on Eqn.(8)

Eqn. 8 assumes that the roof collapses when the total roof load equals a safety factor of 1.67 times the total design load.

The predicted roof load, $(P_r)_{pre}$, is the peak load (i.e., the value anywhere on the leeward side for $\theta > 275\ \beta P_f/\gamma\ W$, the value at the eave for $\theta < 275\ \beta P_f/\gamma W$) from the ASCE 7-98 provisions. In order to present an unbiased comparison between observed and predicted roof loads, the larger of the $(P_g)_{obs}$, and $(P_g)_{50}$ was used as input. This was required since for a few case histories (i.e., #4, #19 and #27) the observed ground snow load was substantially larger than the 50 year value. Note that if this weren't the case, (i.e., $(P_g)_{obs} < (P_g)_{50}$ for all case histories) the appropriate ground snow load for comparison purposes would be $(P_g)_{50}$. That is, one would be interested in comparing the observed load on the roof with the design load from ASCE 7-98.

In applying the ASCE 7-98 provisions, the importance factor I was taken as equal to 1.0 for all cases. Similarly the default value for the thermal factor was 1.0. Structure specific information on alternate values for C_t was used when available. Specifically case history structures #19, #21 and #27 were unheated ($C_t = 1.2$) while reportedly case history structure G was kept just above freezing ($C_t = 1.1$).

Notice that the ratio of observed to predicted roof loads ranges from 1.97 to 0.54 with an average value of 0.996 and a median of about 0.86. In this sense the 7-98 provisions seem appropriate in that "on average" they produce predicted loads which are reasonably close to the observed values.

As one might expect, some of the largest predicted loads, specifically 3.2 and 3.3 kPa (67 and 68 psf) for case histories 28 and C, correspond to larger ground snow loads of 1.9 and 2.4 kPa (40 and 50 psf). However the observed roof loads of 1.7 and 1.8 kPa (36 and 37 psf) for those case histories were not particularly large. That is, the ratio of observed to predicted were unusually low for the two case histories with the largest ground loads. Hence one could argue that although the provisions appear to provide reasonable loads on average, they tend to overpredict for locations with large $(P_g)_{50}$.

4.3 Functional Form

The final issue raised in relation to the 1998 provisions involved the functional form for the drift load equations. Wayne Tobiasson (1999) suggested that the dependence upon roof aspect ratio L/W be eliminated. He also suggested that the increase in drift load with p_g be something less than linear.

In his discussion of the aspect ratio, Tobiasson presents a graph of the gable roof drift parameter β ploted against a normalized roof load (apparently a ratio of the peak roof load to the observed ground load) from case histories in O'Rourke and Auren (1997) paper. Since there was essentially no correlation between β and the roof load ratio, Tobiasson concluded that the aspect ratio and β were not important parameters.

However due to the nature of the case history database and the influence of aspect ratio (i.e., β) upon gable roof drift formation, one would not expect a significant correlation. As indicated previously, β is intended to quantify the likelihood or potential for gable roof drift formation with wind possibly coming from a number of directions. That is, considering a large inventory of gable roof structures with ridgelines orientated in a number of directions, individual structures with a large aspect ratio L/W would be susceptible to drift formation for wind from a relatively wide range of directions. Conversely for individual structures with a small aspect ratio, they are susceptible to drift formation for a narrow range of wind directions (i.e., wind essentially perpendicular to the ridgeline). Hence, if one were to measure drifts on this large inventory of structures over a number of years that is with snowstorms from a number of different directions, gable drifts on roofs with large L/W (i.e., large β) would occur more frequently. Hence because of the increased frequency of occurrence, all other aspects being equal, the 50 year drift value is an increasing function of aspect ratio and β.

The data base developed by O'Rourke and Auren (1997a) was not the result of a multiyear survey of a large number of structures. It was instead a collection of case histories from the technical literature and insurance company files. As such, gable roofs where no drifts formed were not part of the database. In that sense the structures or datapoints were pre-screened. Since the aspect ratio influences the likelihood of drift formation over a number of years as opposed to the size of the drift generated during a single winter storm, the lack of correlation between β and a normalized roof load is not surprising.

Note in this regard that the aspect ratios for the case histories in Table 1 range from 2.25 to 13.86 with an average of 5.1 and a median of about 3.6. Since a gable roof that is square in plan has an aspect ratio L/W of 2, all the case history structures were longer than they were wide. This supports the reasoning behind the β factor, that is, gable roof drifts are more likely to form on large aspect ratio roofs.

Tobiasson also recommended that the functional relationship between the unbalanced roof load and the ground load be changed. Recall that in the current ASCE 7-98 approach, the unbalanced or drift load is a linear function of p_s or p_f and, through eqs. (1) and (2) a linear function of p_g. In essence Tobiasson suggests that β be a decreasing function of P_g such that the unbalanced or drift load increases with P_g but at less than a linear rate.

There are empirical and theoretical arguments in support of such a change. Note that for nine of the twelve case histories in Table 1, the controlling ground snow load (large of $(P_g)_{obs}$ or $(P_g)_{50}$) ranged from 1.2 to 1.7 kPa (25 to 35 psf). As noted previously, for the two case histories with large ground loads (case histories #28 and #C), the current ASCE 7-98 provisions overestimates the observed roof loads. Conversely for the single case history with comparatively smaller ground load (case history #26) the current provisions underestimated the observed load. This empirical evidence suggests that β should in some fashion be a decreasing function of p_g.

From a theoretical perspective, drifts only form on windy days when there is "driftable snow" up-wind of a geometric irregularity. In this context "driftable" refers to snow which has not been subject to freezing, rain, sleet or above freezing temperatures which result in crust formation. Drifts often are the result of blizzard-like conditions which can last a couple of days. In comparatively low P_g areas, the snowfall (driftable snow) associated with a big blizzard is often comparable to the 50 year ground snow load (i.e., most or all p_g is driftable). However in comparatively high p_g areas, the 50 year ground

snow would not result from a single blizzard. The 50 year value is often due to a series of snowfalls over a period of months. These snowfalls may or may not be accompanied by or followed by strong winds. Also crust formation may occur between an individual snowfall and strong wind sometime later (i.e., not all p_f driftable). Since one expects the percentage of available driftable snow (i.e., ratio of driftable to 50 year ground snow value) to be a decreasing function of p_f one could argue from a theoretical perspective that β should be a decreasing function of p_g.

5 SUMMARY

Prior and current approaches for gable roof drift loads in the ASCE 7 load standard were reviewed. An analysis of recent loss experience suggests that gable rood drifting is a significant load mechanism, frequently leading to structural performance problems. On average the overall magnitude of the gable roof drift loads prescribed in the current ASCE 7-98 standard compare favorably with observed loads. Exceptions are apparent underestimation for locations with small ground snow loads and apparent overestimation for locations with large ground loads. This empirical evidence as well as more theoretical considerations support a future enhancement in which the gable roof drift parameter β is a decreasing function of the ground snow load.

6 REFERENCE

American Society of Civil Engineers (1996) "ASCE 7-95, Minimum Design Loads for Buildings and Other Structures", Reston, VA.

American Society of Civil Engineers (1999) "ASCE 7-98, Minimum Design Loads for Buildings and Other Structures", Reston, VA.

O'Rourke, M.and Auren, M., (1997a), "Snow Load on Gable Roofs", J. Struct. Engrg., ASCE Vol. 123, No. 12, Dec. pp. 1645 to 1651

O'Rourke, M. and Auren, M., (1997b), "Unbalanced Snowload on Gable Roofs", Snow Engineering: Recent Advances, Proc. 3rd Intl. Conf. On Snow Engrg., Sendai, Japan, May 1996, AA Balkema, Rotterdam, pp. 201 to 208

Tobiasson, W., (1999), Discussion of "Snow Loads on Gable Roofs", by O'Rourke and Auren, J. Struct. Engrg., ASCE, Vol. 125, No. 4, pp. 470 to 471

Snow Engineering, Hjorth-Hansen, Holand, Løset & Norem (eds)© 2000 Balkema, Rotterdam, ISBN 90 5809 148 1

Roof loading due to rain-on-snow

M.J.O'Rourke
Rensselaer Polytechnic Institute, Troy, N.Y., USA

C.Downey
Zaldastani Associates, Boston, Mass., USA

ABSTRACT: A study of structural loading on gable roofs due to rain falling on an existing roof snowpack is presented. As background, rain-on-snow provisions in various exiting national standards are reviewed and a summary of structural damage in the U.S. arguably related to rain-on-snow is presented. Three separate approaches are then used to quantify the intensity, duration and return period for a rain-on-snow event. The first approach is case history information from a 1996-97 rain-on-snow event (unknown return period). The second is a statistical approach in which the ground snow load plus daily rain with a 50 year return period is determined (known return period, unknown intensity and duration). The third is a two year return period rain (unknown snow condition). Based upon the resulting roof surcharge loads, revised rain-on-snow provisions for the ASCE 7 load standard are suggested.

1. INTRODUCTION

In many regions of the world it is not unusual for rainfall to occur when building roofs are already covered with snow. Winter storm systems which are accompanied by air temperatures that fluctuate around freezing, and spring rains which fall before the winter snowpacks have melted are two situations which lead to such rain-on-snow events. In such cases, rain traveling through the snowpack can temporarily increase the roof load and hence needs to be considered to some degree in building codes and load standards. The overall objective of the study reported herein was the development of a rain-on-snow surcharge appropriate for use in building codes and load standards. In the following sections, current building code approaches and available rain-on-snow case histories are reviewed. Precipitation data from 19 first order weather stations in the U.S. is then used to characterize, probabilistically, the rain-on-snow event. Finally, recommended procedures for incorporating the effect of rain-on-snow are developed in relation to the ASCE 7-98 load standard.

2 CURRENT CODE APPROACHES

2.1 *ASCE 7-98*

The rain-on-snow approach in ASCE 7-98 (1998) assumes that the effects of rain-on-snow will be reflected in the ground snow load measurements in areas with a design ground snow load greater than 0.96 kPa (20psf). In addition, the ASCE 7-98 provisions are related to roof slope. Specifically the surcharge of 0.24 kPa (5 psf) is not required for slopes of ½ on 12 or greater Note that for a given intensity and duration of rain, it takes a shorter amount of time for rain to drain from a steeply sloped, snow covered roof than from a nominally flat snow covered roof.

2.2 *NBCC*

The National Building Code of Canada (1995) requires that a rain-on-snow surcharge load be added to the design snow load for geographic regions which prescribe such a load. The rain-on-snow surcharge is derived from the one-day rain with a 30-year return period, and ranges from 0.1 to 0.7 kPa (2.1 to 14.6 psf).

2.3 *New European code for snow loads*

A background document (Del Corso et al; 1996) describing the new Eurocode provisions for snow loading presents two empirical methods for deriving ground snow loads from ground snowdepth measurements. The empirical formulae are intended to establish an appropriate snow density for design purposes. The code acknowledges that snow density is influenced by "rain falling onto the snow (possibly causing a considerable increase in density)". However there are no actual factors contained within the empirical formulae, which explicitly account for rain-on-snow.

2.4 AIJ recommendations

The Architechnical Institute of Japan (1996) Recommendations fail to mention the effect of rain falling on an existing roof snowpack. Hence it appears that the AIJ assumes that the effects due to rain-on-snow are reflected in ground snow load measurements. The AIJ equivalent snow densities, which relates maximum annual snow depth to maximum annual ground snow load are larger than those for Europe and the U.S. and hence may include the effects of some rain-on-snow.

3 CASE HISTORIES

3.1 Pre-1996

Prior to the winter of 1996-97, information on structural damage due at least in part to rain-on-snow was, with a few exceptions, not available in the technical literature. The authors' literature search yielded only six case histories. The degree to which deterioration, inadequate design and/or poor construction contributed to the pre-1996 collapses is unknown. An exception is the Hartford Civic Center. Although there was a heavy rain which followed snowfall, it is believed that the collapse was due in large part to an error in the initial structural design (inadequate stiffness of the brace for a space truss compression chord). Additional information on the pre-1996 case histories is available in O'Rourke and Downey (1999)

3.2 1996-97 Washington State

A series of mixed precipitation events over the 1996-97 holiday season led to a large number of structural collapses in the Pacific Northwest. A report by the Structural Engineers Association of Washington (SEAW, 1998) presents a tally of structural damage by occupancy, detailed information on a subset of the damaged buildings, and other information. According to the SEAW report, there were a total of 1663 buildings damaged by the holiday storms. The buildings were a mix of engineered and non-engineered structures. There were a number of factors, besides simple overload, which contributed to the collapses.

First of all more than a quarter of the damaged facilities were residential carports or outbuildings, which in all likelihood were non-engineered. Mobile homes often were designed to Housing and Urban Development (HUD) standards which specify lower roof snow loads than the Uniform Building Code (UBC) which governed the design of most other types of engineered structures in the area. Marinas appeared to have been designed for lower loads than UBC, with the expectation that snow removal would be implemented as needed. In 1994, UBC revised downward the allowable stress for various grades and species of wood. Hence the corresponding structural components in older buildings would likely be considered undersized using current codes. Finally, improper initial grading of dimensional lumber and deterioration of wood exposed to the elements were cited as contributing factors for some individual structures.

The senior author was involved in an analysis of two of these collapses. In both cases the structures were cold rooms (intentionally kept at or below freezing) with measured roof snow loads larger than the design roof snow load (2.7 kPa (57 psf) vs 1.5 kPa (31.5 psf) in one case, 1.39 kPa (29 psf) vs 1.05 kPa (22 psf) in the other) and other problems (lack of a stiffener where the end of a hot rolled beam sits atop a column in one case, incorrect weight for roof top cooling equipment in the other). In both cases calculations suggest that rain-on-snow was not a significant consideration (in one case the surcharge at the eave for a rain-on-snow scenario or for the rain simply refreezing in the roof snowpack was about 5% of the total load while in the other case the roof collapsed before it rained).

In terms of quantification of the rain-on-snow event, fairly detailed weather information for Yakima Airport and Sea-Tac Airport is available in a report by Wistar (1997) of Accu-Weather Inc. At Yakima on 12/31/96, we have a ground snow load of about 1.1kPa (23 psf) and rain with an average intensity of 0.287 cm/hr (0.113 in./hr) over a 12 hour period. In the Seattle area on 1/2/97, we have a ground load of roughly 0.96 kPa (20 psf) and rain over an 11 hour time period with an average intensity of 0.297 cm/hr (0.117 in./hr).

4 50-YEAR RAIN ON SNOW EVENT

In this section, daily precipitation and ground snow load data is used to quantify, in a probabilistic sense, the potential magnitude of a 50 year rain-on-snow event. Precipitation data for the period 1974 through 1987 (14 years) was obtained from the National Oceanic and Atmospheric Administration (NOAA) for 19 weather stations across the United States. For this set of weather stations the 50 year ground snow loads from the ASCE 7-98 commentary vary from 0.38 (psf) to 1.63 kPa (34 psf). Extreme value analyses were performed on the data to determine the 50-year mean recurrence interval (MRI) values for both the ground snow load and the rain-on-snow event.

For the ground snowload, the 14 values at each site for the annual maximum daily ground snow water equivalent were fitted to the appropriate extreme value distribution. The 50-year MRI ground snow values for all 19 sites were then determined. Determination of the rain-on-snow event was somewhat more difficult as it involved a number of assumptions. First of all, it was assumed that the snowpack is warm and that a daily rain would flow through the snowpack (as opposed to refreezing within the snowpack) and then directly into the ground below. Furthermore we assume that the daily rain would not have been recorded in the daily ground snow measurement. The daily rain is calculated as the daily precipitation minus the daily snowfall. Since the NOAA data presented daily snowfall in inches of snow as opposed to inches of water equivalent, we assumed that fresh snow has one tenth the density of water which, according to Tabler (1996), is a reasonable value for fresh snowfall. Finally it is assumed that all the daily rain is in (i.e., flowing through) the ground snowpack at the same time.

As with the ground snow load data, the 14 values for the annual maximum ground rain-on-snow load (i.e., daily ground snow load plus estimated temporary daily rain surcharge) for each site was fitted to the appropriate extreme value distribution and 50-year MRI values were determined.

The 50-year rain-on-snow ground surcharge (difference between the 50-year rain-on-snow ground load and the 50-year ground snow load) is shown in Table 1. Note that the surcharge ranges from 0 to 0.33 kPa (7 psf) with an average value of 0.12 kPa (2.5 psf). The zero surcharge values for some sites in the Great Plains (i.e., Grand Island, Omaha and Denver) suggest that these areas do not experience significant winter rainfall.

Note that the ground snow surcharge in Table 1 has a known return period (i.e., 50 years) but the intensity and duration of the corresponding rain event are unknown. That is, the 0.33 kPa (5 psf) surcharge for Boston could be due to 1.22 cm (0.48 inches) per hour for two hours, or 0.24 cm (0.096 inches) per hour for about 10 hours.

Table 1. 50 year MRI rain-on-snow surcharge (kPa) for ground snow.

Station	Surcharge	Station	Surcharge
Raleigh-Durham, NC	.05 kPa	Omaha, NE	0 kPa
Knoxville, TN	.24 kPa	Denver, CO	0 kPa
Newark, NJ	.29 kPa	Missoula, MT	05 kPa
Albany, NY	.34 kPa	Eugene, OR	.19 kPa
Boston, MA	.24 kPa	Portland, OR	.24 kPa
Syracuse, NY	0 kPa	Seattle, WA	.10 kPa
Columbus, OH	.10 kPa	Yakima, WA	0 kPa
Milwaukee, WI	0 kPa	Baltimore, MD	19 kPa
Erie, PA	.14 kPa	Rockford, IL	.10 kPa
Grand Island, NE	0 kPa		

5 TWO YEAR RAIN

In this section, an alternate approach for characterizing the "rain input" for a roof snow model is pursued. In the previous section, the ground surcharges resulted from precipitation which actually fell during winter months. Unfortunately the actual rainfall intensities and durations (needed for determination of roof loads) associated with these winter rains are unknown. In this section, the potential "rain input" is characterized using probabilistic information on annual rainfall intensities and durations. Unfortunately these rains may or may not occur during winter months, and more specifically when roof snow loads are significant.

Rainfall intensities and associated durations, as developed herein, are based on maps in Niedringhaus (1972, 1973). The Niedringhaus maps present 5, 30 and 60 minute duration rainfalls (in inches) for return periods of 10, 25, and 100 years. Site-specific rainfalls for our 19 weather stations was interpolated from these maps. This site-specific data was then extrapolated to determine the values with a return period of 2 years. Table 2 presents the extrapolated 2 year MRI rain intensities (cm/hour) with an associated duration of 60 minutes, for the 19 weather stations.

For the vast majority of the 19 sites the 2-year MRI, 60 minute duration rain in Table 2 is an upper bound for the 50 year rain-on-snow surcharge in Table 1. The intensities in Table 2 are for a known duration (i.e., 1 hour) and known return period (i.e., 2 years). However, they may or may not occur during the winter and specifically when the ground snow load is substantial.

6 ROOF RAIN-ON-SNOW SURCHARGE

In this section, the additional roof load due to a rain-on-snow event is evaluated using analytical relations available in the technical literature. When the snowpack temperature is such that rain flows through the snowpack as opposed to refreezing within the snowpack, the roof surcharge load is composed of two parts. The first is the load associated with rain percolating downward through the upper unsaturated layer of the roof snowpack. The second is the load associated with rain traveling downslope towards the eaves in a saturated snow layer directly above the roof surface. Colbeck (1977) presents detailed analytical relations for the load on the roof due to flow downslope in the saturated layer. Information in the *Handbook of Snow*, Gray & Male (1981) allows a simplified method for estimating the load due to rain percolation through the unsaturated layer. In the development which follows we assume that the roof snowpack temperature is such that none of the rain refreezes in the snowpack.

The *Handbook of Snow* establishes 2 to 60 cm/min as the typical range for the percolation velocity. A midpoint value of v = 30 cm/min is used, hence, assuming steady state flow, the travel time for rainflow through the unsaturated snowpack layer, t_u, is the roof snowpack depth divided by v.

Since the portion of the surcharge load due to vertical percolation through the unsaturated layer is small, the simplified steady state flow assumption does not introduce significant errors. Hence the vertical percolation load is taken as the travel time, t_u, multiplied by the rainfall intensity, i, and the unit weight of water, ρ_w.

$$W_u = t_u \, i \, \rho_w \qquad (1)$$

In the calculation of W_u, the roof snow load is taken equal to the ground snow load, p_g, and the depth is determined from the snow density relation in the ASCE 7-98 load standard. Although these assumptions are conservative, they do not introduce significant errors since, as noted previously, W_u is small with respect to W_s.

The portion of the roof surcharge load due to flow in the saturated layer is mainly dependent on roof slope, θ (in degrees), roof length, w, (e.g., eave-to-ridge distance), rainfall intensity, and rainfall duration.

From Colbeck, the maximum depth of the saturated layer for a sloped roof is

$$d_e = (i/\alpha k_s)^{1/2} w \, \tanh[F(\alpha k_s i)^{1/2}(\tau - t_u)/w\phi]/F \qquad (2)$$

where ϕ is the snow porosity, i is the rainfall intensity, α is a constant equal to $5.47 \times 10^6 \; m^{-1} \; s^{-1}$, k_s is the permeability of the saturated layer, τ is the rainstorm duration and

$$F = 0.1 \, n \, [(1 + \theta/2.2)^{1.06} - 1] + 1 \qquad (3)$$

where $n = (\alpha \, k_s/i)^{1/2}$. The shape of the saturated layer is assumed to be elliptical and the average additional load (i.e., located somewhere roughly midway between the ridge and the eave) from the saturated layer is then:

$$Ws = 0.25 \, \pi \, d_e \, \phi \, \rho_w \qquad (4)$$

Based on Colbeck and the *Handbook of Snow*, a typical value of $k_s = 10^{-9} \; m^2$ will be used herein for the permeability, a somewhat conservative constant value of 0.75 will be used herein for porosity.

Table 2. 60-minute rainfall intensities (cm/hour) for a 2-year return period.

Weather Station	2-year MRI	Weather Station	2-Year MRI
Raleigh Durham, NC	4.72	Omaha, NE	4.29
Knoxville, TN	4.06	Denver, CO	2.34
Newark, NJ	3.58	Missoula, MT	1.57
Albany, NY	2.79	Eugene, OR	1.63
Boston, MA	3.05	Portland, OR	1.55
Syracuse, NY	2.94	Seattle, WA	1.19
Columbus, OH	3.10	Yakima, WA	0.64
Milwaukee, WI	3.56	Baltimore, MD	3.84
Erie, PA	2.92	Rockford, IL	3.75
Grand Island, NE	4.04		

Table 3. Roof surcharge loads (kPa) for 2-year rainfall from Table 2.

Station	W = 6.1 m 1.19°	4.76°	W = 30.5 m 1.19°	4.76°
Ral. Dur.	.343	.253	.365	.359
Boston	.226	.173	.237	.233
Yakima	.048	.043	.049	.048

Table 4. Roof surcharge loading (kPa) for 50 year rain-on-snow from Table 1.

Station	W = 6.1 m 1.19°	4.76°	W = 30.5 m 1.19°	4.76°
Albany	.259	.193	.273	.269
Rockford	.095	.071	.098	.097
Yakima	0.0	0.0	0.0	0.0

Table 5 Average roof surcharge load (kPa) for Seattle/Yakima case histories

E to R Distance W (m)	Roof Slope, θ (deg) 0.60°	1.19°	2.39°	4.76°	9.46°
6.1	.12	.07	.05	.04	.02
15.3	.21	.16	.10	.06	.04
30.5	.24	.22	.17	.10	.06
76.3	.25	.25	.23	.19	.12

6.1 *2-year rain*

The roof rain-on-snow surcharge due to a 2-year MRI rain was determined. The roof snow load prior to the rain is taken, somewhat conservatively, to be the 50 year MRI ground snow load. Again this assumption only affects W_u and hence does not introduce significant errors. The resulting roof rain-on-snow surcharges for Raleigh-Durham, Boston and Yakima (i.e., the largest, median and smallest intensities in Table 2) are presented in Table 3 for a rainstorm duration of 60 minutes, for eave-to-ridge distances, w, of 6.1 and 30.5 m (20 and 100 ft.), and roof slopes of ¼ on 12(1.19°), and 1 on 12(4.76°). Note, as one might expect, the surcharge loads are decreasing functions of the roof slope and increasing functions of the eave-to-ridge distance.

6.2 50-year ground surcharge

The roof surcharge loads due to the 50 year ground surcharge in Table 1 were determined. As noted previously, while the return period for the ground surcharge in Table 1 is known, the intensities and associated durations are not. For the purposes of roof surcharge calculation, the rain intensity is taken as the water equivalent of the 50 year ground surcharge with a duration of 1 hour. For the purposes of calculations W_u, the roof snow load is taken as that listed in the commentary for ASCE 7-95. The resulting roof surcharges are presented in Table 4, for Albany, Rockford and Yakima (i.e., the largest, median and smallest surcharges in Table 1), and the same eave to ridge distances and roof slopes as Table 3. For some sites, such as Yakima, the average roof rain-on-snow surcharges is zero since the ground surcharge is zero. For the other sites, the roof surcharge as one might expect is an increasing function of eave-to-ridge distance.

6.3 Case History Roof Surcharge

The roof rain-on-snow surcharge for weather conditions representative of Seattle and Yakima in the 1996-97 winter was also calculated. The roof snow load was taken as equal to the ground snow load of roughly 1.01 kPa (21psf). Again this assumption only affects the load related to rain percolating through the snow-pack. The rain imput was about 0.30 cm/hr for an 11 hour period, or a total of about 3.3cm. This total rain precipitation is similar to the Boston value in Table 2, that is 3.05 cm/hr for a 1 hour duration.

For this Seattle/Yakima case history, the average roof surcharges for an eave to ridge distance, W of 6.1 m (20 ft.) were 0.0722 kPa (1.5 psf) and 0.04 (0.844 psf) for ¼ on 12 (1.19⁰) and 1 on 12 (4.76⁰) roof slopes respectively. The corresponding values for W = 30.5 m (100ft) are 0.22 kPa (4.6 psf) and 0.10 kPa (2.1 psf). For comparison purposes, the surcharges for the Seattle/Yakima case histories are all less than those for the 2-year rain in Boston (higher intensity, shorter duration, but similar total precipitation as Seattle/Yakima) as shown in Table 3. For a comparatively large eave to ridge distance of 30.5 m (100 ft) and a small roof slope of 1.19⁰ (1/4 on 12) the Seattle /Yakima surcharge is about 90% of the Boston 2 year rain value (.22/.237 = 0.93). For larger roof slopes and/or shorter eave-to-ridge distances, the percentage is smaller. Finally, if all of the approximate 3.3 cm of rain had simply refroze in the roof and ground snowpacks the increase or rain surcharge would have been about 0.32 kPa (6.8 psf). Note that if the rain refreezes it would, presumabley, be recorded as part of any ground snow measurement program.

7 RECOMMENDED PROVISIONS

This section presents proposed modifications to ASCE Standard 7 for incorporating the effects of rain-on-snow. These proposed modifications are based upon the gable roof drift load provisions and the minimum roof load provisions in ASCE 7-98, as well as rain-on-snow information developed herein.

7.1 Gable roof drift load

ASCE 7-98 requires unbalanced loads for gable roofs with moderate slopes in degrees,

$$70/w + 0.5 \leq \theta < 70 \qquad (5)$$

where w is the eave to ridge distance in feet. These provisions account for windward roof snow being blown across the ridgeline and onto the leeward side. The lower bound slope is intended to exclude low sloped roofs, such as membrane roofs, upon which significant unbalanced load have not been observed. For a typical roof the drift surcharge would be on the order of about 60% of the sloped roof design load. As such the drift surcharge would be larger than or at least comparable to the roof rain-on-snow surcharges as determined herein. That is, if unbalanced loading (gable roof drifting) and rain-on-snow were separate load cases, unbalanced loads could serve as an appropriate surrogate for rain-on-snow. The authors believe that provisions requiring rain-on-snow in combination with unbalanced loads on a gable roof, or for that matter roof step drift loads, would be overly conservative. Note in this regard that the probability of the joint occurrence of unbalanced loads (requiring snowfall and wind) in combination with rain-on-snow (requiring an existing snowpack and rain) is less than the probability of either occurring separately.

7.2 Minimum roof load

The minimum roof load provision in ASCE 7-98 are

$$\begin{aligned} p_f &= I \cdot p_g & p_g \leq 20 \text{ psf} \\ p_f &= I \cdot 20 & p_g > 20 \text{ psf} \end{aligned} \qquad (6)$$

For slopes $\leq 70/w + 0.5$. That is, minimum roof load provisions apply only to very low sloped roofs for which gable roof drift loads do not apply. It is the authors view that these minimums are intended to account for cases in low ground load areas where a single storm could result in roof loads comparable to the ground load (i.e., single events where there is insufficient time for wind and thermal effects to significantly reduce roof loads).

In summary, if the slope of a gable roof structure is greater than or equal to 70/W + 0.5 with W in feet, then unbalanced or drift load provisions apply. For lower slopes, the minimum roof loads given by eqn. (6) apply.

7.3 Rain-on-Snow Loading

It is instructive to consider the rain-on-snow surcharge in light of the roof geometry relation in Eqn. (5). Table 5 presents the average rain-on-snow surcharge, for various values of the eave to ridge distance and roof slope θ, resulting from the 1996-97 Seattle/Yakima event.

The solid dark line in the table is the lower bound slope from Eqn. (5). That is, above and to the left of the dark line such the combination of W and θ is small enough such that gable roof drift loads are not required, although minimum roof loads are required. Conversely, below and to the right of the dark line, gable roof drift loads, but not minimum loads, are required.

Note that the rain-on-snow surcharge below and to the right of the dark line tend to be comparatively smaller than those above and to the left. That is, the rain-on-snow surcharges tend to be largest for roof geometries where ASCE 7-98 requires minimum roof loading while the surcharges tend to be smaller for geometries requiring gable roof drift loading.

7.4 Recommendations

Based upon these observations, the authors suggest that a roof rain-on-snow surcharge be incorporated into the minimum flat roof load provisions. The specific recommendation, in English units to match the ASCE 7-98 primary format, is as follows.

For $p_g \leq 20$ psf

$$p_f = \begin{cases} I \cdot p_g \\ 0.7 c_e \ c_t I p_g + 4 \end{cases} \text{ (use the larger of)} \qquad (7)$$

For $p_g > 20$ psf

$$p_f = \begin{cases} I \cdot 20 \\ 0.7 c_e c_t I p_g + 4 \end{cases} \text{ (use the large of)} \qquad (8)$$

where c_e, c_t and I are the standard exposure, thermal and importance factors in ASCE 7. Furthermore, it is recommended that the flat roof minimum apply to the same grouping of roof geometries as suggested for ASCE 7-98. Note that as recommended in Equations (7) and (8) the minimum load would correspond to the larger of two scenarios. The first scenario (i.e., I·pg or I·20) corresponds to the single storm event (no wind, thermal or rain effects). The second scenario (i.e., 0.7 c_e c_t I p_g + 4) corresponds to rain in combination with "normal" wind and thermal effects.

7.5 Acknowledgement

The research described herein was supported by the Metal Building Manufactureres Association and the American Iron and Steel Institute. However the results, conclusions and recommendations are the authors alone and do not necessarily reflects the views of MBMA or AISI

REFERENCES

American Society of Civil Engineers (1998), "ASCE 7-98, Minimum Design Loads for Buildings and Other Structures," Reston, VA.

Architectural Institute of Japan (1996), "AIJ Recommendations for Loads on Buildings," Toyko, Japan.

Colbeck, S.C., (1977), "Roof Loads Resulting From Rain-on-Snow," Cold Regions Research and Engineeering Laboratory Report 77-12, Hanover, New Hampshire.

Del Corso, R. Granzer, M., Gulvanessian, H., Raoul, J., Sandvik, R., Sonpaolesi, L., and Stiefel, U., (1996), "The New European Code on Snowloads," Proc. 3rd Intn'l Conf. On Snow Engrg. Sendai Japan, May, 1996, A. A. Balkema/Rotterdam Pub.

Gray, D. and Male, D. (ed.) (1981), "Handbook of Snow; Principles, Processes, Management and Use," Pergamon Press Canada/Willowdale, Ontario.

National Research Council of Canada (1995), "National Building Code of Canada," Ottawa, Canada.

Niedringhaus, T. E., (1972), "Distribution of Mean Monthly Precipitation and Rainfall Intensities," U.S. Army Engineer Topographic Laboratories Special Report ETL-SR-72-5, Fort Belvoir, Virgina.

Niedringhaus, T. E., (1973), "Rainfall Intensities in the Conterminous United States and Hawaii," U.S. Army Engineer Topographic Laboratories Special Report ETL-SR-74-3, Fort Belvoir, Virgina.

O'Rourke, M. J. and Downey, C. (1999) "Roof Loads due to Rain-on-Snow Surcharge," Research Report, Dept. of Civil Engrg. Rensselaer Polytechnic Institute, Troy, NY pp. 122.

Structural Engineers Association of Washington (1998) "An Analysis of Building Structural Failures due to the Holiday Snow Storms," FEMA Mitigation Directorate/Bothell WA pub.

Tabler R. D., (1994) "Design Guidelines for the Control of Blowing and Drifting Snow," Strategic Highway Research Program SHRP-H-381, National Research Council, Washington, DC.

Wistar, S., (1997), private communication

Snow Engineering, Hjorth-Hansen, Holand, Løset & Norem (eds) © 2000 Balkema, Rotterdam, ISBN 90 5809 148 1

Anchor temperatures of avalanche defence snow supporting structures in frozen ground

M. Phillips, P. Bartelt & M. Christen
Swiss Federal Institute for Snow and Avalanche Research (SFISAR), Davos Dorf, Switzerland

ABSTRACT: Snow supporting structures represent a possible thermal disturbance in frozen ground because they are made of steel and therefore have distinctly different thermal characteristics than the ground they are anchored in. To determine their thermal influences, a test site located in alpine permafrost terrain equipped with experimental snow supporting structures has been investigated since 1997. Temperatures of two types of commonly used structures, snow bridges and snow nets, are measured on their supporting beams and on three types of anchors: micropiles, rope anchors and steel tubes. Reference temperature measurements are made in boreholes located in strategic positions on the test site. The long term effects of the structures on ground temperature are simulated using a two-dimensional finite element program that solves the instationary heat conduction equation. Results of measurements and simulations show that the steel structures do not cause significant modifications to the thermal regime of the ground.

1 INTRODUCTION

Snow supporting structures are commonly used in the Alps to retain snow in avalanche starting zones located above settlements and lines of transport. Structure sites can be located in permafrost areas, i.e. around 3000 m a.s.l., on potentially unstable frozen sediments. The structure foundations reach depths of 3 to 6 m, and penetrate the problematic active layer of permafrost which undergoes substantial temperature fluctuations. In this layer phase changes occur on a seasonal basis. Until present, it was not known whether or not the steel structures, which warm under the influence of direct solar radiation, can conduct heat into the ground via the foundations and contribute to slope instability on steep avalanche slopes. Damages to structures due to creep of frozen sediments have been observed, but the causes were not clear (Stoffel 1995). Snow supporting structures do modify the temporal and spatial distribution of the snowcover; in particular, they induce a delay in spring snow melt. On a long term basis however, this phenomenon has a slight cooling effect on the thermal regime of the ground (Phillips et al., 2000).

2 STUDY SITE

The study site Muot da Barba Peider is located above Pontresina, in the eastern Swiss Alps. It is a Northwest oriented avalanche slope in permafrost terrain at 2980 m a.s.l. The slope angle is 38° and the ground cover consists of rough scree and fines on gneiss bedrock. The volumetric ice content of the scree is 10-20% and there is ice in clefts and faults within the bedrock. The layer of scree is about 1 m thick and is underlain by a layer of impervious fines (clay and silt) of varying thickness. The scree is highly pervious: most water from snow melt or precipitation is immediately evacuated downslope on the fines.

3 METHODS

In 1997 experimental snow supporting structures were erected on the site. Two rows of snow nets and two snow bridges (Fig. 1) were built. Three types of anchors were used: micropiles, cable anchors, and pinched steel tubes. On the upper snow net and upper snow bridge, the anchors were equipped with YSI 46008 thermistors at various depths. The thermistors were attached to the anchors before the latter were inserted in boreholes. Grout was then injected into the boreholes. The thermistors are in direct contact with the steel foundations and the grout. Campbell CR10X data loggers are attached to the supports of the structures and thermistor wires are protected from rock fall by plastic tubes. A control borehole was drilled 50 cm away, parallel to one micropile and equipped with thermistors at depths equivalent to those of the thermistors on the micropile. Like the foundation boreholes, this borehole was filled with injection grout and the thermistor casings are in direct contact with the grout, attached to the outside of a plastic tube containing the thermistor wires.

The structure temperature above the ground surface was monitored on two steel supporting beams: an IPE 220 (double T) profile support on snow bridge 1 (Fig. 4) and a cylindrical, hollow swivel support on snow net 1 (Fig. 5). In both cases, the thermistor was inserted into a steel tube welded to the support 20 cm above the ground surface in a position protected from direct solar radiation and generally free of snow in winter due to the wind scooping effect around the support. The steel tube containing the thermistor was filled with silicone heat conducting paste (Dow Corning 340 silicone heat sink compound). The temperature was measured at two hour intervals. These supporting beams are linked to the underlying foundations with bolts and plates (Figs. 2a, 2b).

The aim of these measurements was to establish whether or not steel snow supporting structures conduct any heat into the underlying ground, to determine whether the type of structure (snow bridge or snow net) can have an effect on the thermal regime of the ground and to verify whether different types of anchor can influence ground temperature in different manners. Whereas the thermal conductivity of the ground at Muot da Barba Peider lies between 0.9 and 3.5 $Wm^{-1}K^{-1}$, the steel components of the structures have a thermal conductivity of approximately 60 $Wm^{-1}K^{-1}$.

It should be noted that although the thermistors were attached to the different anchors at corresponding depths, the nature of the terrain conditions during construction of the experimental snow supporting structures (steep slope, frequent borehole collapses during drilling, uneven ground surface) led to variations in the final location of thermistors after the anchors had been installed in their boreholes. This implies that temperature differences between different anchors must to a certain degree be attributed to the fact that the thermistors are not all at exactly the same depth in relation to the ground surface.

4 RESULTS

4.1 Support temperatures

Average daily temperatures of the snow bridge support and the snow net support are shown in Fig. 6.

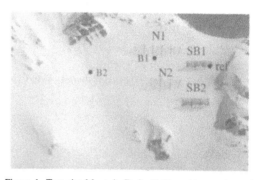

Figure 1. Test site Muot da Barba Peider with experimental snow nets (N1, N2) and snow bridges (SB1, SB2), 20 m boreholes (B1, B2) and a borehole 50 cm away from and parallel to a micropile (ref). Scale: N1 is 15 m long.

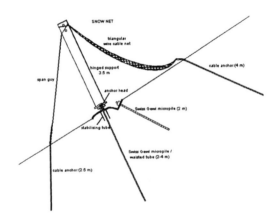

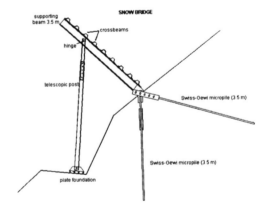

Figures 2a, 2b. Snow net (top) with micropiles and cable anchors; snow bridge with micropiles and steel plate foundations.

246

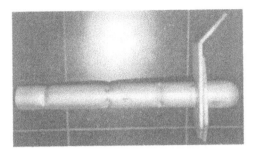

Figure 3. Pinched steel tube (can be used instead of a micropile under snow net support)

Figure 4. Snow bridge supports equipped with thermistors 20 cm above the ground surface. Note the wind-scooping effect in the snow around the supports.

Figure 5. Snow net supports equipped with thermistors 20 cm above the ground surface

The snow net support is generally warmer because it receives a little more direct solar radiation due to the nature of the local topography (potential solar radiation was measured at both locations using a solar compass – Phillips, 2000). Hourly temperature measurements are not shown but indicate that the maximum temperature measured on a support was on the snow bridge with a temperature of 22.6°C on 20.7.1998 at 16.00 hours.

The data illustrates how strongly support temperatures fluctuate on a daily basis. They are strongly influenced by air temperature, cloud cover, solar radiation, wind and precipitation (Phillips, 2000).

4.2 Anchor temperatures

In Fig. 7 the temperatures registered on a snow-bridge micropile (Swiss Gewi, 32 mm diameter) are shown along with the temperatures measured in the parallel reference borehole 0.5 m away, at depths of 2 and 4 m. The temperature differences between the micropile and the reference borehole are shown for both depths. At 2 m depth, the differences are always less than 0.4°C (average 0.03°C). At 4 m, the temperature differences between micropile and ground are even smaller (average 0.02°C). This can be explained by the fact that at 2 m the ground consists of sediments, in which convective heat transfer can occur (locally variable), whereas at 4 m, within the bedrock, only conduction occurs. These measurements indicate that heat is not conducted along the snow bridge supports to the underlying micropiles. The micropiles have virtually the same thermal regime as the surrounding ground. The strong temperature fluctuations registered on the structure supports are not transmitted to the micropile.

Comparison of the temperature evolution on three types of anchor under a snow net (Fig. 8) shows that the cable anchor and pinched tube have similar temperatures, whereas the micropile is warmer in winter and cooler in summer. This can partly be explained by the fact that the micropile thermistor is located slightly deeper than the thermistors on the other two anchors. Heat transfer at that depth is mainly by conduction (smooth temperature curve), whereas the cable anchor and the pinched tube display more sudden changes in temperature, particularly in spring during snow melt – which can be attributed to convective heat transfer (air and water transport). The differences in temperature on the three anchors may also be due to slight variations in local stratigraphy. The average temperature of both the cable anchor and the pinched tube over the entire measurement

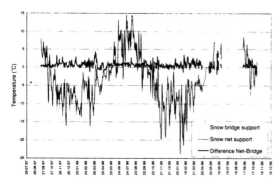

Figure 6. Support temperatures (24 hour averages) on N1 and SB between September 1997 and October 1999.

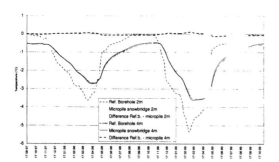

Figure 7. Snow bridge micropile and ground temperatures in a parallel borehole at 2 and 4 m depth. Temperature differences between micropile and ground are shown for both depths.

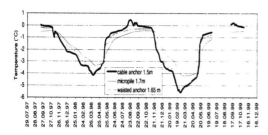

Figure 8. Temperatures of 3 types of anchors (cable anchor, micropile, pinched tube) under a snow net.

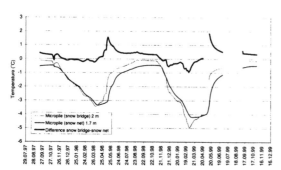

Figure 9. Temperature evolution on micropiles of two different types of snow supporting structure (snow bridge and snow net).

period is −1.9°C and that of the micropile is very slightly warmer : −1.7°C.

The thermistor on the snow net micropile (Fig. 9) is 30 cm nearer the ground surface than the thermistor on the snow bridge micropile. Despite this, the latter experiences greater and more rapid temperature fluctuations particularly during spring snow melt. This has two possible explanations: the ground may be more pervious in the vicinity of this micropile, allowing infiltration of melt water and thus rapid convective heat transfer – or the type of structure may influence ground temperature. As was shown in Fig. 7, heat does not seem to be conducted along the micropile from the snow bridge support to the anchor or from the anchor into the surrounding ground. Snow cover distribution is however more heterogeneous near snow bridges than snow nets as the former are less porous. It is possible that the presence of a reduced snow cover (and therefore reduced insulation) immediately below the snow retaining surface of the snow bridge leads to these differences which are less pronounced in the absence of a snowcover.

5 TEMPERATURE SIMULATIONS USING THE FINITE ELEMENT PROGRAM *HAEFELI*

The program *haefeli* solves the two-dimensional instationary heat transfer equation using the finite–element method. It was developed by the co-authors to simulate the creeping deformation and temperature distribution in phase-changing snowpacks (Bartelt et al., *these proceedings*) and was subsequently extended to treat permafrost. The thermophysical properties of the ground components (density, thermal conductivity and specific heat capacity) have to be specified, as well as the volumetric contents of rock, air, water and ice. Input parameters for boundary conditions on all four sides of the element mesh are ground and anchor temperature data measured in the field. These can be temporally variable (e.g. ground temperature data) or constant (e.g. a geothermal heat flux). The width and height of the mesh can be defined, the number of layers (up to 10) and their individual properties. Three node triangular elements are used in the simulations. The model is axis-symmetric. This is to avoid the anchor being considered as an endless component in the two-dimensional calculation. Phase change is taken into account by the model, which is of particular importance in an active layer containing moisture, which freezes and melts on a seasonal basis.

5.1 Simulation of the effects of anchor temperatures on ground temperatures over 100 years

Average monthly temperatures measured over 12 months (October 1997-1998) were run 100 times, as the snow supporting structures are assumed to have a design life of approximately 100 years. Climatic conditions are assumed to be those prevailing during the time of measurement. The lower boundary condition at 6 m depth is a heat flux of 0.03 W/m². The upper boundary condition is ground temperature measured in a reference borehole at 0.5 m depth. This depth was chosen in order to avoid having to include the problematic top layer of loose scree (in which convective heat transfer can occur) and the complex ground surface. The simulated anchor constitutes several lateral boundary conditions on the left side of the mesh which are derived from temperatures measured at different depths along a snow bridge micropile. Lateral boundary conditions on the right side of the mesh are measured ground temperatures in borehole B2. The thermo-physical properties of the ground were obtained from borehole core data and are shown in Fig. 10. Calculations are effected axis-symmetrically.

The temperature contour plots in figures 11 and 12 display simulated temperature distribution 1 m away from the micropile in year 1 and year 100. When compared with figures 13 and 14, which show temperature evolution 4 m away from the micropile, two differences can be observed:

1 Temperatures are colder nearer to the micropile (i.e. 1 m away) at the end of the winter (the -2°C isotherm reaches about half a metre deeper) than they are 4 m away from the micropile.

2 There is very little difference in temperature distribution between year 1 and year 100. In both cases the ground is slightly warmer at the beginning of the winter in year 100.

Apart from the above mentioned points, very little variation of temperature distribution occurs. Temperature differences observed are probably induced by a certain disturbance caused by the arrangement of the lateral boundary conditions along the micropile (Fig. 10). This also explains the fact that the isotherms are not smoothly rounded.

6 CONCLUSIONS

Temperatures of steel supports of snow supporting structures fluctuate strongly on a diurnal basis but anchor temperatures are not affected because the supports do not get very hot (< 30°C) and warming only occurs for short periods of time (hours) under

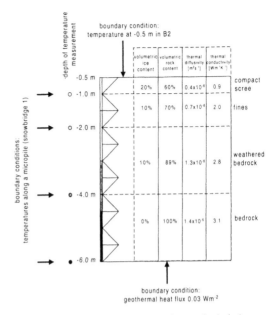

Figure 10. Boundary conditions and thermo-physical characteristics of the ground used for the simulation of the presence of a micropile in the ground.

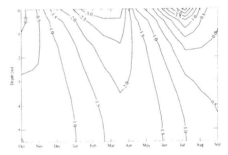

Figure 11. Spatial and tempoal ground temperature distribution at a distance of 1 m from a micropile (year 1).

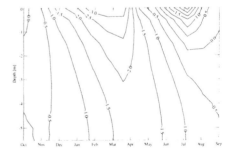

Figure 12. Spatial and temporal ground temperature distribution at a distance of 1 m from a micropile (year 100).

249

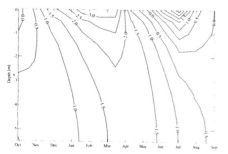

Figure 13. Spatial and temporal ground temperature distribution at a distance of 4 m from a micropile (year 1).

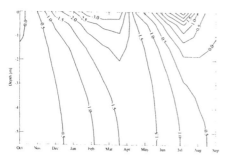

Figure 14. Spatial and temporal ground temperature distribution at a distance of 4 m from a micropile (year 100).

the influence of direct solar radiation. In addition, the contacts between the supports and the underlying anchors are of a complex nature (Fig. 2a,b), preventing effective heat conduction between the two steel elements. Heat does not appear to be conducted along the anchors or into the ground via snow supporting structures.

The different types of anchor have slightly different thermal regimes, but as the thermistors were not installed at exactly the same depths on all three due to technical problems, the interpretation of the temperature data is problematic, as small differences in depth near the ground surface can have significant effects on thermal regime.

Discrepancies in the temperature of a micropile beneath a snow net and a micropile beneath a snow bridge are probably due to the presence of a pronounced gap in the snowcover around the snow bridge supports (Fig. 4). The snow cover is less heterogeneously distributed in the vicinity of snow nets due to their higher porosity.

7 ACKNOWLEDGEMENTS

This project (Permafrost CH-2000) is supported by the BUWAL (Bundesamt für Umwelt, Wald und Landschaft) and by the Swiss cantons Graubünden and Valais. We are grateful to Patrik Thalparpan and Walter Ammann for their role in the organisation of the project and to numerous other people at SFISAR for their dedicated involvement.

REFERENCES

Bartelt P., Christen M., Wittwer S. Two-dimensional numerical simulation of the creeping deformation in a phase changing snowpack. *These Proceedings*.

Phillips M., Bartelt P., Christen M. 2000 (in press). Influences of avalanche defence snow supporting structures on ground temperature in alpine permafrost terrain. *Annals of Glaciology* 31.

Phillips, M. 2000. 1: Influences of snow supporting structures on the thermal regime of the ground in alpine permafrost terrain. Davos, Eidgenössisches Institut für Schnee- und Lawinenforschung.

Stoffel L. 1995. Bautechnische Grundlagen für das Erstellen von Lawinenverbauungen im alpinen Permafrost. *Eidg. Inst. Schnee- und Lawinenforsch. Mitt. 52.*

Snow Engineering, Hjorth-Hansen, Holand, Løset & Norem (eds) © 2000 Balkema, Rotterdam, ISBN 90 5809 148 1

Probabilistic description of ground snow loads for Ukraine

S. Pichugin
Poltava State Technical University, Ukraine

ABSTRACT: This material is based on a large bulk of meteorological data. Considerable probabilistic characteristics of ground snow load were revealed through its study. Ukrainian winters are changeable with little to much snow. Various models of snow loads were developed. Method of parameter comparisons of different snow models was worked out. All necessary probabilistic parameters were obtained for snow districts of Ukraine. It is demonstrated that the design snow loads are considerably lower than the actual ones. It's pointed out that the development of National Codes of atmospheric loads can be considered as one of the main tasks.

INTRODUCTION

Snow load probabilistic research for Ukraine is of great importance now. This is a problem which deals with the necessity for developing National Codes of atmospheric loads on building structures as well as the need for reliability design of buildings and structures, preventing structure failures which routinly occur every winter in Ukraine. Snow loads are described with different probabilistic models by many authors. However, these approaches lack both systematic analysis and model comparison which can cause different results in structure reliability design. This problem has been studied for several years at Poltava State Technical University. The results presented in this paper are the integral part of a structure reliability estimation method developed through the author's work (Pichugin 1995a,b).

1. SNOW LOAD PROBABILISTIC DISTINCTIONS

The results of regular snow measurements for 15 – 40 years at 62 Ukrainian meteorological stations have been taken as the reference statistic material. Ground snow load realisations were obtained, their intervals run 5 or 10 whole days. This statistic analysis has demonstrated some specific characteristics of snow loading in the territory of Ukraine.

1.1 *Snow season cycle*

During winter, snow load has two short transitional irregular parts. The beginning of winter is the period of snow accumulation, the end of winter is the snow melting stage. The main winter period lies between the winter average beginning date t_S and the average end date t_F. The $10^{th} - 15^{th}$ of November is the starting point of winter for Ukraine (t_S) and the 10th of April can be treated as the end of it (t_F). The period $t_W = t_F - t_S$ is the most important stage of stable snow cover. At this time snow load has relatively high values, which are of interest to the structure reliability design. Within this period (t_W) some general rules of snow load can be traced and they will be described later.

1.2 *Season winter period*

Seasonal changes of mathematical expectation $\bar{X}(t)$ and standard $\hat{X}(t)$ of snow load (t – is the time interval which is calculated from the 1st of September) is of skewness nature and its top corresponds to the middle of February (Fig. 1). This trend is described approximately as a polynomial of the 3rd degree. Meanwhile, the district coefficients of variation V and relative skewness S can be roughly treated as constants (Pachinski 1999). In this paper we use the snow loads for different districts of Ukraine in accordance with the Ukrainian Code (Loads 1987).

1.3 *Frequent character of snow loads*

Stable snow load (in the interval $t_W = t_F - t_S$) is of a stationary frequent character. This conclusion is based on the fact that snow normalised correlation function and effective frequency have no significant distinction during winter.

1.4 *Distribution of snow load values*

Experimental snow distributions for Ukrainian territory are of bimodal character. Therefore (Fig. 2), the normal law that satisfactory describes the snow loads of many snow districts cannot be applied in these cases. That's why so-called polynomial-exponential distribution was substantiated and used. Its normalised presentation is as follows:

$$f(\gamma) = exp\left(C_0 + C_1\gamma + C_2\gamma^2 + C_3\gamma^3\right), \qquad (1)$$

where $\gamma = (x - \overline{X}) / \hat{X}$ – normalised deviation of snow load.

If V and S are constant the coefficient of exponential index in expression (1) remains constant and does not depend upon the date (t). As a result snow load is stationary not only in frequency, but in normalised distribution of ordinate (1).

1.5 *Specific characters of snow polynomial-exponential distribution*

This distribution (Fig. 2) has an exponential maximum to the left at the original coordinates which means absence of snow during some winter periods. This is a specific characteristic of changeable Ukrainian winters with little snow. The second top of the curve is determined by the period of stable snow loads. It is interesting to analyse the influence on distribution of coefficients of the polynomial argument in expression (1). Term C_0 determines the original ordinate, C_1 determines down – going exponential part, positive coefficient C_2 – existence and height of a curve top. The last negative multiplier, C_3, suppresses the effect of C_2 to the level of $\gamma = (3 - 3.5)$ and it makes a curve tag down to the X-axis. Thus, it is not by coincidence that the distribution tags (1) are located lower then the normal distribution ones. It is necessary to note that charts of integral functions of snow load distribution $F(\gamma)$, are of more stable and smooth character in comparison with differential functions $f(\gamma)$ (Fig. 2).

2. SNOW LOAD AS A RANDOM PROCESS

2.1 *Outliers of snow random process*

Mentioned above, probabilistic specific characteristics were taken into account while presenting ground

snow load in the form of quasi-stationary differentiated random process. Mathematical expectation, $\overline{X}(t)$, and the standard, $\hat{X}(t)$, of the process change during a seasonal cycle just as the effective frequency, ω, and normalised distribution of ordinate (1) remain constant. The outlier frequency of this process for the moment (t) of a seasonal cycle can be derived from a stationary process (Pichugin 1995a)

$$\nu_+(x,t) = \omega\, f(\gamma) / \sqrt{2\pi} =$$
$$= \omega\, exp\left(C_0 + C_1\gamma + C_2\gamma^2 + C_3\gamma^3\right) / \sqrt{2\pi}, \qquad (2)$$

where $x = \overline{X}(t)(1 + \gamma V)$ is the chosen level of snow load, $\gamma = \left[x - \overline{X}(t)\right] / \left[\overline{X}(t)V\right]$ normalised load deviation.

Let us estimate the probability of snow load excess of level x per year (opposite problem). It is determined by the quantity of outliers of quasi-stationary random process over that level. It is calculated with the help of the integral expression (2) at the interval t_W

$$Q(x, t = 1 year) = N_+(x, t_w) = \int_0^{t_w} \nu_+(x, t) dx. \qquad (3)$$

The mean annual curve of snow load outliers for different districts were obtained by summation of numbers of outliers for the fortnight intervals. These curves are presented in Figure 3. They are characterised by sharp peaks at the beginning and long extended parts at the level of N_+ (1 year) = 0.2 – 1.0. These parts are formed by ground snow storage and the snow melting period during different months. This specific characteristic is typical for rather warm Ukrainian winters. The nature of outlier curves has no simple analytical description, but it is much simpler than the distribution (1).

Let us take into account the proportion between the number of outliers per year and return period $T = 1/N_+$ (1 year). Than, it is possible to determine the loads of different return periods directly using the outlier curves. These loads are shown on the additional low scale in Figure 3. Because of specific characteristics of snow loads an outlier comparatively low normalised snow load value $\gamma = 0.2 - 0.4$ correspond to the return period T (which equals 1 year). These values are located at the beginning of the rise. The transition to T period, which is 5 – 10 years occurs as the shift along the stretch curves to level with $\gamma = 1.5 - 2.7$. The loads of higher return periods (T = 20 – 50 years) are closely grouped, located on descending outlier curve parts $\gamma = 2.5 - 3.0$.

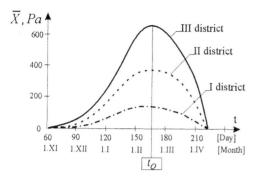

Figure 1. Seasonal change of the snow load mathematical expectation.

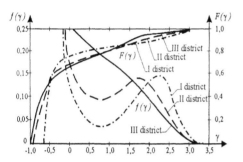

Figure 2. Normalised district snow load distributions.

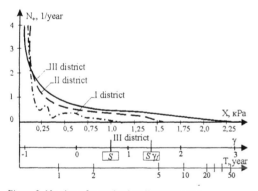

Figure 3. Number of snow load outliers (per year).

2.2 Trend coefficient

The trend coefficient, K_{tr}, was introduced to simplify computation and to get closed decisions. This coefficient equals the proportion of numbers of quasi-stationary random process outliers during the t_w period to the number of outliers during the same stationery process period. It corresponds to the trend top for $t_o = 165$ whole days:

$$K_{tr} = \frac{\int_0^{t_w} v_+(x,t)\,dt}{v_+(x,t_0)t_w}. \qquad (4)$$

Figure 4 illustrates the district values K_{tr}. They are in the intervals between $0.13 - 0.70$ for $T = 1 - 50$ years. If $T > 10$ years, they slowly descend. Instead of expression 3 for probabilistic description of snow load (inverse problem), Ktr application allows one to use the dependence for stationary process in the form of

$$Q(x,t) = \frac{tt_w K_{tr}}{\sqrt{2\pi}} \exp\left(C_0 + C_1\gamma + C_2\gamma^2 + C_3\gamma^3\right), \qquad (5)$$

where $\gamma = \left[x - \bar{X}(t_0)\right] / \left[\hat{X}(t_0)\right]$; t – is the serviceability term in years.

This approach allows, one to derive the normalised level of snow load γ (from 5). It corresponds to a given probability of exceeding [Q (t)] (direct problem). This decision is a root of a cube equation

$$C_1\gamma + C_2\gamma^2 + C_3\gamma^3 + \left\{C_0 + \ln\left[\frac{tt_w K_{tr}\omega}{Q(t)\sqrt{2\pi}}\right]\right\} = 0. \qquad (6)$$

3. SNOW LOAD PROBABILISTIC MODELS

3.1 Absolute maxima of a random snow process

As mentioned above, models in the form of quasi-stationary random process fully characterise ground snow load, but it takes a lot of initial statistical information which is difficult to access. The model, in the form of random snow load maxima, is more laconic and accessible. Distribution of maxima determined by the tag part of this random process is located higher than the level of characteristical normalised maximum γ_0. The letter is derived from the equation $N_+(\gamma_0; 0 \le \tau \le t) = 1$, where $N_+(\bullet)$ – the number of outliers of random process (Bolotin 1969). This distribution for absolute snow maxima has the integral and differential function:

$$F(\gamma,\gamma_0) = 1 - \frac{K_{tr}}{K_{tr0}} exp\left[C_1(\gamma - \gamma_0) + \right.$$
$$\left. + C_2(\gamma^2 - \gamma_0^2) + C_3(\gamma^3 - \gamma_0^3)\right], \qquad (7)$$

$$f(\gamma;\gamma_0) = \frac{K_{tr}}{K_{tr0}}\left(C_1 + 2C_2\gamma + 3C_3\gamma^2\right) \times$$
$$\times \exp\left[C_1(\gamma - \gamma_0) + C_2(\gamma^2 - \gamma_0^2) + C_3(\gamma^3 - \gamma_0^3)\right], \qquad (8)$$

where K_{tr} and K_{tro} are trend coefficients corresponding to the γ and γ_0 levels. It's easy to see that the distribution (8) is normalised (neglecting K_{tr}/K_{tro}). The example of a distribution like that is given in Figure 5.

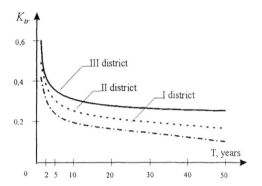

Figure 4. Trend coefficient of snow load.

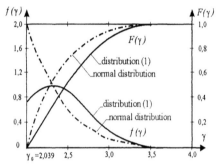

Figure 5. Absolute maxima of snow load random process (II district, t = 5 years).

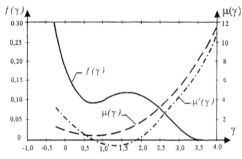

Figure 6. Estimation of the exponential condition for snow body sampling (II district).

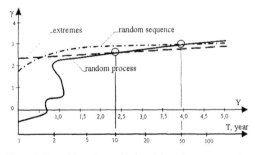

Figure 7. Exposition of snow load models at the extreme scale.

3.2 Random sequence and discrete presentation on snow load

Snow load is presented rather often as a random sequence of independent loads. This model is sometimes called a scheme of independent tests. The load intensity, $\lambda = N / t_\lambda$, is the frequent characteristic here, N is the number of loads; t_λ is the chosen time interval. The answer to inverse problem is derived from the use of the integral function $F(\gamma)$

$$Q(\gamma,t) = \lambda t \left[1 - F(\gamma)\right] =$$

$$= \lambda t \int_\gamma^\infty exp\left(C_0 + C_1\gamma + C_2\gamma^2 + C_3\gamma^3\right)d\gamma . \qquad (9)$$

Direct problem can be reduced to the solution of a cube equation

$$C_0 + C_1\gamma + C_2\gamma^2 + C_3\gamma^3 + ln\left[\frac{\lambda t}{\mu(\gamma)Q(t)}\right] = 0 , \qquad (10)$$

where $\mu(\gamma) = f(\gamma)/[1 - F(\gamma)]$ – distribution intensity of a snow load.

Applying to snow load, this model advantage is the possibility to obtain a priori λ parameter (for example for month or annual intervals) and relative access of meteorological data in the intervals mentioned above. Besides, this integral function $F(\gamma)$, is of a smoother character and is more suitable for computation in comparison to the differential function $f(\gamma)$ (Fig. 2). Discrete presentation of snow load is of the same form, its frequent parameter is mean duration of overloading $\overline{\Delta}(\gamma)$. In this case the design formula are derived from (9) and (10) with the help of substitution $\lambda = \overline{\Delta}^{-1}$.

3.3 Snow extremes

This presentation is widely used and it describes well annual snow maxima. Gumbel's well known distribution of the I-st type (Gumbel 1967) is used.

$$F(y) = exp\left[-exp(-y)\right]; y = \alpha_n(\gamma - u_n), \qquad (11)$$

where u_n and α_n are the characteristical extremum and experimental intensity correspondingly, which depend upon body sampling n_e.

Let's check the possibility of applying experimental presentation for maxima from snow load samples, subjected to polynomial and exponential distribution (1). With this purpose, we'll check the validity of condition:

$$\mu(y) = f(y)/[1 - F(y)] = \mu'(y) = -f'(y)/f(y).$$

Its left part is evaluated numerically, the right part is of closed type $C_1 + 2C_2\gamma + 3C_3\gamma^2$. As it is possible to see at Figure 6, the given condition is carried out from the level $(3.5 - 4)\gamma$. Thus, snow load sample is of an exponential type and the extremum use is quite correct in that case.

For selection procedure simplification of u_n let's connect general extreme proportions with snow load distribution

$$Q(u_n) = n_c^{-1} = f(u_n) / \mu(u_n) =$$
$$= exp\left(C_0 + C_1 u_n + C_2 u_n^2 + C_3 u_n^3\right) / \mu(u_n). \quad (12)$$

Cube equation for u_n estimation is derived from (12)

$$C_0 + C_1 u_n + C_2 u_n^2 + C_3 u_n^3 + ln\left[n_c / \mu(u_n)\right] = 0. \quad (13)$$

If relation $u_n = \gamma - y / \alpha_n$ is substituted into (12) we will obtain expression (14) for the body sampling (by α_n).

$$n_c = \alpha_n \, exp\left[-C_0 - C_1(\gamma - y / \alpha_n) - \right.$$
$$\left. -C_2(\gamma - y / \alpha_n)^2 - C_3(\gamma - y / \alpha_n)^3\right]. \quad (14)$$

Another expression for n_c is derived from (12) by u_n

$$n_c = \mu(u_n) exp\left(-C_0 + C_1 u_n - C_2 u_n^2 - C_3 u_n^3\right). \quad (15)$$

4. COMPARISON OF PROBABILISTIC SNOW MODELS

The connection of models in the form of random process, random sequence, and discrete presentation is described as follows

$$\lambda = \Delta^{-1} = \omega\mu(\gamma)K_{tr} / \sqrt{2\pi}. \quad (16)$$

The selection of corresponding extreme parameters can be performed numerically or with the help of extreme scale (Fig. 7). Normalised deviation of γ load is graphed on the ordinate axis and distribution argument (11) $y = -ln\left[-ln F(t)\right]$ or corresponding probabilities Q (t) of γ level exceeding connected with the return period of T load is graphed on the abscissa axis. Random process transition to the scale (Fig. 7) is carried out by annual outliers, the number of which equals $N_+(\gamma, 1\, year)$

$$y = -ln N_+(\gamma, 1\, year). \quad (17)$$

As it can be seen in Figure 7, a random process chart is of a complicated and irregular character if T < 5 years. The curves which illustrate random sequences and discrete presentations are of a smoother character and approach to random process curves at the top quite closely if T ≥ 5 years. Gumbel's extremes are shown by straight lines in Figure 7. The analysis demonstrated that the most suitable snow extremes are from the 10 year samples. The corresponding line is tangent to the random process curve at point T = 10 years, it exceeds, insignificantly,

snow loads for T = 10 – 50 years and it is higher than the irregular part of the curve for T < 10 years, which is of no importance.

5. MEAN SNOW LOAD PARAMETERS OF UKRAINIAN DISTRICTS

All necessary mean snow load parameters were determined, for the mentioned above probabilistic models of ground snow loads, for districts of Ukraine (Table 1). In accordance with existing Codes (Loads 1987) the territory of Ukraine can be devided into three snow districts. The first district stretches along southern and western Ukrainian parts; the second one is situated in the northern and western parts of Ukraine and the third one is the narrow area in the north-east territory. Analytical expressions, numeric methods, and extreme scales (Fig. 7) were used. Snow load probabilistic model parameters were determined on the basis of equality of normalised load level γ (T = 50 years) (Fig. 7).

There are, in Table 1, the largest values of mathematical expectation for all examined snow load models corresponding to the top of a seasonal trend, coefficient of variations and coefficients of polynomial-exponential distribution of the ordinate (1). In addition, some particular characteristics for every model are tabulated there (Table 1). These data allow one to give a full description of every possible snow model. It is necessary to pay attention to the fact that all the models give close results in normalising snow load and structure reliability estimation. Wind load is described by the author in the same way (Pichugin 1997).

6. PROVISION OF DESIGN VALUES OF GROUND SNOW LOAD

Generalised snow load parameters are tabulated in table 1 and they make it possible to vary the use of probabilistic structure design. In particular, using the obtained results, it was possible to estimate quantitatively the existing Codes of snow load for Ukrainian territory. The obtained results are given in Table 2 where S – is specified snow load; $S_{\gamma f}$ – design snow load; γ_f – load factor; $\gamma_n = (S - \bar{X}) / \hat{X}$ – standard deviation of standard load; $\gamma_d = (S\gamma_f - \bar{X}) / \hat{X}$ – the same of design load; T_n, T_d – standard deviation and return period correspondingly to the specified and design snow load.

The data given in Table 2 demonstrate that specified and design snow loads are of short return periods like T = 2.53 – 3.85 years. It justifies that design loads in accordance with the Codes (Loads 1987) are not ensured enough and are much lower than real ground snow loads for Ukrainian territories. Everything mentioned above gives evidence that the development of a National Ukrainian Snow Load Code is an urgent task. This fact was pointed out by the author ·in his previous works (Pichugin 1994). As a temporary measure the increasing of snow load factors can be proposed i.e. from $\gamma_f = 1.4$ and 1.6 to $\gamma_f = 2.4 - 3.0$ (Pichugin 1994).

Table 1. Probabilistic parameters of snow load

Model, parameter, symbol, unit	Numeric values for snow districts of Ukraine		
	I	II	III
Common parameters Mathematics expectation,			
$\overline{X}$, kPa	0.164	0.344	0.631
Coefficient of variation, V	1.60	1.26	0.920
Coefficients of equation (1)			
C_0	-2.265	-1.736	-1.313
C_1	-3.885	-1.926	-0.725
C_2	3.855	1.885	0.445
C_3	-0.920	-0.506	-0.178
Quasi-stationary random process Effective frequency ω,			
1/24 hours	0.141	0.095	0.073
Absolute maxima of random process Characteristical maximum			
level, γ_0	2.688	2.459	2.113
Random sequence Load intensity,			
λ, year^{-1}	7.90	4.86	3.20
Discrete presentation Mean duration of overload-			
ing, $\overline{\Delta}$, hour	13.55	18.31	18.92
Extremes per year / per 10 years			
Body sampling, n_e	7.14	3.98	2.42
	71.4	40.0	24.2
Characterictical extremum, u_n, kPa	0.738	1.074	1.163
	0.86	1.409	1.858
Extreme intensity,			
α_n, kPa^{-1}	17.883	6.860	3.310

Table 2. Probabilistic provision of the design snow load for Ukraine

Load value	Snow district			
		S, Pa	γ_n	T_n, years
Specified load	I	500	1.29	2.72
	II	700	0.82	2.94
	III	1000	0.64	2.53
		$S_{\gamma f}$, Pa	γ_d	T_d, years
Design load	I	700	2.06	2.43
	II	980	1.47	3.15
	III	1400	1.26	3.85

CONCLUSION

Ground snow load is of a quasi-stationary origin. Its mathematical expectation and standard have a seasonal trend. At the same time snow frequent characteristics and normalised ordinate distribution remain constant during the season. Snow load can be described faithfully by different probabilistic models: quasi-stationery differentiated random process with its absolute maxima, random sequence, discrete presentation, and extremes. System comparative analysis of these snow load models was performed, and parameters for Ukrainian districts were validated. These results can be used for practical design of structure reliability. It is substantiated that existing Codes considerably underestimate real snow loads and they are badly in need of updating.

REFERENCES

Bolotin, V.V. 1969. *Statistic Methods in Structural Mechanic.* San Francisco: Holden Day.

Gumbel, E.I. 1967. *Statistics of Extremes.* New York: Columbia University Press.

Loads and Loadings, 1987. Construction Standards and Rules 2.01.07-85. Moscow: Stroyisdat (in Russian).

Pashinski, V.A. 1999. *Building Structures under Atmospheric Loads in Ukraine.* Kiev: UkrNIIProectstalconstructsia.

Pichugin, S.F. 1994. Design Coefficients of Codes Based on the Reliability Analysis of Steel Structures. *Ukrainian Construction* (1): 18-20.

Pichugin, S.F. 1995a. Probabilistic Presentation of Load Actings on Building Structures. *News for High Educational Establishments. Construction* (4): 12-18 (in Russian).

Pichugin, S.F. 1995b. Reliability Estimation of Steel Elements under Variable Loads. *XLI Konf. Naukowa KILIW PAN i KN PZITB "Krynica 1995".* Krakow Krynica: Poletechnica Krakowska.

Pichugin, S.F. 1997. Probabilistic Analysis on Wind Load and Reliability of Structures. *Proc. 2nd European & African Conf. Wind Eng., Geneva, Italy, June 22-26 1997.* Padova: SGFditoriali.

Snow Engineering, Hjorth-Hansen, Holand, Løset & Norem (eds) © 2000 Balkema, Rotterdam, ISBN 90 5809 148 1

European snow loads research project: Recent results

L. Sanpaolesi, R. Del Corso & P. Formichi
Department of Structural Engineering, University of Pisa, Italy

ABSTRACT: The European Commission DGIII/D-3 sponsored in 1997 a pre-normative research activity aimed at improving the scientific knowledge and models for the determination of snow loads on buildings, by producing a sound common scientific basis, which can be accepted by all European countries involved in the drafting of Eurocodes.

The study, recently concluded, was carried out by eight important research Institutes from six different European countries and lasted three years.

The research programme was in two consecutive phases. Phase I, concluded on March 1998, provided methods and techniques for the determination of ordinary and exceptional snow loads on the ground and was finalised in the production of a new European ground snow load map. Phase II, concluded in June 1999, investigated methods and techniques for determination of ordinary and exceptional snow loads on roofs and defined appropriate criteria for determining the serviceability loads on such roofs.

The papers paper gives an overview of the main achievements of the research activity, which are also widely discussed in many other contributions presented in this Conference.

1 INTRODUCTION

At the end of the activity of the Project Team for the draft of the ENV 1991-2-3 Snow Loads, in 1993, many problems were left unresolved in the field of snow loads on buildings, due to the lack of a sound common scientific background, therefore intermediate provisions were given.

The implementation of the Eurocode on snow loads lead the European Commission DGIII/D-3 to sponsor in 1996 a wide pre-normative research activity aimed at giving a scientific based answer to the principal current open problems.

The study, started in December 1996 and concluded in June 1999, was carried out by the following research Institutes from six different European countries:

1. Building Research Establishment Ltd, Construction Division (United Kingdom)
2. CSTB, Centre de Recherche de Nantes (France)
3. Ecole Polytechnique Fédérale de Lausanne, (Switzerland)
4. ISMES, Structure Engineering Department (Italy)
5. Joint Research Centre, ISIS (EU)
6. SINTEF, Civil and Environmental Engineering (Norway)
7. University of Leipzig, Institute of Concrete Design (Germany)

8. University of Pisa, Department of Structural Engineering (Italy)

The research programme is in two consecutive phases, each one divided in two tasks as follows:

Phase I (Sanpaolesi et al. 1998)

Task Ia: "Development of models for the determination of snow load on the ground"

Task Ib: "Development of models for exceptional snow loads"

Phase II (Sanpaolesi et al. 1999)

Task IIc: "Definition of criteria to be adopted for serviceability limit states verifications"

Task IId: "Analytical study for the definition of shape coefficients for the conversion of ground snow loads into roof snow loads".

The study allowed to collect snow data both from Meteorological Offices and from direct experimental measures, in an format as much homogeneous as possible all over Europe. Based upon such data statistical investigations were performed to define the European ground snow load map, which represents one of the most important outputs of the work, eliminating or reducing the discrepancies arising at borderlines among different countries and harmonizing the different snow loads treating criteria within CEN area.

During the investigations, related to the definition of the map, were studied the "exceptional snowfalls" which have been responsible of many failures mainly in maritime climates. A criterion for the identification of these falls and a proposal on how to treat them were defined.

Serviceability combination factors were investigated starting from the available daily snow data, following various statistical procedures, in order to define the optimal ψ coefficients for SLS verifications. Important and interesting results were obtained in particular with the differentiation of ψ values within the European climates.

Finally were investigated the coefficients for the conversion of ground snow loads into roof snow loads. Both theoretical and experimental studies were carried out. The latter represent an innovative investigation in temperate climates since similar studies were performed only in cold climatic areas.

2 DEVELOPMENT OF MODELS FOR THE DETERMINATION OF SNOW LOAD ON THE GROUND

Usually a probabilistic model is applied to represent variable loads. Similar to other actions, e.g. actions due to wind, temperature or earthquake, the snow load on roofs varies not only with time but also in space (topographical position of the site). In the present research these two influences were treated separately. Firstly the variation of snow loads with time (usually a period of many years) was investigated at pre-defined places represented by the stations of observation. By applying extreme value statistics to these observations, a snow load was derived corresponding to a given probability of exceedence. This was fixed to 0,02, as specified in the Eurocode 1 Basis of Design (CEN, 1994), which corresponds to a mean recurrence interval equal to 50 years.

Secondly the geographical distribution of snow was investigated.

Geographical Information Systems (GIS) were used in order to produce the European ground snow load map. A range of options and computer software for handling, interpreting and visualising the relevant data were explored and appropriate recommendations made.

During the map elaboration phase criteria were defined for regionalisation of data in order to achieve consistency across Europe in deriving a map not substantially influenced by national boundaries.

2.1 Basic snow data and statistical model

The work has reviewed current practice in eighteen European countries and in consultation with the appropriate National Meteorological Offices, has identified the statistical techniques and data that were both available and required for determining characteristic snow load values (Sanpaolesi et al. 1998, Gränzer 2000).

Only a few countries offered water equivalent values (weight of the melted snow cover) which could be used directly. In other countries only depth measurements were available. They had to be transformed into loads by using appropriate densities models, which were defined according to the state of the art knowledge available in each CEN member state.

The analyzed database is made up by 2600 weather stations, where records have been kept for periods usually between 20 and 40 winters, though at some stations for more than 90 winters.

Similarities and differences between individual national approaches were identified, which led to the need to develop a reference model for statistical analysis of ground snow loads, to be adopted all across Europe. This statistical model was defined on the basis of the values of the best coefficients of correlation, obtained during the interpolation of the sampled snow data, according to the following three different cumulative distribution functions: Gumbel (Type I), Lognormal and Weibull.

Notwithstanding slight differences, due to the variation of climatic conditions (Lognormal distribution seems to fit better the samples in continental climates, while Type I fits better in maritime and mild climates) it was agreed to use, as reference procedure, in order to avoid inconsistencies, the Gumbel (Type I) distribution, determining its parameters by means of the least square method.

2.2 Regionalisation, European snow map

Once obtained the 50-yearscharacteristic ground snow loads at each weather station the following step is to investigate the geographic variation of the loads through the regionalisation of the available data.

Ten different homogeneous climatic regions were identified on the basis of geographic and climatic consideration and their influence on snowfalls. These regions are shown in figure 1.

Within a distinct climatic region the snow load is considered to be a function of the altitude of the site, but affected by random deviations (including also topographical effects). This allows a mean altitude function to be defined for each climatic region, except for those areas with show a poor correlation with altitude.

Within each region were identified homogeneous zones, where given parameters for the altitude-load formula hold.

A spatial interpolation analysis was performed with GIS techniques to get the spatial representation of the variation of snow loads on the ground, reduced at sea level, all across the examined climatic area.

Particular attention was paid to the setting of the parameters of the GIS interpolation phase and special devices were employed to reduce the inconsistencies at borderlines between climatic regions.

Figure 2 shows, as an example, the altitude function for the Alpine climatic region. In figure 3 it is shown the ground snow map for the Alps. It can be seen how the borders of the climatic region have been defined irrespective of the administrative borders of the countries in the studied area.

The map was deeply checked by two different validation procedures. Characteristic values obtained at each weather station were compared with the corresponding map values and with the Eurocode values. Generally very good results are found and the map values appear to be very slightly conservative if compared with the calculated characteristic values.

The other check performed was the comparison, at borderlines between climatic regions, of the ground snow loads values obtained at given points on the border, by applying the formulas holding in each neighbouring region.

Due to the use of a particular interpolation technique at borderlines, very small differences are generally registered.

3 DEVELOPMENT OF MODELS FOR EXCEPTIONAL SNOW LOADS

In some regions isolated very heavy snow falls have been recorded resulting in snow loads significantly larger than those that normally occur. Such snowfalls significantly disturb the statistical processing of more regular snowfall data.

Before the present study only the French code gave information on how to identify and treat these exceptional values in the ultimate limit states verifications.

"Exceptional" values were encountered in many areas of Europe, mainly in maritime and coastal areas and in mild climates where snow is intermittent and generally short duration snowfalls are registered. These values were defined in a numerical way and places where such exceptional values were encountered were localised in Europe, in order to find out geographical, meteorological and all other possible sources of influence which should have lead to the registered exceptional snowfalls (Sims, 2000).

The figure 4 shows an example of exceptional snow load according to the given definition: *if the ratio of the largest load value to the characteristic load determined without the inclusion of that value is greater than 1,5 then the largest load value shall be treated as an exceptional value.*

The two lines fitting the sample show the effect of the maximum value when this is included or excluded from the it.

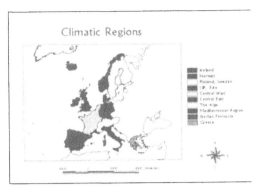

Figure 1. European climatic regions.

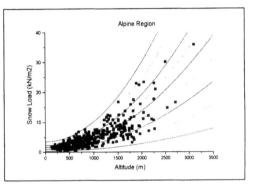

Figure 2. Alpine Climatic Region: Altitude-Load relationship

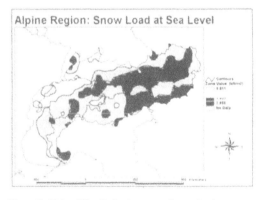

Figure 3. Alpine Climatic Region: ground snow load map.

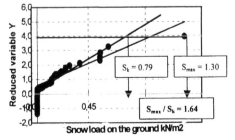

Figure 4. Exceptional snow load at Pistoia (Italy) 58 m a.s.l.

259

Figure5. Location of the European weather station where exceptional snow load were registered.

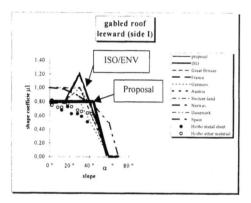

Figure 6 Shape coefficient for the leeward side of a duo-pitched roof.

When encountered, the exceptional values were discarded from the samples to calculate the characteristic ground snow load, used for the definition of the map.

Over the 2600 weather stations analysed, at approximately 160 of them exceptional snow loads were found. The location of such stations is shown in figure 5.

A proposal for the treatment in the Eurocode of the exceptional values was set up via the verification of an accidental design situation, where the design value of the accidental action is defined as:

$$A_d = k\, s_k$$

Were k is a constant value greater than 1,5 , suggested to be equal to 2,0 and s_k is the characteristic ground snow load at the site considered.

4 DEFINITION OF CRITERIA TO BE ADOPTED FOR SERVICEABILITY LIMIT STATES VERIFICATIONS

Serviceability load combination factors were investigated, starting from the available daily snow data and following different statistical procedures, in order to define the optimal ψ coefficients for ULS and SLS verifications. Important and interesting results were obtained, particularly with regard to the differentiation of ψ values within European climates.

4.1 *Derivation of combination factor ψ_0*

In the Eurocode 1 Basis of Design (CEN 1994) the combination factor ψ_0 is applied to the snow load effect when the dominating load effect is due to some other external load, such as wind, to which, in the present studies, investigations were restricted. Accordingly, a derivation of this combination factor strictly requires a refined modelling of both the snow and wind including the modelling of their variation with time. However, procedures based on such a refined model are typically time consuming both with respect to collection of input data, numerical algorithms and computation time. As a consequence, simplified procedures for the derivation of this factor were also investigated.

In particular the following three different approaches, as described in the Eurocode 1 Basis of Design (CEN 1994) and ISO 2394 (ISO 1998) background documents, were employed:
(i)   Simplified methods: the Turkstra's rule and the Design Value Method;
(ii)   the Borghes-Castanheta model;
(iii) the upcrossing-rate analysis model.

These were applied to snow and wind data coming from several weather stations in Germany and Norway, leading to the following main conclusions:
▪ the factors obtained from the Borghes-Castanheta and upcrossing rate models agree fairly well;
▪ the values obtained from the simplified method gives results that are generally on the safe side than those obtained from the other methods.

Due to the great difficulties in getting refined and high quality combined data sets for snow and wind data at each European weather station, it was decided to calculate the combination coefficient ψ_0 by the means of the Design Value Method (which gives good and slightly conservative results in comparison with Turkstra's rule) and extreme value distribution Type I.

One of the most interesting achievement of this task is undoubtedly the calibration of the values of the combination coefficient ψ_0 in the various climatic regions as shown in table 1.

Table 1. ψ coefficients for the European climatic regions

Region	ψ_0	ψ_1	ψ_2	Notes
Alpine	0,6	0,45	0,10	Altitude: > 1000m
	0,5	0,30	0,00	<= 1000m
UK and Eire	0,4	0,04	0,00	
Iberian	0,5	-	-	Altitude: > 500m
	0,4	0,00	0,00	<= 500m
Mediterranean	0,5	0,10	0,00	
Central East	0,55	0,40	0,10	Altitude: > 500m
	0,4	0,20	0,00	<= 500m
Central West	0,4	0,05	0,00	
Greece	0,5	0,00	0,00	
Norway	0,7	0,50	0,20	Altitude: > 300m
	0,6	0,40	0,20	< 300m
Finland-Sweden	0.65	0,60	0,20	
Iceland	0,6	0,20	0,00	Area: South-west
	0,6	0,40	0,10	North-East

The obtained range for the values put in evidence the need to differentiate the combination coefficient as it is recommended in the Eurocode 1, where, at the moment only one value, equal to 0,6, is given for the whole CEN area.

4.2 Derivation of ψ_1 and ψ_2 for the determination of the frequent and quasi permanent values

Starting from the definition given in the Eurocode 1 Basis of Design (CEN, 1994) the frequent value for snow load is determined such that:

1. the total time, within a chosen period of time, during which it is exceeded for specified part; or
2. the frequency with which it is exceeded is limited to a given value.

For snow the duration of load above given thresholds should be used as the only criterion. The frequent value is determined such that the fractile of time during which it is exceeded is chosen to be 0,05.

Always according to the Eurocode 1 Basis of Design the quasi-permanent value is determined such that the time during which it is exceeded, is a considerable part of the reference period of time. This may be set as 0,5 of the reference period.

Any evaluation of ψ coefficients for snow loads has to start from the time series observed at different weather stations.

Snow load on the ground may be considered as a process in time, to which it is possible to associate a probability of occurrence to each snow load level.

The corresponding cumulative distribution function of daily snow loads makes it possible to obtain the load level having a certain probability P of not being exceeded. The ψ_1 and ψ_2 coefficients are the relation between these fractiles of P probability and the characteristic value of the snow load.

In the present research the investigation was based directly on the empirical cumulative distribution function, representing the observed short-term snow loads at 63 weather stations spread all over Europe.

In Eurocode 1 Basis of Design (CEN 1994) the factor ψ_1 is specified as 0,2 and the factor $\psi_2 = 0,0$, but it is written that modification for different geographical regions may be required. The results from the present research clearly show the necessity of this differentiation. In particular, the difference between the maritime and continental (and/or mountainous) climates results evident. As shown in table 1, the factor ψ_1 reaches its maximum values Finland and Sweden (0,6), in the Alpine region (0,45), in Norway (0,5) and in Iceland (0,4), i.e. in regions of continental/mountainous climate. Also ψ_2 values in these regions deviate from zero (the maximum is in Norway and it is equals to 0,2). On the contrary, for areas of a maritime climate (UK, Mediterranean, Iberian, Central West, Greece) the ψ_1 values are very small (maximum 0,10 in the Mediterranean) and ψ_2 values are equal to zero.

5 ANALYTICAL STUDY FOR THE DEFINITION OF SHAPE COEFFICIENTS FOR THE CONVERSION OF GROUND SNOW LOADS INTO ROOF SNOW LOADS

In the majority of the current codes the roof snow load is normally calculated from the ground snow load by multiplying by a conversion factor, which accounts for the roof shape, thermal characteristics, exposure and other possible influencing factors, such as the presence of glass parts in the roof.

Experimental investigations and theoretical modeling on the snow deposition on the roofs were mainly developed in cold countries, such as Canada, USA, Russia and Norway. No relevant information were available for European climatic area. To emphasize the importance of these coefficients in design, their values, in current codes, generally range from 0,8 to 1,6 and occasionally even higher.

Within the present research, it was made an attempt to cover the lack of scientific background for the determination of roof shape coefficients.

As a first step it was reviewed the state of the art about the theoretical modeling of the snow deposition on roofs, with particular reference to the snow drifting, to the metamorphism and to the ablation models.

From the above investigations were obtained many important information, aimed at setting up an experimental measurement to collect in-situ measurements to calculate shape coefficients.

The results of the experimental measurements campaign, which actually lasted for one complete winter and for a part of a second winter, are to be considered a starting point for the assessment of the shape coefficients commonly adopted in the Eurocode.

Measurements were carried out in four CEN countries, representing different European climates: Germany, Italy, Switzerland and United Kingdom.

A total number of 81 roofs were monitored, of which 26 flat and the remaining 55 gabled.

The techniques of measurements adopted were quite different. The simplest was the use of wooden poles installed on the roof to measure the depth of the snow layer on the roof itself, to be compared with the snow layer on the ground; both values were converted into snow loads by the measurement of density.

More complex devices were used in UK and in the Italian Alps. Snow pillows were installed directly on the roofs and on the ground to record the snow load variation with time.

Notwithstanding the scarcity of time available to perform measurements, several goals were met.

For the first time in temperate climates it was directly measured the variation of the ratio between snow load roof and snow load on the ground, which, even if for a very limited set of roof shapes, leads to some important conclusions.

There is evidence that, in particular for gabled roofs with slope equal to 30°, the leeward side shape coefficient does not reach as high value as 1,2 as it is shown in the figure 6. This is also in accordance with the investigations carried out by Høibø in Norway (Høibø 1989).

Another evidence is that for the windward side of a gabled roof a sensible reduction of the shape coefficient is observed for roof angles higher than 15°-30°, which is in line with the present Eurocode provisions.

Considerations were also possible for the evaluation of the exposure coefficient, which is normally set to 1,0 in the Eurocode. From the measurements the following range of values was found: from 0,7 in windswept areas up to 1,1 in sheltered sites.

Another important goal met is the assessment of a sufficiently reliable set of density measurements giving further support to the density models chosen in the first phase of the research (Del Corso, Formichi 1999).

In parallel to the in nature experimental campaign another important experimental activity was carried out at the climatic wind tunnel available at CSTB in Nantes (France).

Within the wind tunnel investigations were tested several roof shapes, to integrate the in-situ data.

Flat, gabled, cylindrical, two levels flat roofs, multipitched roofs of different sizes were reduced to scale and tested in the wind tunnel with different wind directions and intensities.

The in-situ measurements were also used to calibrate the wind tunnel parameters for the reproduction of the natural conditions of snow and of the other environmental features.

The collected information represent a huge database from which important conclusions can be drawn. In particular useful information are now available for a better understanding of drift patterns in complex roof shapes, such as multipitched or multilevels roofs (Sanpaolesi et al. 1999).

6 CONCLUSIONS

The European Commission DGIII/D-3 sponsored in 1996 a wide prenormative research activity which lasted three years and which was carried out by eight European research Institutes from six CEN countries.

The research programme allowed to study on European scale many important open problems in the filed of snow loads on buildings.

One of the main achievements of the research is the new European ground snow load map, which has been derived irrespective to administrative borders of CEN countries and which is based upon a fully updated snow database, which covers 18 CEN member states, with about 2600 weather stations, with time series varying from 20 up to 100 years. This database, now available, allowed to establish a common an homogeneous procedure to get the characteristic snow loads on the ground all over Europe.

The map was drawn by an extensive use of the most updated GIS software, which made possible refined interpolations ad fittings of the data.

During such studies were defined and identified the "exceptional snowfalls", an important phenomenon interesting maritime and temperate climates, where snow is more intermittent and irregular. A normative proposal on how to treat the exceptional snow values was elaborated.

The huge database of daily ground snow loads allowed in-deep studies about snow loads for serviceability verifications and a new and articulated set of ψ coefficients was derived.

Finally, a coordinated both theoretical and experimental activity for the study of roof snow loads was carried out and relevant information for European climate was obtained.

Future research work is still needed, but it can be said that the European snow loads research project represents and important step forward in the knowledge in the field of snow loading in European climatic area and the obtained results are a reliable scientific basis to improve the elaboration of the European CEN code on snow loads.

REFERENCES

CEN, 1994. ENV 1991-1: 1994 Basis of Design and Actions on Structures, CEN Central Secretariat Brussels

CEN, 1995. ENV 1991-2-3: 1995 Actions on Structures – Snow Loads, CEN Central Secretariat Brussels

Del Corso, R., Formichi, P. 1999 *Shape coefficients for conversion of ground snow loads to roof snow loads.* 16th BIBM International Congress. Venice May 25-28 1999.

Ellingwood, B., Redfield, R., 1983. Ground snow loads for structural design. Journal of Structural Engineering, ASCE, Vol. 109 n°4.

Gränzer, M., 2000 *New ground snow loads for an European standard.* Ibid.

Gumbel, E. 1958 *Statistics of Extremes*, N.Y./London, Columbia University Press.

Høibø, H., 1988 *Snow load on gable roof. Results from snow loads measurements on farms buildings in Norway.* First International Conference on Snow Engineering, Santa Barbara – California, US Army Corps of Engineers, Special report 89-6

International Standard Organisation, 1998, ISO/FDIS 4355 Bases for design of structures – Determination of snow loads on roofs. Geneve.

Sims, P. et al., 2000 *European ground snow loads map: irregular and exceptional snow falls.* Ibid.

Sanpaolesi, L et al. 1998. *Scientific support activity in the field of structural stability of civil engineering works: Snow loads – Final Report Phase I*, Commission of the European Communities DGIII/D-3, Brussels

Sanpaolesi, L et al. 1999. *Scientific support activity in the field of structural stability of civil engineering works: Snow loads – Final Report Phase II*, Commission of the European Communities DGIII/D-3, Brussels

Sanpaolesi, L. et al., 1983. *Analisi statistica dei valori del carico di riferimento da neve al suolo in Italia (Statistical analysis of reference ground snow load in Italy).* Giornale del Genio Civile - Roma

Thom, H.C.S., 1966. *Distribution of maximum annual water equivalent of snow on the ground.* Monthly weather review (94): 265-271.

Snow Engineering, Hjorth-Hansen, Holand, Løset & Norem (eds) © 2000 Balkema, Rotterdam, ISBN 90 5809 148 1

Snow distributions on greenhouses

G. Scarascia-Mugnozza & S. Castellano – *University of Bari, Italy*

P. Roux & J. Gratraud – *Cemagref, Montpellier, France*

P. Palier & M. Dufresne de Virel – *Centre Scientifique et Technique du Bâtiment, Nantes, France*

A. Robertson – *Silsoe Research Institute, UK*

ABSTRACT: Greenhouses are very lightweight structures that are particularly sensitive to wind and snow loading Recent development of structural design standards has highlighted the need for specific data that accounts for the special requirements and functions of greenhouses (lightweight, relatively inexpensive, and often very large, structures that are covered with transparent plastic films or glass). One such need for data concerned snow load distributions. To address this issue, a programme of testing has been conducted in the Jules Verne Climatic Wind Tunnel Facility in Nantes (France). Tests have been conducted on a range of large-scale models of plastic-clad greenhouses to evaluate snow load shape factors. The results are presented and are compared with existing codified snow load shape factors.

1 INTRODUCTION

Structural optimisation of greenhouses is important in order to minimise capital costs in response to an extremely competitive market. It is equally important, however, because crop yield is directly linked to light transmission, and commercial profitability is therefore increased by minimising the presence of opaque structural members. Greenhouses are, as a result, very lightweight structures that are particularly sensitive to wind and snow loading. The total area of greenhouses in the EU is in excess of 70 000 hectares. Structural failures are not uncommon and can occur on a large scale. Some 400 hectares of greenhouses (predominantly polytunnels) collapsed under snow in the Perpignan region of France in January 1992 (Roux & Motro 1993). In southern Italy, in Puglia, on January 1993 about 100 hectares of glass and plastic greenhouses collapsed due to snowfall followed by wind action. These have disastrous consequences on the local horticultural economy for several years (Figure 1 illustrates structural damage produced by snowfall on polytunnels). In January 1976, approximately ¼ of all the polytunnels in England and Wales were destroyed by a wind storm. In many cases, the origin of these collapses was deficient design of the structure because of the absence of appropriate codified design procedures.

In response to such problems, standards have been created in several countries that provide minimum design requirements for greenhouse structures. A European initiative followed these national approaches and a European Standard on the design of commercial greenhouses prepared by CEN/TC 284 will be available for the first time in the near future (CEN, 1999). This draft European Standard ('prEN'13031) is based on the Structural Eurocodes, but wherever supplementary information was needed and was available, or wherever dedicated information (derived specifically from reputable tests on greenhouse structures) was available, it has been incorporated into the draft 'prEN'. For example, no appropriate rules were contained in Eurocode 3 for the design of slender monotubular steel arches that are commonly used for 'polytunnel' greenhouses, so structural tests and calculations were conducted on representative frames to formulate rules (Roux & Motro 1994, Roux & Lezeau & Motro 1995, Roux & Robertson & Motro 1997)

The development of this standard has highlighted the need for specific data that account for the special requirements and functions of greenhouses (lightweight, relatively inexpensive, restricted human occupancy, and often very large structures in which environmental control is important in order to maximise plant growth).

One such need for data concerned snow load distributions on single and multispan greenhouses. Currently, snow shape coefficients for greenhouses are extrapolated from data for conventional buildings. However, such data ignore some peculiarities of greenhouses that are likely to produce different snow loading distributions, such as:

- the unusual curved and arched roof shapes of many plastic-clad greenhouses (both single and multi-span),

Figure 1. Snow damages to polytunnels in Perpignan (France).

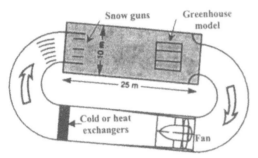

Figure 2. Jules Verne 'Thermal circuit'.

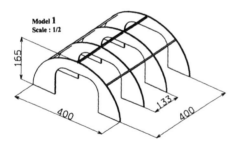

Figure 3. The arched model (dimensions in cm).

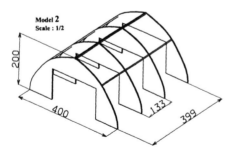

Figure 4. The 'gothic' arched model (dimensions in cm).

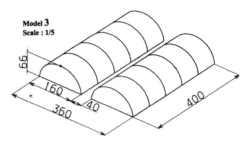

Figure 5. The two arched models (dimensions in cm).

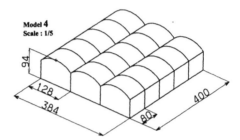

Figure 6. The multi-span model (dimensions in cm).

Figure 7. The full-scale arched greenhouse (dim. in cm).

Figure 8. The snow depth measurement device.

- the effect of different frictional/adherence properties of snow on flexible and semi-rigid plastic film cladding materials,
- the effect of higher rates of heat loss than in conventional buildings,
- the need to evaluate with higher precision the distribution of snow loads in order to improve calculations and design for mass manufacturing.
- the accumulation of snow in gutters in multispan greenhouses is a highly significant load case which may not be represented adequately by available data for conventional buildings.

To address some of these issues, a programme of testing has been conducted within the European research program « Access to large scale facilities » using the Jules Verne Climatic Wind Tunnel Facility at CSTB, Nantes, France.

2 METHODOLOGY

2.1 Climatic tunnel facility

The inner 'thermal circuit' tunnel at the CSTB Jules Verne facility was used. The working section measures 10 m wide by 7 m high by 25 m long. Different climatic conditions can be simulated such as wind (up to 38 m/s), snow, rain, sun, sand storm, large range of air temperatures, etc. In this study, three snow guns were used at the beginning of the working section as shown in Figure 2. A low wind is generated (about 2 m/s) for transportation of the snow from the guns to the test area.

2.2 Experimental models

Five test models were constructed, two at 1:2 scale (made from plywood as shown in Figure 3 and Figure 4), two at 1:5 scale also made from plywood (Figure 5 and Figure 6), and one at full scale (made from steel tubes as shown in Figure 7). The covering of the models (film or semi-rigid plastic sheet) was fixed at ground level on either side of the arches and, with the multi-span model, along the gutters. The covering is supported circumferentially by the arches and longitudinally by tubes (purlins) and polyester wires as is common practice, especially in France (the distance between wires was 15cm near the ridge, and up to 40 cm near the ground).

2.3 Instrumentation and measurements

A set of vertical external rods spaced at 10 cm intervals were used to measure snow depths. The rods, marked by 2 cm wide black and yellow graduations, were mounted on a frame as shown in Figure 8. Initially, the rods were positioned over the arch and each rod was lowered and clamped in position to reproduce the profile of the unloaded arch. The beam clamping the rods was then raised, enabling the whole frame to be removed for the duration of the snow test. The frame was then repositioned over the

arch and the beam lowered to its original position, causing the rods to penetrate into the snow and give a direct measurement of the profile of snow depth over the arch.

The snow depth was measured in the plane of a central structural arch were no displacement of the film can occur and in the plane midway between two central structural arches where maximum displacement of the film occurs due to snow accumulation. At this position, laser was positioned inside the greenhouse to measure the displacement of the film.

The duration of tests depend on the scale, the geometry, the type of cladding on the model, and on the characteristics of the snow.

2.4 Snow characteristics

Most of the tests were conducted using 'dry artificial snow' produced at an air temperature of $-10°C$. Snow samples were collected after each test, from the ground around the structure and on from the roof of the structure. The dry snow had an average density of 352 kg/m^3 (standard deviation on 12 samples s = 21.4 kg/m^3). Two tests were conducted with 'wet snow' on model 2 and 3. The wet snow was produced at an air temperature of $-5°C$. It had an average density of 397 kg/m^3 (standard deviation on 7 samples s = 46.3 kg/m^3).

The liquid water content of the snow (LWC) was also evaluated using a device which measured the dielectric constant (epsilon) of the snow. This parameter varies according to the quantity of liquid water in the snow. LWC is the ratio of the 'volume of liquid water' to the 'total volume of snow'. For a dry snow LWC is near 0% (less than 2%), for a wet snow (as used during these tests) LWC is about 5% to 8% and for a very wet snow is up to 10%.

2.5 Wind for snow transportation

Wind was generated to transport the snow from the guns to the test area. The wind speed varied from about 2.5 m/s at the beginning of the test down to 1.5 m/s at the end. This speed variation was due to frost formation on the heat exchanger which increased the pressure drop across it. The velocity of the snowflakes in the vicinity of the model is a combination of the wind speed, snowflake speed generated by the guns, and gravity. The chosen wind speed range (for snow transportation) was a compromise to obtain the most uniform snow distribution on the ground at the position of the model.

3 SNOW DEPTH ON THE GROUND IN THE CLIMATIC TUNNEL

The first test consisted of generating snow without any models and to measure the snow depth on the ground in the testing zone (see Figure 9). Measurement were made every 15 minutes up to 80 minutes.

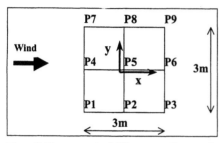

Figure 9. Measurement point for snow on the ground.

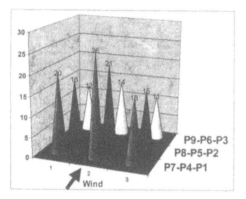

Figure 10. Snow depth on the ground after 80 min.

Figure 11. Snow distribution on Model 1 (covered with plastic film) with dry snow

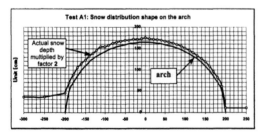

Figure 12. Measured snow distribution on Model 1 (covered with plastic film) at the position of a structural arch.

The amount of snow on the ground was higher on windward side of the test area (due to the trajectory of the guns and the wind-induced transportation of the snow) and some asymmetry was observed between the left and right sides as shown in Figure 10.

The measurements on models were made along the centre-line of the tunnel, so the important measurement positions in the present study were P4, P5, P6. Linear regression of the snow fall on the ground allows a 'theoretical' snow depth on the ground H_K (in cm) to be evaluated (Table 1). This is used in the evaluation of the snow load shape coefficients μ.

4 EXPERIMENTAL RESULTS

Results are presented as plots of the measured snow distribution on the central structural arch, and between central structural arches (where film displacements occur). For the latter plots, the influence of the 3 longitudinal tubes (purlins) is apparent (the purlins restrict displacement of the film which leads to greater snow depth on the ridge and on the purlins to either side). Supplementary photographs are included to give a qualitative appreciation of the snow distribution, including accumulations on the sides where appropriate.

4.1 Experimental results for arched model (model 1) covered with a film

The duration of this test (A1) was 75 min with dry snow. The snow depth was 18 cm on the wind-ward side and 5 cm behind the greenhouse. The average density was 310 kg/m³. In Figure 12, the snow depth has been multiplied by a factor of 2 to improve the clarity of the plot.

4.2 Experimental results for arched model (Model 1) covered with semi-rigid plastic sheet

The duration of this test (A2) was 75 min with dry snow. The snow depth was 30 cm on the wind-ward side and 14 cm behind the greenhouse. The average density was 350 kg/m³.

For this test, displacement of the semi-rigid cladding-sheet (Soplachim Ondex®) were insignificant.

Table 1 : Snow depth on the ground in wind direction.

Snowfall duration	Linear regression*
30 min	$H_K = -0.02(x) + 9.0$
45 min	$H_K = -0.02(x) + 13.3$
65 min	$H_K = -0.0267(x) + 16.7$
80 min	$H_K = -0.04(x) + 20.3$

* where x is the coordinate in cm along x axis centred on point P5 (negative x on windward side).

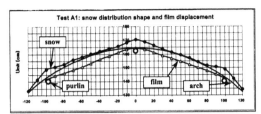

Figure 13. Measured snow distribution on Model 1 and film displacement measurement at a position between two structural arches.

Figure 14. Snow distribution on Model 1 covered with semi-rigid plastic sheet.

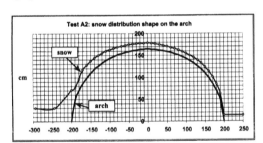

Figure 15. Measured snow distribution on Model 1 covered with semi-rigid plastic sheet.

4.3 Experimental results for gothic model (Model 2) covered with a plastic film

4.3.1 Gothic arch with dry snow
The duration of this test (A3) was 75 min with dry snow. The snow depth was 28 cm on the wind-ward side and 10.5 cm behind the greenhouse. The average density was 347 kg/m^3.

4.3.2 Gothic arch with wet snow
The duration of this test (A6) was 55 min with wet snow. The snow depth was 40 cm on the wind-ward side and 40 cm behind the greenhouse. The average density was 408 kg/m^3.

Figure 16. Snow distribution on Model 2 with dry snow.

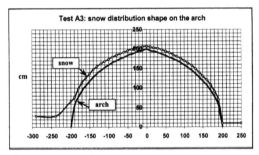

Figure 17. Measured snow distribution on Model 2 (covered with plastic film) at the position of a structural arch.

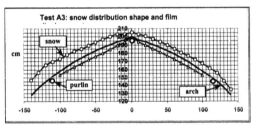

Figure 18. Measured snow distribution on Model 2 and film displacement measurement at a position between two structural arches.

4.4 Experimental results for two arched models (Model 3) covered with plastic film

4.4.1 Model 3 with dry snow
The duration of this test (A4) was 45 min with dry snow. The snow depth was 14 cm on the wind-ward side and 11 cm behind the greenhouse. The average density was 381 kg/m^3.

4.4.2 Model 3 with 'wet' snow
The duration of this test (A5) was 44 min with wet snow. The snow depth was 21 cm on the wind-ward

side and 9 cm behind the greenhouse. The average density was 383 kg/m³.

4.5 Experimental results for 'multispan' (model 4) covered with a film

The duration of this test (A7) was 105 min with dry snow. The snow depth was 16 cm on the wind-ward side and 9 cm behind the greenhouse. The average density was 384 kg/m³.

4.6 Experimental results for full-scale arched greenhouse (Model 5) covered with a plastic film

The duration of this test (B0) was 45 min with dry snow. The snow depth was 20 cm on the wind side and 7 cm behind the greenhouse. The average density was 357 kg/m³.

5 PRELIMINARY RESULTS ANALYSIS

5.1 Snow load shape coefficient and snow on the ground

In codes such as Eurocode 1 (CEN, 1994), snow loads on a roof (denoted 's') are assumed to act vertically and to refer to a horizontal projection of the area of the roof. Snow loads 's' are then determined from : $s = \mu\ C_e.C_t.s_k$ where s_k is the characteristic value of the snow load on the ground and (C_e,C_t) are coefficient that take into account exposure and thermal effect (generally both close to 1.0). From that formula, the snow load shape coefficient μ can be defined as the snow load on the roof divided by the snow on the ground.

If β is the angle formed by a tangent to the arch with the horizontal, then values of μ can be determined from the measurement of snow depth over arches as shown below (see Figure 32):

$$H_s \cdot (\ell \cdot \cos\beta) = H_{roof} \cdot \ell \ \Rightarrow \ H_s = \frac{H_{roof}}{\cos\beta}$$

by definition $\mu = \dfrac{s}{s_k} = \dfrac{H_s}{H_k}$

, where H_k is the snow depth on the ground

If this is applied, for instance to the results of test A3 (Gothic arch with dry snow), with H_K evaluated using the linear regression presented in Table 1 (for test A3, a linear regression for a snowfall duration of 65 min was used), then the snow load shape coefficient μ can be presented as function of the roof angle β as shown in Figure 33, or as a function of the position along the arch (Figure 34).

Snow load shape coefficients given in Eurocode 1 (EC1) for cylindrical roofs are presented in Figure 35. These rules assume that snow is not prevented

Figure 19. Snow accumulation on wind-ward side of Model 2 with wet snow (partially due to slipping on the roof). The snow in the foreground has been removed to reveal the cross-section of the snow distribution.

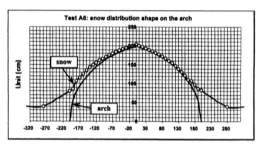

Figure 20. Measured snow distribution on Model 2 (covered with plastic film) with wet snow at the position of structural arch.

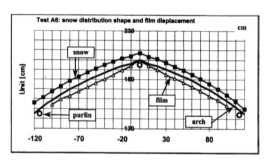

Figure 21. Measured snow distribution on Model 2 and film displacement measurement with wet snow at a position between two structural arches.

from slipping off the roof.

The present results (Figure 34) give very different μ coefficients from those contained in Eurocode 1 (Figure 35). This is primarily because the μ coefficient is not well suited to describing snow accumulations on the steep sides of arched structures. According to EC1, there is no snow load where the tangential roof angle exceeds 60° because of snow sliding off the roof, but this was not found to be the case in the present tests. Instead, as shown in Figures 19, 29 and 30, snow commonly slid from a position close to the ridge (where the tangential roof angle is very small) all the way down to the ground, producing a significant accumulation. If the snow accumulation zones in Figure 33 and Figure 34 are eliminated, μ then varies from about 0.4 to 1.7 which is much closer to the uniform distribution given in EC1, but none of the present experimental results resemble the asymmetric distribution given in EC1 (Figure 35).

It may be that snow accumulation data are not included in EC1 (Figure 35) because the provisions relate specifically to roofs constructed in the traditional manner on top of vertical walls. When snow slides from such roofs, it would simply fall to the ground and so exert no load on the roof structure, and arguably little load on the vertical walls. The EC1 data also presumably relate to traditional roofing materials (concrete, steel, etc) which will have different frictional/adherence properties from the plastic film and sheet investigated here. It should also be noted that the EC1 data apply only to 'cylindrical' roofs whereas the majority of polytunnels are of semi-elliptical or gothic-arch profiles.

5.2 Snow sliding and side accumulations

In all cases where the cladding was resistant to large local displacements (ie the semi-rigid cladding and the plastic film supported by tensioned polyester wires), the snow appeared to slide down the roof once a certain snow thickness was reached, as shown qualitatively Figure 36. This thickness depended on the snow characteristics and, on the friction or adhesion between the snow and the covering material. This may mean that whatever the duration of the snow fall, there is a limit to the depth of snow that can be sustained on the roof of an arched structure. Once the thickness reaches approximately 10 cm (Figure 36, step 1), the snow slides from the roof (see Figure 29 and 30) leaving either a central narrow strip of snow along the ridge (Figure 36, step 2a) or no snow at all over the top of the arch (Figure 36, step 2b). Snow then collects on the arch until again reaching a depth of approximately 10 cm (Figure 36, steps n_a and n_b) when it slides again (Figure 36, steps $n+1_a$ and $n+1_b$). Eventually, the accumulations due to sliding snow grow sufficiently far up the sides of the arch to restrict any further sliding down the roof (Figure 36, final step, accumulation height H_{acc}) and deposition then increases unabatedly over the entire arch. It is likely that this final scenario arose in severe snow fall such as in Perpignan in 1992 (ROUX & MOTRO 1993).

Figure 22. Snow distribution on Model 3 (two small tunnels) with dry snow.

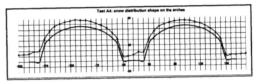

Figure 23. Measured snow distribution on Model 3 (covered with plastic film) with dry snow at the position of structural arch.

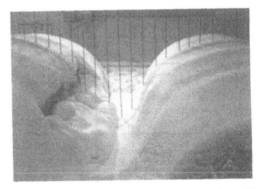

Figure 24. Snow distribution on Model 3 (two small tunnels) with dry snow.

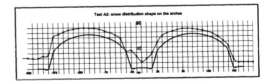

Figure 25. Measured snow distribution on Model 3 (covered with plastic film) with wet snow at the position of a structural arch.

Figure 26. Snow distribution on Model 3 (multispan) with dry snow.

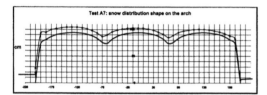

Figure 27. Measured snow distribution on Model 4 (covered with plastic film) with dry snow at the position of a structural arch.

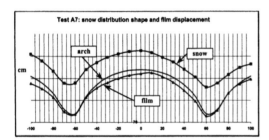

Figure 28. Measured snow distribution on Model 4 and film displacement measurement with dry snow at a position between two structural arches.

Figure 29. Snow distribution on windward side of Model 5 (full scale arched greenhouse) with dry snow.

Figure 30. Snow distribution on leeward side of Model 5 (full scale arched greenhouse) with dry snow, just after a snow slide.

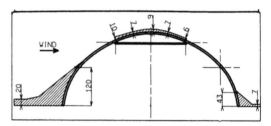

Figure 31. Measured snow distribution on Model 5 (covered with plastic film) with dry snow at the position of a structural arch.

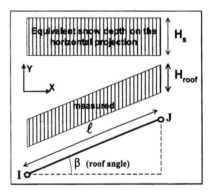

Figure 32. Snow loads on roofs and equivalent value on horizontal projection.

5.3 Film displacement effect

The effect of film displacement can be significant since it can lead to an unstable 'ponding' situation.

If not adequately supported to resist out-of-plane loadings, the film cladding deforms under an initial snow load to form a local pocket, or snow pond, which encourages further snow accumulation and deepening of the pocket etc.

The load amplification due to snow ponding can be quantified by comparing the snow load in the plane of a structural arch with that midway between

272

structural arches (see Figure 37) for those test cases where the film displacement was measured.

Table 2 presents a summary of the ratio of the accumulated snow volume with ponding (between structural arches) and without ponding (directly above a structural arch) for four test cases. This ratio is based on the hatched areas shown in Figure 37 which reflect the greatest differences, but the ratio ignores the fact that there is gradual transition from one distribution to the other. On this basis, the volume (load) is amplified by 10 to 22 % with dry snow, and by 178% with wet snow.

Currently, no rules are given in codes for the calculation of film supporting systems. It is clear that this is an important design consideration because of its consequences on snow load amplification and snow sliding. Further investigations and calculations of the effectiveness of supporting systems under snow loading are needed to provide design procedures.

5.4 Effect of snow characteristics

Only two artificial snow types were used, both produced at air temperature below 0°C. Comparisons with the distribution produced by natural snow would be of interest and value, particularly in relation to the sliding behaviour on the covering.

5.5 Scaling effect

It seems that model scaling has no direct influence on snow distributions up to the onset of any sliding. Thereafter, it appears that the tendency for snow to slide reduces as the scale of the model reduces. The likely physical explanation for this is that as the scale size reduces so the arc length of the snow patch that is inclined to slide shortens; for a given depth of settled snow, the weight of the patch thus reduces, rendering it less inclined to slide. Tests with small scale models are therefore likely to underpredict the snow sliding and snow accumulation behaviour that has been observed in the present tests on polytunnels (ie to fail to reproduce the intermediate stages of snow deposition portrayed in Figure 36).

Table 2 : Effect of film displacement on snow volume on the roof.

Model	SNOW quality	Test n°	Ratio : $V_{ponding} / V_{arch}$
Model 1 (arched model)	Dry	A1	1.22
Model 2 ('Gothic')	Dry	A3	1.20
Model 2 ('Gothic')	Wet	A6	2.78
Model 4 (Multispan)	Dry	A7	1.10

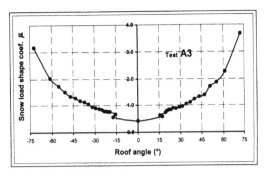

Figure 33. Snow load shape coefficient as function of roof angle β for Model 2 (Gothic arch).

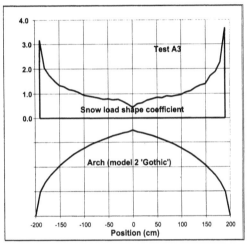

Figure 34. Snow load shape coefficient as function of position across Model 2 (Gothic arch).

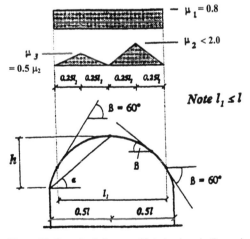

Figure 35. Snow load shape coefficients given in Eurocode 1 for cylindrical roofs.

273

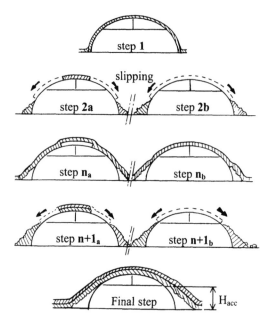

Figure 36. Qualitative analysis of snow deposition on arches during the tests.

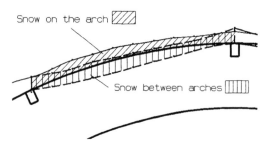

Figure 37. Snow accumulation (ponding) due to film displacement (test A1).

6 CONCLUSION AND PERSPECTIVES

The aim of this preliminary study was to supply data to the scientific community on snow distributions on greenhouses in order to derive codification proposals.

It is concluded from the present analysis that is premature to derive proposals for codification for two main reasons.

Firstly, assessments of the artificial conditions of the Jules Verne climatic tunnel are required particularly in relation to snow characteristics. To address this, the decision was taken to erect the full scale model (Model 5) in the French Cevennes during the winter of 1999/2000 to monitor the snow distributions produced under natural conditions.

Secondly, the suitability of the shape coefficient μ is questioned, particularly in relation to side ac-

cumulations. A complementary study is planned to determine the significance of side accumulations on the global resistance of arches. This will require non-linear calculations including the effects of imperfections as described in a recently proposed methodology for arch calculations (Roux & Robertson & Motro 1997). If side accumulation loads do influence global resistance significantly then appropriate design provisions will be required.

Film displacement has been shown to be a very important parameter both in relation to snow distributions (snow ponding), and the occurrence of sliding. Further work is required to determine the requirements for support systems that effectively avoid film displacement effects. It seems likely that future codification proposals for snow loads will need to reflect whether or not an adequate film supporting system is present.

7 REFERENCES

Roux, Ph., Motro, R., 1993. Etude du comportement global d'arcs monotubulaires: Pathologie des serres tunnels sous l'effet de la neige (Study of the global behaviour of monotubular arches: pathology of tunnel-shaped greenhouses under snow action). *Construction Métallique*, N°1-1993, 23-39

European Committee for Standardisation (CEN), 1999. Draft EN 13031-1 Greenhouses: Design and construction. Part 1: Commercial production greenhouses. CEN, Brussels, November 1999

Roux, Ph., Motro, R., Clavier, L. 1994. Incidence du comportement d'ancrages superficiels sur la résistance globale d'arcs monotubulaires: Application aux serres tunnels (Influence of the behaviour of soil anchorage on the global resistance of monotubular arches: application to tunnel-shaped greenhouses). *Construction Métallique*, N°2-1994, 27-41

Roux, Ph., Lezeau, S., Motro, R., 1995. Evaluation de la résistance d'un arceau de serre tunnel: Incidence du modèle de comportement des tubes (Evaluation of the resistance of a tunnel greenhouse arch: The significance of the moment-rotation behaviour of steel tubes). *Construction Métallique*, N°4-1995, 53-62

Roux, Ph., Robertson, A.P., Motro, R., 1997. The design of slender, monotubular steel arches. *The Structural Engineer* 75 (9) 143-151, 6 May 1997

European Committee for Standardisation (CEN), 1994. ENV 1991-2-3 Eurocode 1: Basis of design and actions on structures – Part 2-3: Snow loads, Brussels, 1994

Snow Engineering, Hjorth-Hansen, Holand, Løset & Norem (eds) © 2000 Balkema, Rotterdam, ISBN 90 5809 148 1

European ground snow loads map: Irregular and exceptional snow falls

P.A.C. Sims – *Building Research Establishment Limited, Garston, Watford, UK*

D. Soukhov – *Institut für Massivbau und Baustofftechnologie, Universität Leipzig, Germany*

D.M. Currie – *Formerly: Building Research Establishment Limited, Garston, Watford, UK*

C. Sacré – *Centre Scientifique et Technique du Bâtiment, Centre de Recherche de Nantes, France*

ABSTRACT: Infrequent, irregular, short duration snowfalls and single isolated very heavy snowfalls need to be taken into account when determining the characteristic values of ground snow loads for use in structural design. They require analysis techniques different from those for determining the characteristic values of frequent and long duration snow cover. This paper outlines work to determine loads corresponding to irregular and exceptional snow falls carried out as part of the European Snow Loads research project. It gives definitions for irregular and exceptional snow loads, develops criteria for identifying them, and outlines the statistical methods needed in their analysis. The methods are illustrated by examples taken from northern Germany and the locations of climatic stations within the eighteen CEN countries with exceptional snow loads are identified on a map of Europe.

1 INTRODUCTION

Some maritime European countries, e.g. UK, Spain, France, Italy, and part of Germany, experience irregular short duration snowfalls in contrast to continental and alpine Europe, where more continuous and longer lying periods of snow result. These irregular snowfalls are typified by the snow melting completely between successive snowfalls so that no residual accumulation of snow results on the ground.

Also throughout Europe, isolated very heavy snowfalls have occurred resulting in snow depths significantly larger than those normally encountered. When the annual maximum values of snow depth at meteorological stations are examined, they sometimes contain one or two very large values which unduly influence the data processing. Moreover, statistical techniques appropriate for analysing more continuous and longer lying periods of snow, are inappropriate for irregular, short duration snowfalls.

This paper outlines the approaches developed in the European Snow Loads research project (Sanpaolesi 2000) for treating both irregular and exceptional snow loads.

2 IRREGULAR AND EXCEPTIONAL SNOW LOADS: DEFINITION

Snow depth records at some meteorological stations may contain significant gaps, either during, or for complete, winter seasons. Either no snow fell, or the snow depth had not been registered because it was less than a prescribed measurement threshold. Thus there are 'snow-less winters' or 'zero-snow years' and this results in shorter and less populated data sets on which to carry out statistical processing. For this reason, statistical techniques for determining characteristic snow load values for regions with long-duration snow cover are not appropriate for regions with irregular snowfalls.

Accordingly irregular snow loads can be defined as: *Infrequent and usually short duration snow falls resulting in loads which may have long periods, sometimes years, between their occurrence.*

For the data record lengths typically available, including the data associated with isolated very heavy snowfalls may significantly disturb the statistical processing of the more regular snowfall data. Accordingly exceptional snow loads can be defined as: *Isolated and very infrequent snowfalls where the resulting snow load is significantly greater than the loads in the general body of snow load data and its inclusion in that data set distorts the statistical analysis.*

3 TREATMENT OF IRREGULAR GROUND SNOW LOADS

Two principal options are available for the analysis of data sets which include a significant number of 'zero-snow' years or have short data records. These approaches: a mixed distribution approach and an event-based maxima approach, are outlined in the following.

3.1 *Mixed distribution approach.*

Irregular snow loads have been treated previously, e.g. in France, using mixed distributions combining the probability of occurrence of snow within a winter (P_{snow}) with the probability of not exceeding a given value of snow load, S_k, when there is snow on the ground during a snow-winter. For the latter, annual snow load maxima were used.

The probability for snow load not to exceed a given value S_k is given by:

$$F (X < S_k) = P_{snow} \times F_{snow} (X < S_k) + (1 - P_{snow})$$

If the total number of observation years is N, and the total number of years with recorded snow load is n, then the probability that snow is present in a given year, P_{snow}, is equal to n/N. The cumulative distribution function of non-zero annual maxima, F_{snow}, is fitted by extreme value distribution type I (Gumbel).

Thus the probability that the snow load will exceed the value S_k (i.e. $F(X > S_k)$) is $(1 - F(X < S_k))$. If the value S_k has a return period (mean recurrence interval, MRI) of 50 years, then P_{snow} equals 0.98 and, correspondingly, the probability of exceedance of the value S_k in any one year is equal to 0.02. The value S_k, called in building standards as the characteristic value, can be then found by means of Equations 1 and 2:

$$F_{snow} (X < S_k) = [0.98 - (1 - P_{snow})] / P_{snow} \quad (1)$$

$$S_k = \{ -\ln[-\ln(1 - 0.02 / P_{snow})] / c \} + u \quad (2)$$

where c and u are the parameters of the Gumbel distribution.

This approach is similar to that proposed in Thom (1966) but the non-zero snow values are fitted to a Gumbel instead of a Log-Normal distribution.

3.2 *Event-based maxima approach.*

The basis for this method is the analysis of extreme wind speed data described in Mayne & Cook (1980) in which data sets are augmented by using maximum wind speeds from individual storms instead of annual maxima. In maritime regions, snow falls are separated by periods during which snow melts and disappears completely and, therefore, extreme loads are generated from single snow events. This is similar to wind storm events.

The use of independent event maxima makes efficient use of the data since it does not automatically discard values which are less than the maxima within any particular year. The method does not specifically identify annual maxima nor 'zero-snow' years. However, for design it is necessary to express the characteristic value obtained as the probability of

exceedance in any one year. This is done by defining the average number of snow events per year. This automatically takes into account the occurrence of years without snow.

Since events, rather than annual maxima are being considered, it is necessary to reflect this in the statistical processing of independent extremes. We seek the probability that all of the observations of a given variate in a given period are less than a specified value. If there are m such observations then the probability that all such observations are less than X is $P^m(X)$. This is the same as the probability that the largest among m independent observations is less than X. Thus, the cumulative distribution function of the extreme value distribution is the cumulative distribution of the parent distribution raised to the power of the sample size. A fuller explanation of this is given in Cook (1985). The sample size is the average number of events in the year and the reduced variate in the Gumbel plot needs to be raised to this power.

In the application of this method, it is important to ensure that the events are inspected for statistical independence. For multiple events occurring very close together in time, it may not be possible to separate them and, therefore, it may be necessary to treat them as a single event and retain only the largest value of the aggregated events. It is also required that the process of occurrence of the events in time will be a stationary one.

The procedure is broadly similar to that using annual maxima (Gränzer, 2000) and is summarised as follows:

1 Extract, and rank in order, independent snow event maxima above a selected measurement threshold (e.g. 0.5 cm depth or corresponding load equivalent).

2 Determine the plotting position for these events based on the total number of snow events n and the total length of record N:

$$Z = -\ln\{ -\ln[(m/n + 1)^{n/N}] \} \quad (3)$$

3 Fit Gumbel to the tail of the distribution, i.e. to the right of the mode.

4 Determine the value of the reduced variate for 50 year MRI:

$$Z = -\ln[-\ln(0.98)] \quad (4)$$

For steps 2) and 4) the following alternative steps can be used:

2) Determine the plotting position for these events:

$$Z = -\ln[-\ln (m/n + 1)] \quad (5)$$

4) Determine the value of the reduced variate for 50 year MRI from the Equation 6:

$$Z = -\ln[-\ln(P)] \quad (6)$$

where: $P = 1 - N / (50 \times n)$

4 EXCEPTIONAL SNOW LOADS

Figure 1, illustrates a typical extreme value plot in which a high value appears. Assuming the authenticity of the high value, is it part of the same population as the main body of snow data for that station or is it from a different population?

Evidence of a different population needs to be sought from meteorological considerations. The fact that it is simply a much higher value than the remainder of the observed values is not sufficient to make that decision. It could easily be part of the same population and simply be the early occurrence of a snowfall with a very long mean recurrence interval (MRI).

If the high value is part of the population, then, as data collection and evaluation continues into the future and the meteorological processes are stationary, values would be expected to appear between the high and next highest values in the extreme value plot as well as within the main body of data. The plotting positions will change with the increased number of data points. The high value will tend to move along the reduced variate axis further away from the origin, move closer to the best fit line and its influence on the parameter determination will diminish. Eventually this high value may be moved sufficiently so that no longer would it be regarded as exceptional and excluded from the statistical analysis. By then, its effect on the parameters of the distribution will be insignificant and indeed little different from neglecting it. Mayne and Cook (1980) illustrate this by showing what happens to a single high value as the sample size is increased progressively by factors of two, three and four.

In the absence of 'future' data, and recognising the length of records that would be required to furnish this information, what do we do now? The previous paragraphs suggest that the distribution should be determined with the exclusion of the high value.

By examining the snow load annual maxima for Hamburg, Germany, using the extreme value distribution type I, we can illustrate the effect on the statistical processing. In Figure 1, the vertical axis Z is the reduced variate depending on P, the probability that the corresponding value of snow load will not be exceeded. The parameters of the best fitting line are determined using the least squares method.

If the exceptional value of 1.82 kN/m^2 is included in the statistical processing (i.e. resulting in the best fit line, line A) then the 50 year MRI (characteristic) value of snow load is 1.23 kN/m^2. However, it is clear that line A is not a good fit to the main body of the data. If the exceptional value is excluded (line B) then a better fit to the main body of data results and the characteristic value is equal to 0.71 kN/m^2. The ratio of the maximum value to the (line B) characteristic value is 2.56, this is much larger than the

partial safety factor for snow load, 1.5 (CEN, 1994).

The next section considers the criteria for identifying exceptional ground snow loads and how they are accounted for in the analysis.

5 TREATMENT OF EXCEPTIONAL GROUND SNOW LOADS

The definition of exceptional snow loads in the previous section does not give the quantitative information necessary to classify types of snow loads prior to data analysis. Moreover, as seen from Figure 1, values that might be regarded as exceptional can only be identified by examination of the annual, or event, maxima plots for each meteorological station.

An objective criteria is needed against which specific values can be compared in order to determine whether they should be included or excluded from the statistical processing. The 1996 revision of the French Code of Practice for the estimation of snow loads on structures N84 (Sacré, 1997) considered exceptional snow loads according to the following criteria derived from snow depths:

$$(H_{50X} - H_{50}) / H_{50X} > 0.5 \qquad (7)$$

and

$$H_{max} > 1.50 H_{50} \qquad (8)$$

where:

H_{50X} the 50yr MRI value with the maximum value of snow depth included,

H_{50} the 50yr MRI value with the maximum value of snow depth excluded,

H_{max} the maximum value of snow depth,

1.50   a factor deduced from the safety factors in the French structural design codes.

However it should be noted that none of the codes of practice of the other 17 CEN countries included such criteria.

The French approach suggested that a generalised criterion should take the form:

$$S_m = k \times S_k \qquad (9)$$

where:

S_m the largest snow load (depth) value registered,

S_k the characteristic value determined by excluding the largest value,

k a constant coefficient.

To determine k two approaches are proposed in the present work, one from statistical considerations and the other from the design equations governing loads and resistance in the structural Eurocodes.

For the statistical approach, the P-fractile, x_p, of the extreme value distribution type I (Gumbel) of a general variate, X, is calculated as:

$$x_p = u - \ln(-\ln P) / c \qquad (10)$$

277

The distribution parameters, u and, c are determined using the method of moments:

$$u = m - 0.57722 / c \qquad (11)$$

$$c = 1.2825 / s \qquad (12)$$

where:
m mean value of the sample
s standard deviation of the sample
V coefficient of variation of the sample

Following substitution, Equation (10) can be written as:

$$x_p = m \{1 - 0.78 \times V [0.577 + \ln(-\ln P)]\} \qquad (13)$$

Now consider the high value to be an 'accidental action', and associate with it an appropriate MRI. The Structural Eurocodes do not stipulate a MRI for accidental actions, however when designing for such actions on nuclear structures a value of 10000 years is often taken. This corresponds to the probability of a specified load not being exceeded during any one year of 0.9999. Now the characteristic value for variable actions is defined as a value with a MRI of 50 years (i.e. $P = 0.98$). The ratio of the loads corresponding to these two probabilities gives the value of k, viz:

$$k = x_{0.9999} / x_{0.98}$$

Inserting the appropriate values into Equation 13 results in the following:

$$k = (1 + 6.73 \times V) / (1 + 2.59 \times V) \qquad (14)$$

k depends on V, the coefficient of variation of the data set.

This value will be different for different data sets and not a single value for all registered data. For example, Figure 2 shows the coefficient of variation for the German data. Though there is a wide deviation, the value of 0.6 is calculated as the average from all the data. Additionally, Annex C in CEN (1995) advises that, in calculations to determine return periods for ground snow loads different to that of the characteristic, a value of 0.5 may be assumed for the coefficient of variation.

Substituting $V = 0.6$ in Equation 14 results in $k = 1.98 \approx 2.0$; whilst for $V = 0.5$, $k = 1.90$.

For a return period of 1000 years, the fractile value has a probability of non exceedance of 0.999. For $V = 0.6$, $k = 1.55 \approx 1.6$ and for $V = 0.5$, $k = 1.51 \approx 1.5$.

These calculations suggest that high load values which satisfy the criterion $k \geq 2$ can be considered to have a MRI of the order of 10,000 years or more and those which satisfy the criterion $k \geq 1.5$ have a MRI of the order of 1,000 years.

The second approach is to use the CEN Basis of Design document (CEN 1994) which defines snow loads as variable actions. This generally relates to snow loads determined by methods appropriate to the processing of regular snow loads. However, some actions, for example seismic actions and snow loads, can be considered as either accidental and/or variable actions, depending on the site. This permits exceptional snow loads to be treated as accidental loads.

Therefore, if the snow event is exceptional, i.e. related to the characteristic load by equation (9), then, in design, two cases need to be considered: persistent/transient (P/T) situations and accidental (A) situations. The coefficient k needs to be determined by taking into account not only consideration of the actions, but also the resistance of structural components.

Following CEN (1994) Section 9, it shall be verified that:

$$E_d \leq R_d \qquad (15)$$

where:
E_d the design value of the action,
R_d the design value of resistance

For simplification, consider the case when snow load is the only variable action and there is only one permanent action (e.g. self-weight). Then, according to Clause 9.4.2 "Combinations of actions" (CEN 1994), there are two cases to consider:

1 persistent and transient design situations for ultimate limit states verification other than those relating to fatigue

$$E_d = \gamma_G \times G_k + \gamma_Q \times Q_k \qquad (16)$$

2 accidental design situations

$$E_d = \gamma_{GA} \times G_k + A_d \qquad (17)$$

where:

G_k	the characteristic value of permanent action
Q_k	the characteristic value of variable action (snow load)
$A_d = \gamma_A \times A_k$	is the design value of accidental action
A_k	the characteristic value of accidental action (snow load)
$\gamma_G = 1.35$	the partial safety factor (PSF) for permanent action
$\gamma_Q = 1.5$	the PSF for variable action (snow load)
$\gamma_{GA} = 1.0$	the PSF for permanent action for accidental design situation
$\gamma_A = 1.0$	the PSF for accidental action (snow load)

According to CEN (1991) for P/T situations the design value for concrete is defined as:

$$R_d = R_{k,c} / \gamma_C \qquad (18)$$

and the design value for steel reinforcement or prestressing tendons is defined as:

$$R_d = R_{k,s} / \gamma_S \qquad (19)$$

For Accidental situations the design value for concrete is defined as:

$$R_d = R_{k,c} / \gamma_{CA} \qquad (20)$$

and the design value for steel reinforcement or prestressing tendons is defined correspondingly as:

$$R_d = R_{k,s} / \gamma_{SA} \qquad (21)$$

where:

$R_{k,c}$ the characteristic value of concrete strength
$R_{k,s}$ the characteristic value of steel reinforcement strength

$\gamma_C = 1.5$ the PSF for concrete (P/T situations)
$\gamma_S = 1.15$ the PSF for steel reinforcement (P/T situations)
$\gamma_{CA} = 1.3$ the PSF for concrete (Accidental situations)
$\gamma_{SA} = 1.0$ the PSF for steel reinforcement (Accidental situations)

Using Equations 15, 16, and 18 it is possible to write for P/T situations for concrete:

$$1.35 \times G_k + 1.5 \times Q_k \le R_{k,c} / 1.5 \qquad (22)$$

Considering as an unfavourable case $G_k = 0.5 \times Q_k$ and noting that $Q_k = S_k$ then:

$$3.26 \times S_k \le R_{k,c} \qquad (23)$$

Similarly for Accidental situations, Equations 15, 17 and 20 and $A_t = S_m = k \times S_k$ from Equation 9 give:

$$1.0 \times G_k + k \times S_k \le R_{k,c} / 1.3 \qquad (24)$$

Taking again as an unfavourable case $G_k = 0.5 \times Q_k$ then:

$$(0.65 + 1.3 \times k) \times S_k \le R_{k,c} \qquad (25)$$

Because the right hand sides in Equations 23 and 25 are the same (the characteristic value of concrete strength), the left hand sides (the design value of action effect) must also be the same, independent of the design situation. Then k can be calculated as:

$$k = (3.26 - 0.65) / 1.3 = 2.0 \qquad (26)$$

Thus in this illustration, for $k \ge 2.0$ the governing design load case will be the Accidental situation

Other ratios of G_k / Q_k can also be taken into consideration. Assuming $G_k = Q_k$, results in $k = 2.28$, and for $G_k = 1.5 \times Q_k$, $k = 2.56$.

Considering the case $G_k = 0.5 \times Q_k$ as unfavourable, isolated snow events with $k \ge 2.0$ should be designated as accidental events. From a pragmatic point of view events with $1.5 \le k < 2.0$ may be also considered as accidental with $k = 2.0$ for subsequent calculations.

Consider again Equations 16 and 17 with, for example, $k = 2.0$ then for P/T situations:

$$E_d = 1.35 \times G_k + 1.5 \times S_k \qquad (27)$$

and for Accidental situations:

$$E_d = G_k + 2.0 \times S_k \qquad (28)$$

The more unfavourable of these two situations (and therefore the decisive one) depends on the ratio of the characteristic value of permanent action to the characteristic value of the snow load. Manipulating Equations 27 and 28 indicate that if this ratio is less than 1.42 then the Accidental design situation is the decisive one.

If the snow load is not the only variable action, then the characteristic values of the other variable actions multiplied by appropriate partial safety factors and combination coefficients should be added into the Equations 16-17 as stated in CEN (1994), Chapter 9. In these cases the snow load is considered to be the dominant action and both design situations will need to be verified.

The foregoing discussion on design situations in the Eurocodes confirms the view that snow load values that are more than twice the characteristic value, calculated by excluding the largest value from the statistical processing, should be regarded as exceptional and considered in design as accidental actions. It is less conclusive for load values of 1.5 times the characteristic value and this needs to be evaluated further in specific design situations. However the earlier statistical considerations indicate that removing such values from the statistical processing is valid if the interpretation ascribed to those values is that they have MRI values in excess of approximately 1,000 years.

Hence, these considerations result in the following conservative criterion for identifying 'exceptional load' values in the observations:

"If the ratio of the largest load value to the characteristic value determined without the inclusion of that value is greater than 1.5 then the largest load value shall be treated as an exceptional value."

The methodology for identifying and handling exceptional snow loads can now be summarised. Whilst it assumes that the annual maximum snow load values are fitted by an extreme value distribution type I, and that the parameters of this distribution are determined by the Gumbel Least Squares Method, other distributions e.g. Log-Normal or Weibull could be used where appropriate. The methodology is:

1 For the annual (or event) maxima data set rank the maxima in ascending magnitude excluding the largest value. The reduced variate Z is determined from:

$$Z = -\ln [-\ln (m/(N))] \qquad (29)$$

where N is the number of maxima, and m is the rank.

2 Best fit a straight line using an appropriate technique, e.g. LSM.

3 Evaluate the parameters of the best fit line and determine the 50 year MRI value, excluding the

largest value, using Equation 30:

$$Z = -\ln[-\ln(0.98)] \qquad (30)$$

4 Determine the value of k which is equal to the largest value divided by the 50 year MRI value from step (3). If $k < 1.5$ then repeat steps 1) to 3) with the inclusion of the largest value in the statistical processing.

5 Record the 50 year MRI value and the largest value if it has been classified as an exceptional value. Annotate the appropriate meteorological station records accordingly.

6 APPLICATION OF METHODOLOGY TO EUROPEAN SNOW DATA SETS

The application of the methodologies outlined previously resulted in exceptional snow loads being identified in: Austria, Belgium, Eire, France, Germany, Greece, Italy, Portugal, Spain and UK, i.e. in 10 of the 18 CEN countries. In total, 159 stations were identified, 47 of which are in Spain and 44 in the UK. The individual stations are shown in Figure 3, more detailed data are given in (Sanpaolesi et al 1998).

For the countries with a maritime climate, either the mixed distribution approach e.g. in France, Italy, or the event based approach, e.g. in the UK, were used to identify the characteristic values taking into account exceptional loads.

Twelve stations in Northern Germany provide an illustration of exceptional loads determination, Table 1. The date of the occurrence is given to aid meteorological explanation, from which it is observed that the all the exceptional loads resulted from a single weather system that persisted in that area for about a fortnight. The characteristic values of snow loads for these stations were obtained for two cases:
1 the exceptional value included in the statistical processing, and
2 the exceptional value excluded from the consideration.
Results are shown in Table 2.

The effect of using distributions other than Gumbel, e.g. Log-Normal and Weibull, for German data are discussed in (Sanpaolesi et al 1998).

If the extreme value distribution type I is used for fitting the data then all 12 of the German climatic stations in Table 2 have k greater than 1.5 and eight of these have k greater than 2.0.

Thus these twelve stations have been identified as having exceptional snow loads and treated according to the procedure for accidental actions.

Of the 159 stations identified throughout Europe, 75 have k values greater than 2.0. Whilst the majority of values lie in the range $2.0 < k < 3.0$, there are a small number with $k > 4.0$. It is worth noting, how-

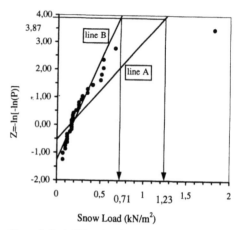

Figure 1. Probability plot of extreme value distribution type I for Hamburg, Germany

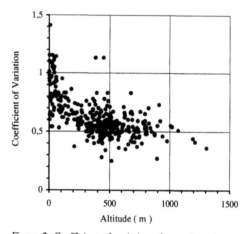

Figure 2. Coefficient of variation of annual maximum snow load for 331 climatic stations in Germany

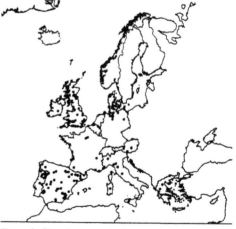

Figure 3. Climatic stations in Europe where exceptional snow load values have been identified.

280

Table 1. Stations in Germany with exceptional snow loads

N	Station	Number of record years	Maximum observed value of snow load Q_{exc} (kN/m²)	Occurrence date of maximum observed value
1	Bederkesa	18	1.52	16.02.1979
2	Bremen	32	1.53	18.02.1979
3	Cuxhaven	31	1.48	01.03.1979
4	Hamburg	33	1.82	15.02.1979
5	Hannover	33	1.23	24.02.1979
6	Kiel	33	1.78	15.02.1979
7	Norden	12	1.07	19.02.1979
8	Norderney	18	1.56	22.02.1979
9	Schleswig	33	2.37	19.02.1979
10	Soltau	25	1.26	19.02.1979
11	Tostedt	27	1.47	16.02.1979
12	Visselhoevede	27	1.54	17.02.1979

Table 2. Characteristic values of snow load and values of plot correlation coefficients for extreme value distribution type I

N	Station	A: Exceptional value is included		B: Exceptional value is excluded		Ratio $k = Q_{exc}/S_B$
		50 year return period value S_A (kN/m²)	Plot correlation coeffi cient.	50 year return period value S_B (kN/m²)	Plot correlation coeffi cient	
1	Bederkesa	1.42	0.886	0.81	0.987	1.88
2	Bremen	1.05	0.824	0.59	0.981	2.59
3	Cuxhaven	1.11	0.851	0.66	0.990	2.24
4	Hamburg	1.23	0.828	0.71	0.976	2.56
5	Hannover	0.90	0.893	0.61	0.995	2.02
6	Kiel	1.22	0.822	0.71	0.981	2.51
7	Norden	1.13	0.770	0.39	0.954	2.74
8	Norderney	1.51	0.860	0.84	0.910	1.86
9	Schleswig	1.60	0.822	0.93	0.967	2.55
10	Soltau	1.07	0.917	0.75	0.982	1.68
11	Tostedt	1.16	0.877	0.75	0.971	1.95
12	Vissel- hoevede	1.17	0.837	0.66	0.975	2.35

ever, that a large number of these stations have small characteristic loads and relatively small maximum values, e.g. Morecambe (UK),for which the maximum registered load value is 0.30 kN/m² and the calculated characteristic value is 0.12 kN/m².

At the time of writing the stations with exceptional loads have not been incorporated into the new European ground snow loads map. They remain as marked on a separate map, Figure 3, in order to facilitate consideration of how exceptional snow loads should be included in the work of the Project Team charged with conversion of CEN (1995) into full EN status.

7 CONCLUSIONS

This paper has defined irregular and exceptional snow falls (loads) and outlined the methodologies that were developed during the European Snow Loads research project for determining the corresponding characteristic snow load values.

For irregular snowfalls, two methodologies have been discussed: mixed distribution and event maxima distribution. The latter is more effective than annual maxima for less well-populated data sets because it does not automatically discard data that is below the annual maximum.

Criteria have been defined for identifying exceptional snow loads by considering both statistical and limit state design arguments. For ratios of the highest observed snow load value to the characteristic value, determined without inclusion of the highest value, greater than 2.0, the observed load value is classified as exceptional. For such ratios between 1.5 and 2.0, the situation is less clear as it depends on the particular structure under design and its location. However it is proposed currently that snow load with ratios falling in this range are also classified as exceptional loads.

The effects of exceptional values on the statistical processing of the more normal processing have been illustrated with respect to sites in Northern Germany. These criteria have been applied throughout the 18 CEN countries and 159 stations with exceptional snow loads have been identified. The locations of these sites have been presented currently in a separate map to the European ground snow loads map.

The codification of these exceptional snow load values remains to be decided but notwithstanding this, such loads should be treated in design according to the procedures for accidental actions in the emerging Eurocodes.

REFERENCES

CEN, 1991, ENV 1992 Design of Concrete Structures, Part 1-1 General Rules and Rules for Buildings. CEN Central Secretariat, Brussels

CEN, 1994. ENV 1991-1 Basis of Design and Actions on Structures, Part 1 Basis of Design. CEN Central Secretariat, Brussels

CEN, 1995. ENV 1991-2-3 Basis of Design and Actions on Structures, Part 2-3 Snow Loads. CEN Central Secretariat, Brussels

Cook, N.J. 1985, *The designer's guide to wind loading of building structures – Part 1: Background damage survey, wind data and structural classification*. London: Butterworths.

Gränzer, M. 2000, New ground snow loads for the European Standard, Proceedings of the 4th International Conference on Snow Engineering, Trondheim.

Mayne, J.R. & Cook, N.J. 1980, Extreme wind analysis. *Wind Engineering in the Eighties, CIRIA conference*, London, 12-13 November 1980.

Sacré, C. 1997. *Analysis of snow load data – A – France and B – Belgium and Luxembourg.* Report EN-AEC 97.87 C, CSTB, Nantes.

Sanpaolesi, L et al. 1998. *Scientific support activity in the field of structural stability of civil engineering works: Snow loads –Final Report Phase I*, Commission of the European Communities DGIII-D3, University of Pisa, Pisa

Sanpaolesi. L et al., 2000. European snow loads research project: Recent results. Proceedings of the 4th International Conference on Snow Engineering, Trondheim.

Thom, H.C.S, 1966. Distribution of maximum annual water equivalent of snow on the ground. *Monthly weather review* (94): 265-271.

Snow Engineering, Hjorth-Hansen, Holand, Løset & Norem (eds) © 2000 Balkema, Rotterdam, ISBN 90 5809 148 1

Analysis of roof load data for the European Snow Load Program 1997-1999

Ulrich G. Stiefel
Gruner Limited, Consulting Engineers, Basle, Switzerland

S. Wunderle
University of Berne, Switzerland

ABSTRACT: During the years of 1997 to 1999 8 European institutes were involved in a European Snow Load Research Project. During the phase II one of the subtasks was the investigation of roof snow loads to determine roof shape coefficients for normative use. These shape coefficients are multiplied with the characteristic ground snow load to result in the relevant roof snow load. Depending on several factors the relevant shape and the magnitude of the roof snow load are determined. The procedures of the measurements and analysis methods the during the winter 98/99 are described. The reasons for the selected models for the multiple regression analysis, the relevant programs and the results from the investigation are shown. As one of the results the measurements of the roof snow loads are continued in Switzerland in the winter 1999/2000 and hopefully in future winters in the coming years.

1 INTRODUCTION

In many areas of Europe the structural design of roofs is governed by the snow load which can be expected to accumulate on the roof. The structure should be capable of withstanding the severest load which will be imposed upon it during its lifetime. On the other hand economics dictate that the building should not be grossly overdesigned.

The procedure to determine the roof snow load, s, in the CEN/TC250/SC1-Code, most national codes on snow loads as well as the ISO-Code 4355 is to multiply the characteristic ground snow load s_k with certain "shape coefficients" μ which take into account some of the effects influencing the roof snow load. To emphasise the importance of these coefficients in design, their values generally range from 0.8 to 1.6 and occasionally even higher.

In order to provide further information towards the harmonisation of roof shape coefficients through Europe, task IId - analytical study for the definition of shape coefficients - was included in the programme of work.

2 MODELS

2.1 *Snow drift models*

Falling snow is deposited on roofs in uniform layers only if the wind speed is low. It is known that with wind speeds in the range of 4 to 5 m/s, much of the snow is deposited in areas of 'aerodynamic shade'. If the wind velocity increases above this range snow particles can be picked up from the snow cover, leading to depletion of the snow cover in areas of high wind speed and re-deposition of the snow on the lee sides of peaked or arched roofs, on lower roofs in the lee of higher roofs, or behind obstructions on the roof.

The amount of transported snow is a function of the wind velocity, the duration of the high wind velocity, the composition of the snow surface, the depth of snow at the source (i.e. upper roof), the topographic relief, the exposure, the size of snow grains, the temperature and humidity of the wind.

Several models exist for the determination of the mass of drifted snow. However none of them can be applied directly for the determination of roof snow load. The final report of the European Snow Load Research, 1999 describes empirical models for the mass of drifted snow and a finale area element method to calculate the mass of drifted snow.

2.2 *Metamorphism*

Metamorphism describes the changes inside the snow pack and starts immediately after the accumulation of snow flakes. After a short time the small needles and points of the flakes are rounded due to the transport of water vapour. The result is an increase of density and a decrease of snow depth. The time for this process depends on the air temperature and the temperature gradient in the snow pack. The temperature gradient could induce a transfer of heat and water vapour in the snow column. Destructive metamorphism becomes apparent with a gradient less than 0.1 °K/cm. There is a growth of rounded grains with a development of bonds between the grains. The number of grains increases. The snow cover is compact and stable.

With a temperature gradient higher than 0.2 °K/cm constructive metamorphism takes place. This gradient builds new facettes without any bonds and the snow cover is mechanically very weak.

The metamorphism of a snow cover is an important process for the transformation of the snow surface, in particular for the shape and characteristic of this surface. A frequent thaw-freeze cycle will transform the new fallen fluffy snow of the surface to a hard and icy crust, which after some cycles may cause the surface to be too hard for snow transport by wind.

2.3 Ablation

Ablation is the process of melting. The process is controlled through energy transfer from the atmosphere to the snow cover. The main energy input is from radiation and heat fluxes although additionally the energy input from rain should be considered.

There are two basic approaches to the prediction of the rate of snowmelt: the degree-day model and the full energy balance method. Both of these methods contain some empiricism; however the degree-day model is entirely empirical and generally site specific, whereas the physical basis of the energy balance approach allows the parameters found by experiment to be widely applicable. The two models are described in the final report of the European Snow Load Research, 1999.

2.4 Conclusions for the model for the roof snow load

In the sections 2.1 to 2.3 the three processes of the roof snow load are described: drifting, metamorphism and ablation. In order to determine the roof snow load all three influences are treated as additive load parts.

According to ISO 4355 the snow load on the roof is given by:

$$s = s_b + s_d + s_s \quad \text{where } s_b = \text{balanced}, s_d = \text{drift}, s_s = \text{slide} \quad (1)$$

The multiplicative approach takes into account the physical behaviour of snow loads

$$s = s_k * \mu = s_k * C_e * C_t * \mu_b * \mu_d \quad (2)$$

where
s_k = characteristic value of the snow load on the ground
C_e = exposure factor, which usually has the value 1.0
C_t = thermal coefficient, which usually has the value 1.0
μ_b = balanced factor, describes the differences between windward and lee
μ_d = drift factor

These two approaches, one additive and one multiplicative, are used in civil engineering design but from the scientific point of view the equations are only a rough approximation to the natural conditions. The application of regression analysis is used to improve the results. The linear regression may have the following form:

$$s = s_k * \mu \quad (3)$$

where $\mu = a + b * \alpha_1 + c * T_{env} + d * T_{env} + e * T_{build} + f(u-x)^3 * t$ (4)

and
$b * \alpha_1$ = term for the slope α_1
$c * T_{env}$ = term to take into account the metamorphism using the temperature of the environment
$d * T_{env}$ = term to take into account the ablation using the temperature of the environment
$e * T_{build}$ = term to take into account the insulation of the roof
$f(u-x)^3 * t$ = term for the snow drift with wind velocity u during the time t

The logarithmic regression differs in the following way

$$\mu = m * \alpha_1^{n1} * T_{env}^{n2} * T_{env}^{n3} * T_{build}^{n4} [(u-x)^3 * t]^{n5}$$

and
α_1^{n1} = term for the slope α_1
T_{env}^{n2} = term to take into account the metamorphism using the temperature of the environment
T_{env}^{n3} = term to take into account the ablation using the temperature of the environment
T_{build}^{n4} = term to take into account the insulation of the roof
$[f(u-x)^3 t]^{n5}$ = term for the snow drift with wind velocity u during the time t

Only the second approach was used to determine the snow loads on roofs using the measurements obtained during the 1998/99 winter.

2.5 Implications for the measurement campaign

From the investigation of the different models to treat drift, metamorphism and ablation the following implications are relevant for the measurement campaign:

- The buildings for the measurement campaign should be near to meteorological stations. Additional automatic weather stations may be helpful for further investigations.

- The location of the buildings and their neighbourhood should be described precisely. The dimensions of the buildings (incl. height) should be determined.

- For continental climates the depth of the snow cover and the density of the snow volume should be measured at least every 2 weeks on the roofs and near to the building in an undisturbed area.

- Some of the buildings on which the roof snow loads are measured should be used for verification of the simulation of snow drift in the wind tunnel. It is important that the recorded meteorological parameters should be used as an input for the investigations in the wind tunnel.

3 SELECTED SITES

The following table 1 shows an overview of the locations of the roof measurements.

4 DATA ANALYSIS

The principles for the data analysis are described in the final report of the European Snow Load Research, 1999. As far as possible regression analysis is performed on the data of every climatic region. From this analysis basic information on the shape coefficient can be drawn in EXCEL-program to analyse the data uses simple and multiple regression analysis considering the mathematical expression of equation 5. The EXCEL-datasheet with information about the statistics used is treated in detail in Gruner, 1999.

The following coefficients are used in the regression analysis:

T_{env} = Average temperature in Switzerland of an early morning hour
(7 - 8 a.m.), others average daily values

T_{build} = 20 °C if the construction is heated
T_{env} if the construction is not heated

DD = The number of the days with temperatures higher than 0 °C since the previous observation resp. since start of snow (only if snow lying) will be multiplied by the number of degrees

u = Average wind speed values in Switzerland of an early morning hour
(7 - 8 a.m.), others average daily values

u_{high} = Average high wind speed values above 4 m/s (if possible hourly mean values) since last observation [m/s]; data from meteorological station

α = Roof slope

5 MULTIPLE LINEAR REGRESSION ANALYSIS

The following tables 2, 3 and 4 show the results of the data analysis. Since the most measurements are for Switzerland and the meteorological data could be gathered quite completely the first analysis is performed for these data. If for one of the other climatic

regions more than five, or at least more than three data sets (partly incomplete with respect to meteorological data) were available, a separate analysis is worked out. In addition reasonable combinations are performed to analyse the differences between the results for different climatic regions. Of particular interest are the possible differences between the continental continuous snow falls and the maritime single snow falls.

5.1 *Flat roofs*

The following remarks can be made from these data analysis on flat roofs:

- The correlation factor for the Swiss data set with $R^2 = 0.55$ is quite high compared with other roof snow load measurements in the United States. The effective and estimated values are plotted in the final report of the European Snow Load Research, 1999. From the t-test only the parameter T_{env} is shown to be sufficiently reliable to be used as a parameter in the regression analysis. This means that from a statistical point of view the data must be partly improved and additional investigations are necessary. Nevertheless impor-

Table 1: Overview of all sites of roof snow load measurements

| Partici-pants | Number of roofs | | | Material | Personnel | Met. Stations |
	Total	Flat roofs	Gabled roofs			
Swit-zerland	35	14	21	Wooden poles	Mainly employees of meteo stations	Yes
Italy	13	3	10	Poles	Dept. Personnel – local employees	Yes (Abetone)
Great Britain	25	9	16	Poles	Met. Office	Yes
ISMES	5	0	5	Pressure transducers	Automatic stations	Automatic stations, every 3 h per day
Germany Leipzig	3	0	3	Wooden poles	Geographic faculty	Yes
Total	81	26	55			

Table 2: Correlation coefficient R^2 for the roof snow load data on flat roofs

Climatic region	N° sites	Correlation coefficient R^2
Switzerland	10	0.55
Italy Apennine	3	0.20
United Kingdom	2	Remarks see below
Switzerland + Italy Apennine	13	Remarks see below
Switzerland + United Kingdom	18	Remarks see below

Table 3: Correlation coefficient R^2 for the roof snow load data on gabled roofs side I (lee)

Climatic region	N° sites	Correlation coefficient R^2
Switzerland	17	0.12
Italy Apennine	4	-
Italy Dolomite	5	-
Germany	3	-
United Kingdom	2	-
Switzerland + Italy Dolomite +(Alps)	22	0.04
Switzerland + Italy Apennine	21	0.0001
Switzerland + Italy Dolomite + Germany (Continental)	25	0.12
Switzerland + Italy Dolomite + Italy Apennine	26	0.0001
Climatic region	N° sites	Correlation coefficient R^2
Switzerland + Italy Dolomite + Italy Apennine+ Germany	29	0.02
Switzerland + Italy Dolomite + Italy Apennine + Germany + United Kingdom	31	00.1

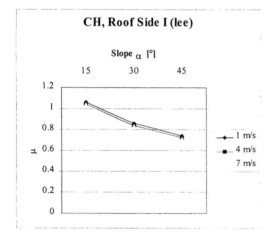

CH, Roof Side I (lee)

Figure 4: Calculated roof shape coefficients for the roof snow load on gabled roofs side I (lee) for Swiss data for different slopes of the roof and different wind speeds

tant parameters to characterise the roof snow load can be determined.

- The correlation factor for the Italian data with R^2 = 0.20 is not as high as that for the Swiss measurements; this is probably due to the small quantity of data. From the statistical point of view three data sets are insufficient for reasonable analysis. To introduce all of the the available information into the analysis it was decided to use all the measured data.

- For two of the UK sites BRE has the required meteorological data and only on two flat roofs was there enough snow to be measured. This is not sufficient for a statistical treatment.

- The combination of the Swiss data with the data from the Apennines is not reasonable because the same meteorological data does not exist.

- The combination of the Swiss data, representative for a continental climate, with the data from United Kingdom is not reasonable due to few data from United Kingdom.

5.2 Gabled roofs side I (lee)

From the analysis of the roof snow load data for the leeward side of the gabled roofs (side I) in table 3 the following conclusions can be drawn:

- The correlation coefficient R^2 = 0.12 for the Swiss data is low. The reasons for this fact might be bad data collection or not considering all the relevant influences. To eliminate the first reason several improvements of the data were performed, such as elimination of data that was gathered after snow removal due to extremely heavy snow falls or wrong measurements after discussion with local observers. The t-test shows values with an unusually high probability of 15 to 20 % for a wrong decision (5 % is normal). All the parameters therefore are only of limited use for the determination of the roof shape coefficients.

In figure 4 the roof shape coefficients for different slopes of the roof are given, using the regression equation. As shown in this figure the shape coefficient is reduced remarkably for higher slopes whereas different wind speeds play only a minor role for the shape coefficient.

- There was an insufficient number of data sets for Germany and for the Italian Apennines and Dolomites in order to calculate reasonable correlation coefficients. The number of data sets is approximately equal to, or smaller than, the number of independent variables of the multiple linear regression analysis, therefore correlation coefficients of 0.75 and 1 result.

- The same correlation coefficient for the Swiss data as for the combination of the Swiss, German and Italian Dolomites data might be a hint that for these climatic regions – all with continuous snow fall and a build up of the snow layer during several weeks and months – the same regression equation might be valid. Unfortunately there are insufficient data for this to be a final conclusion.

- The combination of the Swiss data and the Italian Dolomites with the data from the Italian Apen-

Table 5: Correlation coefficients R^2 for the roof snow load data on gabled roofs side II (windward)

Climatic region	N° sites	Correlation coefficient R^2
Switzerland	18	0.27
Italy Apennine	4	
Italy Dolomite	5	
Germany	3	
United Kingdom	1	
Switzerland + Italy Dolomite (Alps)	23	0.13
Switzerland + Italy Apennine	22	0.0001
Switzerland + Italy Dolomite Germany (Continental)	26	0.19
Switzerland + Italy Dolomite + Italy Apennine	27	0.002
Switzerland + Italy Dolomite + Italy Apennine + Germany	30	0.02
Switzerland + Italy Dolomite + Italy Apennine + Germany + United Kingdom	31	0.01

Climatic region	N° sites	Env Temp Corr. Coeff. R^2	Wind Speed Corr. Coeff. R^2	High Wind Speed Corr. Coeff. R^2	Degree Days Corr. Coeff. R^2
Switzerland	10	0.54	0.08	0.69	-
Italy Apennine	3	-	-	-	0.20
Italy Dolomite	0	-	-	-	-
Germany	0	-	-	-	-
United Kingdom	2	-	-	-	-
Switzerland + United Kingdom	12	0.004	-	-	-

Figure 7: Correlation coefficients R^2 for simple linear regression analysis for the parameters with influence on the roof shape coefficient of flat roofs

5.3 Gabled roofs side II (windward)

From the analysis of the roof snow load data for the windward side of the gabled roofs (side II) the following conclusions can be drawn:

- The correlation coefficient $R^2 = 0.27$ for the Swiss data is higher than for the leeward side but is still not satisfactory (table 5). The reason can be the same as described above. The t-test shows better values than for the leeward-side. The probability of an erroneous decision is for wind speed less than 5 % and for the other two parameters, slope of the roof and temperature, approx. 15 %.

- In figure 6 the roof shape coefficients for the different slopes of the roof are given, using the regression equation. As shown in this figure the shape coefficient is reduced remarkably for the higher slope values whereas different wind speeds play only a minor role for the shape coefficient. Comparing these values with figure 4 the values for the lee side of the roof are about the same as for the windward side, this is not expected. Other investigations have always shown a remarkable difference between the two roof sides.

- The combination of the continental data shows a smaller correlation coefficient, $R^2 = 0.19$, than for the Swiss data. The reason for this fact can not be determined from the existing data basis.

- The combination of the Swiss data with the data from the Apennine follows the same pattern as for the leeward side.

- The combination of the continental data with the data from United Kingdom shows again a low correlation.

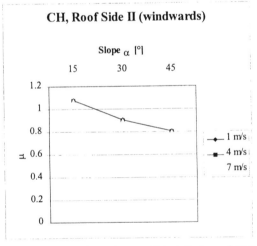

CH, Roof Side II (windwards)

Figure 6: Calculated roof shape coefficients for the roof snow load on gabled roofs side II (windward) for Swiss data for different slopes of the roof and different wind speeds

nine shows an extremely low correlation coefficient. This might be sign that the Italian Apennine follows a different law for shape coefficients than continental regions.

- The combination of the continental data with the data from United Kingdom again seems to have no correlation. If this data is significant from a statistical point of view cannot be judged due to few data.

6 SIMPLE LINEAR REGRESSION ANALYSIS

Since the multiple linear regression analysis in chapter 5 shows only partially satisfactory results, the influence of each single parameter on the roof shape coefficient is investigated in this section. The following table gives details of the simple linear regression analysis.

Table 8: Correlation coefficients R^2 for simple linear regression analysis for the parameters with influence on the roof shape coefficient of the leeward side of gabled roofs side I

Climatic region	N° sites	Env Temp. Corr. Coeff. R^2	Wind Speed Corr. Coeff. R^2	High Wind Speed Corr. Coeff. R^2	Degree Days Corr. Coeff. R^2
Switzerland	17	0.001	0.02	0.40	0.10
Italy Apennine	4	-	-	-	0.592
Italy Dolomite	5	0.026	0.079	-	0.006
Germany	3	0.967	0.967	-	0.642
United Kingdom	2	-	-	-	-
Switzerland + Italy Dolomite (Alps)	22	0.007	0.001		
Switzerland + Italy Dolomite + Germany	26	0.017	0.007	-	
Switzerland + Italy Dolomite + Italy Apennine	26	-	-	-	
Switzerland + Italy Dolomite + Italy Apennine + Germany	29	-	-	-	
Switzerland + Italy Dolomite + Italy Apennine + Germany + United Kingdom	31	-	-	-	

6.1 Flat roofs

From these results the following conclusions can be drawn:

- For the Swiss data the high wind speed parameter with a correlation coefficient $R^2 = 0.69$ and environmental temperature with $R^2 = 0.54$ describe quite well the influences on the shape coefficient. The t-test gives normal results; the probability of an erroneous decision is normally small. This correlation is coherent with the discussion of the influencing parameter for drift, metamorphism and ablation in chapter 2.

- The correlation coefficient for the degree days parameter of the data from Italy Apennine is with $R^2 = 0.20$ not as high as the R^2-value for the environmental temperature parameter of the Swiss data. Further investigations would be necessary to conclude from a statistical point of view the significance of the differences.

6.2 Gabled roofs side I (leeward)

From these results the following conclusions can be drawn:

- The two parameters, environmental temperature and wind speed, have for the Swiss data no correlation with the shape coefficient. Whereas high wind speed and roof slope have similar effects from a statistical point of view on the determination of the roof shape coefficients with relatively high correlation coefficients of 0.4 and 0.1.

- From the data no higher values for a slope between 30 ° and 45 ° can be observed.

- The Italy Apennine values show high correlation for the roof shape ($R^2 = 0.59$).

- High correlation coefficients for other data sets are mainly due to few data sets.

- All combinations show no relevant correlation or significant differences.

6.3 Gabled roofs side II (windward)

From this analysis the following conclusions can be drawn:

- As for the Swiss data on the leeward side of the roofs the parameter 'environmental temperature' has no correlation with the shape coefficient.

- For sites with low wind speeds far larger shape coefficients are determined than for sites with high wind speeds.

- The correlation coefficient for the roof slope of the Swiss data is only very small, for the Italy Apennine with $R^2 = 0.3$ small. The t-test values show a big probability for the Swiss data and an increased probability for the Italy Apennine data of an erroneous decision. It seems that the slope of the roof on the windward side does not have a dominant influence.

- High wind speed is strongly correlated to shape coefficients, the correlation coefficient of 0.77 as well as the t-test statistics support this fact for the Swiss data.

- Other comments are similar to the ones for the gabled roof side I (leeward).

7 ANALYSIS FOR WIND EXPOSURE CLASSIFICATION

This section treats the influence of the wind exposure on the shape coefficients. Similar investigations have been performed by O'Rourke, Koch, Redfield (1983) and by Ellingwood (1985). The following figures show the mean and standard deviation for all data from the different types of roofs. As expected the shape coefficients for windy sites are (much) smaller than for sheltered sites as are the coefficients of the windward side. The standard deviation of the data becomes larger for more windy sites.

There are no significant differences between the climatic regions. However it must be mentioned that since the statistical basis excludes Switzerland it is rather limited.

Compared to the investigation by O'Rourke, Koch, Redfield (1983), the European mean values are rather higher, and the standard deviation smaller than the American values.

8 ANALYSIS FOR ROOF SLOPE CLASSIFICATION

The roof shape coefficients are subdivided into the following categories:
- $0 - 7°$
- $8 - 22°$
- $23 - 37°$
- $38 - 52°$

From the following figures 13, 14 the shape coefficients for the different roofs dependent on the roof slope can be seen.
From these figures the following conclusions can be drawn:

- A decrease of the roof shape coefficient with an increase in the slope of the roof is obvious, as expected.

- For the leeward side the shape coefficient is constant up to 30 °.

- The shape coefficients for the windward side are smaller than for the leeward side, as expected.
  For the leeward side no increase for 30 ° can be determined.

Table 9: Simple linear regression analysis for the parameters with influence on the roof shape coefficient of the windward side of gabled roofs side II

Climatic region	N° sites	Env Temp.	Wind Speed	High Wind Speed	Degree Days
		Corr. Coeff. R^2	Corr. Coeff. R^2	Corr. Coeff. R^2	Corr. Coeff. R^2
Switzerland	18	0.002	0.24	0.77	0.03
Italy Apennine	4	-	-	-	0.305
Italy Dolomite	5	0.093	0.02	-	0.023
Germany	3	0.991	0.991	-	0.723
United Kingdom	1	-	-	-	-
Switzerland + Italy Dolomite (Alps)	23	0.025	0.078	-	0.41
Switzerland + Italy Dolomite + Germany	26	0.043	0.084	-	0.063
Switzerland + Italy Dolomite + Italy Apennine	27	-	-	-	0.002
Switzerland + Italy Dolomite + Italy Apennine + Germany	30	-	-	-	0.02
Switzerland + Italy Dolomite + Italy Apennine + Germany + United Kingdom	31	-	-	-	0.01

9 INFLUENCE OF WIND FOR DRIFTING ON GABLED ROOFS

The influence of wind on the drifting of snow on gabled roofs is investigated, based on the data gathered in Switzerland during the 1998/99 winter. From these investigations the following conclusions can be drawn:

- The wind direction during snow storms influences directly the distribution of the snow on the gabled roof: the roof slope on the windward side has less snow than the leeward side, as the following summary shows:

- 20 roofs with both roof slopes investigated
- 5 roofs without wind during snow fall
- 6 roofs with no difference between the shape coefficients of the two roof slopes
- 8 roofs where the windward roof slope has the smaller shape coefficient and the leeward roof slope the larger shape coefficient, and 1 roof with reverse results to these.

- Two roof configurations were selected with the eaves directions at right angles: a one-storey office building in Bern-Liebefeld and a barn in Davos. The shape coefficients for the office building were equal due to little snow. The difference between the shape coefficients for the roof slopes of the barn perpendicular to main wind direction is slightly greater than the difference between the shape coefficients of the roof slopes in the main wind direction.

- For the results of the correlation analysis see chapter 5 and 6. From this analysis the higher velocity gusts (> 4 m/s) are shown to have a very important influence on the formation of the drift.

10 CONCLUSIONS FOR FUTURE SNOW MEASUREMENTS

The following aspects can be concluded from the roof snow load measurements during the 1998/99 winter:

- Satisfactory roof snow load measurements need time; measurements must be performed throughout several winters.

- For each climatic region at least 5 to 10 roofs of each type must be equipped and measured.

- A meteorological station or wind and temperature measurements in the vicinity of the roofs must be guaranteed.

- Close support for the observers of the measurements is necessary. Several visits to the sites, especially after the first snow falls, need to be planned. The observers should be as reliable as possible.

11 CONCLUSIONS, RECOMMENDATIONS

From the roof snow load measurements in nature numerous data are available for the determination of roof shape coefficients in European climatic regions. Provisionally the roof shape coefficients as shown in figure 15 for different roof slopes are proposed. The different curves in several codes of European countries are also shown in this figure.

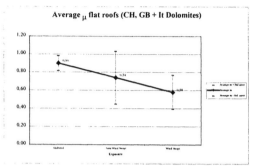

Figure 10: Mean and standard deviation of the roof shape coefficient for flat roofs for Swiss, Italian Dolomites, United Kingdom

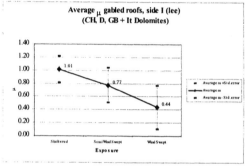

Figure 11: Mean and standard deviation of the roof shape coefficient for gabled roofs side I (lee) for Swiss, Italian Dolomites, United Kingdom and German sites

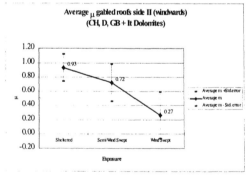

Figure 12: Mean and standard deviation of the roof shape coefficient for gabled roofs side II (windward) for Swiss, Italian Dolomites, United Kingdom and German sites

The exposure coefficients are provisionally proposed as follows:

- Sheltered 1.1
- Semi wind swept 0.9
- Windswept 0.7

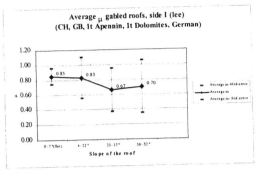

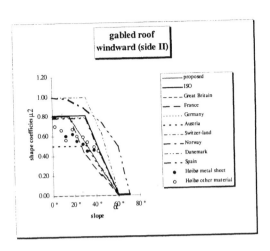

Figure 13: Mean and standard deviation of the roof shape coefficient for gabled roofs side I (lee) with different slopes for Swiss, Italian Apennine, Italian Dolomites, United Kingdom and German sites

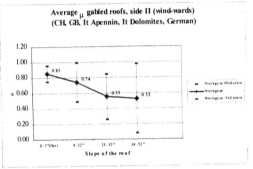

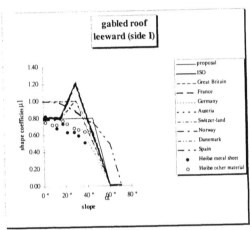

Figure 14: Mean and standard deviation of the roof shape coefficient for gabled roofs side II (windward) with different slopes for Swiss, Italian Apennine, Italian Dolomites, United Kingdom and German sites

Figure 15: Provisional proposal for roof shape coefficients depending on the slope of the roof for gabled roofs

A more detailed investigation with more measurements in the different climatic regions should improve the results of the multiple linear regression analysis. From this investigation different values for the climatic regions might be determined. From today's knowledge the scatter of the data due to natural influences is far greater than the possible influences from climatic regions, since the main influence for the roof snow load is considered by calculating the roof snow load based on the ground snow load.

From the wind tunnel tests a general confirmation of the roof shape coefficients for flat roofs and for roofs with a small roof slope can be determined. However the values are generally larger than those obtained from the measurements in nature. For large roof slopes different shape coefficients result. The measurements in nature confirm the results from previous measurements, that the roof shape coefficients are reduced for larger roof slopes, whereas the wind tunnel tests suggest a slight increase of the shape factors for both the leeward and windward sides of gabled roofs.

12 LITERATURE
DETAILLED LISTE OF LITERATURE SEE FINAL REPORT OF THE EUROPEAN SNOW LOAD RESEARCH, 1999

Ellingwood B, O'Rourke M.: 2 (1985). Probabilistic models of snow loads on structures, Structural Safety, 291 - 299

Gruner: June 1999. Data analysis for roof snow load investigation of the European Snow Load Project, Gruner Ltd, Bale

O'Rourke, Koch, Redfield: 1983. Analysis of roof snow load case studies, uniform loads, M. O'Rourke, P. Koch, R. Redfield, CRREL Report 83-1, US Army Corps of Engineers, Cold Regions Research & Engineering Laboratory

University of Pisa: Sept. 1999. Final report of the
European Snow Load Research

Pomeroy, J. W. & D. M. Gray: 1990. Saltation of
Snow. – Water Res. Research, 26, No.7, 1583-1594

DeQuervain, M.R.: 1972. Snow structure, heat and
mass flux through snow. – In: The role of snow and
ice in hydrology, Proc. Of the Banff Symposia,
Vol.1, 203-224

Snow Engineering, Hjorth-Hansen, Holand, Løset & Norem (eds) © 2000 Balkema, Rotterdam, ISBN 90 5809 148 1

Statistical investigation on snow load in Germany for verification of serviceability limit states

D. Soukhov
Institut für Massivbau und Baustofftechnologie, University of Leipzig, Germany

ABSTRACT: According to design philosophy of Eurocodes some representative values (not only characteristic one) of variable actions are to be used. These values (e.g. frequent and quasi-permanent ones) have a larger probability of occurrence in comparison with the characteristic value. To find out these values the specific statistical data (e.g. daily snow depth and/or water equivalent) should be used. The procedure for the determination of these values (with the help of the coefficients ψ_1 and ψ_2) for snow load is explained on the basis of German snow data. The calculated values of ψ_1 and ψ_2 for Germany are considered as proposed ones for new European standard EN 1991-1-3 "Snow Loads".

1 INTRODUCTION

The document CEN (1994) defines the limit states as the states beyond which the structure not longer satisfies the design performance requirements. Two groups of limit states are distinguished: ultimate limit states (ULS) and serviceability limit states (SLS).

Ultimate limit states concern the safety of the structure and the safety of people, and include:
– loss of equilibrium of the structure or structural element,
– loss of stability of the structure or structural element,
– failure due to fatigue.

Serviceability limit states concern the functioning of the construction work, the comfort of people and the appearance, and include for example:
– deformations and displacements which affect the appearance or use of the structure,
– vibrations which cause discomfort to people or limit the functional effectiveness of structure.

Ultimate limit states and serviceability limit states shall be verified for different design situations and different combinations of actions.

The following combinations of actions shall be considered for ULS:
1 combination for persistent and transient design situations,
2 combination for accidental design situations,
3 combination for seismic design situations.

For SLS the following combinations shall be taken into account:
4 characteristic (rare) combination,
5 frequent combination,
6 quasi-permanent combination.

According to CEN (1994) some representative values of variable actions are used in these combinations:
– characteristic value,
– combination value,
– frequent value,
– quasi-permanent value.

Characteristic value has normally a return period of 50 years and is the main representative value. Combination value takes into account a reduced probability of simultaneous occurrence of the most unfavourable values of several actions. It is obtained by multiplication of the characteristic value by factor ψ_0. The frequent value is determined such that:
- the total time , within a chosen period of time, during which it is exceeded for a specified part, or
- the frequency with which it is exceeded,
is limited to a given value. It can be obtained by multiplication of the characteristic value by factor ψ_1. The quasi-permanent value is determined such that the time during which it is exceeded is a considerable part of the reference period of time. The time during which it is exceeded may be set as 0.5 of the reference period. The quasi-permanent value may also be determined as the value averaged over the reference period of time. The quasi-permanent value can be obtained by means of multiplication of the characteristic value by factor ψ_2.

The characteristic value of a dominant variable action is used in combinations (1) and (4); combination values of non-dominant variable actions are used in combinations (1) and (4); frequent value of dominant variable action is used in combinations (2) and (5); quasi-permanent values of variable actions are used in combinations (2), (3), (5) and (6).

During the prenormative research work

"European Snow Loads Research Project" (CEC 1998) the snow data of some hundreds German climatic stations were collected with the help of Deutscher Wetterdienst (German meteorological office). On the basis of these data the statistical analysis of snow load was undertaken and a new snow load map for Germany as a part of European snow load map was elaborated. Because in this stage the characteristic values of snow load (values with return period of 50 years) were of interest, the extreme values (annual maxima) were taken into consideration. For determination the frequent and quasi-permanent values of snow load these data are not appropriate and therefore other procedures and original snow data should be used.

2 INVESTIGATION ON CLIMATIC STATION LEINEFELDE

2.1 Snow climate of Leinefelde

The climatic station Leinefelde is located in the central part of Germany, in the north-west of the province Thuringia, and has an altitude of 356 m above sea level. This station is chosen for investigation due to its representativeness, considering location and climate. The last one can be considered as mixed, and this is typical for the most of regions in Germany.

The complexity of climate in Leinefelde can be seen in Figures 1-2, where two different winters are shown. The first winter can be considered as the one with continental climate (although with some features of maritime climate). The second winter represents the maritime weather (but with some features of the mixed one). Considering all winters (not shown here) it is possible to conclude that the climate of this station is the mixed one.

The available data are the daily snow depth measurements from 1957 to 1993 with some gaps. The total record length is 26 winters.

2.2 Determination of snow event

Let us define for further investigation the term "snow event". Under snow event we will understand the snow pack on the ground during the time from the first occurrence of snow till it will be fully melted.

For all winters the snow occurs between the 1st November and the 30th April. It means that the possible time period for snow layer is 180 days. But in reality there are a lot of days without snow (we call them "zero values"). If only the days with snow depth greater than 1 cm are taken into account (this threshold was chosen because the discrete observations made with step of 1 cm) than we will have 1294 non-zero snow days during 26 years or 50 non-zero snow days per winter on average. The total number of events will be 164 with mean value of 6.3 per winter.

If we look at each winter then we will find that the number of events varies from 2 to 14 per winter. It can be shown that the distribution of the number of events per winter can be fitted well by the Poisson distribution. This distribution represents the number of events k which occur over equal time intervals (one winter), assuming that events occur independently at a constant average rate.

The probability density function of the Poisson distribution:

$$f(k,\lambda) = \lambda^k \exp(-\lambda) / k! \qquad (1)$$

with expected value $E(k) = \lambda$ and variance $Var(k) = \lambda$; where $\lambda > 0$ and $k = 0, 1, 2...$

Parameter λ can be obtained as the ratio of the observed number of events occurred to the number of intervals (winters). For Leinefelde: $\lambda = 164 / 26 = 6.3$ (the mean value of events per winter).

At the next step the duration of events is considered. All 164 events have been ranked in order regarding their duration. These values vary between 1 day and 117 days. The histogram of event duration can be seen on Figure 3.

Some theoretical probability distribution functions were verified whether they fit the observed event duration: exponential, normal, log-normal, Gumbel, Weibull. It was found that exponential distribution fits the data best. Also log-normal distribution can be considered as a candidate, however does not fit the data in region of small durations (one day).

The exponential distribution has the probability density function:

$$f(t,\lambda) = \lambda \exp(-\lambda t) \qquad (2)$$

with expected value $E(t) = 1/\lambda$ and variance $Var(t) = 1/\lambda^2$; $t > 0$ is the time between events and parameter $\lambda > 0$ is the average rate of events.

It has to be noticed that in this approach t is the time between events and therefore if $\lambda = 6.3$ then: $E(t) = 1 / 6.3 = 0.1587 \times$winter $= 0.1587 \times 180 = 28.6$ days (mean time between events). But for duration of the event itself the mean value $E(\tau) = 7.9$ days. The explanation can be seen in Figure 4, here t is the time between events and τ is the duration of event.

As it was already mentioned the climate of Leinefelde is typical for Germany and represents the mixed one with influences of both maritime and continental weather systems. This is reflected in the exponential probability plot for duration of events (Fig. 5). The function of this plot can be subdivided into 3 parts.

The first part with event duration from 1 to 20 days represents the influence of the maritime climate, the second part with event duration from 21 to 34 days can represent the mixed climate (although the fitting is not quite well) and the third part with event duration from 35 to 117 days represents the influence of the continental climate.

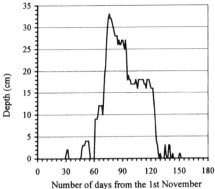

Figure 1. Snow depth of Leinefelde for winter 1978-1979

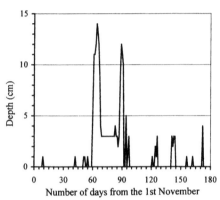

Figure 2. Snow depth of Leinefelde for winter 1979-1980

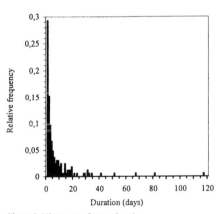

Figure 3. Histogram of event duration

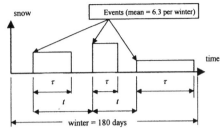

Figure 4. Duration of events and time between events

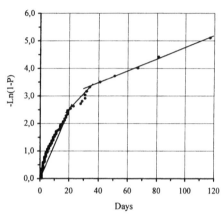

Figure 5. Exponential probability plot for event duration

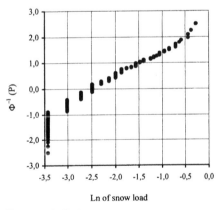

Figure 6. Distribution of event snow load maximum, log-normal probability paper

2.3 *Maximum of event snow load*

The maximum of snow load for each event is found by means of conversion of snow depth to snow load using the snow density model (so called "snow load factor") of DWD (German meteorological office) given in the report of Caspar & Krebs (1974). All 164 event maximum values are then ranked in order and cumulative distribution function (CDF) is obtained. The last one is verified to be fitted by some distributions: exponential, normal, log-normal, Gumbel, Weibull. For example a fit of log-normal distribution can be seen in Figure 6. But the exponential distribution fits the data best. It can be confirmed by Figure 7. The probability distribution function of exponential distribution is:

$$F(x) = 1 - \exp(-\lambda x) \qquad (3)$$

For probability plot the reduced variate Z can be found as:

$$Z = -\ln(1 - P) = \lambda x \qquad (4)$$

Parametr λ is obtained by least squares method (LSM) and equal to 6.67 $(1/(kN/m^2))$.

Because the mean value of the number of events per year is equal to 6.3 then the number of events per 50 years will be $N = 6.3 \times 50 = 315$. The characteristic value is defined to be exceeded one time per 50 years on average. Therefore the probability not being exceeded for this value defined by means of consideration of event distribution will be:

$$P = 1 - 1/315 = 0.9968 \qquad (5)$$

The reduced variate then will be equal to:

$$Z = -\ln(1 - 0.9968) = 5.7526 \qquad (6)$$

On Figure 7 it can be seen that the value of snow load which corresponds to this value of Z is equal to 0.86 kN/m^2. This value is very close to characteristic value of snow load for Leinefelde which was obtained using annual maximum data (CEC 1998) and equal to 0.90 kN/m^2. This confirms the possibility of application of the event approach to define the representative values of snow load.

2.4 Frequent and quasi-permanent values of snow load

To define the frequent value of snow load the first possibility from the two approaches given in chapter 1 "Introduction" is used: "The frequent value is determined such that the total time, within a chosen period of time, during which it is exceeded for a specified part is limited to a given value". It means that duration of snow load above the given level should be considered (we call it "duration over the threshold" curve).

The ratio of this duration to the chosen period of time (normally one year) is taken mostly as a value of 0.95, but other values from 0.9 to 0.99 can be also used depending on construction works considered.

If we consider N years of observations then we will have $365 \times N$ values of snow (for Leinefelde $N = 26$ and $365 \times N = 9490$). But only k days during N years the snow covers the ground. k is the number of snow days (1538 for Leinefelde). Therefore $p = k / (365 \times N)$ is the probability of snow presence and $q = 1 - p$ is the probability that no snow is present. All $365 \times N$ values are ranked in order. Then the maximum value can be found (for Leinefelde this is 38 cm of snow depth). The range between zero and the maximum value is subdivided into m intervals ($m = 100$ was chosen and interval is equal to 3.8 cm). Then the number of days with snow depth which belongs to each interval is counted. The

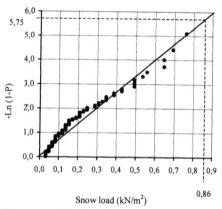

Figure 7. Distribution of event snow load maximum, exponential probability paper

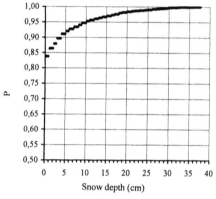

Figure 8. Probability that the snow depth is less or equal to a given level

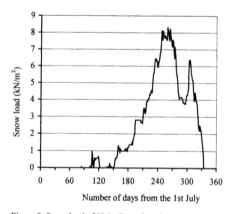

Figure 9. Snow load of Fichtelberg for winter 1964-1965

ratio of this number to the total number of days $365 \times N$ (9490 days for Leinefelde) corresponds to the probability that snow depth will be present in this interval. Thus the cumulative distribution function

(CDF) showing the probability that snow level is less or equal to a given value can be obtained. For Leinefelde this function is shown in Figure 8.

The function begins with the value 0.84. This corresponds to q, the probability that no snow is present. This "observed" CDF is checked whether it can be fitted by one of the theoretical distribution functions. Five distributions are considered: exponential, normal, log-normal, Gumbel, Weibull. Non of these distributions is found to be a good fit. The best result shows the normal distribution but its fitting of the upper tail of data is weak. Therefore it is decided to work directly with the obtained CDF. The fractiles of probabilities from 0.9 to 0.99 for this distribution can be seen in Table 1. The snow depth values are converted to the load values by means of the snow density model (snow load factor) elaborated by the German meteorological office DWD (Caspar & Krebs 1974).

As it mentioned above the characteristic value of snow load for Leinefelde was calculated in the report CEC (1998) and is equal to 0.90 kN/m^2. The ratio of fractile to the characteristic value gives the value of factor ψ_1 used for determining the frequent value because the last one is:

$$Q_{frequent} = \psi_1 \times Q_{characteristic} \qquad (7)$$

Normally the fractile of probability 0.95 is used for the most construction works. Then the factor ψ_1 will be equal to 0.20 (see Table 1).

The quasi-permanent value of snow load defined as 0.5 fractile of CDF results in zero because the CDF begins with 0.84 as it was already mentioned. It is decided to use the second procedure for determination of quasi-permanent value (see chapter 1 "Introduction"): to define the value averaged over one year (365 days). According to this method the quasi-permanent value for snow depth is 1.38 cm and for snow load is 0.022 kN/m^2.

The quasi-permanent value is defined as:

$$Q_{quasi-permanent} = \psi_2 \, Q_{characteristic} \qquad (8)$$

The ratio of the found average-over-one-year value to the characteristic value gives the value of factor ψ_2 used for the determination of the quasi-permanent value. This factor for Leinefelde is:

$\psi_2 = 0.022 / 0.90 \approx 0.03$

3 INVESTIGATION ON OTHER CLIMATIC STATIONS

According to above described procedure two other German climatic stations are also investigated. For both stations the original data are directly measured water equivalent. Station Fichtelberg located in Saxony in the mountains Erzgebirge at altitude 1213 m represents mountain/continental climate with very heavy snow load. The record period is from 1951 to 1993. As an example the winter 1964-1965 is shown on Figure 9.

Station Potsdam located near Berlin at altitude 81 m represents maritime climate. The record period is from 1893 to 1993. It is the longest record period among all European climatic stations which were available for "European Snow Loads Research Project" (CEC 1998).

Applying the described method the factors for frequent and quasi-permanent values result in the following values for Fichtelberg:

$\psi_1 = 0.39$

$\psi_2 = 0.08$

For Potsdam these values become equal to:

$\psi_1 = 0.12$

$\psi_2 = 0.02$

These calculated values for German climatic stations are included in the report CEC (1999) of European prenormative research for Eurocode 1, Part 2-3 "Snow Loads".

4 CONCLUSIONS

- For specific combinations not only the characteristic value but other representative values of snow load shall be used. These values have a larger probability of occurrence. To determine these values the annual snow maxima are not appropriate. The other snow data (preferably daily ones) should be used.
- Investigations on snow data of the climatic station Leinefelde with mixed climate show that the snow event approach can be helpful for the consideration of snow as a time loading process.
- The distribution of the snow events number per winter is fitted well by the Poisson distribution. The exponential distribution fits best a distribution of snow event duration as well as a distribution of snow event maxima.
- By means of the "duration over the threshold" function the frequent and quasi-permanent values for snow load can be obtained. The coefficients ψ for these values are proposed for Germany as follows:

$\psi_1 = 0.40$
$\psi_2 = 0.10$
for altitude greater than 500 m,

and

$\psi_1 = 0.20$
$\psi_2 = 0.00$
for altitude less than 500 m.

Table 1. ψ_1 values for different fractiles

Probability that given value is exceeded	Probability that given value is not exceeded	Snow depth m	Snow density kN/m³	Snow load kN/m²	ψ_1
0.1	0.9	0.04	1.65	0.07	0.08
0.09	0.91	0.04	1.65	0.07	0.08
0.08	0.92	0.05	1.66	0.08	0.10
0.07	0.93	0.07	1.69	0.12	0.14
0.06	0.94	0.08	1.70	0.14	0.16
0.05	0.95	0.10	1.72	0.17	0.20
0.04	0.96	0.12	1.74	0.21	0.24
0.03	0.97	0.16	1.79	0.29	0.33
0.02	0.98	0.19	1.82	0.35	0.40
0.01	0.99	0.25	1.88	0.47	0.52

5 ACKNOWLEDGMENTS

The author would like to express his appreciation for financial support made by the Commission of the European Communities DGIII-D3 to carry out the reported work.

REFERENCES

Caspar, J.W. & Krebs, M. 1974. Auswertung langjähriger Beobachtungen über Schneehöhen und Schneelasten. *Abschlußbericht, Deutscher Wetterdienst, Zentralamt.* Offenbach/Main.

CEC, 1998. Scientific Support Activity in the Field of Structural Stability of Civil Engineering Works, Snow Loads. Commission of the European Communities DGIII-D3. *Final report, Phase I. University of Pisa. Department of structural engineering.* Pisa.

CEC, 1999. Scientific Support Activity in the Field of Structural Stability of Civil Engineering Works, Snow Loads. Commission of the European Communities DGIII-D3. *Final report, Phase II. University of Pisa. Department of structural engineering.* Pisa.

CEN, 1994. ENV 1991-1: 1994 Basis of Design and Actions on Structures, Part 1: Basis of Design. CEN European Committee for Standardisation. Brussels.

Snow Engineering, Hjorth-Hansen, Holand, Løset & Norem (eds) © 2000 Balkema, Rotterdam, ISBN 90 5809 148 1

Statistical investigation on ground snow load in Germany

D. Soukhov
Institut für Massivbau und Baustofftechnologie, University of Leipzig, Germany

ABSTRACT: The present work is carried out as a part of the prenormative research "European Snow Loads Research Project". The snow information is collected from a plenty of German climatic stations and investigated using statistical procedures. The best fitting probability distribution functions are identified for different geographical regions.

1 INTRODUCTION

The prenormative research work "European Snow Loads Research Project" was initiated by the Commission of the European Community "DG III - Industry" in 1997. In the first phase the snow data were collected from the meteorological offices of all 18 CEN countries. These data are either water equivalent of the snow layer or (in the most of the countries) the values of snow depth on the ground. In the latest case the depth should be converted into load values by mean of appropriate snow density model. In the next stage the characteristic values of snow load (the values with a return period of 50 years) on the ground were calculated using the same statistical procedures for all countries. Afterwards the homogeneous European snow load map was elaborated (CEC 1998).

2 THE GROUND SNOW LOAD MODEL

The snow load is a variable action with a very high variation in space and time. Snow conditions vary between different geographical regions and can also depend on the local climate and on the altitude of the site. The process of snow accumulation and depletion on the ground is the very complex one and depends on different meteorological and geographical parameters (air and ground temperature, wind, humidity, exposure to the sun, geographical surroundings, distance from the sea etc). According to the prevailing climatic conditions the processes of snow accumulation and depletion can be divided into two basic groups. In the regions with continental and mountainous climate the snow layer increases steadily during the winter and then melts during a relatively short period. The maximum snow load occurs normally in the late winter. In the regions with maritime climate the snow cover is

intermittent during the winter. The snow layer can exist only some days or hours and then disappears. Thus the winter snow load maximum can be achieved even by a single snow fall. During some winters there may be no snow at all. In the regions with the mixed climate the snow loading is the combinations of the two above mentioned processes.

The occurrence of snowfalls, the duration and intensity of snow layer (and therefore snow load) are of random nature. Thus investigation of snow should be undertaken on the stochastic basis. To describe the problem exactly time-dependent modes are to be used (see for example CIB (1991)). But according to the design philosophy of Eurocodes the European snow load map should represent the extreme values of snow load. To find these extreme values it is enough to consider the yearly maximum values of the snow load on the ground. The sample should cover at least 30 or preferably 50 winters, so that the statistical analysis can result in reliable characteristic values.

Characteristic value of snow load Q_k is defined as a value with a given probability P of not being exceeded during one year. Return period is defined as a period of time during which the characteristic value Q_k can be exceeded once on average. The return period T and the probability P are connected by Equation 1:

$$T = 1 / (1 - P) \qquad (1)$$

The annual maxima of snow load can be considered as stochastically independent. Thus the value of return period automatically defines the probability of not exceeding during one year by means of conversion the Equation 1 into Equation 2:

$$P = 1 - 1 / T \qquad (2)$$

Then the characteristic value can be obtained as a fractile of this probability.

According to CEN (1994) T is equal to 50 years for common structures. This corresponds to the probability $P(Q_k) = 0.98$.

The result of this approach (i.e. the characteristic value of snow load) is very sensitive to the choice of the probability distribution function used for fitting statistical data. Which theoretical distribution fits the data best depends on the climatic and geographical conditions of site. Also statistical parameters (e.g. size of random sample) are important.

3 PREVIOUS INVESTIGATIONS

Some different conclusions about mentioned problem can be found in the research literature. Investigations in USA (Ellingwood & Redfield 1983) show that the log-normal distribution fits the observed values of the annual maximum snow load better than any extreme value distribution for the most of the weather stations. In Ellingwood & Redfield (1983) it is pointed out, however, that the type of distribution is geographically and climatically dependent. The use of the log-normal distribution is also recommended in ANSI (1982). In the work of Akerlund (1989) the snow depth and snow density were considered as mutually independent, random events. Then the 50 year Mean Recurrence Interval (MRI) value of snow load was calculated by means of the Pearson distribution. It was noted that the log-normal distribution fits the data also well but gives a larger characteristic value of snow load.

Only few investigations were dedicated to this problem in Germany. In the reports of Caspar & Krebs (1974) and Gränzer & Riemann (1980), as well as by Gränzer (1983), the extreme value distribution type I was used for determination of characteristic values of snow load. In the report of Luy & Rackwitz (1978) the snow depth measurements in Germany were analysed. For 60% of all climatic stations considered the gamma distribution was found to be the best fitting one, second was a log-normal distribution, only a small number of stations had Gumbel or normal distributions as the best fitting distribution. The gamma distribution is also recommended for snow loads in the documents CIB (1991) and JCSS (1996).

The choice of the fitting distribution in the report of Luy & Rackwitz (1978) was made basing on the likelihood probability criterion. The distribution which gives the maximum of the likelihood probability for a given sample is declared as the best fitting one. The parameters of the distribution were determined by means of the maximum likelihood method. Only snow depth was analysed in the work of Luy & Rackwitz (1978), therefore, all conclusions should be actually related only to snow depth. It is well known that conversion of snow depth into snow loads is very complex and depends on different parameters. For Germany the model for snow density was elaborated by the German Meteorological Office in the report of Caspar & Krebs (1974). The snow density (called "snow load factor") is a non-linear function of the snow depth. Thus the best fitting distribution for the depth is not necessarily the best fitting one for the load. The documents CIB (1991) and JCSS (1996) propose to use the gamma distribution for snow load and for all geographical regions, although the analysis of Luy & Rackwitz (1978) was based only on the West German data.

A main disadvantage of the gamma distribution is the lack of a theoretical probability distribution function. The calculation of probability and fractiles (and also the probability plot) is possible only by means of numerical integration. This was pointed out in the recent work (Bachmann et al. 1997) and it was proposed there to use the Gumbel distribution instead of the gamma distribution for rough calculations because Gumbel is the asymptotic extreme value distribution for the gamma distribution.

The latest investigations in Russia use only the extreme value distribution type I (Otstavnov 1996). But it is necessary to note that in Russia a very continental climate dominates. Izumi & Mihashi & Takahashi (1989) considered the Japanese snow data. It was pointed out that the extreme value distributions type I, type III and log-normal are the most applicable distributions in Japan.

4 THE INVESTIGATION ON THE CLIMATIC STATION ADELMANNSFELDEN

First of all the snow data from the climatic station Adelmannsfelden (Germany) was statistically investigated. The period of measurements was from 1935 to 1968 (totally 30 winters). The station is located in Baden-Württemberg at altitude of 470 m. This investigation was done before the water equivalent data were obtained from the German meteorological office (DWD). Thus the snow depth measurements were considered and then converted into snow load according to the model of the DWD (Caspar & Krebs 1974) which was completed by Gränzer & Riemann (1980). This model considers the snow load factor (i.e. snow density) D (kN/m³) as a function of snow depth h (m). For depth less than 1.53 m the load factor is:

$$D = 1.5981 + 1.2982 \times h - 0.8109 \times h^2 + \\ + 0.59907 \times h^3 - 0.20652 \times h^4 \qquad (3)$$

For depth greater or equal to 1.53 m the load factor D is equal to 2.7 kN/m³.

The model of CEB (1976) was also considered:

$$\rho = 300 - 200 \times \exp(-1.5 \times h) \qquad (4)$$

where:
ρ is the snow density (kg/m³)
h is the depth of the snow layer (m)

In a paper of Gränzer (1989) another formula was proposed:

$\rho = \rho_e \times \mu \times \ln\{1 + \rho_0 \times [\exp(h/\mu) - 1]/\rho_e\}/h$ (5)

where:
ρ is the mean density (kg/m^3)
ρ_e is the extreme density = 500 kg/m^3
ρ_0 is the minimum density = 170 kg/m^3
h is the depth of the snow layer (m)
μ is a scaling factor for depth = 0.85 m

The graphs of these models can be seen in Figure 1. All three models were used for conversion the snow depth data of Adelmannsfelden into snow load. It was found that using of all three models lead to the similar results regarding the choice of the best fit distribution. In research group of the "European Snow Loads Research Project" (CEC 1998) it was decided to use further only the model of DWD (Caspar & Krebs 1974).

Five different probability distribution functions are considered to be the best fit for observed data:
– Extreme value distribution type I for maximum (Gumbel)
– Extreme value distribution type II for maximum
– Weibull (extreme value distribution type III for minimum)
– Log-normal distribution
– Normal distribution

During the work on "European Snow Loads Research Project" it was decided to consider the least squares method (LSM) and moment's method (MM) for the determination of distribution parameters. The using of the maximum likelihood method leads to the necessity to solve the system of non-linear equations for most types of distributions. This was considered not to be reasonable for many hundreds of stations across Europe (CEC 1998).

One can see on the probability plot for the extreme value distribution of type I that the line with parameters defined by LSM (Fig. 2) fits the data better than the line with parameters defined by moment's method. Moreover, the second line gives the smaller 50 year MRI (mean recurrence interval) value i.e. underestimates a characteristic value of snow load (Fig. 3). Thus, it was decided to use for further investigations only LSM to obtain homogeneous results across Europe. The results for all five distributions can be seen in Table 1.

Two criteria for the choice of the distribution are used in this table:
1 Coefficient of correlation (CoC) between reduced variable (according to probability paper) and the original values of snow load (or logarithm of snow load). The larger this coefficient (in the ideal case it would be equal to 1.0) the better the probability distribution function fits the data.
2 Maximum likelihood probability (MLP). The larger the likelihood probability the better the probability distribution function fits the original data (this criterion is very seldom, used only by Luy & Rackwitz (1978)). The choice of parameters is based on LSM.

According to both criteria only three theoretical distributions can be considered to be the best fit for snow load data:

Table 1. Comparison of different distributions

Type of distribution	Snow load value with return period of 50 years		Coefficient of correlation	Maximum likelihood probability
	Least square method	Method of moments		
	kN/m^2	kN/m^2		
Extreme value distribution type I	1.96	1.79	0.9885	$3.4 \cdot 10^{-7}$
Log-normal distribution	2.72	1.88	0.9856	$7.3 \cdot 10^{-7}$
Weibull distribution	1.83	1.75	0.9862	$16.1 \cdot 10^{-7}$
Extreme value distribution type II	5.00	1.79	0.9468	$0.5 \cdot 10^{-7}$
Normal distribution	1.67	1.56	0.9723	$0.5 \cdot 10^{-7}$

– Extreme value distribution type I for maximum (Gumbel)
– Log-normal distribution
– Weibull (extreme value distribution type III for minima)

If the criterion of maximum likelihood probability is applied then Weibull will have the first place. If the coefficient of correlation is used then Gumbel will have the first place, Weibull will be at the second one, the log-normal distribution will be at the third one. The difference between values of the coefficient of correlation for all three functions is very small. This means that these three distributions fit the data almost with the same accuracy, that can be confirmed by means of the probability plots (Figs 2, 4, 5). Further only this criterion, the maximum of coefficient of correlation, will be used.

5 ANALYSIS OF GERMAN DATA

Considering the background information of building standards of CEN members one can find that almost all countries use the extreme value distribution type I (Gumbel), apart from Denmark where Weibull distribution is applied. The problem of the best fitting distribution is considered in "European Snow Load Research Project" (CEC 1998) regarding the different climatic regions in Europe. In this chapter the German investigations are presented.

Original data which are the water equivalent, measured three times per week, and daily snow depth are collected from 331 stations. The most stations from Western Germany have a record period of 30 years, the most stations from Eastern Germany have a record period of 45 years (including two stations with about 100 years record period).

The mean value, standard deviation and coefficient of variation of annual maximum of snow load versus altitude for all climatic stations can be seen in Figures 6-8.

301

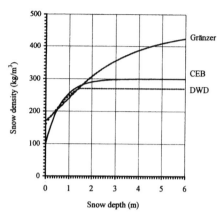

Figure 1. Different snow density models

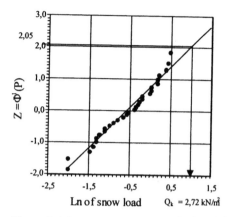

Figure 4. Probability plot for log-normal distribution, LSM

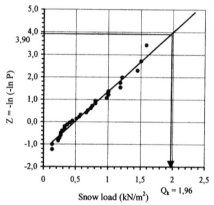

Figure 2. Probability plot for extreme value distribution type I, least squares method (LSM)

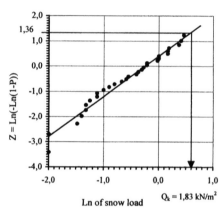

Figure 5. Probability plot for Weibull distribution, LSM

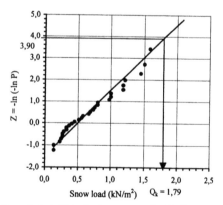

Figure 3. Probability plot for extreme value distribution type I, method of moments (MM)

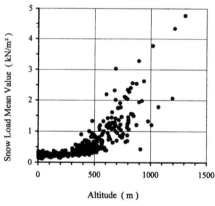

Figure 6. Mean values of annual maximum of snow load versus altitude

302

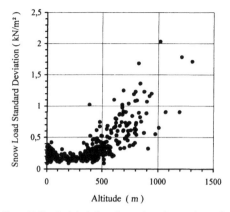

Figure 7. Standard deviation of annual maximum of snow load versus altitude

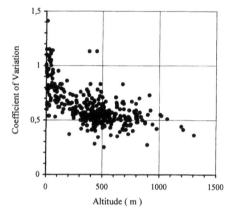

Figure 8. Coefficient of variation of annual maximum of snow load versus altitude

Figure 9. Location of the climatic stations with the different best fitting probability distribution functions in Germany

As it can be seen from these figures, the mean value and standard deviation increase with altitude while the coefficient of variation has a tendency to decrease with altitude. Figure 8 shows that the scatter of coefficient of variation is very large: from 0.2 up to 1.4. The largest values of the coefficient of variation are observed at low altitudes. This is statistical reason that different probability distribution functions fit best the data at different stations.

Another reason is the geographical (climatic) conditions. Middle and southern Germany have a rather heterogeneous relief, many of the middle high mountains are cut by valleys. This causes very strong differences between local climatic conditions.

The application of the above mentioned approach (probability plot, LSM, coefficient of correlation between reduced variable and snow load values) shows which distribution is the best one for 331 climatic stations in Germany:

– Gumbel distribution -   78 stations
– Log-normal distribution -   171 stations
– Weibull distribution -   82 stations

Thus the log-normal distribution is best in 52%, Gumbel in 23%, Weibull in 25% of all stations. The locations of the stations with the best fitting functions can be seen in Figure 9.

The colours of this map mean the following:
– light grey: stations with Gumbel distribution as the best fitting one,
– black: stations with log-normal distribution as the best fitting one,
– dark grey: stations with Weibull distribution as the best fitting one.

It can be seen that northward the middle high mountains (51-52 degree of north latitude) a log-normal distribution dominates. The Northern low plain (Norddeutsche Tiefebene) and almost the whole of eastern Germany (the former DDR) belong to this region.

Southward this latitude the map is very heterogeneous. In the middle high mountains Hessisches Bergland, Taunus, Fichtelgebirge and also in the flat land of the Lower Rhein (Niederrhein) the log-normal distribution dominates again.

Log-normal distribution prevails also in more than 50% of the stations in Rothaargebirge (mixed with Gumbel and Weibull), Fränkische Schweiz (with Gumbel and Weibull), Frankenwald (with Weibull), Schwarzwald (with Gumbel), Rhön (with Gumbel).

Weibull distribution dominates in Odenwald, Böhmerwald, Bayerischer Wald (mixed with log-normal), Fränkische Alb (with Gumbel), in Harz and in the high western Alps (Allgäuer Alpen).

Gumbel distribution dominates in Sauerland, Weser-Bergland (mixed with log-normal), Donau Ebene (mixed with Weibull), Oberschwaben (mixed with log-normal) and in the high eastern Alps (Chiemgauer Alpen).

In some mountain regions the best fitting distribution depends obviously on the main wind direction. So for example in the northern Eifel the log-normal distribution dominates, Gumbel is a prevailing one on the east-southern slopes. There is a very complicated situation in Schwäbische Alb: Gumbel dominates in the North, log-normal on west-northern slopes and in the West, Weibull on south-eastern slopes. A mixed picture can be observed also in Taubergrund, where only in the East the log-normal distribution dominates.

6 CONCLUSIONS

The analysis performed above confirms the complexity of the choice of the probability distribution function for a large region. For future research it can be recommended to consider this problem together with meteorological facilities taking into account the local climatic conditions. For Germany the following rough conclusions can be done:

– In the regions northward the middle high mountains (51-52 degree of north latitude) it appears that the log-normal distribution is the best fit.
– In middle and southern Germany the extreme value distribution type I (Gumbel) can be used for fitting, even in the regions where Weibull distribution dominates. This is because in comparison with Weibull distribution the Gumbel one gives slightly larger 50 years MRI values (i.e. it is conservative).

ACKNOWLEDGMENTS

The author would like to express his appreciation for financial support made by the Commission of the European Communities DGIII-D3 to carry out the reported work.

REFERENCES

Akerlund, S. 1989. Snow depth and Snow Loads in Umea. *First International Conference on Snow Engineering; Santa Barbara, California, July 1988*: 3-7. Hanover, New Hampshire: CRREL Special Reports 89-6.

ANSI, 1982. ANSI A58.1: Minimum Design Loads for Buildings and Other Structures. *American National Standards Institute*. New York.

Bachmann, H. & Rackwitz R. & Schueller G. 1997. Tragwerkszuverlässigkeit, Einwirkungen. *Der Ingenierbau* (8). Ernst & Sohn.

Caspar, J.W. & Krebs, M. 1974. Auswertung langjähriger Beobachtungen über Schneehöhen und Schneelasten. *Abschlußbericht, Deutscher Wetterdienst, Zentralamt*. Offenbach/Main.

CEB, 1976. Common unified rules for different types of construction and material. Comite Euro-international du Beton, CEB. *Bulletin 116-E*. Paris.

CEC, 1998. Scientific Support Activity in the Field of Structural Stability of Civil Engineering Works, Snow Loads. Commission of the European Communities DGIII-D3. *Final report, Phase I. University of Pisa. Department of structural engineering*. Pisa.

CEN, 1994. ENV 1991-1: 1994 Basis of Design and Actions on Structures, Part 1: Basis of Design. CEN European Committee for Standardisation. Brussels.

CIB, 1991. Action on structures. Snow loads. *Report by Commission W81*. First edition.

Ellingwood, B. & Redfield, R. 1983. Ground Snow Loads for Structural Design. *Journal of Structural Engineering, ASCE* 109 (4): 950-964.

Gränzer, M. & Riemann. H. 1980. Statistische Auswertung langjähriger Schneemessungen zur Ermittlung der Schneelastverteilung im Bundesgebiet. *Schlußbericht, Landesstelle für Baustatik Baden-Württemberg*. Tübingen.

Gränzer, M. 1983. Zur Festlegung der rechnerischen Schneelasten. *Bauingenieur* 58: 1-5.

Gränzer, M. 1989. Angabe von Schneelasten, geografisch nach Zonen gegliedert, für den Eurocode "Lasten, Teil 7". *Schlußbericht, Landesstelle für Baustatik Baden-Württemberg*. Tübingen.

Izumi, M. & Mihashi, H. & Takahashi, T. 1989. Statistical Properties of the Annual maximum Series and a New Approach to Estimate the Extreme Values for Long Return Periods. *First International Conference on Snow Engineering; Santa Barbara, California, July 1988*: 25-34. Hanover, New Hampshire: CRREL Special Reports 89-6.

JCSS, 1996. Probabilistic Model Code, Part 2: Actions on structures, 2.12 Snow Load.

Luy, H. & Rackwitz, R. 1978. Darstellung und Auswertung von Schneehöhenmessungen in der Bundesrepublik Deutschland. *Berichte zur Zuverlässigkeitstheorie der Bauwerke* (31). München.

Otstavnov, V. 1996. Elaboration of Draft Map for Snow Loads in Russia. *Research Report* 16-13-135/96 (in Russian). Moscow: CNIISK.

Snow Engineering, Hjorth-Hansen, Holand, Løset & Norem (eds) © 2000 Balkema, Rotterdam, ISBN 90 5809 148 1

Proposal of chart and table to estimate roof snow accumulation on the ground after sliding down

M.Takita & M.Watanabe
Hachinohe Institute of Technology, Japan

ABSTRACT: This paper presents a simple analytical model that modeled roof snow as a rigid body. This model contains some parameters: eaves height, pitch of roof, horizontal roof length, kinetic friction constant of roofing material and so on, and describes horizontal distance from the roof edge to the impact point of roof snow. Based on the model, the chart and the table were proposed, that enable to evaluate roof snow accumulation on the ground without special program codes.

1 INTRODUCTION

Considering on the planning of dwelling houses in heavy snow fall regions, it is an important theme to make dispose snow that lay on the roof or on the ground. Japanese designers have been interested in the theme because more than half of the regions of Japan is heavy snow fall region. Considering on the method of disposing roof snow by sliding, the method is very effective on the reduction of snow load, on avoiding human injury by snow drop, on the maintenance of dwelling houses and so on. Then it has been required to establish the technology of planning dwelling houses in heavy snow fall region.

The authors have been studied on the factors related to the subjects on controlling roof snow by sliding. In the previous study the authors indicated that the necessity to develop the technologies for estimating behavior of snow and counter measuring to the snow (Watanabe, M. & Takita, M. 1997). It is particularly important to establish the methods to evaluate numerically the behavior of roof snow; namely, how far the roof snow fall from the roof edge, how the roof snow accumulates on the ground. For example, Paine studied on the building design for heavy snow fall region (Paine, J. C. 1988), but there are few methods that evaluate the roof snow accumulation, then an experimental formula developed by Nakamura (Nakamura, H. 1978) has been used in Japan.

The authors have been developed a method that based on a simple model of roof snow and evaluates numerically roof snow accumulation on the ground. The authors' method contains some parameters, for example, the pitch of roof, the eaves height, the depth of snow on roof, the kinetic friction constant of roof, the viscous force and so on, that have not been introduced to Nakamura's formula (Takita, M. & Watanabe, M. 1994a, b). The authors' method contains the parameters that are related to the planning of dwelling houses and is simple, but it is not easy to use our method because it requires special computer program.

The objective of this study is to propose the chart and the table. These are based on the authors' method and enable to evaluate the accumulation of roof snow on the ground without computer program. In the first stage of this paper, the authors' method will be presented and modified to develop the chart and the table. In the second stage, the chart and the table will be proposed.

2 FORMULATION OF ROOF SNOW BEHAVIOR

2.1 *Derivation of formulas*

The behavior of a snow block on roof will be described by the models shown in Figure 1 and Figure 2. The model of Figure 1 defines sliding behavior of a snow block on roof. Using this model the velocity of a snow block at roof edge will be derived by the model. The model of Figure 2 defines a fall of a snow block, and the horizontal distance from roof edge to the impact point of the snow block will be calculated by the model.

In these models, the following assumptions will be applied:
(a) roof snow is a collection of snow blocks,
(b) a snow block is rigid and dose not deform,
(c) a snow block is not affected by air friction nor wind force,
(d) a snow block impacted on the ground dose not splash nor slide on the ground,
(e) surface of roof is plain, smooth and homogeneous.

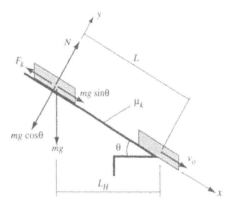

θ : pitch of roof
L : sliding distance of a snow block
L_H : horizontal sliding distance of a snow block
μ_k : kinetic friction constant of roof
v_o : original velocity of a snow block at roof edge
m : mass of a snow block
g : gravity acceleration
N : normal force on roof
F_k : kinetic friction force

Figure 1. Sliding model of roof snow

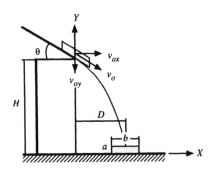

v_o : original velocity of a snow block at roof edge
v_{ox} : horizontal component of v_o
v_{oy} : vertical component of v_o
a : height of a snow block
b : horizontal length of a snow block
H : eaves height
D : horizontal distance from roof edge to the impact point

Figure 2. Falling model of roof snow

Furthermore, the viscous force between the roof and the snow will be ignored.

2.1.1 Basic formulas
Studying on the x-y coordinate system that is defined in Figure 1, the equation of motion of a snow block on roof is found to be

$$m\ddot{x}(t) = mg\sin\theta - F_k \qquad (1)$$

where $F_k = \mu_k N$ and $N = mg\cos\theta$.

The snow block begins to slide when $mg\sin\theta > F_k$. Then integrating Equation 1 by time t and applying

the initial conditions $x(0) = 0$ and $\dot{x}(0) = 0$, where $t = 0$ means the time when the snow block begin to slide, the velocity of the snow block at the edge of roof will be led to

$$v_o = \sqrt{2g\alpha L} \qquad (2)$$

where $\alpha = \sin\theta - \mu_k\cos\theta$. If the velocity of the snow block at the roof edge was decided, the horizontal distance from roof edge to the impact point of the snow block will be evaluated by the model of Figure 2.

On the X-Y coordinate system that is defined in Figure 2, the equation of motion of a snow block, that fall from the roof edge with the original velocity v_o, is found to be

$$m\ddot{Y}(t) = -mg. \qquad (3)$$

In Equation 3, $t = 0$ means the time when the snow block leaves the roof edge.

Integrating Equation 3 by time t, and applying the initial conditions $Y(0) = H$ and $\dot{Y}(0) = -v_{oy}$, the position of the snow block at time t $Y(t)$ will be led to

$$Y(t) = -\frac{1}{2}gt^2 - v_{oy}t + H.$$

Then the time T when the snow block impacts on the ground will be expressed as

$$T = \frac{1}{g}\left(\sqrt{v_{oy}^2 + 2gH} - v_{oy}\right), \qquad (4)$$

and the horizontal distance from the roof edge to the impact point of the snow block, D, will be led as

$$D = \frac{v_{ox}}{g}\left(\sqrt{v_{oy}^2 + 2gH} - v_{oy}\right) \qquad (5)$$

where $v_{ox} = v_o\cos\theta$ and $v_{oy} = v_o\sin\theta$.

2.1.2 Modification of formulas
As a preparation for making the chart and the table, Equation 2 and Equation 5 will be modified as follows. Replacing L and α to L_H and α_d respectively, Equation 2 will be modified as

$$v_o = \sqrt{2g\alpha_d L_H} \qquad (6)$$

where $\alpha_d = d - \mu_k$ and $d = \tan\theta$. Here, constant α_d means that the snow block begins to slide when the pitch of roof d is greater than the kinetic friction constant μ_k.

In Equation 6, the horizontal distance D is affected by the eaves height H. This is not appropriate to evaluate snow accumulation on the ground, because when snow lies on the roof it usually lies on the ground, too. Namely, it will be required that the eaves height H must be changed to include the snow cover on the ground. Then, introducing a new parameter H_S that means the snow cover on the ground, and modifying Equation 5 by using parameters α_d, z and L_H,

$$D = 2\alpha_d z L_H \left(\sqrt{1 + \frac{H_d}{d}\frac{1}{\alpha_d z L_H}} - 1 \right) \qquad (7)$$

where $z = d/(1+d^2)$.

The dependency among the parameters related to the calculation of the horizontal distance D will be summarized as Figure 3. In the figure the parameters surrounded by the rectangle are independent, and arrows show the dependency between the parameters.

2.1.3 *On the evaluation of the range of accumulation of roof snow*

To evaluate the range of accumulation of roof snow on the ground, it will be required to evaluate the behavior of the snow block after the impact, namely splashing of the snow block, sliding on the ground or on the snow and so on. As Yamaguchi mentioned, evaluating the range of splashing of snow, it is required to clarify the impact force (Yamaguchi, M. et al. 1992).

This means that the authors' method is not appropriate to study on the roof snow sliding and its accumulating on the ground, because the authors' method has been ignored splashing and sliding on the ground. But the studies by Yamaguchi (Yamaguchi,

M. et al. 1992) and Abe (Abe, O. 1996) show that the horizontal distance from roof edge to the impact point of roof snow that is calculated by some numerical methods gives larger distance than the actual phenomenon, even if the kinetic friction force or viscous force would be evaluated.

Then, except for the case that the snow block splashes in wide range by the impact to the ground with high impact force, it will be expected that the authors' way, that is treating the accumulating range as the sum of the horizontal distance and the half length of snow block, gives safety results.

2.2 *Effect of parameters*

Figure 4 shows the relationship between the pitch of roof d and the horizontal distance normalized by the horizontal sliding distance L_H.

$$\frac{D}{L_H} = 2\alpha_d z \left(\sqrt{1 + \frac{\beta}{d}\frac{1}{\alpha_d z}} - 1 \right) \qquad (8)$$

where $\beta = H_d/L_H$. Here, D/L_H and β are the horizontal component and the vertical component respectively, that are normalized by L_H. Figure 4 (a), (b) and (c) show that the horizontal distance D become

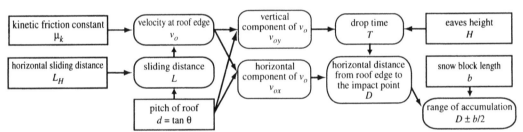

Figure 3. Relationship between the parameters that are related to the evaluation of the accumulation of roof snow (the parameters surrounded by the rectangle are independent parameters. Arrow means relationship among the parameters)

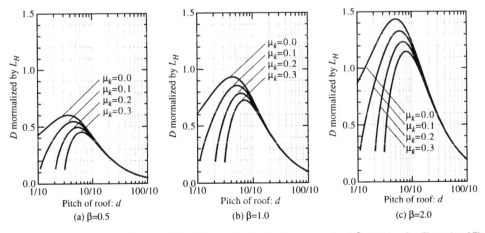

Figure 4. Relationship between the pitch of roof d and D normalized by L_H (Here $d = \tan\theta$ and $\beta = H_d/L_H$. See Figure 1 and Figure 2 on the parameters.).

maximum when the pitch of roof is in the range from 3/10 to 7/10, and that the horizontal distance D is in inverse proportion to the kinetic friction constant μ_k. Furthermore when μ_k is constant, the horizontal distance D extremely increases according to the increment of drop distance H_d.

From these results, it is clear that the degree of effect to the horizontal distance D is great in the order of β, μ_k and d.

2.3 Comparison with an experimental formula

Equation 9 is an experimental formula that gives the width of roof snow accumulation. This equation was formulated by Nakamura from field works at Shinjyo in Japan (Nakamura, H. 1978).

$$l_1 = \frac{M + \sqrt{M^2 - 1.12(l_2 - 0.7l_3)h_0 - 0.6l_2{}^2}}{0.56} \quad (9)$$

where $M = -0.3h_0 - 0.7l_2$; $h_1 = h_0 + 0.7l_1$ and $h_2 = h_0 + 0.7l_1 - 1.2l_2$

In Japan, this formula has been used to evaluate roof snow accumulation, but this formula doesn't contain parameters related to the roof, for example the eaves height H_d, the kinetic friction constant of roof, the pitch of roof and so on (see Figure 5).

Figure 6 is a comparison of the Nakamura's formula and the authors' method, where D is the horizontal flying distance given by the authors' method and l_1 is the range of snow accumulation given by Nakamura's formula. In Nakamura's formula, l_1 was calculated for $h_0 = 1, 2, 3, 4$ m and $l_2 = 0.5$ m. In the authors' method, D was calculated for $\mu_k = 0.1$ and 0.2, $H = 5$ m, $d = 4/10$, and H_d was applied as the drop distance. In the Nakamura's formula, horizontal roof length l_3 was assumed as the horizontal sliding distance of the roof snow block. In the authors' method, it is impossible to evaluate accumulation range of snow on the ground, then it was assumed as the maximum value of the horizontal distance D is caused by a snow block sliding from the top of the roof.

From Figure 6, it is clear that D becomes longer than l_1 when the drop distance H_d becomes longer. Whereas, when H_d is short or horizontal sliding distance became long, authors' method gives shorter distance than the range of accumulation given by Nakamura's formula.

Considering on the difference between the Nakamura's formula and the authors' method, Nakamura's formula gives the result of snow accumulation for a long term, and the authors' mthod gives the behavior of roof snow in a short term. Then if the authors' formula will be applied to evaluate long term behavior of roof snow, the formula should be applied several times considering on the vary of snow fall.

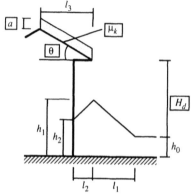

l_1 : width of roof snow accumulation
l_2 : eaves depth
l_3 : horizontal roof length
h_0 : snow conver on the ground
h_1 : height of roof snow accemulation
h_2 : height of roof snow accumulation beside of wall

Figure 5. Parameters on the Nakamura's experimental formula (the parameters surrounded by the rectangle have not been included in the formula)

3 CHART AND TABLE

In this section, the chart and the table for estimating roof snow accumulation on the ground will be presented. The chart enables easy and visual estimation of roof snow accumulation for the specified parameters. The table enables to estimate generally the accumulation for various combinations of the parameters.

3.1 Chart for evaluating roof snow accumulation

The chart is composed by four figures as shown in Figure 7. Figures <1>, <2> and <3> are used to decide the velocity of a snow block at roof edge, the drop time of the snow block and the horizontal distance from roof edge to the impact point of a snow block respectively, and Figure <4> is used to plot the result. As presented in Figure 3, the proposed chart depends on the pitch of roof, the horizontal sliding distance, the drop distance and so on, then if these parameters were changed, the another chart must be prepared.

For example, defining the pitch of roof d, the kinetic friction constant μ_k and the drop distance H_d as $4/10, 0.1$ and 500 cm respectively, the usage of the chart will be explained as follows.

(a) Check the horizontal sliding distance L_H of a snow block that will be evaluated. L_H is given by the summation of the sum of the horizontal length of snow blocks that slid previously and the half of the horizontal length of the snow block that will be evaluated.

(b) On Figure <1>, read horizontal velocity v_{ox} and vertical velocity v_{oy} at roof edge. If L_H is 300 cm and L_H is 0.1, v_{oy} and v_{ox} will be decided as point A and B respectively.

(c) Draw line from point A on Figure <1> to Figure <2>, and decide the drop time T at the intersection C on the curve corresponding to the value of L_H. If H_d is 500 cm, point C will be decided.

(d) Draw line from point C on Figure <2> to Figure <3>, and decide the horizontal distance D. The line of v_{ox} will be drawn by the horizontal axis at the top of Figure <3>.

(e) Draw line from point D on Figure <3>, and plot snow block figure according to the length and the depth of the snow block. Here, point D means the horizontal center point of the snow block.

By the procedures from (a) to (e), the range of accumulation of a snow block will be decided. Then applying these procedures to snow blocks remained on the roof, the shape of the accumulation of roof snow will be plotted.

3.2 Table for evaluating roof snow accumulation

The proposed chart is very useful as far as the required parameters are included on the chart, but the generality of the chart is restricted because the combination of the parameters will be huge. Then we propose the table that enables to evaluate generally roof snow accumulation.

In order to derive the table, the parameter D of Equation 7 wil be redefined as

$$D = 2P_L\left(\sqrt{1 + \frac{P_H}{P_L}} - 1\right) \qquad (10)$$

where $P_L = \alpha_d z L_H$ and $P_H = H_d / d$. Here, P_L is a variable parameter that depends on the horizontal sliding distance L_H, and P_L is a constant. Using Equation 10, the table will be proposed as Figure 8. The table is composed by two components; Table <1> is the summary of parameters that define the problem, and Table <2> is the work sheet for evaluating roof snow behavior.

For example, defining the eaves height, the pitch of roof, the snow cover on the ground and the kinetic friction constant of the roofing material as 500 cm, 4/10, 100 cm and 0.1 repectively, the usage of the table will be explained as follows.

(a) Check these parameters on Table <1>, and calculates the other values of H_d, α_d, z, $\alpha_d z$ and P_H.

(b) Devide roof snow to the blocks tht have appropiate horizontal length. In the example, the roof snow was devided to five blocks in one meter length in each.

(c) On the work sheet of Table <2>, calculate the values of L_{Hi}, P_{Li}, D_i and $D_i \pm b_i / 2$ for each snow block.

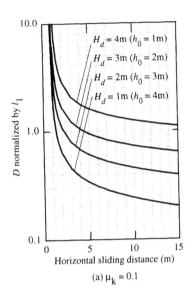

(a) $\mu_k = 0.1$

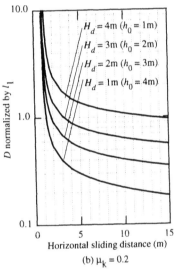

(b) $\mu_k = 0.2$

Figure 6. Comaprison of the range of accumulation given by Nakamura's experimental formula and the horizontal distance from roof edge to the impact point calculated by the authors' method (in the Nakamura's formula: $l_2 = 0.5\ m$, $h_0 = 1\ m$, $4\ m$, in the authors' method: $H = 5\ m$, $d = 4/10$)

(d) Plot the results.

As presented previosly the proposed table is formulated by the simple arithmatic formulas, then the table shown in Figure 8 will be evaluated easily on a spread sheet application program by a personal computer.

309

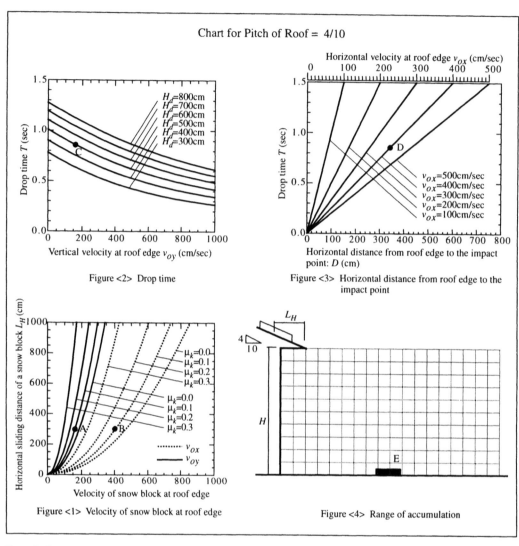

Figure 7. Exmple of the proposed chart (In Figure <1>, check horizontal sliding distance L_H of a snow block, and decide velocity components v_{oy} and v_{ox} at the intersections A and B. In Figure<1> μ_k is the kinetic friction constant of roof. Draw line from point A on Figure <1> to Figure <2>, and decide the drop time T at point C. Draw line from point C on Figure <2> to Figure <3>, and decide the horizontal distance D at point D. On Figure <3> v_{ox} that was decided on Figure <1> will be required. Finally draw line from point D on Figure <3> to Figure <4>, decide the horizontal center point E of the snow block and plot the figure of the snow block on Figure <4>.)

4 CONCLUSIONS

In this study, roof snow was treated as a collection of snow blocks, and the behavior of a snow block, sliding on the roof and accumulating on the ground, was formulated by a rigid body model. The characteristic of the authors' method was clarified by the study on the relationship among the parameters and by the comparison of Nakamura's experimental formula.

Then the chart and the table that enable to evaluate roof snow accumulation on the ground were proposed. The chart visualizes snow accumulation easily without

calculation. The table enables to solve the problem for any combination of parameters by arithmetic operations.

To simplify the chart and the table the viscous force has been ignored in this study. The development of the chart and the table included the viscous force should be the next theme. Furthermore, the authors' formula is based on a rigid body model of roof snow block, then the introduction of the compression of snow, splashing of snow at the impact on the ground and so on should be the theme for further study.

Table <1> Summary of parameters

Eaves height H (cm)	500	Snow cover on the gorund H_S (cm)	100	Drop distance $H_d = H - H_S$ (cm)	400
Pitch of roof $d = \tan\theta$	4/10	Kinetic friction constant of roof μ_k	0.1	$\alpha_d = d - \mu_k$	0.3
$z = d / (1 + d^2)$	10/29	$\alpha_d z$	3/29	$P_H = H_d / d$	1000

Table <2> Work sheet for calculating horizontal flying distance of snow blocks on roof

Snow Block No i	Horizontal length of snow on the roof (cm)	Horizontal length of a snow block b_i (cm)	Thickness of a snow block a_i (cm)	Horizontal sliding distance L_{Hi} (cm)	$P_{Li} = \alpha_d z L_{Hi}$	Horizontal distance from roof edge to the impact point D_i (cm)	Range of accumulation	
							$D_i - b_i/2$	$D_i + b_i/2$
1	500	100	50	50	150/29	133.9	83.9	183.9
2	400	100	50	150	450/29	220.0	170.0	270.0
3	300	100	50	250	750/29	274.0	224.0	324.0
4	200	100	50	350	1050/29	315.0	265.0	365.0
5	100	100	50	450	1350/29	348.3	298.3	398.3

$$D_i = 2P_{Li}\left(\sqrt{1 + P_H/P_{Li}} - 1\right)$$

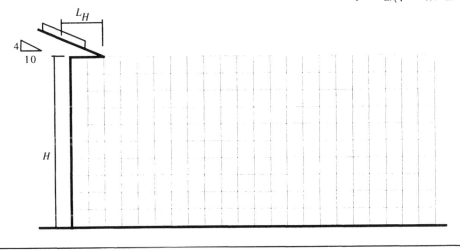

Figure 8. Example of the proposed table (In Table <1>, check the parameters of the parameters of the eaves height, the pitch of roof, the snow cover on the ground, the kinetic friction constant and the drop distance from roof edge, and calculate the values of z, α_d, $\alpha_d z$ and P_H. Devide roof snow to some blocks. In Table <2>, for each snow block, check the horizontal length of snow on the roof, the horizontal length of a snow block, the thickness of a snow block and the horizontal sliding distance, and calculate the values of P_{Li}, D_i and the range of accumulation. Finally plot the result on the figure at the bottom).

REFERENCES

Abe, O. 1996. Shape of banks formed under the very large sloped roof – mesuremenet and estimation –, *Proc. of 13th symposium on snow engineering, Journal of snow engineering. Japan society for snow engineering* 159-164 (*in Japanese*)

Paine, J. C. 1988. Building design for heavy snow area, *Proc. of 1st international conference on snow engineering, CRREL special report* 86-6: 483-492

Nakamura, H. 1978. Shape of the snow bank formed under eavas by the fall of snow on roofs. *Journal of the Japanese society on snow and ice* 40 (1): 37-41 (*in Japanese with English abstract*)

Takita, M. & Watanabe, M. 1994a. Simulation of sliding and accumulation of roof snow, *Proc. of 10th symposium on snow engineering, Journal of snow engineering. Japan society for snow engineering* 144-158 (*in Japanese*)

Takita, M. & Watanabe, M. 1994b. Simulation of sliding and accumulation of roof snow (Part 2 Effect of viscous drag), *Proc. of 11th symposium on snow engineering, Journal of snow engineering. Japan society for snow engineering* 39-42 (*in Japanese*)

Yamaguchi, H., Tomabechi, T., Yamada, T., Nakajima, H., Ito, T. and Hoshino, M. 1992. Study on the property of snow when sliding off membrane structure roof, *Journal of structural and construction engineering, Architechtual institute of Japan* 437: 91-96 (*in Japanese with English abstract*)

Watanabe, M. & Takita, M. 1997. Fundamental study on control of snow sliding, *Journal of snow engineering. Japan society for snow engineering*, 13 (1): pp.3-12 (*in Japanese with English abstract*)

Snow Engineering, Hjorth-Hansen, Holand, Løset & Norem (eds) © 2000 Balkema, Rotterdam, ISBN 90 5809 148 1

Developing ground snow loads for New Hampshire

Wayne Tobiasson, James Buska & Alan Greatorex
Cold Regions Research and Engineering Laboratory (CRREL), Hanover, N.H., USA

Jeff Tirey, Joel Fisher & Steve Johnson
Structural Engineers of New Hampshire Incorporated (SENH), Laconia, N.H., USA

ABSTRACT: Because of New Hampshire's hilly landscape, mapped values of ground snow load are not available for much of its area. We conducted snow load case studies to establish ground snow loads for a specific elevation in each of the 140 towns where no values are currently available. That work was done by three researchers and three structural engineers practicing in New Hampshire. While our methods of analysis varied somewhat, our results were comparable and the feedback we received from each other was quite valuable. We then established an elevation correction factor to transfer our snow load answers to other elevations in each town. We did not do case studies for the 102 towns in New Hampshire where mapped values are available. We are now planning to do that, as we believe that case studies improve snow load design criteria. We suggest that similar studies be conducted for other places in the United States.

1 INTRODUCTION

The primary resource document for the design of structures in the United States is American Society of Civil Engineers (ASCE) Standard 7, "Minimum design loads for buildings and other structures" (ASCE 1996). It is commonly referred to as ASCE 7-95. The first step in determining design snow loads is to determine the ground snow load at the place of interest. ASCE 7-95 contains a map of the United States overlaid with that information. That map was made by Tobiasson and Greatorex of CRREL using data from 226 "first order" National Weather Service (NWS) stations, where snow depths and snow loads are measured frequently, and data from about 11,000 other NWS "co-op" stations, where only the depth of snow on the ground is measured frequently. In some areas, extreme local variations in ground snow loads preclude mapping at a national scale. In those areas the national map contains the designation "CS" instead of a value. CS indicates that case studies are required to establish ground snow loads in these areas. Figure 1 presents the information from the ASCE 7-95 map on a larger map of New Hampshire, showing county and town boundaries. The zoned values in Figure 1 are ground snow loads with a 2% annual probability of being exceeded (i.e., the 50-year mean recurrence interval value). As can be seen in Figure 1, all of New Hampshire is either in a "CS" area or the zoned values have elevation limits (the numbers in parentheses) above which case studies are needed. Thus, case studies are needed to determine ground snow loads for many buildings in New Hampshire. ASCE 7-95 requires that, in these situa-

tions, ground snow loads "shall be based on an extreme value statistical analysis of data available in the vicinity of the site using the value with a 2% annual probability of being exceeded (50-year mean recurrence interval)".

At CRREL a methodology has been developed to conduct snow load case studies. It and the data used are described in the paper, "Database and methodology for conducting site specific snow load case studies for the United States," which was presented at the Third International Conference on Snow Engineering (Tobiasson & Greatorex 1997). That database also contains information from an additional 3300 locations across the United States where ground snow loads are measured a few times each winter by other agencies and companies.

Figure 2 shows New Hampshire overlaid with town boundaries and the location of each station in the database. There is 1 NWS "first order" station, and 89 NWS "co-op" and 91 "non-NWS" stations in New Hampshire. First order stations in adjacent states within 50 miles (80 km) of the border and other stations within 25 miles (40 km) of the border were also used in our analysis. They are also shown in Figure 2. Shading in that figure and its legend indicate towns we studied and others we did not.

Structural Engineers of New Hampshire, Inc. (SENH), is a non-profit professional association of structural engineers. Their members expressed interest in using the CRREL database and methodology to develop ground snow loads for each town in New Hampshire. Several volunteered their time to conduct case studies. All prior case studies had been done by two or three CRREL personnel familiar with the database and methodology. To see how well the methodology could be used by others to determine ground snow loads, CRREL trained five practicing licensed SENH engineers in the case study methodology and 20 case studies were done by both groups. This pilot study showed that comparable results could be achieved when the groups shared ideas. CRREL and SENH then entered into a Cooperative Research and Development Agreement (CRDA) to determine ground snow loads for the 140 New Hampshire towns in the "CS" zone; 17 other towns in that zone in portions of the White Mountain National Forest where little or no construction is to be expected were not studied. We did not do case studies for the remaining 102 towns where, as shown in Figure 1, ground snow load values up to a limiting elevation are available on the map in ASCE 7-95. We reasoned that we did not wish to develop values that might contradict mapped values in ASCE 7-95. We have subsequently changed our minds on this point, as will be discussed.

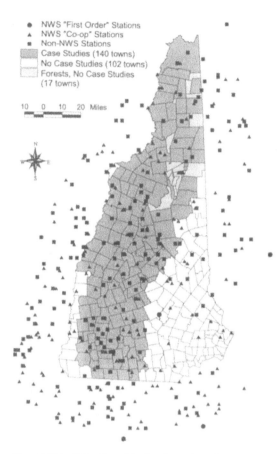

Figure 2. State of New Hampshire showing stations where ground snow load information is available and where our case studies were and were not done. (To convert miles to km, multiply by 1.609.)

2 ESTABLISHING CASE STUDY LOCATIONS

United States Geological Survey (USGS) 1:24000 scale topographic maps of the state were used to determine the coordinates of the geographical center, not the population center, of each town to the nearest minute of latitude and longitude. Those maps show town boundaries as well as roads and buildings. We did not use the elevation of the geographical center as the case study elevation but, instead, determined six elevations for each town: (1) lowest land; (2) lowest building; (3) lower limit of most buildings; (4) upper limit of most buildings; (5) highest building; and (6) highest land. Significant elevation differences exist within most towns. Thus, each ground snow load answer would not be a single value for all places in a town but a value at the case study elevation and an elevation factor for correcting that value to other elevations in that town.

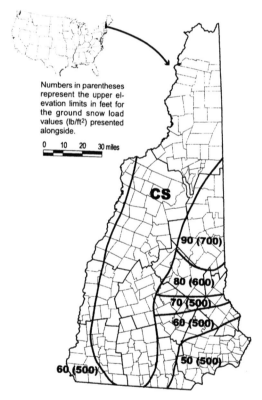

Figure 1. State of New Hampshire showing town and county boundaries overlaid with the ground snow load information in ASCE 7-95. (To convert lb/ft² to kN/m², mulitply by 0.0479, for miles to km, multiply by 1.609, and for ft to m, multiply by 0.3048.)

We chose an elevation near the upper limit of most buildings as our case study elevation. Had we done these case studies at lower elevations, failure to apply the elevation correction factor would have resulted in inappropriately low design loads for some of the buildings in each town.

3 CASE STUDY FORMS AND GUIDELINES

Case study forms were computer-generated for each town. Figures 3 and 4 present such forms for the town of Salisbury. The first page (Fig. 3) contains the data available in the vicinity. For many towns, that tabulation contains data from neighboring states. For Salisbury, periods of record range from 4 to 44 years; about half the stations are NWS and half non-NWS, and ground snow loads are available in the vicinity at elevations from 350 ft (107 m) to 1500 ft (457 m), bracketing the 900 ft (274 m) elevation chosen for Salisbury.

The final page (Fig. 4) of each case study contains two plots of ground snow load vs. elevation. The upper plot contains just the data from the nearest six to eight stations, while the lower plot contains all the data available within a 25-mile (40-km) radius, plus any NWS first order data within 50 miles (80 km). As shown in Figure 4, the elevation of interest is highlighted on the plots, as is a straight line of best fit using least squares and the best fit value of the ground snow load at the elevation of interest. For some towns the ground snow load "answer" is similar on the upper and lower plots but for other towns it is quite different.

Ground snow loads generally increase at higher elevations up to the tree line. Above the tree line, they may decrease because of wind action. The upper plot in Figure 4 has a negative "slope" (i.e., elevation correction factor) of -1.67 lb/ft^2 per 100 ft (-0.26 kN/m^2 per 100 m). The few data points on the "nearest 6" plot result in an unrealistic slope and thus the ground snow load answer of 68 lb/ft^2 (3.3 kN/m^2) is not to be trusted. The lower "all values" plot in Figure 4 contains enough data points to generate a physically more realistic slope of 2.5 lb/ft^2 per 100 ft (0.39 kN/m^2 per 100 m) and, thus, a believable ground snow load of 80 lb/ft^2 (3.8 kN/m^2).

Data from near the 6288-ft (1917-m) summit of Mt. Washington created problems. The tabulated ground snow load there is only 56 lb/ft^2 (2.7 kN/m^2), which is far below the ground snow load at many other places at elevations well below 1000 ft (305 m). The high winds on that treeless summit result in ground snow load measurements that are much too low to be used for our purposes. Several plots containing the Mt. Washington value have a negative slope and the ground snow load answer suffers as a result. While Mt. Washington and a few other stations frustrated us, their implications were worth considering. Mt. Washington's

redeeming value was to remind us that we should not apply our elevation correction factor above the tree line.

Each of the three CRREL researchers and the three SENH structural engineers involved was provided with a copy of the "data and methodology" report mentioned previously (Tobiasson & Greatorex 1997), several representative case studies done by CRREL previously, and written suggestions by Tobiasson and Greatorex for conducting case studies, a copy of which can be obtained from CRREL.

We began by working on 40 towns, about half of which were in the rugged northern portion of the state and the rest in the rolling hills of southwestern New Hampshire. We each conducted our analysis in our own way and forwarded our "preliminary" ground snow load answers to a third party at CRREL, who tallied them without divulging the author of each value, and then sent the tally to us. We then reassessed our answers in light of those of the five others, and then sent in our "semi-final" answers, which were tallied in a similar fashion, then returned to us. We met shortly thereafter to discuss our various methods of analysis and our answers and to arrive at a final answer for each of the 40 towns. As a result of our first meeting, we each made some changes to our method of analysis. We then repeated the process for the remaining 100 towns being studied.

4 DIFFERING WAYS OF ARRIVING AT ANSWERS

The three individuals representing CRREL had done many case studies and were comfortable with the case study forms and the guidelines for analysis. They closely followed the instructions, giving more weight to closer stations and stations with longer periods of record. They gave little weight to stations with less than about 15 years of record and they gave little weight to stations where the ratio of the 50-year ground snow load (i.e., P_g on the case study tabulation) to the largest ground snow load ever measured there (i.e., the Record Max value, P_{max}, on the case study tabulation) was greater than 1.6. They flagged such stations on the upper plot and added a few stations somewhat further away, but with longer periods of record, to replace them. Often, more stations were added than were eliminated. Then they either "eyeballed" or calculated a new line of best fit in their quest for that case study's answer. When "eyeballing" in a line of best fit, they gave it a slope of between 2 and 2.5 lb/ft^2 per 100 ft (0.31 to 0.39 kN/m^2 per 100 m), based on the written suggestions mentioned above. Two of them found it valuable to bound the good data by upper and lower lines at one of these slopes. Their answer was usually somewhat above the midpoint of the upper and lower bounds at the case study elevation. The third individual devoted additional attention to the geographical position of stations used in his analysis. He plotted this for some case studies.

The three SENH practicing structural engineers had participated in the pilot study. Each had developed a slightly different way of doing case studies. They chose not to work on the case study plots, believing them to contain too much information of limited value, which hides trends of interest. Instead, they reanalyzed only the better stations in the data tabulation. One of them felt that the NWS co-op information, since it is based on measurements of the depth of snow on the ground, not measurements of the weight of that snow, is inferior to the non-NWS values, which are measurements

of the weight. The other five individuals felt that both the NWS and non-NWS data sets were of comparable value, each having its own strengths and weaknesses. The individual who focused on the non-NWS data only included NWS information when few non-NWS data were available. He attempted to have 6 to 8, and occasionally 10, stations with 20 or more years of record in his analysis. He did not use stations where the P_g/P^r_{max} ratio was greater than 1.5. He re-plotted the P_g values selected vs. elevation and used a straight line, least squares fit to establish a preliminary answer. That answer was modified with consideration given to the slope of his trend line and the scatter of points.

SNOW LOAD CASE STUDY FOR

Salisbury, New Hampshire

Latitude 43° 23' N   Longitude 71° 46' W   Elevation 900 ft

Station	Radius (mi.)	Azimuth (from site)	Elev. (ft)	P_g (psf)	Record Max. (psf)	Years of Record Total	Years of Record No Snow
NWS FIRST ORDER							
CONCORD (W.E.)	18	125	350	63	43	40	0
CONCORD WSO AP ("DEPTH")	18	125	350	44	38	44	0
NEW HAMPSHIRE (NWS co-op)							
BLACKWATER DAM	5	143	600	69	59	44	0
FRANKLIN	7	56	390	83	94	13	0
FRANKLIN FALLS DAM	8	54	430	72	67	44	0
SOUTH DANBURY	10	311	930	101	85	22	0
NEW LONDON	11	279	1340		51	9	0
BRADFORD	14	236	970	75	73	39	0
BRISTOL 2	14	9	590		27	8	0
WEST HENNIKER	16	201	500		59	5	0
GRAFTON	16	315	840	101	67	25	2
MOUNT SUNAPEE	16	261	1260	132	78	18	2
GILMANTON	18	79	1030	86	55	16	0
LAKEPORT	19	61	560	69	68	34	0
LAKEPORT 2	19	61	500	67	28	11	2
ALEXANDRIA	19	339	1370		38	5	0
GILMANTON 2 E	20	83	800		23	4	0
WEARE	21	174	720	50	32	20	0
NEWPORT	21	270	790	78	57	39	1
NORTH CHICHESTER	21	109	360		27	8	0
DEERING	22	201	1010	83	41	16	0
EAST DEERING	22	189	790	77	65	26	0
SOUTH WEARE	23	171	700	82	71	18	0
ALTON	25	84	800		28	5	0
NEW HAMPSHIRE (NON-NWS)							
SALISBURY	1	90	760	72	54	40	0
ANDOVER	4	315	700	76	61	32	0
BLACKWATER	5	166	620	69	56	40	0
FRANKLIN FALLS	7	45	400	73	54	39	0
SOUTH DANBURY	10	315	800	74	62	40	0
DAY POND	12	218	780	83	62	29	0
LITTLE SUNAPEE	15	287	1490	93	59	31	0
NEW LONDON	15	287	1170	86	75	26	0
CHASE VILLAGE	16	180	700	81	59	29	0
GRANLIDEN	17	276	1220	89	60	31	0
SADDLE HILL	18	33	1020	73	69	41	0
GILFORD	18	49	1000	90	71	40	0
CARDIGAN MOUNTAIN	19	336	1500	72	64	15	0
NEW HAMPTON	19	24	560	76	62	41	0
GRAFTON CENTER	19	317	900	69	60	24	0
NELSON BROOK	20	78	770	89	55	11	0
EVERETT DAM	22	159	460	78	53	29	0
WASHINGTON	22	236	1500	88	64	22	0
MEREDITH	22	43	880	80	62	40	0
WASHINGTON	22	237	1340	90	61	11	0
WEIRS BEACH	23	54	520	50	38	27	0
HOYT HILL	24	360	950	72	73	41	0
SALMON BROOK	25	223	1300	88	57	22	0

Figure 3. Case study data tabulation for the town of Salisbury. (To convert lb/ft^2 to kN/m^2, mulitply by 0.0479, for miles to km, multiply by 1.609, and for ft to m, multiply by 0.3048.)

316

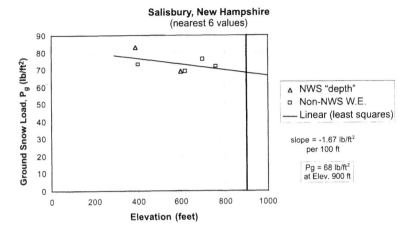

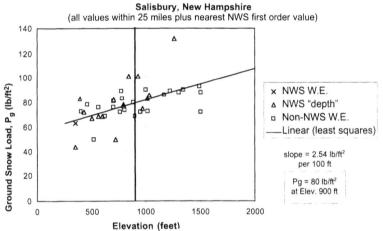

Figure 4. Case study plots for the town of Salisbury. Note that the scales on the two plots differ. (To convert lb/ft^2 to kN/m^2, mulitply by 0.0479, for miles to km, multiply by 1.609, and for lb/ft^2 per 100 ft to kN/m^2 per 100 m, multiply by 0.1572.)

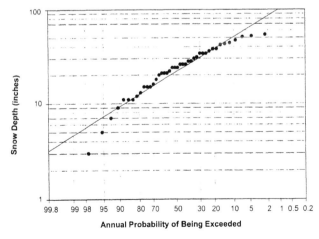

Figure 5. Log-normal probability plot for Milford which has a high P_g/P_{max} ratio of 1.76. (To convert inches to meters multiply by 0.025.)

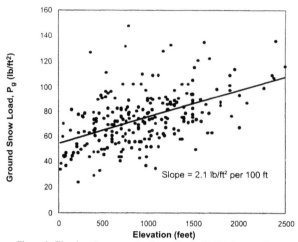

Figure 6. The elevation correction factor for the 236 highest quality stations used in our analyses was 2.1 lb/ft² per 100 ft (0.33 kN/m² per 100 m). (To convert lb/ft² to kN/m², mulitply by 0.0479, and for ft to m, multiply by 0.3048.)

When several points at about the elevation of interest fell above the trend line, he increased his preliminary answer.

The other two SENH structural engineers considered both NWS and non-NWS data, but one of them gave more weight to the non-NWS information because it eliminated the step of having to relate snow depths to snow loads (see equation 1 in Tobiasson & Greatorex 1997). Both of these individuals developed selection criteria that eliminated from consideration a number of the stations on the case study form. The acceptance criteria of one individual were (1) at least 15 years of record, (2) less than 15 (sometimes 20) miles (24, sometimes 32 km) away and (3) P_g/P_{max} ratio no more than 1.75 for non-NWS stations and no more than 1.5 for NWS stations. The other individual's acceptance criteria were (1) at least 20 years of record, (2) less than 15 miles away, and (3) P_g/P_{max} ratio no more than 1.5. Both then adjusted each selected ground snow load to the case study elevation by using an elevation correction factor of from 2.0 to 2.5 lb/ft² per 100 ft of elevation difference (0.31 to 0.39 kN/m² per 100 m). Both then determined the average value of the ground snow load at that elevation for all the stations selected. In the vicinity of Mt. Washington, where a station or two had a value quite different from this average, a second average was often calculated, eliminating the outliers. One individual developed separate averages for all data and for "non-NWS" data and gave more weight to the "non-NWS" average. He always plotted all the data he analyzed and frequently referred back to the case study plots before finalizing his answer.

A review of each individual's final answers indicates that no one's approach caused them to be consistently much lower or much higher than the group's final answer. Thus, quantitatively, it appears that the process we developed to arrive at answers tended to bring each of us to about the same answer. We expect that if any one of us had used our method of analysis alone, without receiving feedback from the others along the way, we may have arrived at significantly different answers for some towns. Thus, we conclude that there is merit in involving several individuals in a way that they periodically receive anonymous feedback from each other. This process allowed the group to determine most answers before our meetings and precluded the need to discuss many of the case studies at those meetings. When we met, we concentrated on the few case studies on which we had remaining concerns or disagreements. This left time for us to explore ways of improving the process, ways of simplifying our findings, and ways of incorporating them into the national standard (i.e., ASCE 7-95) and into practice within New Hampshire. It also allowed us time to discuss our increasing understanding of ground snow loads in New Hampshire.

5 ADDITIONAL INVESTIGATIONS

For 69 of the 302 stations shown in Figure 2, where a 50-year ground snow load is available, the P_g/P_{max} ratio exceeded 1.5. Often, the 50-year ground snow load at such stations greatly exceeded other ground snow loads in the vicinity. For example, the upper outlier in the lower plot in Figure 4 has a high P_g/P_{max} ratio of 1.7. Responding to this complication proved to be the most controversial aspect of our analysis. To better understand what was happening, we examined probability plots of several of these stations and determined that, for them, the log-normal distribution used to generate the ground snow load values on the case study forms does not fit the actual trend in lower probabilities very well. Figure 5 illustrates this for Milford, where the P_g/P_{max} ratio is 1.76 and the log-normal value at a 2% annual probability of being exceeded (50-year mean recurrence interval) greatly exceeds the data trend there. With this evidence, we gave little weight in our analysis to stations with high P_g/P_{max} ratios.

Once we had all 140 case study answers, we compared them to the answers on the upper and lower plots on the last page of the case study form. The upper "nearest 6" plot answers did not agree with our answers well at all. Only 59 of the upper plot answers were within 5 lb/ft² (0.2 kN/m²) of our 140 case study answers. For 50 stations the upper plot answers were from 10 to 38 lb/ft² (0.5 to 1.8 kN/m²) away from our answers. The lower "all values" plot answers were within 5 lb/ft² (0.2 kN/m²) of our answers for 116 of the 140 case studies (i.e., 83% of the time). However, for eight stations, the "all values" answers were from 10 to 20 lb/ft² (0.5 to 1.0 kN/m²) away from our answers. Thus, while the "all values" answers provide good indications of the "correct" answers most of the time, further study will occasionally result in significantly different, better answers.

The elevation correction factor can also be examined on the upper and lower plots. On the upper plot that factor varied widely between 13.5 lb/ft² per 100 ft (2.12 kN/m² per 100 m) and minus 9.0 lb/ft² per 100 ft (minus 1.41 kN/m² per 100 m). The average value of this widely divergent and physically unrealistic set of numbers was 1.8 lb/ft² per 100 ft (0.28 kN/m² per 100m). We place little value on this average, as it is significantly influenced by some values that are physically unrealistic. Stations like Mt. Washington create these inappropriate values. On the "all values" plot, the slopes make somewhat better physical sense, but Mt. Washington and a few other stations still create problems. Slopes vary from 5.3 lb/ft² per 100 ft (0.83 kN/m² per 100 m) to minus 3.0 lb/ft² per 100 ft (minus 0.47 kN/m² per 100 m) and average 2.4 lb/ft² per 100 ft (0.38 kN/m² per 100 m).

We further examined the elevation correction factor by studying each station in our database. We eliminated stations with less than 15 years of record, others with an elevation above 2500 ft (762 m), and others with P_g/P_{max} ratios less than 0.9 or greater than 1.7. For the remaining, high quality stations, the line of best fit of their elevation to their 50-year ground snow load, P_g, produced a slope of 2.1 lb/ft² per 100 ft (0.33 kN/m² per 100 m), as shown in Figure 6. While we expect that the elevation correction factor varies from place to place in New Hampshire, we do not have enough data to support such differences. Thus, we have used this elevation correction factor for all New Hampshire towns.

6 FINDINGS

Our answers for the 140 towns are presented in Table 1. Some of the towns listed in Table 1 are only partially in the CS zone. At this time for those towns, we recommend that the ground snow load be determined using the information in Table 1 rather than from the map in ASCE 7-95. The case study process is a more detailed and thus, in all likelihood, a more accurate assessment of the ground snow loads in these towns. This is consistent with the guidance in the commentary attached to ASCE 7-95, which states that "detailed study of a specific site may generate a design value lower than that indicated by the generalized national map. It is appropriate in such a situation to use the lower value established by the detailed study. Occasionally, a detailed study may indicate that a higher design value should be used than the national map indicates. Again, results of the detailed study should be followed"(ASCE 1996).

After discussing the pros and cons of having a portion of New Hampshire defined by the ASCE 7-95 map and the remainder defined by our case studies, we concluded that it would be best to expand our case studies to cover the entire state. We have agreed in principle to do that and will revise the CRDA between CRREL and SENH to increase the scope of work accordingly. We expect that once we have done the entire state and examined all of our answers, some of the values in Table 1 may change a little. Thus, we advise readers to consider those values as interim in nature.

To determine the ground snow load at elevations other than those listed in Table 1 (i.e., at elevations other than those where the case studies were conducted), the values in Table 1 should be increased or decreased by an elevation correction factor of 2.1 lb/ft² per 100 ft (0.33 kN/m² per 100 m). For example, in Hanover where the Table 1 value is 75 lb/ft² at 1300 ft (3.6 kN/m² at 396 m), at an elevation of 900 ft (274 m) the answer would be $75 + (2.1/100)(900-1300) = 75 - 8 = 67$ lb/ft² (in SI units: $3.6 + (0.33/100)(274 - 396) = 3.6 - 0.4 = 3.2$ kN/m²).

We have not fully investigated the upper limit above which our elevation correction factor does not apply. At this time it seems safe to use it up to an elevation of 2500 ft (762 m) in New Hampshire. At higher elevations a larger elevation correction factor may be needed.

7 CONCLUSIONS AND RECOMMENDATIONS

The current case study plots contain some data of limited value that mask rather than define trends. Perhaps stations with fewer than about 14 years of record should be eliminated from the plots on the case study forms and perhaps stations with P_g/P_{max} ratios exceeding about 1.7 should also be eliminated from those plots.

Table 1. Case study findings for the 140 towns studied.

Town	Case study elevation (feet)*	Ground snow load, P_g (lb/ft^2)**
Acworth	1500	90
Albany	1300	95
Alexandria	1100	85
Alstead	1300	80
Andover	900	80
Antrim	1000	80
Ashland	800	75
Bartlett	1200	105
Bath	1000	65
Bennington	1000	80
Benton	1600	90
Berlin	1600	95
Bethlehem	1800	105
Boscawen	700	75
Bradford	1200	85
Bridgewater	1000	80
Bristol	1000	80
Campton	1300	85
Canaan	1200	85
Carroll	1700	95
Center Harbor	900	80
Clarkesville	2000	90
Colebrook	1600	80
Columbia	1600	80
Croydon	1200	90
Dalton	1300	80
Danbury	1000	85
Deering	1200	90
Dixville	1900	90
Dorchester	1400	80
Dublin	1600	90
Dummer	1400	90
Easton	1400	85
Ellsworth	1400	90
Enfield	1300	85
Errol	1600	90
Fitzwilliam	1300	75
Francestown	1100	80
Franconia	1700	95
Franklin	700	75
Gilsum	1200	85
Gorham	1400	105
Goshen	1400	90
Grafton	1400	90
Grantham	1400	90
Greenfield	1100	80
Green's Grant	1700	105
Greenville	1000	75
Groton	1200	80
Hancock	1300	85
Hanover	1300	75
Harrisville	1500	90
Harts Location	1300	100
Haverhill	1200	75
Hebron	900	80
Henniker	1000	80
Hill	1100	85
Hillsboro	1000	80
Holderness	1000	80
Hopkinton	800	80
Jackson	1800	115
Jaffrey	1300	80
Jefferson	1700	100
Keene	900	70
Laconia	900	80
Lancaster	1300	70
Landaff	1300	80
Lebanon	1200	80
Lempster	1600	95
Lincoln	1400	95

Table 1 (cont'd).

Town	Case study elevation (feet)*	Ground snow load, P_g (lb/ft^2)**
Lisbon	1100	75
Littleton	1200	75
Lyman	1200	75
Lyme	1100	70
Lyndeborough	1000	80
Marlborough	1300	80
Marlow	1600	90
Martin's Loc.	1300	100
Mason	1000	75
Meredith	1000	80
Milan	1500	100
Milford	600	70
Millsfield	1700	90
Mont Vernon	900	75
Moultonborough	900	80
Nelson	1500	90
New Boston	800	80
New Hampton	1000	80
New Ipswich	1300	80
New London	1400	95
Newbury	1300	90
Newport	1200	85
Northumberland	1200	75
Odell	1800	90
Orange	1500	90
Orford	1100	70
Peterborough	1000	75
Piermont	1400	75
Pittsburg	1700	80
Plainfield	1300	90
Plymouth	900	75
Randolph	1900	110
Rindge	1300	80
Roxbury	1300	80
Rumney	1300	85
Salisbury	900	80
Sanbornton	1000	80
Sandwich	1100	85
Second College Grant	1500	90
Sharon	1300	80
Shelburne	800	90
Springfield	1500	95
Stark	1200	75
Stewartstown	2000	90
Stoddard	1600	90
Stratford	1100	70
Success	1600	100
Sugar Hill	1600	90
Sullivan	1400	90
Sunapee	1400	90
Surry	1100	80
Sutton	1100	85
Swanzey	800	65
Temple	1300	80
Thornton	1200	85
Tilton	900	80
Troy	1300	75
Unity	1500	90
Warner	800	80
Warren	1300	80
Washington	1700	100
Waterville Valley	1800	105
Weare	900	80
Webster	700	80
Wentworth	1200	80
Whitefield	1400	75
Wilmot	1200	90
Wilton	900	75
Windsor	1200	85
Woodstock	1200	85

*To convert feet to meters, multiply by 0.3048.
**To convert lb/ft^2 to kN/m^2, multiply by 0.0479.

Most of us think that the NWS and non-NWS databases are of comparable value and both should be used when developing ground snow loads.

The "all values" plot provides a good indication of the "correct" answer in most cases, but in a few cases it is not a very good indication. Thus, simply using the "all values" answer is not recommended.

The three structural engineers involved chose to somewhat modify the analytical procedure developed by CRREL, each in his own way. Nonetheless, when coupled with our anonymous feedback process, it was easy for us to reach a consensus in almost all cases.

Stations with P_g/P_{max} ratios greater than about 1.5 were given little weight and those with ratios above about 1.7 were largely discounted in our analysis. We determined that the log-normal distribution does a poor job of predicting extreme values for such stations. Stations with P_g/P_{max} ratios less than about 0.9 appear to create similar problems.

An elevation correction factor of 2.1 lb/ft^2 per 100 ft (0.33 kN/m^2 per 100 m) works well for New Hampshire to an elevation of about 2500 ft (about 762 m). This factor may increase at higher elevations. It should not be assumed to apply in other parts of the country.

Based on what we learned by conducting the 140 case studies in the CS zone, we think it is important to do case studies for the 102 New Hampshire towns not in that zone. We will begin that work in the near future.

The case study process involves a more detailed examination of an area than was achieved some years ago when the national snow load map was made by two of us. Thus, the case study process, in all likelihood, produces a more accurate ground snow load.

In "CS" areas on the national map, case studies are required. In other areas where mapped values have elevation limits or change rapidly within short distances, case studies are recommended.

8 ACKNOWLEDGEMENTS

Financial support was provided by 14 New Hampshire structural engineering companies, by the New Hampshire Building Officials Association, and by the U.S. Army Corps of Engineers. Over 75% of the work reported here was done by volunteers.

REFERENCES

American Society of Civil Engineers, 1996. *Minimum design loads for buildings and other structures*. ASCE Standard 7-95. New York, NY.

Tobiasson, W. & Greatorex, A,. 1997. Database and methodology for conducting site specific snow load case studies for the United States in Snow Engineering, Recent Advances. In *Proceedings of the Third International Conference on Snow Engineering), Sendai, Japan, 1996*. Rotterdam: A.A. Balkema.

Snow Engineering, Hjorth-Hansen, Holand, Løset & Norem (eds) © 2000 Balkema, Rotterdam, ISBN 90 5809 148 1

Wind effects on snow accumulation on two-level flat roofs

Manabu Tsuchiya & Takeshi Hongo
Kajima Technical Research Institute, Japan

Tsukasa Tomabechi
Hokkaido Institute of Technology, Sapporo, Japan

Hiroshi Ueda
Chiba Institute of Technology, Japan

ABSTRACT: When estimating snow accumulation on a building roof in a heavy-snowfall district, for structural and architectural planning, it is important to know its volume and drift.
Snow accumulation around buildings is usually investigated by snow wind tunnel tests. However, it is not easy to conduct such tests. Snow accumulation is closely related to the wind flow pattern around a building.
The purpose of this study was to clarify the relationship between snow accumulation and wind flow around buildings, and to thus predict the snow accumulation from wind flow patterns from wind tunnel tests. This paper investigates snow accumulation and wind flow around two-level flat-roofed buildings. Snow accumulation patterns on roofs were demonstrated from snow and wind tunnel test results and field measurement results. Furthermore, wind flow patterns around buildings were clarified in detail from wind tunnel test results. Finally, the relationship between snowdrifts on roofs and velocity fields around buildings was obtained.

1 INTRODUCTION

At the snow resistant design stage for buildings in heavy snowfall districts, one of the most important requirements is to estimate the snow accumulation patterns caused by wind flows as well as the snowfall volume. It is difficult to estimate these patterns accurately because of the effects of structural shapes and different wind characteristics. At present, snow accumulation is investigated by snow wind tunnel tests and field measurements. However, snow wind tunnel tests are not easy to conduct due to the requirement for an exclusive wind tunnel and special technology. Furthermore, it cannot truly be said that the similarity law for snow wind tunnel tests has been satisfactorily established yet. Field measurement investigations cannot be carried out in general, because arbitrary building shapes cannot be chosen because of the utilization of the existing buildings, and field measurements are greatly affected by wind and snow conditions (O' Rourke M.J. et al. 1981, Hashimoto S. et al, 1986).

The authors noticed that snow accumulation patterns on roofs and building surroundings are greatly influenced and varied by wind flows in their peripheral areas. The relationship between the snow accumulation pattern and the wind flow has been investigated for basic building shapes. This paper studies the relationship between snow accumulation patterns on two-level roofs of a building obtained from field measurement during a snowstorm, and snow and the wind flow around this building obtained from wind tunnel tests (Tsuchiya M. et al, 1998, 1999).

2 SNOW ACCUMULATION PATTERNS

2.1 *Outline of Field Measurement*

Field investigations into the characteristics of snowdrifts on two-level roofs of a building were carried out on the campus of the Hokkaido Institute of Technology in the west of Sapporo City, Japan from December 1999 through February 2000. The shape of the model with two-level flat roofs subject to the investigation and the installation conditions for this model on the ground are shown in Fig.1 and Photo.1, respectively. The lower level roof is 5.4m(6H) wide, 4.5m(5H) deep and 0.9m(H) high. The upper level is 5.4m(6H) wide, 1.8m(2H) deep and 0.9M(H) high. The model was installed 0.5m above the ground to maintain this shape even if snow fell. Both roofs and walls of the model were made of wood and the entire surface was painted. Two models were made. One was installed with the lower level roof facing windward of the NW wind, which blows mainly during the winter in this area. The other was placed so that the upper level roof facing the same direction.

The snow accumulation patterns on the roofs were measured at 0.1m intervals using a snow scale at the central section. To obtain the snow accumulation patterns under a constant wind direction, the snow on the roofs was removed every day. Furthermore, the surfaces of fallen snow around the models were smoothed.

Regular measurements of basic wind direction, wind velocity and snowfall were carried out using a wind vane, a three-cup anemometer and a sonic snow depth gauge, which were installed at a location where they were not influenced by the presence of the models.

2.2 Observation Results

Typical snow accumulation patterns on the lower level roof are shown in Fig.2 using the value of S/H for the snow depth (S) obtained by transferring the height (H). Fig.3 illustrates the variation in wind direction, velocity and snow depth observed over 18 hours during which this snow accumulation was formed after removing the snow on the day preceding the field measurement. Wind direction at right angles to the model was set at 0°. Fig.3 indicates that the wind direction angle during this period was 22.5°~ 45°. The mean velocity during a period of 10 minutes was 3 ~ 5m/sec. The maximum instantaneous wind velocity was 12m/sec and the outdoor temperature was about -2°C.

The snow accumulation patterns on the lower level roof for the upwind facing it show that the snow depth S/H at the end point A on the windward side is about 0 (Fig.2 (a)). In the area B ~ C (X/H=0.2 ~ 1.1), the snow accumulation is almost constant with light snow. In the area C ~ D (X/H=1.1 ~ 2.5), the snow accumulation rapidly changes and at point D the snow depth S/H is 0.32. Over the entire area of D, E and F, the snow accumulation is convex with the minimum value at point E. At point F, where the lower roof meets the vertical wall, the maximum snow depth (S/H) of 0.33 is shown.

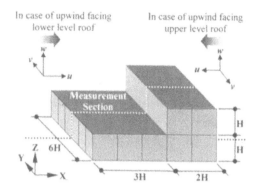

Field test : H=900mm, Wind tunnel test : H=45mm.

Figure 1. Size of the two level flat roof model.

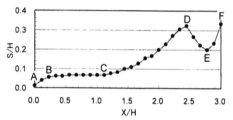

(a) In case of upwind facing lower level roof.

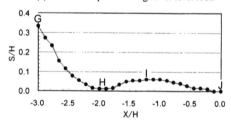

(b) In case of upwind facing upper level roof.

Figure 2. Snow Accumulation on lower level roof.

Photograph 1. Field test model on the campus of the Hokkaido Institute of Technology.

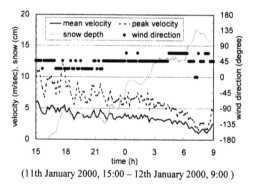

(11th January 2000, 15:00 – 12th January 2000, 9:00)

Figure 3. The records of wind and snow at field test area.

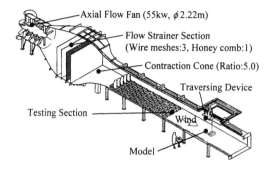

Figure 4. Outline of the wind tunnel.

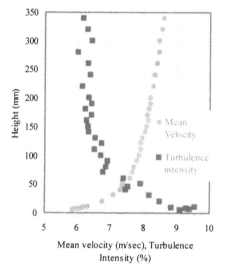

Figure 5. Profile of mean velocity and turbulence intensity.

For the upwind facing the upper level roof, the snow accumulation patterns on the lower level roof show a maximum snow depth S/H of 0.33 at point G where the vertical wall meets the upper level roof. The snow accumulation rapidly decreases from this point through point H (X/H=-2.0), the value of S/H approaches 0. The depth slightly increases in the area of H ~ I and then gently decreases again in the area of I ~ J. At point J, at the end on the windward side, S/H reaches 0.

From comparison between these two cases, it can be seen that the maximum snow depth is almost the same (S/H=0.33). However, the total snow accumulation on the roof is greater for the upwind facing the lower level roof.

3 WIND FLOW CHARACTERISTICS

3.1 Wind Test Outline

In order to investigate the characteristics of wind around the two-level roofed building which was the subject of this study, wind tunnel tests were carried out using a model. The Eiffel type boundary-layer wind tunnel belonging to the Kajima Technical Research Institute was used for the tests. Fig.4 shows the outline of this wind tunnel. A 1/20 scale model of the specimen made for the field observations was used (Fig.1). The height (H) of the lower level roof was 45mm and the height ($2H$) of the upper level roof was 90mm. The wind flow used in the wind tunnel tests was a boundary-layer flow obtained by installing spires and a saw on the windward side of the measurement area. Fig.5 illustrates the vertical profile of this flow. The vertical profile of the mean wind velocity can approximate the exponential distribution of the power exponent α=1/7. The mean velocity at a reference height H of the lower level roof was 7.4m/sec and the turbulence intensity was 7.5%.

The velocity was measured at the central section (X-Z surface) of the model. The velocity of the u-w component was measured by installing a triple split fiber probe (DANTEC 55R92) as a traversing device. The minimum interval of the measurement points on both the model wall surface and the roof surface was 5mm. The measurements were carried out by A/D transformation over 30min. The sensor output was set at 500Hz at each measurement point.

3.2 Profile of Mean Wind Velocity

Figs.6 (a), 6 (b) and 6 (c) illustrate the mean flow vector distribution, the mean velocity contour u/U_H and the mean velocity contour w/U_H, respectively, for the upwind facing the lower level roof. These wind flow characteristics for the upwind facing the upper level roof are shown in Figs.7 (a), 7 (b) and 7 (c). The aforementioned characteristics are discernible only when the wind direction angle is 0°.

Each velocity component was obtained by transferring the mean velocity U_H at the height H of the lower level roof into a non-dimensional form assuming that there is no model.

When the upwind faces the lower level roof, the flow against the lower level wall is divided into two types. One forms a vortex in the vicinity of the ground on the windward side and the other separates from the edge of the lower level roof. On the lower level wall, a downward flow $\overline{w}/U_H \approx -0.2$ caused by the vortex formed in he vicinity of the ground is discernible. A reverse flow of $\overline{u}/U_H \approx -0.2$ can be seen near the ground. The flow separated from the windward edge of the lower level roof blows against the upper level wall again. This separated flow is then divided into the following two types of flow: a downward flow that forms a vortex on the lower level roof and an upward flow that separates from the windward edge of the upper level roof. Due to this flat vortex formed on the lower level roof, a reverse flow occurs over the entire area of the lower level roof. The maximum value of $\overline{u}/U_H \approx -0.4$ is shown in the proximity of *X/H*=2.5. The downward flow near the upper level roof surface is $\overline{w}/U_H \approx -0.2$. On the upper level roof, no reverse flow caused by the separation bubble is discernible. The comparatively fast mean velocity of $\overline{u}/U_H \approx 0.5$ can be seen.

When the upwind faces the upper level roof, the flow against the upper level wall on the windward side descends from a height of about 1.5*H* and is divided into a flow that forms a vortex on the windward side of the building and a flow that separates upward from the windward edge of the upper level roof. This vortex induced downward flow is $\overline{w}/U_H \approx -0.4$. The reverse flow in the vicinity of the ground shows a value of $\overline{u}/U_H \approx -0.5$, which is larger than the flow value for the upwind facing the lower level roof. A large wake is formed on the leeward side of the building. This is caused by the separated flow from the windward edge of the upper level roof. Due to this wake, all flows on both the lower and upper level roofs, as well as in the vicinity of the ground on the leeward side of the building, are shown as reverse flows. The reverse flow value on both the upper and the lower level roofs is $\overline{u}/U_H \approx 0.3$. At the leeward edges of the upper and the lower level roofs, small separated flows induced by this reverse flow can be observed.

3.3 Profile of Fluctuating Wind velocity

Figs.6(d) and 6(e) show the contour of fluctuating velocity with u direction components and that with w direction components, respectively, for the upwind facing the lower level roof. Figs.7 (d) and 7 (e) show the velocity contour for the upwind facing the upper level roof. These fluctuating velocity components were obtained by transferring the mean velocity U_H at the height H of the lower level roof into a non-dimensional form when the wind direction is 0°.

For the upwind facing the lower level roof, the u direction component greatly varies on the windward side on the lower level roof and in the vicinity of its center, and shows a value of $u'/U_H \approx -0.3$. At the center and in the vicinity of the leeward side, the w direction component greatly fluctuates. A value of $w'/U_H \approx -0.2$ is indicated. As a result, it is clear that the vortex (separation bubble) induced by the separation from the windward side of the lower level roof is not a stable flow and that it changes its flow state constantly. For the upwind facing the lower level roof, both the u direction component and w direction component greatly fluctuate in the vortex area near the windward ground surface. The values of both u'/U_H and w'/U_H are about 0.3. In the windward portion of the wake caused by the separation from the windward edge of the upper level roof, the fluctuating velocity in the u direction shows a large value of $u'/U_H \approx 0.4$. In the leeward portion,

the value of the fluctuating velocity in the w direction is $w'/U_H \approx 0.25$.

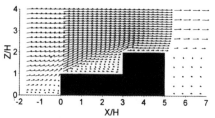

(a)　Distribution of mean flow vector.

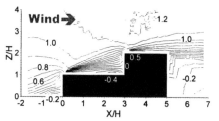

(b)　Contour of mean velocity $\overline{u}/U_H$.

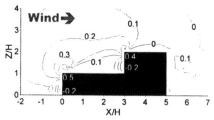

(c)　Contour of mean velocity $\overline{w}/U_H$.

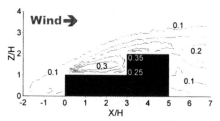

(d)　Contour of r.m.s. velocity u'/U_H.

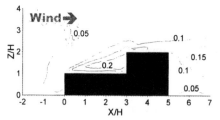

(e)　Contour of r.m.s. Velocity w'/U_H.

Figure 6.　Characteristics of wind flow around the model in case of upwind facing lower level roof (wind direction $\theta = 0$ degrees).

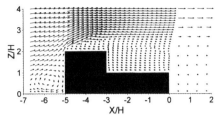

(a) Distribution of mean flow vector.

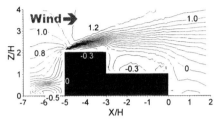

(b) Contour of mean velocity $\bar{u}/U_H$.

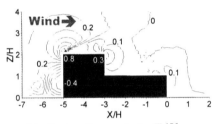

(c) Contour of mean velocity $\bar{w}/U_H$.

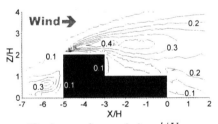

(d) Contour of r.m.s. velocity u'/U_H.

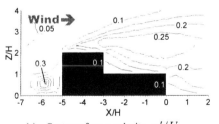

(e) Contour of r.m.s. velocity w'/U_H.

Figure 7. Characteristics of wind flow around the model in case of upwind facing upper level roof (wind direction θ = 180 degrees).

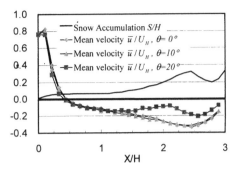

Figure 8. Comparison of mean wind velocity for angled wind direction in case of upwind facing lower level roof.

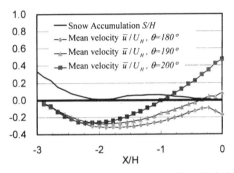

Figure 9. Comparison of mean wind velocity for angled wind direction in case of upwind facing upper level roof.

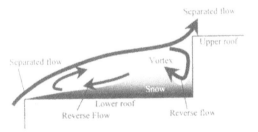

Figure 10. Typical wind flow and snow accumulation pattern in case of upwind facing lower level roof.

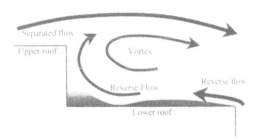

Figure 11. Typical wind flow and snow accumulation pattern in case of upwind facing upper level roof.

3.4 Variation in Mean Wind Velocity for Angled Wind Direction

The u component of the velocity 5mm above the lower level roof surface where the wind blows at an angle to the wall was investigated (Figs.8 and 9). A flow that is opposite to the main flow is indicated as minus (-). When the wind blows from the lower level roof side, variation in the velocity is hardly discernible for wind directions $\theta = 0°$ and $\theta = 10°$. Although for $\theta = 20°$ the current of the reverse flow weakens slightly in the vicinity of $X/H=1.5 \sim 2.5$, no great variation in velocity distribution can be seen. For wind blowing from the upper level roof side, the velocity varies greatly with wind direction. For $\theta = 180°$, restoration of the wind velocity can be recognized at $X/H=0 \sim 1.0$ on the leeward side of the lower level roof. However, the wind blows in the opposite direction. In the vicinity of $X/H = 0.5$ for $\theta = 190°$, in which the wind direction is oblique to the wall, and in the vicinity of $X/H = 1.0$ for $\theta = 200°$, the flow reattaches to the lower level roof. It is thus clear that the flow direction on the roof is the same as that for the main flow.

4 RELATIONSHIP BETWEEN FLOW CHARACTERISTICS AND SNOW ACCUMULATION PATTERNS

The above results, were used to investigate the relationship between the characteristics of flow and snow accumulation patterns during snowfall, shown in Fig. 2.

For the upwind facing the lower level roof, both the snow falling near the windward side of the lower level roof and the snow accumulated on the roof are blown away by the widely fluctuating separated flow occurring at the windward edge of the lower level roof and accumulates on the leeward side. The downward flow that descends from the stagnation point of the lower level wall surface as well as the strong reverse flow induced by this downward flow drive the snow to the leeward side. It can be assumed that the snow accumulation pattern having a peak in the proximity of point D, shown in Fig.2 (a), is formed by the repetition of these flow patterns. However, for $\theta = 20°$, a lag is discernible between the wind velocity distribution and the snow accumulation pattern. Further investigations must be carried out into the effects of the velocity distribution with the w direction component as well as the velocity distribution with variations in wind directions. Fig.10 is a typical figure showing the relationship between the wind flow and the snow accumulation pattern.

For the upwind facing the upper level roof, both the falling snow and the snow accumulated on the roof are pushed toward the upper level wall surface by the reverse flow on the lower level roof, which is due to a large wake formed by the strong separation from the windward edge of the upper level roof. Furthermore, a strong ascending flow occurs near the wall of the upper level. It can be assumed that the snow accumulation pattern shows a minimal value in the vicinity of point H indicated in Fig.2 (b) because a strong reverse flow is induced here by these two flows. Moreover, it can be thought that the variation in snow depth from point I through point J occurs because the snow is blown away toward the leeward side due to the reattachment of the flow. This is the same phenomenon as for $\theta = 200°$ shown in Fig.9. A typical example of this is shown in Fig.11.

5 CONCLUSION

Field observations have been carried out of snow accumulation patterns on a two-level roofed building. At the same time, characteristics of the wind flow around this building were clarified by wind tunnel tests. The relationship between the snow accumulation pattern and the flow was then investigated. The results obtained from the investigations into the relationship between the snow accumulation pattern and the wind flow characteristics were described. Furthermore, the possibility of estimating snow accumulation patterns from mean wind velocity and fluctuating wind velocity was indicated.

However, there are still problems awaiting solution in the relationship between these two phenomena. This relationship, which has yet to be dealt with as a general problem, should be further investigated.

REFERENCES

O' Rourke M.J., Speck R.S. and Von Bradsky, P., 1981. Uniform snow load on structures, *American Society of Civil Engineers*, Vol.108, No.ST12, pp.2781-2798.

Hashimoto S. et al, 1986. Relation between the snowfall on roof and characteristics of the wind, *Summaries of Technical Papers of Annual Meeting Architectural Institute of Japan*, Structures 1, B-1021

Tsuchiya M. et al, 1998, 1999, Wind effects on snow accumulation on two-level flat roofs, *Annual report Kajima technical research institute*, vol. 46, pp.63-68, vol. 47, pp.161-164

4 Housing and residential planning

Snow Engineering, Hjorth-Hansen, Holand, Løset & Norem (eds) © 2000 Balkema, Rotterdam, ISBN 90 5809 148 1

Lessons to be learned from indigenous architecture

H. Bull
BSA Architects, San Francisco, Calif., USA

ABSTRACT: Out of necessity, early man used available materials to create shelter from the elements. Building forms evolved which responded practically to local climate conditions. These early builders achieved a timeless architecture rooted in solving problems. In recent times, advanced communication, transportation and technology have allowed us to be impractical with the knowledge that artificial means are available to overcome inefficiency. Modern buildings often use inappropriate architectural forms and materials based on fashions originating in other climates. This results in unnecessary problems which are difficult and costly to remedy. Thoughtful study of indigenous architecture could lead to the design of contemporary buildings which would be suited to their climates.

1 INTRODUCTION

My architectural education occurred during a period when traditional architecture was looked down upon as not being functional. "Form follows function" was the watchword of the new architecture. The study of architectural history was limited to European classical styles and did not include early architecture of other continents. Indigenous or vernacular buildings were not considered "architecture".

Over a period of more than forty years. I have specialized in architecture in areas of deep snow and cold weather. During that time, I have observed many buildings that looked functional, but in reality reflected architectural fashions rather than answering problems brought about by severe climate. Increasingly, I have noted that vernacular architecture of buildings throughout the world responded

practically to the particular climates in which they were located. These early builders dealt with the laws of physics instinctively through their day-to-day experience. Their designs were simple and direct. Today, the typical engineer or architect has computing power which was inconceivable forty years ago. Perhaps as a result, our building designs and building codes have become unnecessarily complex. We believe that any problem can be solved, sometimes ignoring the fact that a simpler design would be a better solution.

It is my belief that if we study and apply the lessons of indigenous architecture, we would learn to build more sensibly and economically. We would also then be designing buildings where form truly follows function.

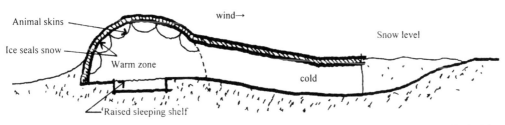

Figure 1 – Inuit igloo, northern Canada. Entrance to lee side of wind, below the living level, so the warm air (which rises) does not escape. Ice and snow act as insulators against the below freezing temperatures and harsh winds. Heating the igloo after construction forms an ice layer that seals cracks. Skins hung inside further insulate structure.

2 INDIGENOUS ARCHITECTURE

Historically, indigenous architectural forms have been derived from a combination of response to climate and need to use readily available materials. An extreme example of such sensible design is the traditional igloo of the Inuit people of far northern Canada. The design uses snow as insulation as well as structure. The hemispherical form maximizes volume while minimizing surface area. The lower entry tunnel and raised sleeping shelves use convection to preserve heat. Such a shelter can be erected and made habitable in a matter of hours by an experienced person using only a long knife. (See Figure 1)

The limitations of available materials and the need to seek shelter from the elements, resulted in sensible and architecturally harmonious buildings and communities that have given an identifiable "personality" to the architecture of different regions of the world. These architectural designs came about from a process of trial and error over many centuries.

2.1 Roof design responding to the climate

Roof slopes also vary according to climate and available materials. Flat roofed buildings evolved in hot and arid climates, where the roofs could be used for sleeping on relatively cool nights. Steep roofs shed water quickly in hot and rainy climates and walls are often quite open to encourage air circulation. In very cold climates, snow was found to be an insulator from temperatures below the freezing point, and roofs typically have a more moderate slope to keep the snow on the roofs, but steep enough to drain away the rain and melted snow.

Roofing materials were chosen on the basis of practicality and availability. To prevent snow from sliding, roof slopes were kept moderate. Stones and logs were often secured to the roofs to further ensure that the snow would not move. Entries were typically located facing the sun under the gable end of the roof to avoid dripping water on the inhabitants and visitors. In high elevations near or above timberline, flat stones were laid up shingle fashion to shed water. Where timber was more available, wood planks were lapped side by side or made into shingles. Sod roofs (earth with grass) were also used to provide insulation and hold snow, which in turn provided further insulation.

Various strategies were employed to keep the snow on the roof from melting. Often animal fodder was stored in lofts above the dwelling quarters. This created insulation for the inhabitants against the cold and also prevented heat from escaping to melt the roof snow. In some regions, farm animals were housed at the ground floor and their body heat helped warm the building's inhabitants above by convection.

The early mountain dwellers learned of the destructive power of avalanches. They did not want roof avalanches. It is interesting to note that typical roof slopes in the high Alps of central Europe and in rural Scandinavia correspond to generally accepted slopes for snow stability. In the book, *The Avalanche Enigma*, Colin Fraser notes that a 22° slope is considered generally safe from avalanche origination and slopes greater than 30° are considered hazardous. In deep snow areas, vernacular roofs seldom exceed 30°, except for religious structures where the roof form is designed to "reach for heaven". These roofs are typically greater than 45° and designed to shed snow. Entrances are under the gable end of the roof where entering worshippers are not endangered. (See Figure 3)

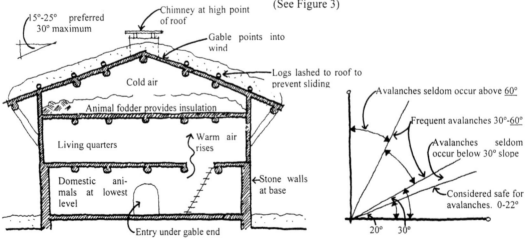

Figure 2 – Traditional high alpine home

Figure 3 – Avalanche danger as a function of terrain slope

In vernacular snow country architecture roof forms were kept simple, usually symmetrical gables. Roofs seldom intersected, thereby avoiding valleys which would concentrate water and increase problems of leaking. Whole villages in the high Alps were composed of buildings with similar roof forms oriented in the same direction. Typically, roof ridges pointed in the direction of the prevailing wind and down the slope of the terrain. Broad roof overhangs protected walls from water and consequent weathering problems. In summer, when the sun is higher in the sky, the overhangs also provided shade. (See Figure 2).

2.2 *Available building materials*

Because building materials had to be carried by people or domesticated animals, it was necessary to use those materials that were located nearby. In mountainous regions, stone and timber were usually available. At lower elevations where timber was plentiful, it was used for walls, roof framing and sometimes for the roofing surface. In timber buildings made of stacked logs, the joints were chinked to prevent movement of cold air. In some regions, particularly Scandinavia, sawn boards were later added to either (or both) exterior or interior walls to further increase insulation and create a more finished appearance. At higher elevations, timber was scarce and reserved for roof structure where tensile strength was important. Stone was used for walls but needed to be thick to provide an equivalent insulation to log walls. Being resistant to decay, it was almost invariably used for foundations and lower walls where there was likely to be large amounts of water. Stone walls were laid in mortar made from locally available materials such as mud strengthened with animal dung. Sometimes they were also plastered for appearance or to achieve increased weather tightness.

3 SNOW COUNTRY DESIGN TODAY

With increased ease of transportation in recent years, it has been possible to ship building materials from one region to another. Styles of architecture have become similar worldwide as communication has improved internationally. The result has been metal and glass office buildings being built all over the world, regardless of climate, which is a topic written about in a number of books. These buildings may be suitable in temperate climates, but not in tropical or subarctic climates.

In mountainous regions and other areas with heavy snowfall, building forms are also often based on current architectural fashions rather than a response to problems created by severe climate conditions. Thus, we often find that entries to buildings are subject to ice and snow falling from steep roofs. This creates a maintenance problem as well as a potential life safety hazard. Steep roof slopes and metal roofing materials are often employed in the mistaken belief that it is desirable to get the snow off the roof. While it is certainly true that roof snow can cause structural failure and water infiltration, it is important to recognize that these problems should be addressed sensibly and directly. Roof structure must be engineered to hold the maximum credible snow load in the particular geographical area. Often it is not. Even in the California, Oregon, and Washington mountains where snow loads can be 1500 to 2000 kg/m2 (300-400 pounds/square foot), it is better to design for the vertical loads than attempt to solve the problems of sliding snow. Many buildings in these areas have collapsed because it was assumed that the snow would slide off and reduce the roof load. Because of previous slides, there was no place for the new snow to be deposited. Collapses have also occurred when sliding snow has broken exterior walls. Balconies, decks, chimneys,

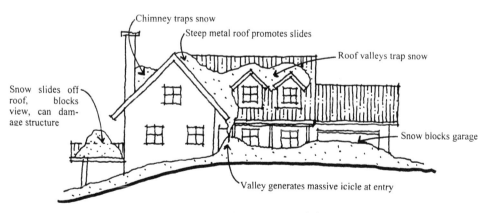

Chimney traps snow
Steep metal roof promotes slides
Roof valleys trap snow
Snow slides off roof, blocks view, can damage structure
Snow blocks garage
Valley generates massive icicle at entry

Figure 4 – Inappropriate snow country design

and lower buildings are regularly destroyed by sliding snow. Most important, many people have been killed by the enormous forces of snow avalanching from roofs.

3.1 *Ice dams and roofing materials*

Most leaks are due to ice dams, which can in large part be avoided by proper design. Ice dams are principally caused by heat from inside the building melting the snow on the roof. The resulting water flows down the roof slope until it meets the colder eave. Here it freezes, building up ice and forming a barrier that traps the water flowing down the roof. The water then leaks into the building, especially if it has overlapping roofing such as tile, slate, shingles or shakes. Sometimes continuous metal sheet roofing is chosen to avoid this problem. The low coefficient of friction of metal relative to snow increases the tendency of the snow to slide. However, this sliding is unpredictable. Snow has been known to slide on slopes less than two percent and adhere temporarily to 60° roofs. Snow often sticks to a metal roof when the temperature is cold or when held by roof penetrations such as chimneys or plumbing vents. Snow will also be trapped in valleys where snow will slide from two opposing directions. It accumulates over time and often forms a massive ice dam at the low end of the valley. When snow slides off a metal roof, it does so unpredictably and fast, often landing far away from the building. Signs are available in all the languages of countries with heavy snowfall which warn of the danger of sliding snow and ice falling from the roof.

Such signs identify buildings which have not been properly designed for the winter. Many devices have been developed to prevent snow sliding on roofs. These snow retention devices are necessary with any roof slope if the roofing material has a low coefficient of friction such as metal, tile or slate. They may also be necessary with high friction roofing materials such as granular faced asphalt composition if the roof slope is greater than 30°. In any case, entries or any areas where there might be people, should be protected if there is any possibility that they would be subject to sliding snow.

3.2 *Cold roofs*

Much as been written about cold roofs as a cure for ice dams and consequent leakage. Many do not work as hoped. Often the roof form is too complex, with complicated intersections, valleys, dormers and other interruptions which impede the smooth flow of air. Many times the space between the two roofs is too constricted for air to move by convection. The intake and exhaust apertures are often too small or unbalanced. Deep snow can also block the upper exhaust vent eliminating the effectiveness of the cold roof. Even when the cross section of the cold space is generous, the insulated warm envelope is often compromised by penetrations such as for utilities or recessed light fixtures. Vapor retarders, if present at all, are often discontinuous. Such openings will leak warm moist air into the cold roof void and will waste energy, cause condensation and ice dams.

Figure 5 – Modern village using traditional form. All roofs point down slope into wind.
Entries under gable ends. (Verbier, Switzerland)

4 CONCLUSIONS

How can we apply the lessons of traditional indigenous snow country architecture to improve the performance of modern buildings?

1 Design roofs to retain snow. Use high friction roofing materials to prevent sliding. Keep roof slopes less than 30°.
2 Avoid pitching roofs toward where there may be people. Locate entries under gables.
3 Keep roof forms simple, avoiding valleys and complicated intersections. Keep roof penetrations to a minimum and locate them at the highest part of the roof if possible.
4 Use "cold roof" detailing to avoid ice dams. This can be achieved by double roofs with cold air in between. Alternatively, use an extremely high value insulation to keep the roof surface uniformly cold.
5 Use broad roof overhangs to protect walls and windows from water and sun exposure.
6 Orient buildings and roofs to respect sun and wind.
7 Choose materials that are practical for the local climate and will weather gracefully.

REFERENCES

Alexander, C. 1979. *The Timeless Way of Building*. New York. Oxford University Press
Bugge G. & Norberg – Schultz, Christian. 1969. *Stav og Laft i Norge*. Byggekunst, Norske Arkitekters Landforbund
Bull, Henrik. 1986. *Lessons from Traditional Mountain Building*. Ski Area Management Magazine: May. pp. 97-99
Bull, Henrik. 1999. *Designing Roofs in Snow Country*. Fine Homebuilding. June-July. pp. 96-103
Cereghini, M. 1957. *Building in the Mountains*. Milano. Edizioni del Milione
Dorward, Sherry 1990. *Design of Mountain Communities*. New York. Van Nostand Reinhold
Fraser, Colin. 1966. *Avalanche Inigma*. London. John Murray
Holan, Jerri. 1990. *Norwegian Wood*. New York. Rizzoli
Kirk, Ruth. 1978. *Snow*. New York. Morrow Press
Olgyay, V. 1963. *Design with Climate*. Princeton. Princeton University Press
Rudofsky, Bernard. 1965. *Architecture Without Architects*. New York. Doubleday & Co. Garden City
Taylor, John S. 1983. *Commonsense Architecture*. New York. W.W. Norton
Von Frisch, Karl. 1974. *Animal Architecture*. New York. Harcourt Brace Jovanovich

Snow Engineering, Hjorth-Hansen, Holand, Løset & Norem (eds) © 2000 Balkema, Rotterdam, ISBN 90 5809 148 1

Information on building in cold climates – Results from research in Northern Norway

A. Nielsen
Building Science, Narvik Institute of Technology, Norway

ABSTRACT: Building houses in cold climate makes it necessary to take snow into account. The research group in Narvik has since 1992 been working with analysis of snow around buildings using computational fluid dynamics (CFD) and with experiments with building models in Svalbard. The information on the research and the results has to get out to the builders. We have made a web-site on Internet, where we give advise to people building in cold climates. Here we have information on placing the house on the site taking into account sun, wind and snow and other important factors. It is possible to calculate solar radiation and snow loads for different places. This information site is mostly written in Norwegian as solutions are based on Norwegian building tradition, but the result can be useful in other countries.

1 INTRODUCTION

Many areas in Norway are much influenced by snow and wind. In the coastal areas the problems with wind are most severe. High wind speed in storms has made damage especially on houses, that are not build in accordance with the building regulations. The wind loads can be severe in cases, where the house has a roof with a large overhang and placed on a hill to get a fine view from the house. Some winters in northern Norway with much snow gives many problems for the inhabitants as seen in Figure 1. It is for instance damage from snowloads, extra work for snow removal or snowdrift into the roof with later moisture problems in the building.

It is of great practical value to be able to make planning of single houses or groups of houses so that we can reduce the snow problems. In old days buildings ware made by local people who had the experience to know where to build and how the house should be constructed. Today we see the same house types all over Norway and not enough information on how to build. The Norwegian State Housing Bank does something with this problem. It gives extra loan for houses that are planned for cold climate and they have made an information book Husbanken (1994).

2 RESEACH IN NORTHERN NORWAY

At the start of "Building Science" at Narvik Institute of Technology it was decided to make research on snow and wind effects on buildings. This was based on the fact that this was of interest for all buildings in Northern Norway and also in other parts of Norway. Most previous investigations have been based on wind tunnel measurements done on a model of the building and terrain. That is very expensive method. It will be much easier to make calculations on a computer with CFD (Computational Fluid Dynamics), but calculations of this type were in 1992 still difficult as good models were lacking and the calculations were time-consuming.

In the meantime the capacity for computers has increased tremendous and the model that was done on a UNIX system can today by made on a high class PC. The research has been described first in Bang et al. (1994). The result from Narvik for numerical snow-simulation is found in Sundsbø (1997) and for wind tunnel experiments and wind simulation in Wiik (1999). Experiments are done with a building model on Svalbard as seen in Figure 2.

The snow distribution around the building model was measured with surveying equipment. The result is seen in figure 3. The same case was calculated in a CFD simulation of snowdrift as seen in Figure 4. Comparing the results from Figure 3 and Figure 4 show similar distributions. This looks rather promising for being able to predict snow distributions based on CFD calculations. There are still some problems for the lee zone and the scaling factors. More details on this and other cases can be found in Thiis (1999), Thiis & Gjessing (1999). Model experiments have been made on a model of a new mess building in Ny-Ålesund, Thiis et al. (1999). Numerical modelling and experiments are described in Thiis & Jaedicke (2000).

Figure 1.Buildings in Fuglenesdalen in Hammerfest in April 1997. This was a day with a snowstorm as can be seen on the windows.

Figure 2. Model building with snowdrift from right (Nielsen & Thiis 1999).

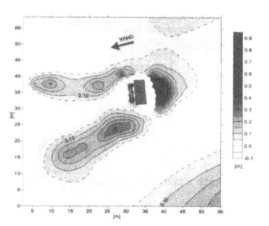

Figure 3. Measured snow height in m from the experiment in figure 2.

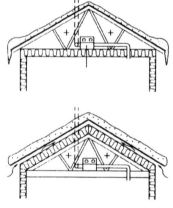

Figure 4. Calculated snow distribution with CFD for the experiment in figure 2.

Figure 5. Index-page of web-site on buildings adjusted to sites with wind and snow. Address see Nielsen (2000).

Figure 6. Roof with a ventilation system in the attic. In the top picture heat loads melt the snow. At the bottom is the system inside the thermal envelope and there is a ventilated air space between the insulation and the roofing material. From BRDS 525.002.

3 INFORMATION ON BUILDING DETAILS

Getting information on new research to the building industry is a slow process. New knowledge usually only leads to changes in practice after it is incorporated in standards and regulations. The building regulations in the Nordic countries are based on performance requirements. They are often phrased in very general terms and are at times difficult to interpret. There is a need for a more practical description of how to build without risk of mistakes. In many countries it is necessary to find this information in many types of publications as articles and research reports. In Norway it is different. The Norwegian Building Research Institute gives a high priority to the applications of building research. The most important information is the Building Research Design Sheets (BRDS 2000). There is approximate 900 sheets each consisting of 4 to 12 pages. The sheets are presented in a "cookbook" format, so the recommendations can be easily and successfully implemented. The sheets are not formally approved by official authorities, but are regarded as a semi-official interpretation of the regulations. Both practitioners and authorities turn to the Building Research Design Sheets for examples of how the regulations can be satisfied. The whole series is sold to subscribers (approx. 10000) and single sheets for direct sale.

These Building Research Detail Sheets contains much information on all the aspects that have to be looked at for different parts of the building. That includes also areas related to wind and snow. An example is calculation of glass thickness in roof glazing under snow and wind load based on the Timoshenko formula (BRDS A525.583).

4 WEB-SITE ON INTERNET

The WWW (World Wide Web) is a very good supplement to the Building Research Detail Sheets to spread the information on how to build in areas with wind and snow. We have started to build at web-site on snow- and wind-problems as this gives us an opportunity to present our research results and also show pictures as Figure 1, Figure 10-13 of some of the problems with information on what to do. As this is written for architects, engineers and builders we have decided to make the main part of the text in Norwegian. It includes pictures, text and some small practical calculations written as JAVA-applets.

We have not made the final address for web-site, but you will find a link to it from my web page Nielsen (2000). The front page, that is using frames, is seen in Figure 5. There will be links to the following subjects:

– Climate
– Design

– Economy
– Research
– More knowledge
– Pictures
– Calculations
– Other constructions

I will describe three of these points in more detail; that is Climate, Design and Calculation methods.

4.1 Climate

This part contains descriptions on climate and how it affects buildings. For each climate parameter is a number of pages that describe the effect and how it must be handled.

– Solar radiation
– Wind
– Snow
– Outdoor temperature
– Moisture
– Indoor temperature

We have included more than snow and wind effects as many decisions include the other factors. For instance is it important to get solar radiation through the windows in cold climates, as this will mean energy savings. The information will have a link to calculation methods for solar radiation for a free site and for solar path with mountains around.

4.2 Design guide

This part gives a description of how to make buildings or parts of buildings in climates with wind and snow. The information is from Børve (1987), Husbanken (1994), Strub (1996) and many building design sheets BRDS (2000). It contains drawings and pictures of details and buildings. Figure 6 shows the problems with snow melting on the roof due to internal heat sources.

Another problem is related to show in detail, how to make wall constructions without getting snow into the ventilated air gap on the outside of the construction. Many practical problems are illustrated.

The geometry on the building is important and placing the building in the right orientation to the normal wind and snowdrift direction is very important. Figure 7 is a drawing from Strub (1996) of a building form designed to control snowdrifting.

4.3 Calculations

This part contains calculation programs for different effects. Each calculation is written as JAVA-applets on a single page. The page has cells that have to be filled with input numbers for the calculation. Pushing a bottom on the page starts the calculation that can give results in the form of numerical values in some cells or plotted as a drawing.

The calculations are:

- Solar radiation trough a window
- Solar path and the horizon
- Solar radiation inside a room
- Indoor thermal comfort
- Thickness of glazing in a glassroof
- Snowloads on roof

Some of these calculations will be described more in detail. The first is: Solar path and the horizon. This is useful to find out which time on a selected day the sun is above the real horizon at a site. Information on the month and day and the position of the site are given as longitude and latitude. The horizon is given as a number of points – height in degrees in the different orientations. The horizon is drawn on the page each time a new point is added. This makes it easy to check the results. After the horizon is defined the calculation button is clicked. The solar path is calculated with the method in Lund 1976. This gives a drawing of the solar path in the same diagram as the horizontal values, se figure 9. It is easy to see if the sun is above the horizon. Making calculations for other days is easy.

Calculation can also be done of the radiation trough a window. In that case it is of interest to get information where the solar radiation will hit the floor or wall during the day. Figure 10 is the result from a calculation with a south facing window and the solar radiation each hour during June 15 in Northern Norway. It is possible to size the window and the room and decide the orientation of the window and latitude and longitude for the site.

Another calculation is the snow load on a roof after NS3479 (1997) and BRDS 471.041. The snowload depend on the form factor for the roof and the snowload on the ground at the actual site. The standard contains data for all communities in Norway. In the web these data is included, so that you can select the name of the town or community and get the correct snowload on the ground.

For glass roof the load is less than for ordinary roofs, as the snow will melt if the internal temperature is above the outdoor air temperature Dreier et al. (1985), Nielsen (1989) and BRDS G471.051. The load will also depend on the slope of the roof. A calculation on a web page can find the necessary thickness of the glass depending on the size - length and width - of the pane and the load from snow, wind and gravity of the pane, see BRDS A525.583.

It is also possible to calculate the energy consumption of a building giving the information on geometry, insulation, ventilation etc.

We think that these calculation pages are very important as they could by used by architects and engineers.

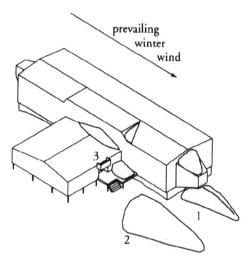

Figure 8. Small snowdrift (1) in the lee of wing aligned with prevailing wind. Shallow drift (2) displaced downwind of wing raised above ground. Deflector (3) clears snow from leeside entrance.

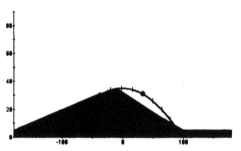

Figure 9. Horizontal shading and the solar path in Marts in Northern Norway. Values in degrees.

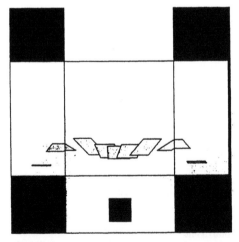

Figure 10. Fold out of floor and walls of a room. The solar radiation hit the wall in the morning and the afternoon on most of the day is the radiation on the floor. There is a calculation for each hour.

Figure 11. Snow accumulation behind a fence around a kinder-garten at Longyearbyen in Svalbard.

Figure 12. Facade with horizontal laths on the administration building at Longyearbyen in Svalbard.

Figure 13. Glaas roof in Trondheim with snow traped in the gutter.

4.4 *Pictures*

This part contains pictures of the different problems described in the other parts of the web-site. The idea is that the library of pictures can have much more examples than in for instance the design part. It can be useful for architects to see different solutions.

Figure 11 is a fence around a kindergarten in Longyearbyen (Svalbard). You can see that it works as a very good snowfence with nearly optimal spacing of the sticks. The fence was made to keep the children inside, but they can easily walk out. The snow coming from the left gives an accumulation of many tons of snow pr meter fence.

Figure 12 is the exterior of the new administration building in Longyearbyen. Note the horizontal laths in front of the window. After a snowstorm the snow will be on the laths. This is not very practical in areas with snowdrift.

Figure 13 is snow on a glassroof with a gutter at the lower end. The gutter works as a snow trap with an extra snow load on the lower part of the glass-roof. A normal glassroof would have no gutter and the snow would slide down along the wall and the roof would be free of snow. The problem could then be to get space below for the sliding snow.

5 CONCLUSION

This is some example of the web-site on buildings in areas with wind and snow. We expect that this can be a meeting place people who are interested in getting solutions for building in cold climate. Results from our research at the Narvik Institute of Technology are presented at the web site. The most important thing that many people forget for a web-site is that it must be updated or visitors will not come back. It is therefore important to have a plan for updates and having some part that is useful for visitors so that they come back. In this case it is the calculation pages, that visitors will return to. We have more ideas of small programs that can be useful.

We would like to thank the Norwegian State Housing Bank for their financial support to build this information web-side. Information on extra loans for climate adapted houses is included on the site.

REFERENCES

Bang, B. & Nielsen, A. & Sundsbø, P.A. & Wiik, T. 1994: Computer simulation of windspeed, windpressure and snowaccumulation around buildings (Snow-Sim), *Energy and Buildings* vol.21 no 3: 235-43
BRDS 2000, Building research design Sheets (1958-2000) in Norwegian from the Norwegian Building Research Institute, Oslo, Norway with numbers:
471.041 Snow- and windloads on roofs, 1995
G471.051 Snowloads on glassroofs , 1990
525.002 Roof construction, Selection of construction types and materials, 1996
A525.583 Roofs, Principles of construction, Calculation of glass thickness, 1988
Børve, A.B. 1987. The design and function of single buildings and building clusters in harsh, cold climates (In Norwegian with English summary), Doctoral Thesis, Oslo School of Architecture.

Dreier, C. et al. 1985. Glass roofs. Constructions. Climate and solutions for Nordic conditions (In Norwegian), Norwegian Building Research Institute, Handbook no 36, Oslo, Norway – second edition 1986

Husbanken, 1994. Wind and weather. Handbook in climate adaptation of buildings in wind exposed areas in Norway (In Norwegian), The Norwegian State Housing Bank, Oslo

Lund, H. 1976. Program BA4 for calculation of room temperatures and heating and cooling loads, Users guide, Report no 44, Thermal Insulation Laboratory, Technical University of Denmark

Nielsen, A. 1989. Snow melting and snow loads on glass roofs. In special report 89-6 from Cold regions research & engineering laboratory, USA, Proceedings from the First international conference on snow engineering, July 1988, Santa Barbara, USA: 168-177

Nielsen, A. & Thiis, T. 1999. Arctic Building Technology, 5th symposium on the Building Physics in the Nordic Countries, September 24-26, 1999 at Göteborg, Sweden

Nielsen, A. 2000 Web-page address: http:\www.hin.no\~an\ , Anker Nielsen, Narvik Institute of Technology, Narvik, Norway

NS 3479, 1997. Design of structures - Design loads, 1990, 3ed. With amendments NS3479/A1:1994, Norwegian Standard Association (in Norwegian), Prosjektering av konstruksjoner - Dimensjonerende laster, Norges Standardiseringsforbund, Norway

Strub, H. 1996. Bare Poles – Building Design for High Latitudes. Carleton University Press, Canada, ISBN 0-88629-278-6

Sundsbø, P.A. 1997. Numerical Modelling and Simulation of Snow Drift. Applications to Snow Engineering. Doctor Thesis from Narvik Institute of Technology and Norwegian University of Science and Technology, Trondheim, Norway, ISBN 82-471-0047-9

Thiis, T. K., Gjessing, Y. 1999. "Large-Scale measurements of snowdrifts around flat roofed and single pitch roofed buildings", Cold Reg. Science and Tech., 30 1-3. Pp. 175-181

Thiis, T. K., Jaedicke, C., Dahl-Grønvedt, M., Johnsen, J., 1999. "The new mess building in Ny-Ålesund – Effects on snowdrifts." Meteorological report series no.5, University of Bergen

Thiis, T., K., 1999. A comparison of numerical simulations and full-scale measurements of snowdrifts around buildings, Wind and Structures (in press)

Thiis, T., K., Jaedicke, C. 2000. Changes in The Snowdrift Pattern Caused by a Building Extension-Investigations Through Scale Modelling And Numerical Simulations, Forth International Conference on Snow Engineering, Trondheim, Norway, 19-22 June 2000

Wiik, T. 1999. Wind loads on low rise buildings. A numerical and experimental study of effects caused by building details on the external surface pressure. Doctor Thesis from Narvik Institute of Technology and Norwegian University of Science and Technology, Trondheim, Norway, ISBN 82-471-0478-4

Snow Engineering, Hjorth-Hansen, Holand, Løset & Norem (eds) © 2000 Balkema, Rotterdam, ISBN 90 5809 148 1

Utility of Gangi and new design of public apartment houses in cold and snowy region in Japan

Takahiro Noguchi & Osamu Itoge
Hokkaido University, Sapporo, Japan

Toshiei Tsukidate
Hachinohe Institute of Technology, Japan

ABSTRACT: There are many apartment houses equipped with Gangi which is like chained corridor and connecting many buildings, in northern Japan. That is arranged mainly for winter use. We analyzed the continued and the open form relating the snow protection. From the resident's point of view, our research clarified how to use Gangi in the real life and how to remove the snow and also what would be required for housing. Finally we tried to clarify the ideal image of Gangi and the common amenity space for the apartment house.

1 INTRODUCTION

In recent years, it's increasing to equip the chained corridor with the roof called 'Gangi' mainly for the public apartment house in snowy regions. Originally Gangi means chained wooden eaves in front of the shop to keep the shopping street off the snow in winter season. They are seen in snowy regions like Hokkaido and Tohoku, where are in the northern region of Japan. It's traditionally called Gangi or Komise **(photo 1)**. Here we will talk about Gangi for the collective housing, different from such a traditional Gangi. Current Gangi is the fireproofed and different form, but it has the same purpose to avoid the snow as origins. More over Gangi for the apartment house is used every day for a link between the apartment and the common facilities. It's also used for the storage and community space through the year. It has an important role as the indoor life or the half-indoor life in snowy regions.

From this point of view, we focused on Hokkaido and Tohoku regions where many apartment houses equipped with modern Gangi are constructing. Hokkaido is a prefecture on the north of Japan, it snows

much and it's extremely cold. It's similar to the climate in the Northern European countries.

Our research tried to analyze the continued and the open form relating the snow protection. Next, from the resident's point of view, our research clarified how to use Gangi in the real life and how to remove the snow and also what would be required for housing and Gangi. Through this research, finally we tried to clarify the ideal image of Gangi and the common amenity space for the apartment house focusing on protecting against snow and activating community life for residents.

2 THE METHOD OF RESEARCH AND SURVEY

For the first step, we collected drawings for public apartment houses equipped with Gangi. They are collected from the total 85 of apartment housing estates in Hokkaido and 3 in Aomori prefecture. They are drawings of the site plan, floor plan, elevation, and section. The form of Gangi was analyzed by these drawings. First the feature of the continuity and the form of housing with Gangi were investigated. They

Table 1. Outline of housing estates and research of questionnaire

Region (prefecture)	Housing estates	City	Total house-holds (homes)	Average family size (persons)	Average house-hold age (years)	Questionnaire (homes)	Answer (ratio) (homes)
Hokkaido	Ooasa	Ebetsu	127	2.7	60.8	109	69 (63.3%)
	Hinode kita	Iwamizawa	64	3.0	39.6	45	28 (62.2%)
	Hinode Minami	Iwamizawa	54	1.8	57.7	30	21 (70.0%)
Aomori	Daisan Toyama	Aomori	213	3.7	33.9	71	59 (83.1%)
	Miyazono	Hirosaki	204	3.3	34.7	77	67 (87.0%)
	Jousai	Hirosaki	200	3.1	43.8	76	61 (80.3%)
Total			862	3.1	36.4	408	305 (74.8%)

343

are sorted into 2 groups-(Gangi connecting buildings), which is connecting between a building and another building and (Gangi attached to a building) on the ground floor of an apartment building. Still more we classified them by the feature of the form and level of open. Based on the previous classification, the questionnaire and observation were conducted for each group on typical 6 collective housing estates (3 in Hokkaido and 3 in Aomori) **(table 1.)**.

3 THE FEATURE OF THE CONTINUITY OF GANGI EQUIPPED WITH THE APARTMENT HOUSE

Gangi equipped with the public apartment house sorted into 4 by the level of continuity and into 2 by it's chained form **(figure 1.)**. Talking about the level of continuity, non-continued type is more than half of all and 15 collective housing estates have a complete continued type, it's about half the number of the collective housing estates which have the chained Gangi.

Next is regarding the feature of its chained form. If it's connected in a line, only 2 to 4 buildings can be connected. In case the building is connected in the right angle with Gangi, it is not only connecting buildings but also it goes through the space like the park and the courtyard where people gather. This linear type is difficult to connect all buildings in the collective housing estate that has many buildings. 2 to 4 buildings are connected each, but they are separated from the next. The rectangular type is easy to connect all buildings. Many of them are perfectly connected.

4 THE FORM OF GANGI AND THE FEATURE OF OPENING

First, we are talking about Gangi connected to the buildings. We researched 16 collective housing estates

of total 30 continuous types. The linear type, which is regarded as the extension of Gangi attached to a building, is excluded. They are classified into 4 types by the level of open at the space created by Gangi.

The both-side open type is major in Hokkaido. It's more than half number. On the other hand, the one side open type only appears in Aomori, not in Hokkaido. In Aomori we found the park or the courtyard in buildings are surrounded by combinations of the one-side open Gangi attached to a building and the right-angle connection of Gangi.

Next, regarding Gangi attached to a building; we chose 76 collective housing estates to fit this survey. They are classified into 3 groups by the level of open of Gangi space. Most of Gangi attached to a building is set in north side of the building because the sunlight from the south is important for the housing. For this reason, many of them are cold and dark. The one-side open with skirting board is the major type at more than half number.

5 GANGI AS THE PASSAGE

The both-type of Gangi connecting buildings (in 4 collective housing estates) and Gangi attached to a building (in 6 collective housing estates) are well used for the passage. For Gangi connecting buildings, the questionnaire shows, "often use" and "sometimes use" make up more than 70% altogether. There are no big differences between Hokkaido and Aomori. Daisan Toyama is not used so much. However that are why there are many buildings that are not connected to Gangi compared with other collective housing estates. According to the questionnaire, "When you use it", the local residents answer like this-70% of them are always using and 30% are using in raining and snowing. It means they are using at any time, even though this facility aims to avoid snow.

Table 2. Use of passage in Gangi (connecting buildings)   (household)

Housing estates	Use often	Use sometimes	Not use much	Unused	Total
Ooasa	36 (52.2%)	21 (30.4%)	5 (7.2%)	7 (10.1%)	69 (100.0%)
Hinode kita	15 (53.5%)	5 (17.9%)	4 (14.3%)	4 (14.3%)	28 (100.0%)
Daisan Toyama	16 (27.1%)	21 (35.6%)	6 (10.2%)	16 (27.1%)	59 (100.0%)
Jousai	36 (46.8%)	15 (19.5%)	11 (14.3%)	15 (19.5%)	77 (100.0%)
Total	103 (44.2%)	62 (26.6%)	26 (11.1%)	60 (18.0%)	233 (100.0%)

Photo 1. Traditional wooden type of Gangi (Komise) in Aomori.

Photo 2.People cannot go out to the adjoining park from Gangi because of the low board (Hinode Kita).

Photo 3. People are chatting, sitting on bench arranged in Gangi (Daisan Toyama).

Photo 4. Children are playing around inside Gangi in winter (Daisan Toyama).

6 GANGI AS THE LIVING SPACE

Gangi is not only used for the passage **(table 2.)**. Both 2 types of Gangi are used in the same way of "chat with neighbors" and "children play". The eaves create neighbor's good community space **(photo 3 and 4)**. Gangi connecting buildings of Daisan Toyama is used for the children's play, it's high at 50% of total use. The space of Gangi is good enough and suitable for the children's play. In case of Hinode Kita it's used much for the children's play. One of main reasons is that there is a park beside Gangi **(photo 2)**. Particularly Gangi connecting buildings is used for "taking a walk", "enjoying the cool of evening or break" at high percentage. "Taking a walk" is 30% to 40%" and "enjoying the cool of evening or break" is 10% to 20% at any collective housing estates.

On the hearing survey, we often heard it's good to enjoy walking in a rainy day and for children to play outside. On the other hand, Gangi attached to a building is used much for the "bicycle parking". Particularly 3 collective housing estates in Aomori, it's 50 to 70% for using as "bicycle parking". It's depend on that they were planned as the bicycle parking space from the first, keeping 2.4m width **(photo 5)**. On the contrary, 3 collective housing estates in Hokkaido are only 0 to 20%. It comes from they have the bicycle parking space independently and Gangi is narrow 2m's width compared with Aomori. In case of Ooasa, Gangi is setting indoors.

7 HOW TO REMOVE THE SNOW FOR THE SPACE OF GANGI

Gangi has the purpose to keep the passage in the period of the snowy season, but as long as it's not an indoor type, it has problems of the snow coming into Gangi **(photo 6 to 8)**. The questionnaire shows; with

Gangi connected to buildings, over 90% answer "snow blow in" **(table 3.)**. "Snow blow in and stay" is also outstanding at high percentages. Particularly at 2 estates in Hokkaido, it's over 80%. It depends on much these 2 estates have only the both-side open Gangi without any protection against the snow. On the contrary, "Snow blow in and stay" is about 50% in Aomori, less than Hokkaido. This is coming from the form of Gangi connecting buildings in Aomori. It's different from Hokkaido. In Aomori, the one side of Gangi is closed. The one-side closed type keeps off the snow more than the both-side open type. In Aomori, the direction to be closed is also confirmed considering the wind in winter.

In case of Gangi attached to a building, "Snow blow in" is about 90% of total, and "Snow stay" is about 60% of them. As the one side is a building, the percentage of "Snow blow in and stay" is a little less than Gangi connecting buildings. In this case, there's no big deference between Hokkaido and Aomori. Both types of Gangi need clearing the snow because the snow blow in and stay **(photo 7 and 8)**. For Gangi connecting buildings, the necessity of clearing the snow is 90% to 100% in Hokkaido and 20% to 60% in Aomori, it's difference depends on estates. The need for clearing the snow is higher in Hokkaido because of its open form. They have further problem of the frozen floor caused by snowing. Gangi currently has a big problem to keep the safe passage in winter.

8 RESIDENT'S REQUIREMENTS FOR GANGI

Gangi is often used for various ways, but residents are not always satisfied. "Satisfaction" is only 50% for both types of Gangi. In case of Gangi connected to buildings, 3 main reasons of dissatisfaction are "snow blow in", "snow stay" and "cold" in the both of regions. For Gangi attached to a building, the reasons

Table 3. Condition in snowy season for Gangi (connecting buildings)   (household)

Housing estates	Snow blow in and stay	Snow blow in and gone	No snow	Total
Ooasa	60 (86.9%)	9 (13.0%)	0 (0.0%)	69 (100.0%)
Hinode kita	24 (85.7%)	4 (14.3%)	0 (0.0%)	28 (100.0%)
Daisan Toyama	31 (52.5%)	26 (44.1%)	2 (3.4%)	59 (100.0%)
Jousai	41 (53.2%)	31 (40.3%)	5 (6.5%)	77 (100.0%)
Total	156 (67.0%)	70 (30.0%)	7 (3.0%)	233 (100.0%)

Photo 5. People are often taking their children for a walk in Gangi (jousai).

Photo 6. Open spaces are all closed in winter season (Hinode kita).

Photo 7. Gangi is used as a valuable passage in snow season (Ooasa).

Photo 8. Residents have to walk carefully on snowing and slippery Gangi (Hinode kita).

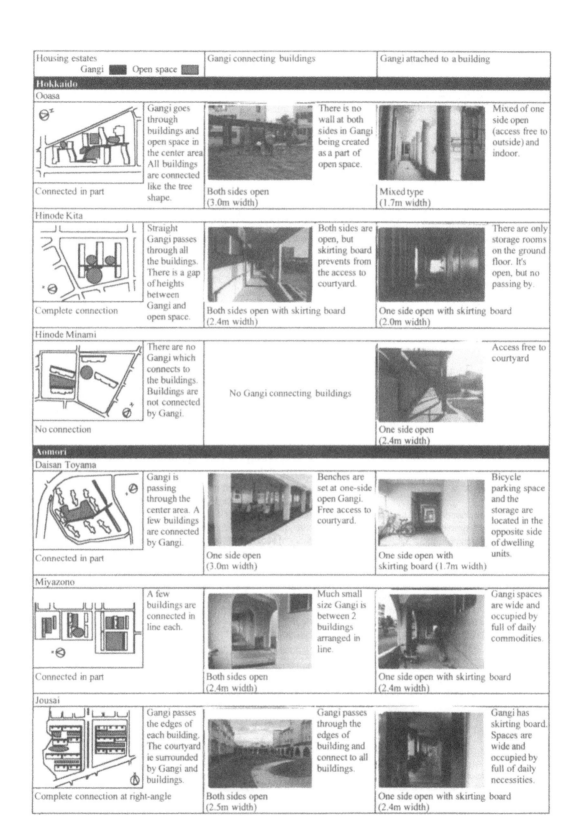

Housing estates — Gangi ■ / Open space ▨	Gangi connecting buildings	Gangi attached to a building
Hokkaido		
Ooasa — Gangi goes through buildings and open space in the center area. All buildings are connected like the tree shape. *Connected in part*	There is no wall at both sides in Gangi being created as a part of open space. — Both sides open (3.0m width)	Mixed of one side open (access free to outside) and indoor. — Mixed type (1.7m width)
Hinode Kita — Straight Gangi passes through all the buildings. There is a gap of heights between Gangi and open space. *Complete connection*	Both sides are open, but skirting board prevents from the access to courtyard. — Both sides open with skirting board (2.4m width)	There are only storage rooms on the ground floor. It's open, but no passing by. — One side open with skirting board (2.0m width)
Hinode Minami — There are no Gangi which connects to the buildings. Buildings are not connected by Gangi. *No connection*	No Gangi connecting buildings	Access free to courtyard — One side open (2.4m width)
Aomori		
Daisan Toyama — Gangi is passing through the center area. A few buildings are connected by Gangi. *Connected in part*	Benches are set at one-side open Gangi. Free access to courtyard. — One side open (3.0m width)	Bicycle parking space and the storage are located in the opposite side of dwelling units. — One side open with skirting board (1.7m width)
Miyazono — A few buildings are connected in line each. *Connected in part*	Much small size Gangi is between 2 buildings arranged in line. — Both sides open (2.4m width)	Gangi spaces are wide and occupied by full of daily commodities. — One side open with skirting board (2.4m width)
Jousai — Gangi passes the edges of each building. The courtyard ie surrounded by Gangi and buildings. *Complete connection at right-angle*	Gangi passes through the edges of building and connect to all buildings. — Both sides open (2.5m width)	Gangi has skirting board. Spaces are wide and occupied by full of daily necessities. — One side open with skirting board (2.4m width)

Figure 1. The feature of the site plan in the housing estates and the form of Gangi

are "snow blow in", "snow stay" and others are "narrow", "cold", and "noisy". They come from most of Gangi attached to a building locates at north side of building and it is not always wide. And also the windows of some dwelling units face to Gangi (photo 5). Compared to each region, the value of Gangi connected to buildings in Hokkaido is significantly low. As the one side is not closed like Gangi in Aomori, snow is easy to blow in and accordingly Gangi gets the frozen and slippery floor.

On the contrary, the value of Gangi attached to a building is lower in Aomori. It caused by Gangi using as a bicycle parking space. Resident's requirements for Gangi are "should be more comfortable for the children's play"(average 31% of total), "need some benches for the rest"(28%), "should be indoor to avoid the snow"(25%), and "get the sunshine more"(19%).

9 CONCLUSION

The feature of continuity is important for Gangi. However there are rather few Gangi that connects all of buildings in collective housing estates. If Gangi falls at right angles with a building, it's easy to make a compact and calm space. Particularly in Aomori, the one-side open Gangi which connects and be arranged at right angles with building are making the preferable space between buildings. In Hokkaido, there are no one-side open Gangi connected to buildings. The both-side open without the wall is major. For this reason, there are no spaces by making up with Gangi like Aomori has. Most of Gangi attached to a building are equipped at north side of building. It's dark and cold in many cases. Many of their form are open with the skirting board.

Gangi is well used not only the passage space but also the common space for residents. The feature of Gangi is appearing for the purpose of going to work or school in a raining or snowing day, the children's play and going for a walk. The form of Gangi is different in Hokkaido and Aomori. For this reason, the characteristic differences in the way of use and resident's satisfaction are seen. Residents in the both regions are not satisfied with Gangi so far.

It should be much improved. We try to improve Gangi for the apartment house as a living space to suit the snowy region. For the resolution, we should develop to design the form that the snow is not blow in much or stop to blow in. We also research the proper relationship between the building (dwelling unit) and Gangi considering for the problem of noise and privacy. Further more we should plan such as flexible types, open in summer time and indoors in wintertime. We need to survey for the plan of the new common space with the combination of the living space and the chained passage to avoid the snow and cold.

Here are our proposals in this conclusion. The first proposal is an idea that Gangi should be placed in the south against in the north where Gangi was usually located. It must be warm with much sunlight even in winter and people would like to gather there. According to this, we need to reconsider the plan for the dwelling unit and the apartment building. Secondly, we should use Gangi in a multi propose as a living space. A warm space like an atrium could be suitable for the purpose as an activity space of chattering, playing, gardening or making some, adding the current purpose for the passage and storage. Thirdly, we need more common facilities connecting to Gangi like the children's playroom and the community room of multi purpose. Further more it's more important for the large collective housing estates to be arranged with the shopping area and the welfare facilities for the aged. The collective housing in the north regions should be changed in a new style. We think it's important to establish a new theory for the house planning and design.

REFERENCES

T.Noguchi, F.Adachi, 1993. "Gangi and Snow Removing System of House-in case of Takada, Joetsu City- Study on the Dwelling Style and Housing Design of Urban Area in Snowy Region", pp.93-103, Journal of Archit. Plann. Environ. Engng, AIJ, No.451.

T. Tanaka, T.Noguchi, 1996. "Daily Activities of Aged and Disabled People and Snow Removal from the Access Areas to their Houses in Winter in Hokkaido, Japan", pp.59-66, Proceedings of Japan Society for Snow Engineering, Vol.13

T.Noguchi, O.Itoge, 1999. "Study on Form and Way of Using of Gangi(Chained Corridor)-in Case of Public Apartment House in Hokkaido and Aomori", pp.173-178, Proceedings of Japan Society for Snow Engineering, Vol.16

T.Noguchi, 1999. "Form of Gangi and Snow Removing of Public Housing in Hokkaido and Aomori- Study on the Common Space Design of Apartment houses in Snowy Regions Part 1", pp.113-120, Journal of Archit. Plann. Environ. Engng, AIJ, No.525.

Snow Engineering, Hjorth-Hansen, Holand, Løset & Norem (eds) © 2000 Balkema, Rotterdam, ISBN 90 5809 148 1

Evaluation of residential areas in harsh climates, Hammerfest, Norway

Harald Norem, Sverre Nistov & Alf Ivar Oterholm
Norwegian University of Science and Technology (NTNU), Trondheim, Norway

Lene Edvardsen
Norwegian Housing Bank, Hammerfest, Norway

ABSTRACT: In areas with no trees and strong winds drifting snow may be a major problem for the daily life in residential areas The paper describes an evaluation of two living areas in Hammerfest, Northern Norway, at a latitude of 70° 40' North.

The paper deals with the evaluation of the areas, to critically analyse the background for the planning and to gain experience for new projects. The use of climatological data to find the critical wind directions and to estimate the magnitude of the snow drift problem is described. The experiences concerning the design and the location of garages and entrances to the houses due to the wind directions and to the topography are also described, as well as the need for an outer sheltering of the whole area. Further, recommendations for the location and the design of access roads and places for storing removed snow are given.

1 INTRODUCTION

In northern and mountainous areas with no trees and strong winds, drifting snow may be a major problem for the daily life, and should thus be an important concern for the planning of residential areas. The most important factors to take into account are the design of the access roads, the location and the design of the houses and the landscaping of the area.

The paper describes an evaluation of two residential areas in Hammerfest, Northern Norway. The areas were developed in the 1980's and during the planning and construction period special attention was paid to the severe climatic conditions in the area. Despite this, there has been considerable problems related to blowing snow. During the winter 1996/97 snow depths well above average and strong winds made it extremely difficult to maintain the access to the area and to the houses.

This paper presents the main conclusions of a study made by SINTEF and NTNU to investigate the reasons for the problems encountered, and to present recommendations to improve the living conditions. The study was financed by the Norwegian Housing Bank with the aim to gather and evaluate these experiences in order to transfer them to the planning of new areas.

2 SHORT DESCRIPTION OF THE AREA

Hammerfest is a major fishing port in Northern Norway with approx. 9000 inhabitants. It is situated at a latitude of 70° 40' North, and is the northernmost town in Norway. In the 1970's the further development of the town was extended to the area Fuglenesdalen, Area 1, 2 and 3, at an altitude of 100-130 m a.s.l, Fig 1.

The climatic conditions at this altitude proved to be extremely difficult compared to the conditions at sea level. The result was severe snow problems by blocked access roads and entrances and houses totally covered by snow. It was thus decided to pay special attention to the climatic conditions when planning the next two areas to be developed, Areas 4 and 6. Photo and maps of the areas are presented in Figs 2 and 3.

The planning process of the area is the basis of the doctoral thesis of Børve (1987). The thesis contains the results of the climatic investigations and model tests with artificial blowing snow as well as the principles for the location and the design of the access roads and the houses.

The main aim of the project was to develop new types of housing keeping the rough climate in mind.

Special attention was made on the sheltering of the areas by use of snow fences and tree planting. These measures were used both as an outer sheltering and inside the residential area. It was also important to develop new dwelling types based on the strong climatic conditions found in the area. A further description of the planning process is found in Edvardsen (2000).

3 INVESTIGATION OF THE CLIMATIC FACTORS

3.1 Critical climatic factors

The climatic conditions both in summer and winter are special in the far north. During the summer there is midnight sun for two months, but even with the sun shining day and night the average summer temperature is only 10° C. In the winter, the sun is missing for two months and there are frequent periods with strong winds and snow falls.

The harsh climatic conditions thus set special requirements to the planning of the residential areas, where the most important are:

Summer conditions
- To create areas for outdoor activities that both has sun and is sheltered for cold winds
- To have windows faced toward the sun in the living rooms

Winter conditions
- To have limited snow collecting on the roof and around the windows
- To have limited amount of blowing snow deposited close to the entrances and the garages
- The houses should be well insulated toward the cold winter wind directions
- The collector and access roads should be easy to keep open during snow storm periods

The main climatic factors that need to be investigated to plan for these essential requirements are snow precipitation, wind, and sun.

3.2 Snow

The snow depths are a designing factor for the dimensioning of roofs and balconies, for the design of access roads and for the size and location of areas for depositing removed snow. Snow falling in calm weather seldom causes considerable problems, but the real problems start when snow is combined with strong winds. This is due to the effect of the wind to erode, transport and deposit snow particles. In areas with strong winds there will be erosive sections with no snow, compared with leeward areas that may have snow deposits several meters of depth.

To have substantial amounts of blowing snow it is necessary to have an open area covered by snow where snow particles may be picked up by the wind. The necessary length to obtain the very worst conditions is approx. 500 m. The length should be measured from a borderline where the snow particles start to be eroded. This borderline may be open water, forests, and creeks or dense developed areas.

The snow depths in Hammerfest vary significantly from year to year. The maximum snow depths on open ground varied between 16 cm and 183 cm in Fuglenesdalen in the 1957-1982 period.

3.3 Wind

Both gradient and topographical winds influence the wind system in Hammerfest. The gradient winds are a result of the pressure gradients between the high- and low-pressure zones. These winds are mainly westerly winds often bringing precipitation from the ocean. The topographical winds follow the directions of the main valleys, and are caused by air temperature differences in the mountains and at the coast. In the winter there is cold air in the mountains compared to mild air along the cost, and the flow has a Southeast direction down the valleys. In summer time the wind direction is opposite, but then the wind speed is generally lower.

The analyses of the wind have to be done separately for the summer and winter seasons. The main wind direction in the summer to be taken into account for the planning of the residential areas are the winds bringing cold air from the ocean, thus making all kind of outdoor activities unpleasant. In the wintertime, it is the wind directions causing snowdrifts that are most important for the planning of the areas.

Heavy snowdrift deposits are formed during periods with winds above a certain value. Generally, the amount of snow in motion is higher when strong winds are combined with simultaneous precipitation. Thus the analyses of the wind need to include information of the wind directions for both high wind speeds and those bringing precipitation.

For the present evaluation the point-value method introduced by Norem (1975 and 1994) has been used. Each wind observation is given a certain value due to the equation:

$$P = \Sigma(V_t - 9) + (V_s - 5)$$

V_t = Recorded wind speed (m/s) without simultaneous precipitation
V_s = Recorded wind speed (m/s) with simultaneous precipitation

The observations for each wind direction is summarised, and a wind rose may be made to present the wind directions that cause most snowdrifts and thus problems for the snow clearing of the roads and entrances, Fig 4. The benefit with the presented procedure is that the only wind observations taken into account are the ones with wind speed above the critical value that gives snowdrift. In addition the value of an observation increases with the wind speed and for days with simultaneous precipitation. It is thus assumed that the wind roses based on the present procedure represents a good basis for the further planning for the winter conditions.

The wind in the Fuglenesdalen area is highly influenced and channelled by the mountainsides at both sides of the valley floor. Fig 4 presents wind roses based on the point-value method, and clearly show that wind from the Southwest is the dominant wind direction in the winter season.

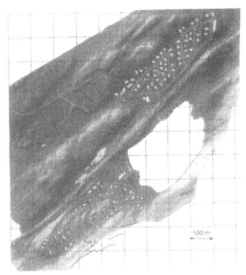

Figure 3   Map of Areas 4 and 6. The map is based on the GIS-method and the darkness represents the slope gradient

Figure 1   Map of the centre of Hammerfest with the developed Areas 1,2,3,4, and 6

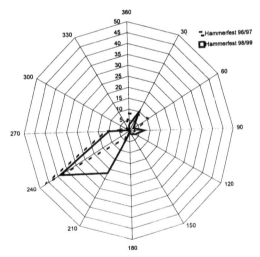

Figure 4   Wind rose for Fuglenesdalen, Hammerfest for the 96/97 and the 98/99 winters based on the point value method

3.4 Sun

The sun is both an important energy source as well as a comfort factor. Even with sun both day and night in the summer it is important to remember that the sun height in the far north is much lower than further south. The maximum sun height in Hammerfest is 43,5°, compared to Oslo with 53,5° at 60°-north. The low sun height sets important require-

Figure 2   Photo of Area 6 and Area 4 in the background

351

ments to the sight hindrances in front of the houses, such as neighbouring houses, trees, fences etc.

Because of the vicinity to the cold Atlantic Ocean really sunny days are rare. During the summer season, only 10% of the days are characterised as bright. This makes it even more important to give the residents opportunity to take the advantage of the sparse sunny hours.

4 CLASSIFICATION OF CLIMATIC EFFECTS DUE TO THE LEVEL OF PLANNING AND THE CLIMATIC FACTORS

The evaluation of the area is easier when important factors and parameters etc. are classified into certain groups. In the present case it was decided to classify the area into three levels of planning; the master-plan level, the structure plan level and the site plan level. For each level the area was evaluated due to snow, wind and sun.

The table below gives the main effects that are studied for each level and for each climatic factor.

4.1 *Master plan level*

Area 4 is located on the lee side of a longitudinal ridge toward Northwest and has a brook depression toward the Southeast. For both of these wind directions the central part of the area is sheltered from strong winds and blowing snow. There is, however, one line of houses located close to the ridge. These houses are in the area of the natural snow deposits from the ridge caused by westerly winds.

The natural sheltering from the wind directions from the Southwest and Northeast is, however, fairly poor. Especially the dominant wind from the Southwest brings a lot of drifting snow to the area.

The wind conditions in Fuglenesdalen seem to be comparable to other residential areas in Northern Norway, but these face less snow problems. The main reason for the more severe winter conditions in Fuglenesdalen is obviously the huge open areas on the windward sides. The need for a natural or artificial sheltering is thus extremely important. Such sheltering may consist of natural ridges or depressions collecting snow, developed living areas close to the new areas, or artificial fencing and planting.

Extensive uses of fences have been implemented in Fuglenesdalen, with limited success. The fences located normal to the dominant wind direction work well if they are located on top of a ridge or on flat ground. The advantage of the fences is mainly by creating shelter and reducing the snow transported into the area. Less effect has been observed on reducing the size of the natural snow deposits in lee of ridges etc.

The only houses having structural damages are those located in major natural snow deposits. In those areas snow fences have been constructed to reduce the size of the snow deposits, but with only limited success. The main lesson learned is thus that houses should not be planned in areas with major natural snow deposits.

Ahead of the development of the area some experiments with tree planting were done. When the planting was combined with snow fences and the use of 1-2 m high earth dams it has been relatively successful. The fences and dams offer shelter and improved living conditions for the trees and herbs. A threat to the plants is the grazing of reindeer, and fences now protect recent plantings.

The quality of a residential area is highly dependent on access during snowstorm periods. The best snow removal conditions are found on roads having no houses on the windward side of the road. To facilitate the snow removal, sufficient space for storing snow should be reserved. These areas should preferably be located at lower levels than the access roads, so the snow deposits are not causing the collecting of new snow on the roads in snow drifting periods.

The width of the collector roads was selected to 10 m in Fuglenesdalen. The large road width was explained by the need for storing snow on the road in order to increase the time between the removal of snow from the road edges. The additional width cause higher speeds on the roads and thus reduces the traffic safety. The group recommended that the collector road should be paved for only 6 m. At one side there should be a pedestrian lane. The pedestrian lane and the area next to the road should be constructed for the use of rotary snow blowers.

The density of the development should be open for discussion. On one side, a high density gives a high sheltering effect, and thus improved wind conditions inside the area. On the other side, the high density gives concentrated snowdrifts in the peripheral parts of the area, limited space for storing snow and less sun in the summer. The group concluded that at least the peripheral areas should be planned relatively open. This gives the planner the opportunity to locate the houses relative to each other in a way so that one house does not collect snow next to the neighbouring houses. In the central part of the developed area, there is most often less transported snow and smaller snowdrifts, and thus the density may be increased.

4.2 *Structure plan level*

As a first step of the development, and simultaneously with the road and water supply, a system of earth dams and snow fences were constructed. The idea behind these systems was to shelter the area and to create improved growing conditions for the plants. After some years the plants should take over the effect of the snow fences, which could then be removed.

Level	Snow	Wind	Sun
Master plan level	Snow drifts in the peripheral zones The effect of snowfences and tree planting Snow removal conditions along the collector roads The effect of snow due to the density of the development	Effect of natural terrain to create sheltering Sheltering effect of snow fences and planting Effect of cold summer winds	Exposition and slope gradients Sunshine
Structure plan level	Effect of the in-area snowfence system and planting Snow drifts due to the relative position of the houses Snow removal conditions along the access roads	Sheltering effects due to the relative positioning of the houses Channelling of the wind due to houses and roads Sheltering effects of the in-area snow fence system on the winter and summer winds	Shadows from one house to another Shadows caused by the in-area snow fence system and planting
Site plan level	Snow drifts around entrances and garages. Effects of fences and walls Snow drifts on roofs and terraces	Wind conditions on areas for outdoor activities Effects of cold wind on the heating of the house	Location of windows and terraces Design of the areas for outdoor activities

Until the whole area was fully developed the earth dams and the snowfences system were very important. After this time there are diverted opinions about the effect. These 1-2 m high dams occupy some area, and in addition leave open channels in the area. On the other hand trees thrive very well close to them. In the opinion of the group such dams should not be higher than 1 m, and to obtain the necessary height of 2-3 m one should use snowfences. The snow fences can then be removed when the plants have reached the necessary height.

The width of the access roads was also selected to 10 m to give space for storing snow. The extra width reduces the traffic safety in the summer season and increases the area to be maintained. The recommendation for further developments is to select access roads only 6 m wide. In very harsh climate 6 m wide paved road and one extra meter on both sides for storing snow may be selected.

The wide access roads have the effect that residents often use the road for parking, especially in snowing periods, when they are afraid that their own entrances are going to be blocked by snow. A good solution would probably be to establish common parking lots, or even additional garages, in the peripheral zones and close to the collector roads. In snowstorm periods when entrances and the access roads are blocked by snow, the residents may then still use their cars.

4.3 *Site plan level*

The two areas evaluated were both planned for extreme climatic conditions. A major part of the constructed houses in addition have a special design for the conditions in Fuglenesdalen. The report of Edvardsen (2000) describes the procedure and the intention of the planning process.

Most of the houses are connected directly to the garages to obtain an easy access from the house to the garage. Usually, the entrance of the garage is close to the access road and the garage doors are located on the windward side of the garage.

Despite all good intentions there have been major problems in using the garages in the wintertime. The main reasons are:

- Walls and fences are located close to the entrances
- Snow walls due to the removal of snow are found close to the entrances
- The entrances are located in the leeward zone from neighbouring houses.

These effects are clearly demonstrated by the photos in Fig 5, showing both the summer and the winter situation.

The garages are located close to the access road and with some angle to the road in order to have a strong eroding wind on the door side. In connection with the gardening the height of the terrain has been increased by a wall and fences have been erected on top of them. To reduce the amount of snow collecting at the entrance, the fences are removed every winter. The wall collects a lot of snow and the resulting snow walls even more. An area intended to be almost free of snow has thus developed to be a local snow drift area.

Figure 5   Picture of an entrance in Area 6. The garage is close to the access road, but the
entrance is on the leeward side of the
fence and the wall. Note that the fence is
removed every winter to reduce the size
of the snowdrifts.

Despite all the problems faced at the entrances
the group is positive to the procedures for the planning of garages and entrances. But a lot of work is
still needed to explain to the house owners how to
design the fences and walls to keep the good intentions from the planning stage by the development of
the garden. It might also be a good solution to have a
reindeer fence surrounding the whole area to prevent
the use of separate fences.

Special attention was paid to the design of areas
for outdoor activities to obtain both sheltered and
sunny areas. Some houses also have an additional
balcony at the first floor level to enable outdoor use
in nice spring days that might occur before the snow
has melted in the garden. These parts of the special
designed houses seem to be very attractive and were
appreciated by the residents.

5 REFERENCES

Børve, Anne Brit (1987)
Houses and groups of houses exposed to harsh climates in
cold areas. (Thesis) (In Norwegian)
Institute of Architecture, Oslo

Edvardsen, Lene (2000)
Dwelling and areas adapted to the climate
Winter Cities 2000, Luleå

Norem, Harald (1975)
Design of highways in areas of drifting snow (Thesis)
Draft Translation 503, CREEL, Hanover, New Hampshire

Norem, Harald (1994)
Snow engineering for roads
Håndbok 172. Norwegian Public Roads Administration,
Oslo

Norem, H., Nistov, S., Oterholm, A. I., (1999)
Evaluation of the climatic conditions for the Area 4 and 6,
Fuglenesdalen, Hammerfest. (In Norwegian)
SINTEF, STF22 A99455, Trondheim

Snow Engineering, Hjorth-Hansen, Holand, Løset & Norem (eds) © 2000 Balkema, Rotterdam, ISBN 90 5809 148 1

Wind tunnel modeling of roof snow accumulations and drift snow on gable roofs

Toshikazu Nozawa, Jiro Suzuya & Yasushi Uematsu
Tohoku Institute of Technology, Sendai, Japan

ABSTRACT: The snow is considered to be drift on leeward side of gable roof, and cause the unbalanced snow load on roof. But in wind tunnel modeling experiments on gable roof, sometimes we obtain the roof snow accumulations in which snow accumulation on leeward side is smaller than that on windward side. Experiments on model buildings with typical shape of roof are carried out, using two kinds of model snow particle. In this paper, we discuss the process of forming of roof snow accumulation.

1 INTRODUCTION

Though the wind tunnel modeling of snow drifts is a successful mean to predict the snow accumulations on a building, it is difficult to assure that the result of wind tunnel modeling precisely reproduce the actual snow accumulations on a roof. To make the condition, in which both the flow around the building and the motion of snow particles are much the same with the actual state is no simple business, and to make some similarity conditions less serious is unavoidable.

Figure 1 (a) shows a result of wind tunnel model test, it well describes the actual roof snow accumulations photographed in Figure 1 (b). In order to compare the predicted snow accumulations with full scale, and improve the reliability of wind tunnel modeling of drift snow, more full-scale case histories are needed. In fact, the full-scale measurement of roof snow involves some difficulties. The balloon system was used to take the air photos of the roof snow shown in Figure 1, and the values of snow depth were obtained through the technique of an aerial survey.

The wind tunnel experiment making use of crashed wheat as model snow obtains the model snow accumulations shown in Figure1. Another conditions for the experiment are decided in a similar manner to the experiments described in following chapters.

In this paper, the wind tunnel modeling of snow accumulations on flat roofs, on pent roofs and on gable roofs are discussed. To simulate the motions of snow around a building, both the similarity of airflow around the building and the similarity of the movement of snow particles near the snow bed are necessary between model and prototype. Various shapes of snow accumulations have been obtained in the wind tunnel modeling experiments on snows on gable roofs. Especially the shape of the accumulations on leeward side roof is considered to be influenced by experimental conditions such as wind velocity, model snow particles, size of model building and so on. In the following chapters, various shapes of model snow accumulation brought by the difference of model snow and by the difference of model size are shown, and the experimental conditions appropriate for wind tunnel modeling of roof snow are discussed.

2 EXPERIMENTS

A large and a small size model of flat roof house, pent roof house and gable roof house are made. The shapes and size of model building are listed in Table 1 with the sketch of each model. Two kinds of particles are used as model snow, one is clashed wheat and the other is activated clay. The physical properties of the particles are shown in Table 2.

Open circuit wind tunnel, with 1.0 m x 1.0m testing section, was used in the experiment, and it has particle dispersing system on its roof. Wind speed was controlled by automatic control system, and was measured at the beginning and at the end of an experiment by hotwire anemometer. The experiments, in which the crashed wheat was used as model snow, was performed at the wind speed of 2.88 m/s that is the value at the threshold of particle movement. The wind speed of the experiment for the activated clay particle was 1.3 m/sec, and it was decided so that the particles fall at the same angle with that observed in crashed wheat cases.

3 EXPERIMENTAL RESULTS

Figure 2 shows the model snow accumulations on the middle of the roof for each model. And Figure 3 shows the distributions of model snow depth on gable roofs. Both figures give the pictures of the ratio to the average depth of whole accumulations on roof.

Model Snow Accumulations on Flat Roof (FL, FS).

The shapes of the model snow accumulations are varies with the kinds of model snow particles, and with the size of model building. In the case of crashed wheat model snow, the depth of the particle accumulation is increasing smoothly towards leeward of the roof. Average depth of the roof snow is about one half of the grand snow. In the case of activated clay, it is seen on both the large model and small model roof, that the depth of the particle accumulation increase suddenly at about 50mm from the windward edge. Average depth of the roof snow is nearly equal to grand snow.

Model Snow Accumulation on Pent Roof (PL, PS).

In the experiments performed with crashed wheat, the shape of model snow accumulations weren't affected by the size of model building. The bow-shaped particle accumulation on the large model and that on the small model have a great resemblance. While in the other hand, the activated clay particles accumulated almost uniformly on the roof.

Model Snow Accumulation on Gable Roof.

Two pairs of building model were made one pair has the roof sloped at degree of 15°, and the other has degree of 30°. The former is named G1-L (large) and G1-S (small), and the latter is named G2-L and G2-S.

The experiments for the crashed wheat particles give a similar figured particle accumulation on windward roof regardless of size of model building. And particle accumulation on leeward side is influenced by the size of model building.

The results of the experiments for the activated clay particles vary with the size of model building and with the slope of the roof. The particle accumulation on windward side of G1-L is different from that on G1-S, the shape of the accumulations on leeward side is varying with the model snow particles.

4 DISCUSSION

The Reynold's Numbers based on the height of eave of large model are 2.06 for crashed wheat test, and 0.93 for activated clay test, and those of small model are 1.03 and 0.46 respectively.

It is said that the agreement of Froud Number between model and prototype is important for modeling of drifted snow. The Froud Numbers based on the particle size are 7.37 for crashed wheat and 8.97 for activated clay respectively. VTR takes the pictures of moving particles during the experiments. Figure 4 is the pictures of crashed wheat particles around the building model, and Figure 5 is those of activated clay.

It is seen on flat roofs, that the activated clay particles nearby the windward edge are taken off through the air separated from the edge, and accumulate on the leeward part of the roof. Saltation trajectories of crashed wheat particle are observed on flat roof, and height of trajectory is less than 5 mm.

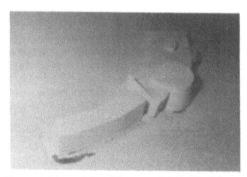

Figure 1 (a)　　Model Snow on Model Building

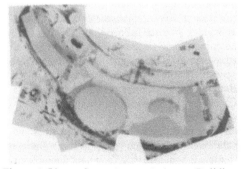

Figure 1 (b)　　Snow Accumulation on Building

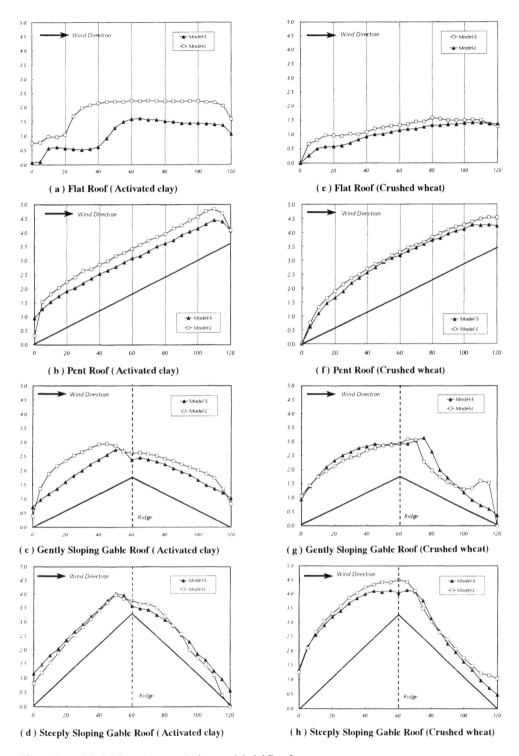

(a) Flat Roof (Activated clay)

(e) Flat Roof (Crushed wheat)

(b) Pent Roof (Activated clay)

(f) Pent Roof (Crushed wheat)

(c) Gently Sloping Gable Roof (Activated clay)

(g) Gently Sloping Gable Roof (Crushed wheat)

(d) Steeply Sloping Gable Roof (Activated clay)

(h) Steeply Sloping Gable Roof (Crushed wheat)

Figure 2   Model Snow Accumulations on Model Roofs

Table 1. Outline of the Model Buildings

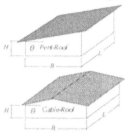

Model type	H	B	L	$\theta°$
F-S Model (Flat roof)	5	10	20	0
F-L Model (Flat roof)	10	20	40	0
P-S Model (Pent roof)	5	10	20	15
P-L Model (Pent roof)	10	20	40	15
G1-S Model (Gable roof)	5	10	20	15
G1-L Model (Gable roof)	10	20	40	15
G2-S Model (Gable roof)	5	10	20	30
G2-L Model (Gable roof)	10	20	40	30

Table 2. Properties of Model Snow Particle

Property	Crushed wheat	Activated clay
Bulk Density (kg/m^3)	436	602
Angle of Repose	50~55°	35~40°
Critical Velocity (m/s)	2.88	4.65
Terminal Fall Velocity (m/s)	0.2	0.4

The air separated from the windward edge of pent roof blow the activated clay particles away to middle part. In the case of crashed wheat particles, saltating particles are observed together with the particles blown away by separated flow.

On the gently sloping gable roofs, the particles taken off through the air separated from windward edge are observed. Those particles are brought to the ridge of small model, and as for the large model; those particles fall on the upper part of windward side. On the steeply sloped gable roofs, the recirculation zone is developed above leeward side, and then the particles drawn into vortexes accumulate on leeward side. The crashed wheat particles, taken off through the air separated from ridge, fall on leeward edge of large model. And on the small model, those particles are blown away to outside of the roof.

5 CONCLUSION

The process of the formation of model snow accumulations on various shapes of roof is examined, according to the results of wind tunnel modeling experiment. Following knowledge on model snow accumulations on gable roof are obtained.

The activated clay particles accumulate on windward side almost uniformly, and the accumulation on leeward side takes shape of arch. The accumulation of crashed wheat particle on windward side is larger than that on leeward side.

The crashed wheat particles blown away by the air separated from ridge fall on leeward edge of large model, and, in the case of small model, those particles blown away to outside of model building.

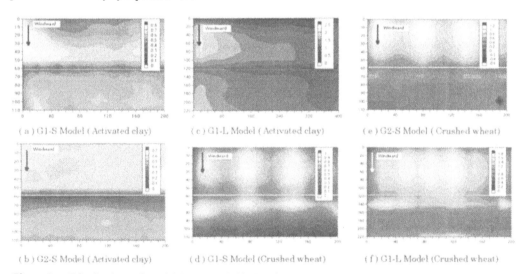

(a) G1-S Model (Activated clay)   (c) G1-L Model (Activated clay)   (e) G2-S Model (Crushed wheat)

(b) G2-S Model (Activated clay)   (d) G1-S Model (Crushed wheat)   (f) G1-L Model (Crushed wheat)

Figure 3   Distributions of Model Snow on Gable Roofs

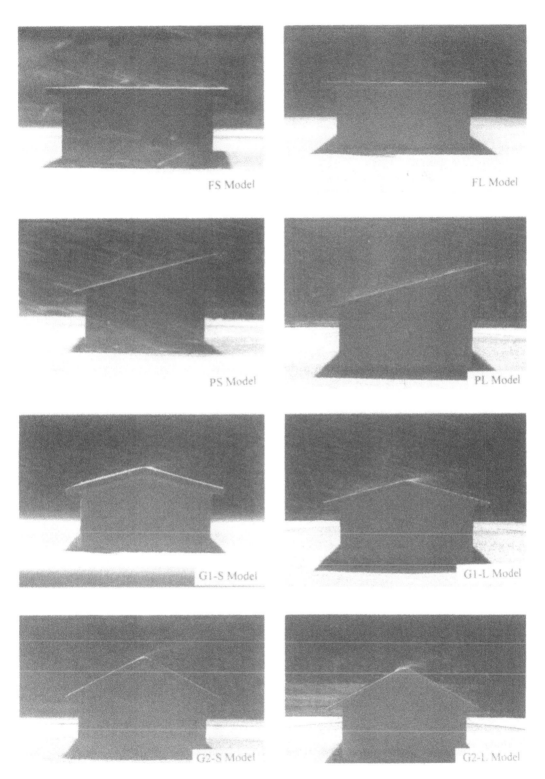

Figure 4   Movement of Model Snow Particles around Model Building   Crashed Wheat Particles

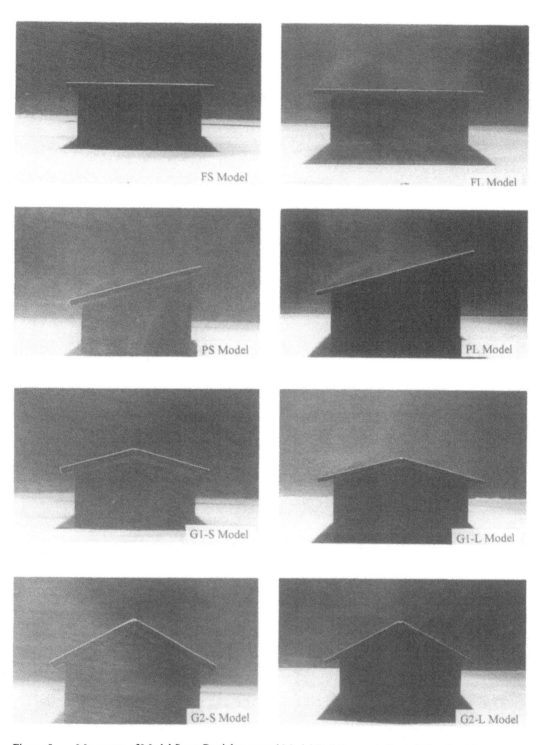

Figure 5   Movement of Model Snow Particles around Model Building   Activated Clay Particles

6 REFERENCES

Damaha Sant'Anna, F. and Taylor, D.A., 1990, Snow Drift on Flat Roofs: Wind Tunnel Tests and Field Measurements, Journal of Engineering and Industrial Aerodynamics, 34, 223-250

Isyumov, N., Mikitiuk, M. and Cookson, P., 1988, Wind Tunnel Modeling of Snow Drifting, Application to Snow Fence, First International Conference on Snow Engineering, 210-226

Isyumov, N., and Mikitiuk, M., 1992, Wind Tunnel Modeling of Snow Accumulations on Large-Area Roofs, Second International Conference on Snow Engineering, 181-193

Kind, R.J., Mechanics of Aeolian Transport of snow and Sand, Journal of Engineering and Industrial Aerodynamics, 36(1990),855-866

Petersen, R.L. and Cermak, J.E., 1988, Application of Physical Modeling for Assessment of Snow Loading and Drifting, First International Conference on Snow Engineering, 276-285

Snow Engineering, Hjorth-Hansen, Holand, Løset & Norem (eds) © 2000 Balkema, Rotterdam, ISBN 90 5809 148 1

Changes in the snowdrift pattern caused by a building extension
– Investigations through scale modelling and numerical simulations

Thomas K. Thiis
Narvik Institute of Technology, The University Courses on Svalbard, Norway

Christian Jaedicke
The University Courses on Svalbard, Norway

ABSTRACT: Snowdrifts around buildings can cause serious problems when formed on undesirable places. The formation of snowdrifts is highly connected to the wind pattern around the buildings, and the wind pattern is again dependent on the building design and the position of the surrounding buildings. The snowdrift pattern around two buildings in Ny-Ålesund, a small village on the arctic island Spitsbergen is investigated. An extension of one of the buildings is planned, and the effects of the new design on the snowdrift pattern is investigated. 1:10 scale models of the buildings, with and without the extension is placed in a valley with uniform wind direction during the winter. The change in the snowdrift pattern around the models, caused by the extension is observed. The upwind snowdrift position of the situation without the extension is compared and validated against measurements of the equivalent snowdrifts around the full-scale buildings. Numerical simulations of the snowdrifting and wind pattern around the different building configurations show, in some areas, a close similarity to the field measurements. The position of the lateral and upwind snowdrifts of both the buildings are well computed, and so is the centre snowdrift between the buildings. The scale-modelling and the numerical method can be applied to other building configurations of limited complexity to determine the optimum position and design of buildings in clusters.

1 INTRODUCTION

When planning buildings in snowy and windy areas, the position and size of the surrounding snowdrifts should be considered to avoid future problems. Also, when significant changes to the exterior design are made, the snowdrift problem should be reconsidered. Several methods have been applied to investigate snowdrift position and size around buildings. A common method is to place a scale model of the buildings in a wind tunnel, apply a granular material on the floor, and let the wind erode the granular material. This has been done with various degrees of sophistication from a full scale wind tunnel with artificial snow (Gandemer et.al., 1997) via boundary layer wind tunnel with a known scaled wind profile and activated clay or glass beads as the granular medium (Anno, 1987) to a simple cross-flow fan and semolina as granular medium. Also outdoor modeling of two dimensional snowdrifts behind snow fences has been tried (Tabler, 1980).

As computer costs are dropping, three dimensional numerical simulations of wind and snowdrift formation around buildings has become a feasible task. (Tominaga and Mochida, 1999; Thiis, 2000; Bang et. al., 1994; Kawakami et. al., 1997; Uematsu et.al., 1991). There are still unresolved problems in this field of work both when it comes to the precision of the wind pattern calculations and the implementation of the different transport modes of snow. The present paper presents data from outdoor modeling and numerical simulations of snowdrifts around two buildings. The results are validated against full-scale measurements of the same buildings.

Ny-Ålesund is a village on the arctic island Spitsbergen on 79° north. The main activity here is research, and because of increased activity, it is necessary to extend the existing mess-building. A CAD (Computer Assisted Design) model of the buildings before and after the extension is shown in figure 1. As can be seen from the figure, the extension will consist of a one level corridor encirceling the old building, a new entrance on the downwind wall and two new wings on the upwind side of the building. Upwind the buildings there is a smooth terrain with no major obstacles. The wind direction in Ny-Ålesund is fairly uniform from ESE and is indicated on the figure.

When planning buildings that will create a substantial change in the perception of the outdoor area, it is today usual to make a 3D CAD model and a computer visualisation of the area. Such a model should be as exact as possible, and if properly made it is possible to use it for other purposes than only the visualisation. In this study, 3D CAD models are imported into an CFD (Computational Fluid Dynamics) solver and the wind pattern and snowdrifting concentration around the buildings are computed.

2 FIELD STUDIES

2.1 Field setup

Tabler (1980) started doing outdoor modeling of snowdrifts around snow fences and snow fences in combination with terrain. He suggests that there will be a self similarity of the wind profile during blowing snow, as long as the geometric similarity exists. This is because the roughness height, z_0, varies approximately as the square of shear velocity, u_*^2, in saltating flow and also that the gusts of the wind will provide a wide range of windspeeds. The scaling of the snow particle paths will to some extent be fulfilled by natural sorting processes over the range of wind speeds during mitigating snow dunes. The wind direction must, however, be the same for both the model and the prototype.

Two 1:10 scale models, the mess-building with the planned new additions and the most upwind building in figure 1 were built in 5mm plywood. The models were placed in a wide valley with uniform wind direction with approximately the same incidence angle to the wind as the buildings in Ny-Ålesund. The frequency distribution of the wind direction for the year 1994 is shown in figure 2. The wind direction 130° was chosen as the representative wind direction. The mess-building model was installed first without any additions and then with all the planned enlargements. During the first experiment, blowing snow was allowed to accumulate around the scale buildings representing the present state. This was also used to evaluate the method, and compared to the snow accumulation around the real, full-scale buildings in Ny-Ålesund. The second experiment included the enlargements and was done to study the possible effects of the new building shape on the snowdrift pattern. Both experiments were started on a plane snow surface. Photographs and land measuring equipment were used to document all experiments. Pictures were taken before and after the experiments to allow direct comparison between the models without and with the accumulated snow. (Thiis et.al.,1999)

2.2 Field results

A contour map of the snowdrift pattern around the models of the buildings before the extension is shown in figure 3. A large center snowdrift between the buildings is clearly visible, together with the upwind and lateral snowdrifts of the two buildings. The center snowdrift causes a lot of problems every winter and develops rather fast. This is why the full

scale center snowdrift is frequently removed, and a comparison of the model and full scale snowdrifts is impossible. The distance between the upwind snowdrift and the upwind wall is therefore used as a validity check for the method. As long as it is the position of the snowdrifts, and not the development rate that is investigated, this should be an appropriate measure for the accuracy of the method. The distance is determined by the dimensions of the building and the vertical wind and turbulence profiles (Thiis and Gjessing, 1999). Table 1 shows the measured distances and the ratios. The ratios are both close to 10 and justifies the outdoor 1:10 scale modelling method.

On this basis it is reasonable to assume that a model run with the extensions on the mess-building, will produce a realistic snowdrift pattern. A contour map of the snowdrift heights after the building extension is shown in figure 4. The main differences in the snow pattern from the original configuration, are the position of the upwind snowdrift of the mess-building and the position of the center snowdrift. Because of the lowered upwind wall, the upwind vortex size is reduced and the snowdrift is positioned a lot closer to the building, approximately 6.5m at the most. The upwind vortex which was originally deflected downwind the sides of the building and creating the well known horseshoe vortex, is in this case too weak to keep the side near the center snowdrift clear of snow. It is not only the weakened horseshoe vortex that causes the burying of the mess-building wing. The wing is extended out in the lateral snowdrift of the upwind building which is causing the blowing snow concentration in this area to increase well above the upwind profile. Since both the snow concentration is increased and the airs ability to carry the snow is decreased near the wall, the snowdrift will grow close to the wall. The center snowdrift extends downwind the wing and buries the downwind wall of the wing. On the other side of the building on the downwind wall of the wing, the snow is accumulating almost up to the top of the roof. Also the most downwind wall on the building is collecting a downwind snowdrift. The length and height of this snowdrift seems to be slightly larger than on the original configuration.

Table 1: Distance from upwind walls to snowdrift on model and in full-scale.

	Full-scale [m]	1:10 Scale model [m]	Ratio
Upwind building	7.7	0.78	9.9
Mess-building	9.1	0.98	9.3

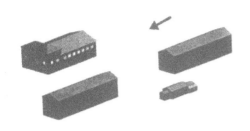

Figure 1 a) building configuration before extension, arrows indicate main wind direction

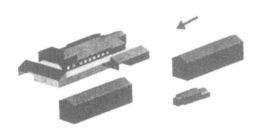

Figure 1 b) building configuration with extensions (CAD model by architect T. Ramberg)

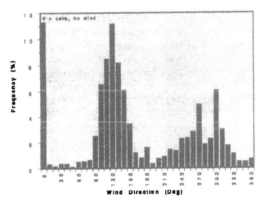

Figure 2 Frequency distribution of the wind direction in Ny-Ålesund in 1994

3 NUMERICAL SIMULATIONS

3.1 Setup

A finite volume CFD solver is used to solve the wind flow and blowing snow concentration around the buildings. The time averaged Navier—Stokes equations are closed with the k-ε turbulence model and solved for a three dimensional grid. The grid size is 90x90x25 cells with grid refinement near the surfaces. The full scale CAD models are converted to the stereolithograpy file format and imported into the flow solver. The inlet wind profile follows the logarithmic expression

$$u(z) = \frac{u_*}{\kappa} \ln\left(\frac{z}{z_0}\right) \qquad (1)$$

where $u(z)$ is the wind speed in height z, z_0 is the roughness of the surface, set to 0.002m, u_* is the friction velocity set to 0.6 m/s and κ is the von karman constant, equal to 0.4. The snow phase is coupled to the wind with the drift flux approximation which is also previously used for this purpose (Thiis, 2000; Flow Science, 1997; Bang et.al., 1994). The snow phase is here allowed to drift through the air phase with the relative velocity

$$u_r = D_f f_1 f_2 \left(\frac{\rho_2 - \rho_1}{\rho_1}\right)\frac{1}{\rho}\nabla p \qquad (2)$$

Here ∇p is the pressure gradient, f is volume fraction, ρ is density and the subscripts 1 and 2 denotes the two different fluids. The total mixture density is $\rho = f_1\rho_1 + f_2\rho_2$. D_f is a coefficient related to the friction between the snow phase and the air. It is found by balancing the drag force on a sphere with the buoyancy force, also known as Stokes form of viscous flow about a sphere.

$$D_f = \frac{2r_0^2}{9\nu} \qquad (3)$$

Here r_0 is the mean radius of the snow particles and ν is the kinematic viscosity of the air.
The viscosity of the air at -20 ° is set to $1.63*10^{-5}$ N*s/m² and the air density is set to 1.38 kg/m³. The density of the snow particles is set to 400 kg/m³. To slow down the snow phase in areas of high snow concentration and/or low shear stress, an apparent viscosity is applied to the snow phase. The apparent viscosity is found from the threshold friction velocity of the steady state snowdrift profile, u_{*t} and the expression (Thiis, 2000)

$$\mu_{app} = \rho_{air} u_{*t}^2 \left(\frac{dz}{du}\right)_{surface} \qquad (4)$$

the threshold friction velocity is set to 0.2 m/s and the inverse vertical wind profile is found from equation 1.

Table 2: Simulated distance from upwind walls to snowdrift, without building extensions.

	5° incidence angle [m]	15° incidence angle [m]
Upwind building	7.0	6.1
Mess-building	10.9	10.7

The method does not consider the building of a new snow surface which will change the wind flow. The snow phase is slowed down, and blowing snow density increased where the surface shear stress is low. The increased blowing snow density will again slow down the wind speed. The presented result must therefore only be considered an initial snowdrift growth area.

The simulations are performed for two directions to show the importance of choosing the right wind direction. A 15° incidence angle represents the 1:10 scale model runs whilst a 5° incidence angle represents the highest frequency in the wind statistic.

3.2 Results

Figure 5 shows the computed area of initial snowdrift growth around the buildings before the extension. It can be seen that it is only the lateral and upwind snowdrifts that are computed. The distances from the upwind walls to the upwind snowdrifts measured at the center of the building are shown in table 2. Compared to the field measurements this distance is slightly under estimated for the most upwind building and over estimated for the mess-building. The difference in snow pattern when changing the wind direction is mainly connected to the center snowdrift. For the 15° incidence angle, this snowdrift is shorter, positioned further upwind and closer to the mess-building compared to the 5° incidence angle simulation. Compared to the 1:10 scale model experiment, the center snowdrift is simulated to be positioned too far upwind.

The results from the simulations of the case including the extensions are shown in figure 6. It can be seen that the upwind snowdrift is positioned closer to the mess-building than for the case without the extensions. The center snowdrift is also positioned closer to the new wing than the old wall. This is also the case for the lateral snowdrift on the opposite side of the building. All these features are recognized in the 1:10 scale modeling. The position of the upwind snow drift of the most upwind building shows, as expected, no major difference from the simulations of the case with no building extensions. The distances from the walls to the upwind snowdrifts are presented in table 3. The 15° incidence angle simulation produces a center snowdrift which is shorter than the 5° incidence angle run. This coincides with the simulations without the building extensions, and is possibly caused by the increased sheltering effect from the most upwind building.

Table 3: Simulated distance from upwind walls to snowdrift, with building extensions.

	5° incidence angle [m]	15° incidence angle [m]
Upwind building	6.9	7.6
Mess-building	8.7	9.0

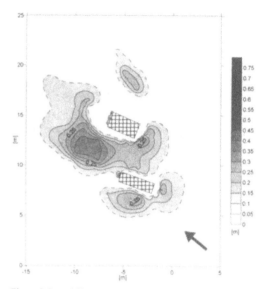

Figure 3 Snowdrift pattern around 1:10 scale model of the mess-building in Ny-Ålesund

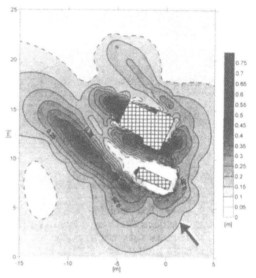

Figure 4 Snowdrift pattern around 1:10 scale model of the mess-building in Ny-Ålesund with extensions

4 CONCLUDING DISCUSSION

The comparison of the field experiments to the full scale measurements supports the theoretical basis of outdoor modeling. It is however crucial that the incidence angle of the wind on the model and the prototype is the same within a few degrees. This can be difficult to obtain in the field because of the relatively short testing time. The models are of such a size and weight that they must be installed in calm weather. It is then difficult to know the exact wind direction of the coming wind episode. Heavier models could be applied to be able to set up the experiment in higher wind speeds, but this would again increase the need for labour and maybe require heavy machinery. Tabler (1980) used 1:30 scale models in his outdoor modeling. In these experiments he used an ice covered lake as surface, which has a smoother surface than the snow surface in the presented experiments. As the roughness increases, the size of the scale models must also increase to maintain similarity of the wind profile.

The numerical simulations computes only the initial snowdrift area and does not consider the new surface of the developing snowdrift. Previous studies have shown that the leeward snowdrifts of the buildings seems to have a slower development rate than the upwind and lateral snowdrifts. This is the reason why the leeward snowdrifts are not computed. The fact that the simulations does not compute any new surface, will also prohibit the burying of any structure, and the development of a new wind pattern around the structure. This is the case for the new wing on the mess building which is closest to the center snowdrift. It is believed that the burial of the wing will reduce the upwind vortex capacity to remove snow, and that at the moment of burial, the upwind snowdrift will advance closer to the wall. The simulated over estimation, compared to the 1:10 scale modeling, of the distance between the upwind wall and snowdrift on the extended mess-building might be due to this phenomenon.

To be able to determine the distances between the

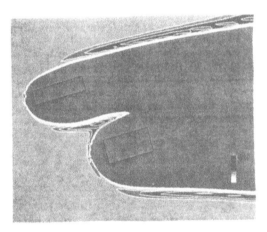

Figure 5 a) Contour map of the blowing snow surface around the buildings without extensions, ρ=1.43 kg/m^3, colour scale in meters, incidence angle is 15°

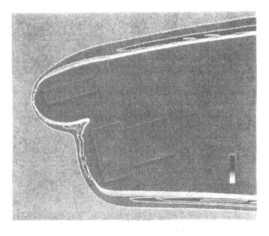

Figure 6 a) Contour map of the blowing snow surface around the buildings including extensions, ρ=1.43 kg/m^3, colour scale in meters Incidence angle is 15°

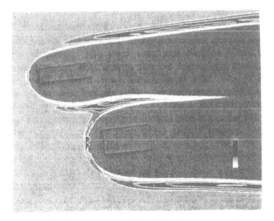

Figure 5 b) Contour map of the blowing snow surface around the buildings without extensions, ρ=1.43 kg/m^3, colour scale in meters, incidence angle is 5°

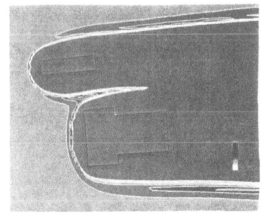

Figure 6 b) Contour map of the blowing snow surface around the buildings including extensions, ρ=1.43 kg/m^3, colour scale in meters Incidence angle is 5°

upwind snowdrifts and the building walls are of great interest when planning buildings in clusters. The presented numerical method of gives a good basis for planning building clusters in relation to the snow deposition. The outdoor modeling is a good, but labour intensive metode of determining the snowdrift pattern around buildings. This is especially the case when the interaction of several buildings is investigated.

REFERENCES

Anno, Y., 1987, CRREL's Snowdrift Wind Tunnel, *Cold Regions Technology Conference '87*

Bang, B., Nielsen, A., Sundsbø, P.A. and Wiik, T., 1994, Computer Simulation of Wind Speed, Wind Pressure and Snow Accumulation around Buildings (SNOW-SIM), *Energy and Buildings*, 21, 235-243.

Flow Science Inc., 1997, Flow 3D User's Manual, *Flow Science Inc., Los Alamos.*

Gandemer, J., Palier, P., Boisson-Kouznetzoff, S., 1997, Snow simulation within the closed space of the Jules Verne Climatic Wind Tunnel, *Snow Engineering: Recent advances, Izumi, Nakamura & Sack (eds)*, Balkema, Rotterdam. ISBN 90 54108657

Kawakami, S., Uematsu, T., Kobayashi, T., Kaneda, Y., 1997, Numerical study of a snow wind scoop, *Snow Engineering: Recent advances, Izumi, Nakamura & Sack (eds)*, Balkema, Rotterdam. ISBN 90 54108657

Tabler, R.D. 1980, Self-similarity of wind profiles in blowing snow allows outdoor modeling, *Journal of Glaciology*, vol. 26, p. 421-432

Thiis, T. K., Gjessing, Y. 1999, Large-Scale measurements of snowdrifts around flat roofed and single pitch roofed buildings, *Cold Reg. Science and Tech.*, 30 1-3. pp. 175-181

Thiis, T. K., Jaedicke, C., Dahl-Grønvedt, M., Johnsen, J., 1999, The new mess building in Ny-Ålesund – Effects on snowdrifts, *Meteorological report series 5-99*, University of Bergen

Thiis, T., K., 2000, A comparison of numerical simulations and full-scale measurements of snowdrifts around buildings, *Wind and Structures*, vol 3.

Tominaga, Y., Mochida, A., 1999 CFD prediction of flow field and snowdrift around a building complex in a snowy region, *J. Wind Eng. and Ind. Aerodynamics* 81 pp.273-282.

Uematsu, T., Nakata, K., Takeuchi, K., Arisawa, Y., Kaneda, Y., 1991, Three-dimensional numerical simulation of snowdrift, *Cold Reg. Science and Tech.*, 20, pp.65-73.

Snow Engineering, Hjorth-Hansen, Holand, Løset & Norem (eds) © 2000 Balkema, Rotterdam, ISBN 90 5809 148 1

The snowdrift pattern around two cubical obstacles with varying distance – Measurements and numerical simulations

Thomas K. Thiis
Narvik Institute of Technology, The University Courses on Svalbard, Norway

Christian Jaedicke
The University Courses on Svalbard, Norway

ABSTRACT: Snowdrifts around buildings can cause serious problems when formed on undesirable places. The formation of snowdrifts is highly connected to the wind pattern around the building, and the wind pattern is again dependent on the building design and the position of the surrounding buildings. The snowdrift pattern around two 2.5m cubes, positioned close to perpendicular to the wind direction is investigated. The spacing between the cubes was varied to investigate the effects of a local wind speedup between the cubes. The measurements show that a large snowdrift forms between the cubes with the largest spacing between them. A smaller spacing between the cubes results in the disappearance of the centre snowdrift. The reason for this is that the small spacing causes the formation of one horse shoe vortex encirceling both cubes and decreasing the amount of snow entering the zone between the cubes. The wind speedup between the cubes will also sweep the area clean of snow. The upwind wind direction, the vertical wind speed profile and the blowing snow flux is measured during the experiments. The experiments were conducted in a valley 3 km wide and 20 km long on Spitsbergen, Norway. The wind in this valley is blowing in the same direction approximately 90% of the time during winter and the site is well suited for studies of snowdrifts and snowdrifting. The different experiment configurations are investigated through numerical simulations, which confirms the measurements. The numerical method can be applied to other building configurations of limited complexity to determine the optimum position of buildings in clusters.

1 INTRODUCTION

The design of buildings in windy and snowy areas requires consideration of the wind transported snow and the snowdrifts formed thereof. The topic of snowdrift formation has previously been studied by the means of full scale experiments, wind tunnel simulations and numerical analysis. Full scale measurements is sparsely reported, however, Haehnel and Lever (1994) have compiled field data from several sources and assessed the properties regarding model validation. Thiis and Gjessing (1999) studied the position and relative size of snowdrifts around cubical obstacles with different rooftops and emphasised on the snowdrifts formed upwind and at the sides of the obstacles.

Numerical analysis of snowdrifts around buildings has been performed by i.a. Tominaga and Mochida (1999), Thiis (2000), Bang et. al. (1994), Kawakami et. al. (1997), Uematsu et.al. (1991). Common for these studies is that the wind field is determined by solving the time averaged Navier-Stokes equations and the blowing snow is then coupled to the wind field. The differences lays mainly in the applied turbulence model and the coupling of the snow phase.

A widely used turbulence model is the k-ε model. Even if this model is known to calculate too high turbulence intensity near impinging surfaces and re-attachment zones, it produces realistic wind patterns around buildings.

Snow can be transported in three different modes; creep, saltation and suspension. The creep mode is because of its limited transported snow quantity, usually neglected when considering mass transport. Saltating particles follows jumping particle paths in the lower tens of centimetres near the surface. Suspension transport mode is initiated when the saltating snow particles are picked up by the wind and transported in the boundary layer without contact with the surface.

A key property of the snow cover when considering blowing snow is the snow threshold friction velocity, u_{*t}. This is the friction velocity of the wind when particle ejection from the surface cease, and it is ranging from 0.07 to 0.25 m/s for fresh, dry snow and from 0.25 to 1 m/s for old, wind hardened snow (Kind, 1981). Assuming that the vertical wind profile follows the widely used logarithmic expression, the friction velocity of the wind, u_*, can be found from

$$u_* = u(z)\kappa / \ln\left(\frac{z}{z_0}\right) \qquad (1)$$

where $u(z)$ is the wind speed in height z, κ is the von karman constant and z_0 is the roughness, deduced from wind speed measurements in two heights, and the expression

The wind pattern around a cube is described by Castro and Robins (1977) and Martinuzzi and Tropea (1993). They describe a recirculation zone up-

$$z_0 = \exp\left(\frac{u(z_2)\cdot\ln(z_1) - u(z_1)\cdot\ln(z_2)}{u(z_2) - u(z_1)}\right) \qquad (2)$$

wind the cube forming between the stagnation point on the upwind wall and the surface. This recirculation zone is extending upwind to the upwind stagnation point and is deflected downwind the sides of the cube, creating a horseshoe shaped system of vor-

$$R = 1.6\sqrt{ah} \qquad (3)$$

tices. Downwind the cube, between the horseshoe vortices, there is a wake region extending downwind to the downwind reattachment point. Beranek (1984) suggested two rules for the determination of the influence area around buildings of rectangular shape. A circle with the radius

can be drawn trough the front and rear stagnation points. The distance from the centre of the circle , M, to the windward face of the building is then found from

$$e = 0.9\sqrt{ah} \qquad (4)$$

here h and a is respectively the height and width of the windward wall. The expressions are based on flow visualisation in wind tunnel experiments, and are valid for $1.22 > h/a > 0.33$ and certain upwind wind and turbulence intensity profiles. It is said that buildings with no overlap of their influence area will not interact. For cubical obstacles this means a distance of $d = 2.5h$. Beranek also suggests that if the distance between the buildings is small, the two buildings begin to act as one building with a long windward wall. They develop a common horseshoe shaped vortex system encirceling both buildings, and not separate systems of vortices between the buildings.

It has been observed that a large centre snowdrift is forming between buildings with a spacing between them. However, if the spacing is under a certain limit, this snow drift does not form. It is believed that due to the funnel shape of the passage between the buildings, the wind speed is increased and is sweeping the passage free from snow.

The phenomenon of centre snowdrift between buildings is in this report studied in detail by the means of field experiments and numerical analysis.

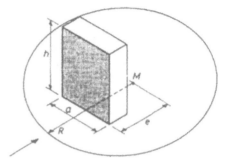

Figure 1. Influence area around buildings of rectangular shape (Beranek, 1984)

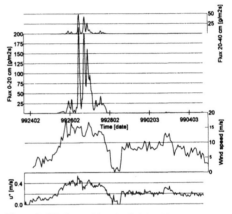

Figure 2. Wind speed in 10m height, friction velocity and snow drifting flux in heights 0-20cm and 20-40 cm during case A

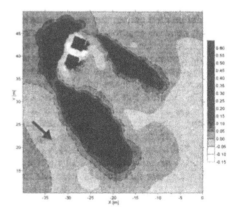

Figure 3. Contour map of snowdrift pattern, and main wind direction in case A

Table 1 Overview of the experiment setup

	Case A	Case B	Case C
Time	10 days	11 days	14 days
Distance btw. cubes	1.7m	4.8m	5.9m

370

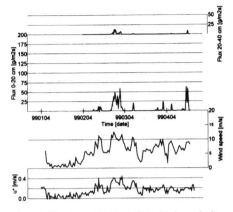

Figure 4. Wind speed in 10m height, friction velocity and snow drifting flux in heights 0-20 cm and 20-40 cm during case B

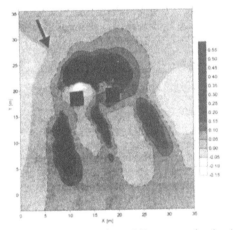

Figure 5. Contour map of snowdrift pattern, and main wind direction around case B

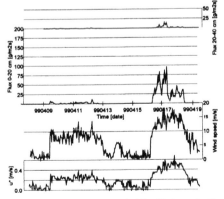

Figure 6. Wind speed in 10m height, friction velocity and snow drifting flux in heights 0-20 cm and 20-40 cm during case C

Since the deposition of the centre snowdrift is, among other factors, dependent upon the wind speed between the buildings, it is evident that the incidence angle of the wind has influence on the formation of the centre snowdrift. In the field experiments of this study, there is a variation of the incidence angle of the wind which makes it difficult to make exact comparisons to Beraneks work, however, the main features of the flow remain.

2 FIELD SETUP

The winter of 1999 a series of experiments were performed in Adventdalen on the arctic island Spitsbergen, Norway. Three of the experiments considered the formation of the centre snowdrift forming between two buildings with varying distance. Two 2.5m cubical shaped obstacles were placed on the flat, uniform surface of a wide valley. The wind pattern in this valley has previously been found to be mainly down the valley during winter. The yearly precipitation in the valley is very low and the thickness of the snow cover was approximately 200 mm of hard wind packed snow. The small amount of precipitation combined with the uniform wind direction makes the area well suited for snowdrift experiments because accumulated, wind transported snow show a very distinct pattern compared to areas with more precipitation. The incidence angle of the wind on the cubes in the tree different experiments was varying from 15 to 26 deg.

After some days of snowdrifting, significant snowdrifts were developed around the cubes, and the experiment was moved upwind to avoid disturbance from the previous experiment. The distance between the cubes was altered, and a new snowdrift was developed. This was repeated again with a new spacing. An overview of the experiments is presented in table 1. After each experiment, the snowdrift's size and shape were measured with a surveying total station. The Kriging method was used to interpolate a surface between the measured points. The deposited snow was very compact with a surface of high hardness.

The upwind wind profile and direction was measured in four heights up to 10m on an automatic weather station. The wind speed sensors were of the cup anemometer type and the sampling interval was ten minutes. The presented plots shows the 1 hour average of the wind speed measurements. The upwind snow drifting flux was measured in two averaged levels, 0-20cm and 20-40cm above the surface. The instrument in use was the FlowCapt sensor from IAV Engineering (Chritin et.al., 1999). The sampling interval was 15 seconds with the 1 hour averaged values presented in the plots. The detection principle of this sensor is based on a mechanical-acoustical coupling. It consists of closed tubes containing electro-acoustical transducers. The signal from the transducers is filtered and a signal proportional to the particle flux is produced.

2.1 Case A

Wind speed, friction velocity and snow drifting flux during case A is shown in figure 2. The main snow flux occurred between 25. and 28. february. It can be seen from the friction velocity plot and the lower snow drifting flux plot that the threshold friction velocity is approximately 0.20 m/s.

Figure 3 shows the snow drift pattern around the two 2.5m cubes in case A. The distance between the cubes was 1.7m and incidence angle was approximately 26°. The upwind snowdrift is continuos transverse of the wind direction, with no local minimum height. This, together with the fact that the centre snowdrift is missing, indicates that there is only one horse shoe vortex, encirceling the two cubes. The maximum distance between the upwind snowdrift and the cubes is found on the most upwind cube and is approximately 1.6m, measured perpendicular to the upwind wall. The distance between the upwind snowdrift and the other cube is approximately 1.35m. Just downwind of the cubes, two leeward snowdrifts have developed. The development rate of the leeward snowdrifts seems to be much smaller than the snowdrifts upwind and on the sides of the building. This is probably because the flux of snow into the wake area is low compared to the surrounding area of the building. The snow is ploughed to the sides and prevented from entering the low surface shear stress area downwind of the cubes.

2.2 Case B

Wind and snow drifting flux data during case B is shown in figure 4. Just ahead of the experiment there was a small snowfall making the threshold friction velocity for snow drifting slightly lower than in case A, approximately 0.18 m/s. The main snow transport occurred between the 2. and the 3. of april. The resulting snowdrift pattern from case B is shown in figure 5. The distance between the cubes was 4.8m and the incidence angle of the wind was approximately 16°. Here, there is a tendency for deposition of a centre snowdrift. The upwind snowdrift have two local maximum heights just in front of each cube, indicating the presence of two horse shoe vortices. However, there is no snow deposition just between the cubes, the centre snowdrift starts just downwind the cubes and extends parallel to the lateral snowdrifts. The centre snowdrift is slightly higher than the lateral snowdrifts, but the downwind extension is smaller than for the two lateral snowdrifts. The distance between the upwind snowdrift and most upwind cube is approximately the same as for the case A, 1.6m and slightly less for the other cube.

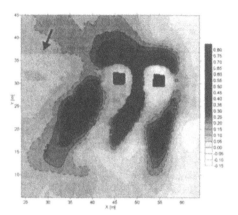

Figure 7. Contour map of snowdrift pattern and main wind direction in case C

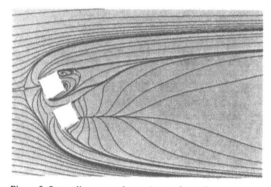

Figure 8. Streamlines around case A near the surface.

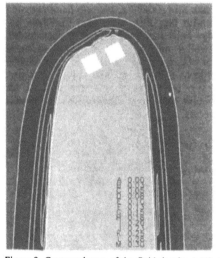

Figure 9. Contoured map of the fluid density 1.41kg/m³, case A.

2.3 Case C

Wind and snow drift flux data during case C is shown in figure 6. It can be seen that the snowdrifting occurred in two episodes, 9-13.apr. and 16-18.apr. the threshold friction velocity was approximately 0.20 m/s for the first episode, and slightly larger for the second episode. The snowdrift pattern around case C is shown in figure 7. Here a distinct centre snowdrift developed during the first wind episode, and which was enlarged during the second episode. The height of the centre snowdrift was 0.80m which is considerably larger than for the two lateral snowdrifts. The presence of two horseshoe vortices is evident in that the snowdrift pattern looks similar to the ones forming around two single cubes with the two adjacent lateral snowdrifts joining together in one large centre snowdrift. The centre snowdrift does, however, not reach as far downwind as any of the lateral snowdrifts. The distance between the cubes was 5.9m and the incidence angle of the wind was 17°. The distance between the cubes and the upwind snowdrift was approximately 1.6m.

3 NUMERICAL SIMULATIONS:

To simulate the wind field and snow drifting around the cubes, a general purpose finite volume CFD code was applied. The CFD code solves the incompressible, time averaged Navier-Stokes equations, using a k-ε turbulence model to close the equations. The turbulence model is known to compute excessive turbulence near the edges of the windward walls, which also can affect the downwind wakes. It is however widely used, and it is capable to produce realistic wind patterns around buildings.

The computational domain is 85x60x15m and consists of 90x89x23 grid cells with grid refinement near the obstacles. The inlet vertical wind velocity profile to the computational domain follows the logarithmic expression from equation 1. The roughness height z_0 is set to 0.002m, which is equal to the upwind measurements, κ is set to 0.4 and u_* is equal to 0.6 m/s. This corresponds to a wind velocity of approximately 12 m/s at a 10 m height. The turbulence intensity on the inlet boundary is set to 0.05. This conserves the vertical wind profile through the computational area which is not affected by the cube. The outlet boundary is continuative, meaning that the normal derivatives of all quantities are set to zero. On the surface boundaries, a law-of-the wall velocity profile is assumed. The lateral and top boundaries have symmetry conditions.

The movement of the snow phase relative to the wind is found using the drift-flux approximation (Thiis, 2000; Flow Science, 1997; Bang et.al., 1994)

$$u_r = D_f f_1 f_2 \left(\frac{\rho_2 - \rho_1}{\rho_1} \right) \frac{1}{\rho} \nabla p \qquad (5)$$

Here ∇p is the pressure gradient, f is volume fraction, ρ is density and the subscripts 1 and 2 denotes the two different fluids. The total mixture density is $\rho = f_1 \rho_1 + f_2 \rho_2$. D_f is a coefficient related to the friction between the snow phase and the air, found by balancing the drag force on a sphere with the buoyancy force.

$$D_f = \frac{2r_0^2}{9\nu} \qquad (6)$$

Here r_0 is the mean radius of the snow particles and ν is the kinematic viscosity of the air.

Other properties used in the simulations are the density and the dynamic viscosity of air at −20 °C which is set to 1.38 kg/m³ and 1.63*10⁻⁵ N*s/m² respectively, the density of snow particles, set to 400 kg/m³ and the apparent threshold viscosity of the snow phase. This property is defined to be

$$\mu_{app} = \rho_{air} u_{*_t}^2 \left(\frac{dz}{du} \right)_{surface} \qquad (7)$$

(Thiis, 2000) where u_{*_t} is set to 0.2 [m/s] and the vertical wind speed gradient is found using equation 1, substituting u_* with u_{*_t}. This results in an apparent viscosity of μ_{app}=2.2*10⁻⁴ N*s/m². The mixture viscosity in the computational cells is evaluated as a volume fraction weighted average of the two constant viscosities. The vertical blowing snow density profile on the inlet is equal to the measured profile of a situation with a 10m wind speed of 12 m/s.

The simulation model calculates the snowdrifting density around the buildings, increasing the viscosity were the snow density is high. A high viscosity slows down the air velocity and increases the density even more. Since the model does not create a new surface, the pattern showed is only the initial snow deposition area which is determined by the surface shear stress and the blowing snow density. It is assumed that the relative size of the snowdrifts can be suggested on this basis.

3.1 Case A

Figure 8 shows the streamlines around case A near the ground surface. It can be seen that the streamlines form only one horseshoe vortex encirceling both cubes. The vortex acts like a canopy and prevents snow from entering the space between the cubes.

This can also be seen on figure 9 where the simulated snowdrifting is shown. The snowdrifts are expressed as iso-surfaces of density. The height of the iso-surface is shown as a contoured map. The two lateral snowdrifts show increased height downwind of the cubes with slightly higher contours of the snowdrift near the most downwind cube. The positions of the downwind maximum heights are however not very accurately calculated and located too far downwind. The reason for this might be the turbulence model calculating too much turbulence near the cubes, thus extending the wake area. The upwind snowdrift shows the same variation as the measured data with a slightly larger distance between the most upwind cube and the snowdrift than the neighbouring cube. Assuming that the edge of the upwind snowdrift is positioned at the upwind edge of the horseshoe vortex, the distances from the cubes to the snowdrift are 2.3m and 1.8m respectively.

3.2 Case B

The simulations of the wind pattern around case B is shown in figure 10.

The streamline visualisation show two separate sets of horseshoe vortices and corresponding wake areas. This corresponds to the assumption made on the basis of the presence of an observed centre snowdrift. The snowdrifting calculations on figure 11 also show increased snow density between the cubes. The position of this increased snow density area is between the cubes, where the measurements show no snow deposition at all. The measured centre snowdrift is positioned further downwind, and the area between the cubes might be submitted to erosion. However, the computed position of the two upwind snowdrifts show reasonable agreement with the measurements. The distances are 1.8m and 1.7m respectively, which is considerably better than for case A. The lateral snowdrifts are calculated too far downwind, again an effect of the extended wake caused by excessive turbulence generation near the upwind walls.

3.3 Case C

The streamlines of the flow around case C is shown in figure 12. As in case B, two horseshoe vortices have developed.

The snowdrifting calculations in figure 13 shows a higher iso-surface with a larger base area between the cubes than in case B. This is probably because while the amount of snow entering the area is somewhat equal for the two cases, the surface shear stress is lower in case C, caused by of the wider opening between the cubes. This corresponds well with the field measurements.

The upwind snowdrifts also show a variation that corresponds to the measurements in that the most downwind cube have a higher upwind snowdrift. The distances from the cubes to the snowdrifts are 2.0m for the most upwind cube and 1.7m for the neighbouring cube. These distances are measured in the field experiment to be approximately equal to 1.6m for both the cubes. The downwind lateral snowdrifts are also here positioned too far downwind, but the variation in height is reasonably well computed in that the snowdrift near the most downwind cube is higher than the other lateral snowdrift.

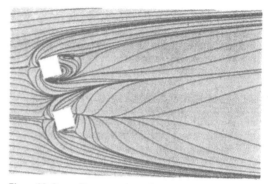

Figure 10. Streamlines around case B

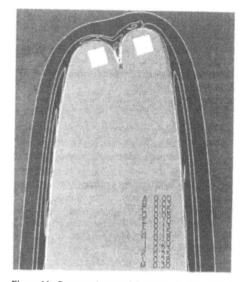

Figure 11. Contoured map of fluid density 1.41 kg/m³ around case B

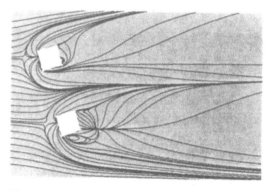

Figure 12. Streamlines around case C.

374

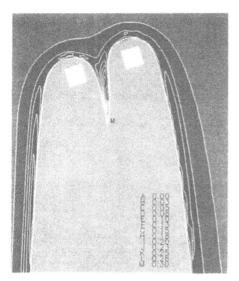

Figure 13. Contoured map of density 1.41 kg/m³ around case C

4 CONCLUDING DISCUSSION:

The measurements show that the distance between buildings are of vital importance when determining the presence of a centre snowdrift. The funnel effect of the passage between the buildings increases the wind velocity in that area, but since the snowdrifting flux also increases, the carried snow exceeds the carrying capacity of the air, leading to accumulation. When the distance between the buildings is under a certain limit, the horse shoe vortex extends around both buildings. The upwind part of the horse shoe vortex obstructs the transport of snow between the buildings and little snow is entrained in the common wake. In this case, there will be no development of a centre snowdrift. This is also confirmed by the simulations.

The measurements and the numerical simulations show some degree of similarity. The position of the upwind snowdrifts are reasonably well computed, and the presence of a centre snowdrift is possible to interpret from the streamline visualisations. However, the differences between simulations and measurements indicates that the wind pattern and the snow phase concentration around the cubes can be more accurately computed. The reason for the inaccuracies may be the previously mentioned inadequacies in the turbulence model, or that the viscosity of the air/snow mixture satisfy a different relation than the applied one.

The numerical method can be used to determine the best configuration of several building configurations, but is also able to suggest the degree of problems concerning snowdrifts around one specific layout.

REFERENCES

Bang, B., Nielsen, A., Sundsbø, P.A. and Wiik, T. 1994, Computer Simulation of Wind Speed, Wind Pressure and Snow Accumulation around Buildings (SNOW-SIM), *Energy and Buildings*, 21, 235-243.

Beranek, W., J. 1984, Wind environment around single buildings of rectangular shape, *HERON*, 29, no.1.

Chritin, V., Bolognesi, R., Gubler, H. 1998 FlowCapt: A new acoustic sensor to measure snowdrift and wind velocity for avalance forcasting, *Cold Reg. Science and Tech.*, 30, pp.125-133.

Flow Science Inc. 1997, Flow 3D User's Manual, *Flow Science Inc., Los Alamos.*

Haehnel, R. B., Lever, J. H. 1994, Field measurements of Snowdrifts, *Proc. ASCE/ISSW Workshop on Physical Modelling of Wind Transport of Snow and Sand.* Snowbird.

Kawakami, S., Uematsu, T., Kobayashi, T., Kaneda, Y. 1997, Numerical study of a snow wind scoop, *Snow Engineering: Recent advances*, Izumi, Nakamura & Sack (eds), Balkema, Rotterdam. ISBN 90 54108657

Kind, R. J. 1981, Snowdrifting, *Handbook of snow, Principles, Processes, Management and use*, pp. 338-359 Gray, D. M., Male, H. ed.

Thiis, T. K., Gjessing, Y. 1999, Large-Scale measurements of snowdrifts around flat roofed and single pitch roofed buildings, *Cold Reg. Science and Tech.*, 30, pp.175-181.

Thiis, T., K., 2000, A comparison of numerical simulations and full-scale measurements of snowdrifts around buildings, *Wind and Structures*, vol 3.

Tominaga, Y., Mochida, A., 1999 CFD prediction of flow field and snowdrift around a building complex in a snowy region, *J. Wind Eng. and Ind. Aerodynamics*, 81 pp.273-282.

Uematsu, T., Nakata, K., Takeuchi, K., Arisawa, Y., Kaneda, Y. 1991, Three-dimensional numerical simulation of snowdrift, *Cold Reg. Science and Tech.*, 20, pp.65-73.

Snow Engineering, Hjorth-Hansen, Holand, Løset & Norem (eds) © 2000 Balkema, Rotterdam, ISBN 90 5809 148 1

Snow drift control in residential areas – Field measurements and numerical simulations

P.A. Sundsbø
Department of Building Science, Narvik Institute of Technology, Norway

B. Bang
Department of Engineering Design, Narvik Institute of Technology, Norway

ABSTRACT: Snow drift in residential areas is mainly a result of an interaction between wind, snowfall, terrain and buildings. Snow drift control is in this context a matter of understanding these parameters and consider them in planning and development. Before going into detailed planning, the dominating snow drifting winds and following main deposition and erosion zones have to be mapped, whether the snow drift problem concerns new development or existing residential areas. This paper considers numerical modelling and field measurements and observations of the dominating snow drifting wind over residential areas in Hammerfest, a town with heavy snow drifts located in the northern part of Norway. The resulting wind map is based on a simulation of the wind conditions over a 17 km^2 area, with hilly and complex topography. A large set of field data consisting of measured snow drift directions, statistical wind data and snow drift observations is used as basis and verification of the simulation. The numerical result was close to observed wind conditions. The three dimensional simulation is performed using Flow-3D, a commercially available computational fluid dynamics (CFD) program.

1 INTRODUCTION

In most snowy regions, the combination of wind and snowfall often leads to unwanted snow depositions in lees created by obstacles or other places where wind reduces its transport capacity. Snow transport is mainly driven by this interaction between wind, topography and vegetation, but also interactions between moving snow particles, humidity, temperature etc. effect the overall transport. In all cases wind is the primary snow drift parameter (Pomeroy & Gray 1997).

Special considerations for snow drift are often necessary during planning of new residential areas in snowy regions. Groups or clusters of buildings seem to be more effective in reducing the wind and collect more drift snow than single, or more widely spread residences. Snow drift control in residential areas is often achieved by a combination of shielding, appropriate house design and well planned localisation. An outer shield with combinations of collector snow fences, embankments and vegetation, can be used to reduce the amount of drifting snow into the residential area. Within residential sections, an inner shelter system consisting of smaller snow drift measures may be effective in preventing undesirable drifts forming. There are also considerations that have to be made regarding snow clearance and storage. Before making any snow controlling efforts it is most important to identify and map the dominating snow drifting winds.

1.1 *Wind map for dominating winter winds*

Winds are heavily affected by local topography and buildings, and it may be appropriate to establish a wind map for the most dominating snow drifting directions. For this purpose all available information about wind and snow drift conditions should be used. In this work we have used computational fluid dynamics (CFD) to calculate a wind map. Statistical wind data is used together with field measurements and field observations to support the numerical approach. A well founded wind map is a good basis for snow drift control of existing settlements or for localising of new residential zones. It is also possible to evaluate fetch distances, the upstream contributing distances where snow is drifting from (Takeuchi 1980). This is important since long contributing distances might lead to large amounts of drifting snow and large snow drifts in downstream lee zones.

This paper presents the method used for making a wind map of dominating snow drift conditions in Hammerfest, which is a part of a larger snow drift analysis for Hammerfest council.

1.2 *Hammerfest*

Hammerfest is a town in the northern part of Norway with a population of about 6000. The region is known for its harsh climate with long and cold winters and therefore the original town centre was naturally located in a bay sheltered from the windy climate. Then the town expanded into regions where winds and snow drift have become a major problem. There is about 6 km of snow fences in Hammerfest and most of them are built as an outer shield for the residential areas in Fuglenesdalen and Baksalen, or to prevent avalanches from Salen, (Fig. 3). Many of the snow fences are in poor technical condition, others seem to be misplaced and/or do not work as intended.

Development of Fuglenesdalen, the valley located north-east of Hammerfest centre, started in 1974 (Børve 1987, Apeland 1997, Norem in press). During the period from 1976-1990 there has been a number of research and cold climate development projects in Fuglenesdalen, but snow drift is still a major problem. It may take about 3 weeks to clear some of these residential areas of snow after a snowstorm, and during that time a new snow storm may occur. Baksalen is another area with large snow drift problems, (Fig. 3).

Annual costs for snow removal and winter maintenance in Hammerfest are high and local authorities are now trying to achieve better snow drift control.

2 FIELD MEASUREMENTS

An analysis of snow drift conditions in Hammerfest was recently performed and four categories of field measurements where evaluated:

1. statistical winter wind data from The Norwegian Meteorological Institute (DNMI) 1958-99
2. measurement of snow drifting wind directions behind 74 snow fences
3. field observations of snow drifting wind directions by Børve (Børve, 1987)
4. field observations of snow drift in residential areas, 1992-99

Statistical wind data has been collected at Fruholmen lighthouse, at the local airport and at Hammerfest radio in the south-east end of Fuglenesdalen. Wind data from Fruholmen is more or less unaffected by local topography and indicates a dominating winter wind direction from south south-east. These results correlate with field measurements at Salen and Baksalen, (Fig. 3). Statistical wind data from Hammerfest airport and Hammerfest radio indicates that the dominating wind turns from south south-east to south-west, as it enters the valley led by the valleysides. This fact is supported by snow drift measurements behind collector fences.

After a considerable amount of field work, a snow shielding plan for Hammerfest council was established, where the current status of the snow fences was documented. Based on this and with the aid of CFD analysis of wind and snowdrift, a recommendation of where to begin efforts of maintenance and construction of additional fences was made.

A. Børve made a thorough analysis of snow drift conditions in Fuglenesdalen and concluded that parts of the valley were affected by winds from different directions. Fuglenesdalen has expanded since Børve made her observations. Her observations regarding the dominating snow drifting wind are in good agreement with measurements behind snow fences made recently.

3 NUMERICAL SIMULATIONS

In this work we have chosen to model the main snow drifting land wind over Hammerfest in three dimensions. The area is about 17 km^2 and has a rather sparse vegetation. There is no specific modelling of variation in surface roughness from vegetation or buildings, or any snow drift effect except that the wind is affected by snow drift roughness.

3.1 Numerical model

A 3D numerical simulation is performed, FLOW-3D from Flow-Science Ltd., a commercially available computational fluid dynamics (CFD) package.

Figure 1. The snow fences at the south side of residential area I and II in Fuglenesdalen do not have enough capacity to shield the residentials.

Figure 2. Snow drift in the south-vest section of residential area IV in Fuglenesdalen during a snowstorm in April 1997.

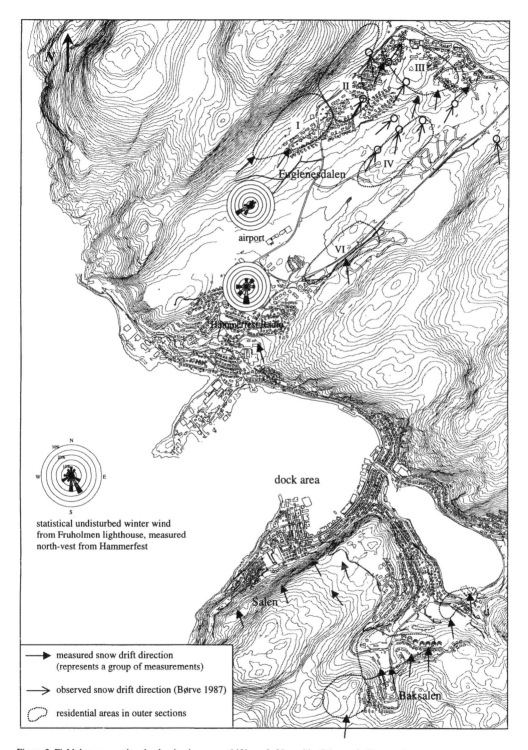

Figure 3. Field data concerning the dominating snow drifting wind in residential areas in Hammerfest.

379

Figure 4. Building covered by drift snow in the south–vest end of residential area IV in Fuglenesdalen, April of 1997. This section is open for snow drift from south-vest.

Figure 5. Summer picture of building in previous figure.

FLOW-3D is a transient fluid code based on a finite-volume technique and the SOLA algorithm (Hirt, 1975). Airflow calculations are performed by numerical integration of the Navier-Stokes equations with the evaluation of turbulence by a renormalized group theory (RNG) model. FLOW-3D uses the Fractional Area/Volume Obstacle Representation (FAVOR) method for surface representation (Hirt, 1993).

3.2 Boundary conditions

Inlet wind velocity is given by the specified logarithmic velocity profile:

$$U(z) = \frac{u_*}{\kappa} \ln(\frac{z}{z_0}) \land z_0 = 0.12 \frac{u_*^2}{2g} \qquad (1)$$

where u_* is the friction velocity; k is von Kármáns constant; $U(z)$ is the horizontal velocity at the height z above the terrain and z_0 is the surface roughness height for snow drifting winds over covered grain fields (Pomeroy & Gray, 1990). The wind speed at boundary is set to 10 m/s at the height of 10 m. The dominating south south-east wind measured at Fruholmen lighthouse is given as inlet wind direc-

tion. Equation 1 is also used as an initial condition for the flow field within the simulation area.

4 RESULTS

Figure 6 shows simulated wind velocities about 5 m above the terrain with statistical wind data, field measurements and observations. It can be seen that the simulated wind velocities near the terrain are in agreement with the collected field data. The dominating wind from south south-east blows over Baksalen and Salen from south south-east and turns south-west into Fuglenesdalen led by the valleysides. The wind direction at the top of the simulation is south south-east according to data from Fruholmen lighthouse.

The old waterfront of Hammerfest seems to be shielded from drifting snow from the current wind direction. This area is shielded either by terrain or by water surfaces, since airborn snow is not likely to travel over open water. The new residential areas in the outer and higher sections of Hammerfest look far more exposed from drifting snow with considerable fetch distances. Especially exposed are the south facing sections of Baksalen and residential area VI, and south-west facing sections of residential areas I, II, III and IV. It should be mentioned that there is practically no snow fence shielding of residential areas VI and IV from the dominating wind, and there is reported a lot of snow drift problems that may be associated with this wind direction, (Figs. 2, 4-5). The shielding of area III is sparse and snow fences south and south-west of areas I and II do not have sufficient capacity.

Hammerfest is exposed to other wind directions that cause snow drift problems, but the purpose of this work was to evaluate the dominating winter wind and we will not go into further discussion about locations of snow fences etc.

5 SUMMARY AND CONCLUSIONS

A numerical simulation of airflow over a residential region has been performed, based on the dominating, snow drifting wind conditions in the region. The airflow near the surface was highly affected by the complex local topography, but the numerical results compared well with field measurements. The project is special in the sense that a large set of field data consisting of measured snow drift directions, statistical wind data and snow drift observations, is used as controlling points within the simulation region.

Snow drift control is of great economical and practical importance for residential areas such as those we find in the outer sections of Hammerfest. The problem is often to predict how snow drifting winds are affected by local hilly topography

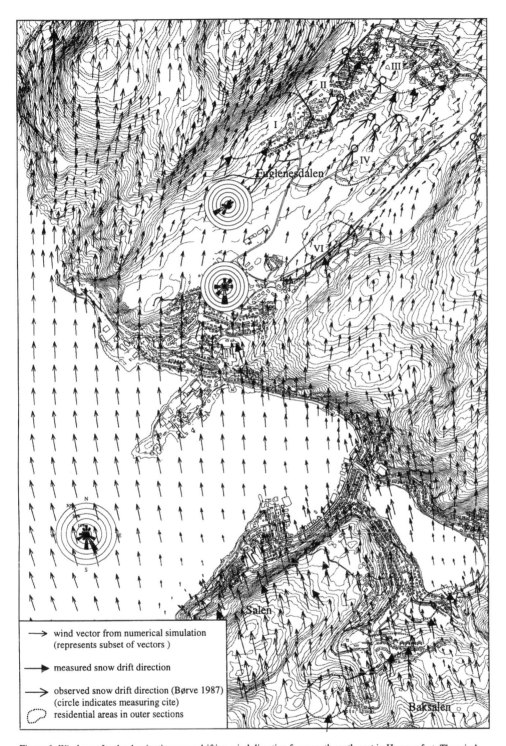

Figure 6. *Wind map* for the dominating snow drifting wind direction from south south-east in Hammerfest. The wind map includes numerical simulated wind vectors at the height of 5 m above the terrain, measured snow drift directions, statistical wind data and snow drift observations.

381

throughout the entire area. Wind map modelling based on computational fluid dynamics is in that case an efficient tool and has the potential of giving fairly comprehensive solutions.

Numerical wind simulations are applicable for providing:
1. additional wind data in complex terrain
2. information about wind conditions between field measurement points
3. evaluations of fetch distances

Besides serving as a foundation for the placement of new buildings and snow fences, wind maps can be used as input for more detailed analysis of snow drift around buildings (Anker et al. 1994, Sundsbø 1997, Thiis in press), or to evaluate wind forces (Wiik 1999) and environmental conditions.

Numerical analysis can probably not replace the traditional methods for snow drift planning based on field measurements and observations, since they still depend on field data to establish boundary conditions and control points within the simulation domain. However, numerical simulations are useful in cases where the wind pattern is complex, driven by local topography or/and buildings. Such conditions are often found in residential areas.

To our knowledge this is the first time CFD has been used in a practical project, where both a full solution of the flow field and snowdrift patterns were achieved. Further work will be to calibrate this model against time dependent snow transport and deposition. If we can manage better predication of snow deposition over time on this scale, it will give us new possibilities for estimating snow build up in avalanche zones as one possible application.

Modelling of snow drift can be found in: Uematsu et al. 1991, Gauer 1998, Liston & Sturm 1998, Naaim et al. 1998 and Sundsbø 1998.

ACKNOWLEDGEMENTS

The authors wish to thank Hammerfest council for giving permission to use field data regarding their snow fences and digital terrain geometry.

REFERENCES

Apeland, K. 1997. Residential planning under heavy snow conditions. *Proc. Of the third international conference on snow engineering*. Sendai, 26-31 may 1996: 483-488.

Børve, A.B. 1987. The design and function of single buildings and building clusters in harsh, cold climates. *Doctoral thesis (ISBN 82 5470 068 0)*. Oslo school of architecture.

Gauer, P. 1998. Blowing and drifting snow in Alpine terrain: numerical simulation and related field measurements. *Annals of Glaciology* 26: 174-178.

Hirt, C.W. 1975. SOLA-A Numerical Solution Algorithm for Transient Fluid Flows. Los Alamos Scientific Laboratory Report LA-5852.

Hirt C.W. 1993. Volume-Fraction Techniques: Powerful tools for Wind Engineering. *Journal of Wind Engineering and Industrial Aerodynamics* 46 & 47: 327-338.

Liston, G.E. & Sturm, M. 1998. A snow-transport model for complex terrain. *Journal of Glaciology 44 (148)*: 498-516.

Naaim, M. et al. 1998. Numerical simulation of drifting snow: erosion and deposition models. *Annals of Glaciology 26*: 191-196.

Nielsen, A. et al. 1994. Computer simulation of windspeed, wind pressure and snow accumulation around buildings. *Proceedings of The International conference on HVAC in cold climate*, Rovaniemi, 15-18 march 1994: 457-467.

Norem, H. et al. Evaluation of residential areas in terrain with heavy snowdrifts, Hammerfest (in press).

Pomeroy, J.W. & Gray, D.M. 1990. Saltation of Snow. *Wat. Resour. Res.* 26 (7): 1583-1594.

Pomeroy, J.W. & Gray, D.M. 1995. Snowcover – accumulation, relocation and management, *National hydrology institute science report no. 7*. Saskatchewan, Canada.

Sundsbø, P.A. 1997. Numerical modelling and simulation of snow drift - Application to snow engineering. *Doctoral thesis (ISBN 82-471-0047-9)*. The Norwegian University of Science and Technology.

Sundsbø, P.A. 1998. Numerical simulations of wind deflection fins to control snow accumulation in building steps. *J. Wind Eng. Ind. Aerodyn.* 74-76: 543-552.

Takeuchi, M. 1980. Vertical Profile and Horizontal Increase of Drift-Snow Transport. *Journal of Glaciology* 26 (94).

Thiis, T.K. A comparison of numerical simulations and full-scale measurements of snowdrifts around buildings. *Wind and Structures* (in press).

Uematsu, T. et al. 1991. Three-Dimensional Numerical Simulation of Snowdrift. *Cold Regions Science and Technology* 20.

Wiik, T. 1999. Wind loads on low rise buildings. *Doctoral thesis (ISBN 82-471-0478-4)*. The Norwegian University of Science and Technology.

Snow Engineering, Hjorth-Hansen, Holand, Løset & Norem (eds) © 2000 Balkema, Rotterdam, ISBN 90 5809 148 1

A study on the modernization of traditional countermeasures against snow damages in snowy town of Japan

Toshiei Tsukidate
Hachinohe Institute of Technology, Japan

Takahiro Noguchi
Hokkaido University, Sapporo, Japan

Abstract

I investigated the traditional houses and towns that were basis on the local climates, and there are many ideas of traditional snow measure. Therefore, it is very important to modernize the traditional ideas. We have to learn how to adapt the nature from traditional house design and towns, snow measure and temporary snow equipment.

Introduction

The snowy and cold area of Japan is located in the north latitude between 35 and 45 degree, but there are various winter climates. Sometimes, big earthquakes hit snowy area in winter. For example, Hanshin-Awaji earthquake in which over 7000 lives lost was occurred in January 17 1995. Houses in snowy area of Japan include many ideas and temporary snow measures.

I think that traditional lives, houses and towns have a lot of ideas to adapt the snowy climate and the nature. We must to research the ideas of traditional snow measure, and modernize the traditional ideas by modern technology.

1. Snow damage and the influence

(1) Snow damages in the present times

We have much snow in Japan, especially it is over 4 m deep. Sometimes snow gives the damage of life. We were attacked by snow everywhere of Japan. The most heavy attacked areas had over 40 times between 15 years (1971-1986).

(2) Pattern of snow damage

Snow falls and cover the land. But snow thaws in spring. Snow affects us on various conditions that snow is in snowfall, snowy, to thaw.

Snowfall and snow that covers the land prevents us to product food and to act in outside. Snow that had slide down on the roof damages the houses, etc. Also thaw of snow make floods and landslides etc

Windstorm overlap snow prevents outdoor action and damages the houses more than snow or wind only. Earthquake in snowy season affects more widespread damages than only snow. Especially in Japan earthquake in winter is the important problem.

2. Snow damages in the present times

The following is the results of questionnaire concerned with snow accident, snow problems and snow processing at local administrations of Tohoku region that is snowy area of Japan.

So traditional urban areas constructed by human scale in snowy region were confused by modern life style, especially by cars, snow processing. And countermeasures are the big economic burden for local administration at snowy area. in Japan

Most important problems are slip accident of cars and violent fall of pedestrians in winter.

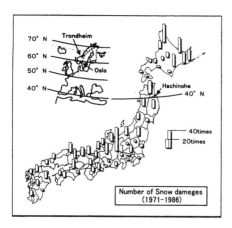

Fig.1 Snow damages in Japan in the present time

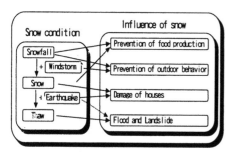

Fig.2. Influence of snow to the life

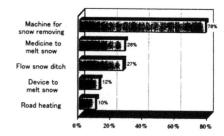

Fig.3. Influence of snow to the life

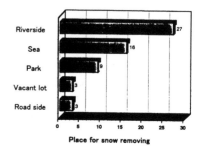

Fig.4 Places of snow removal

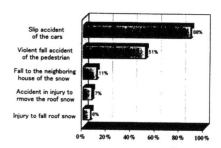

Fig.5 Accidents according to snow

Main snow processing at traditional buildup area are machine snow removing. And we remove snow to the park, sea, and riverside's open space are used as snow desertion place. Average expense of snow processing is about per capital 10~40 million dollars, but most biggest economic burden of local administrations is 10 billion dollars.

Most difficult troubles are to send out each house's snow to roads and to park cars at roadside. For that, the amount of snow processing quantity is increased, those are the important problems in snow processing and town design. Also, there are many complaints by traffic congestion to snow removing by machines.

3. Countermeasures of traditional lives

The countermeasures against snow of traditional lives are the following 6 patterns by methods, technology, temporary equipments included with space.

(1) Storage of food and food processing

As the preparation for snowy, winter we storage food which processes to dry or pickle and stock "Muro" that is a cellar of underground or near the fireplace in the house. We had to storage fresh vegetables in "Muro" not to freeze in traditional life.

Food processing methods for winter are to dry and to pickle. Usually we make dried salmon and dried rice cakes under the eaves. Pickled vegetable makes by salt and some Japanese spices.

(2) Snow measures in traditional houses and temporary equipments

In Japan, there are two types of snow measures. The first is the house design with snow measure. Another one is temporary equipment against snow like wind snow screen. Those snow measures depend on the various snow conditions and many method.

(3) Wind snow screen against snow draft

We set the wind snow screen at the windy side of premise in winter and is usually made by natural materials such reed or straw of rice. We will preserve the reed of wind snow screen and make thatched roof.

(4) "Taka-Happo tsukuri" and snow fence

In snowy area, outside behavior prevents by snowfall and snow on the ground. Because we have to integrate the many functions into the house, we build the multi-floored house or large floored house with work area and storage. Sometimes temporary lavatory is set in the part of work area.

(5) "Gattsho tsukuri " and "Tane"

"Gasho tsukuri" is the large wooden house with large thatched gable roof and multi floor. That house is constructed in snowy, but comparatively warm area as snowy area. The garret was used mainly as silkworm raising space. Roof snow slides and falls to the pond that melts snow. Similar traditional house type with dormer window is called "Taka-Happo tsukuri".

(6) "Chumon tsukuri"

"Chumon tsukuri" which is L or T shape plan is built in more snowy and colder area. Convex part of "Chumon tsukuri" is the entrance hall with stable and storage. External wall and windows are covered with temporary snow fences. This area was attacked sometimes by earthquake, and strong winter wind. Snow type is the middle of wet and dry.

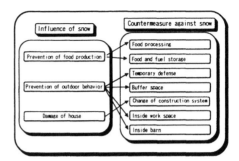

Fig.6 Concept of snow measure of traditional life

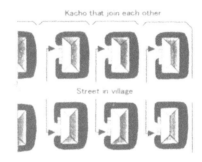

Fig.7 Joined "Kacho" system in plain

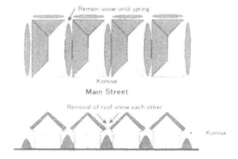

Fig.8 "Komise" and removal of roof snow

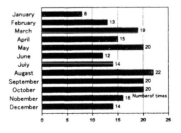

Fig.9 Monthly number of times of earthquakes

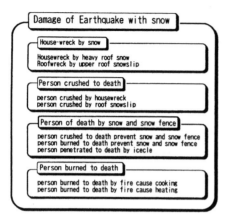

Fig.10 Earthquake damages with snow
by historical grand earthquakes in winter

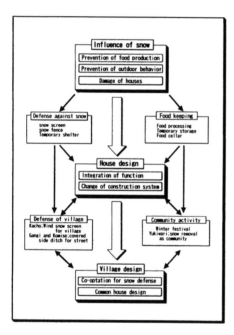

Fig.11 System of traditional snow measure

Table-1 Historical grand earthquakes in winter

A.D.	Damage of earthquake
1/2/1666	1666 "Takada Earthquake" Wrecked house s; About 5,000 Amount of death; About 1,870 Depth of snow : About 4.5m
8/3/1763	1763 "Tsugaru Earthquake" Wrecked houses; About 7,000 Amount of death; About 1,350 (Person burnt to death: about 300) Depth of snow : About 3.0m

The countermeasure against snow of traditional house are consist of a. temporary defense, b. house design and c. construction system. These depend on the natural environment and low energy snow processing methods. The characteristics of snow measures.make unique house styles to adapt to local environment.

4. Snow measures in town

(1) "Kacho" that is the joined snow screen

"Sugoya tsukuri" is popular housing type in Japan included snowy area. But in winter "Sugoya tsukuri" in snowy area is covered with temporary snow fence and wind snow screen made by wood or reed. Wind snow screen that is set up in dry snow area is called "Kacho" or "Shogaki" in Japanese. Each house of "Kacho" joins each other. Therefore village is covered wind snow screen.

(2) "Gangi" and "Komise"

"Gangi" is the traditional covered sidewalk in wet snow area of Japan. People walk safety and comfortably in "Gangi" when heavy snow fall and windy. When it is very deep snow, we had to dig the tunnel to another side of road and "Gangi" that are buried with snow.

"Komise" are the traditional covered sidewalk in dry snow and windy area. "Komise" which is closed by temporary doors in winter is different from "Gangi".

(3) Snow measure system of village and street

Villagers in plain prepare the wind snow screen, which join each other. Pepole in the city build "Gangi" and "Komise" one by one. To join "Komise" each other, street is covered sidewalk connected "Komise".

Those express the cooperation of community and to regard the weak.

5. Earth quake with heavy snow

We have attacked by big earthquakes in winter. The proportion of winter earthquakes is about 38% of total earthquakes; for example, Hanshin-Awaji earthquake was occurred in 17/1/1995.

In history, we had two same scale big earthquakes that one was 1666 " Takada Earthquake ", another one was 1763 "Tsugaru Earthquake " In the two big earthquakes, there were over 5000 houses that were destroyed and over 1000 person who were died. Deep snow increased the damages by the old records.

(1) House wreck by snow

The heavy roof snow had enlarged the wrecked houses and destroyed the roves.

(2) Person crushed to death by snow

Many people were crushed by snow that had slide from roof.

(3) Person crushed to death by snow, snow fence and icicles.

Some people died to prevent escaping by snow or snow fence, and some one died to penetrate by icicles.

(4) Person burned to death

Somebody had burnt to death to prevent escaping by snow or snow fences.

6. Conclusion

According to the historical books, there were many characteristic damages by earthquakes in snowy season.

(1) Traditional snow measure system

Traditional snow measures are consisted with temporary equipments, storage of food and memory of life and the architecture. The concept of traditional one is the co-operation of community and the symbiosis to nature such as ecological and low energy.

(2) The necessity to modernize the raditional snow measure

Recently, there are two types of modern houses in snowy area of Japan. One is the elevated house that is about 2.5 m high like pilotis. Another one is the snow loaded flat roof house. Both house styles are in danger of earthquake with snow. Therefore, we need to learn how to adapt the nature from traditional house design, snow measure and temporary snow equipment.

It is necessary to make present of traditional wisdom of snow countermeasure, for example " Gangi" which are the covered sidewalk and " Yukikirii" which is the snow removing by the area community to decrease the snow burden and accidents.

Photo 1. "Hoshimochi" that is a dried rice cake

Photo 2. Wind snow screen of premise made by straw (Fukushims, Japan)

Photo 3. "Taka-Happo" that is a multi-floor house (ground floor is covered with reed snow fence)

Photo 4. "Gasho tsukuri" and "Tane"
("Tane" is pond to melt snow)

Reference

Tsukidate,T & Noguchi,T. 1997.A study on the design technique of public housing at snowy area in Japan. Proceedings of the third international conference on snow engineering/ Sendai/ Japan/ 26-31 May 1996 etc.

Noguchi,T. & Tsukidate,T 1997.Living style and planning of multifamily housing for snowy, cold region, Proceedings of the third international conference on snow engineering/ Sendai/ Japan/ 26-31 May 1996 etc.

Photo 5. "Chumon tsukuri"
(horses live in the house with people)

Photo 6. "Sugoya tsukuri" and "Kacho"

Photo 7. "Komise" (that is covered with wood door,
"Gangi" is not covered at road side)

5 Transportation

Snow Engineering, Hjorth-Hansen, Holand, Løset & Norem (eds) © 2000 Balkema, Rotterdam, ISBN 90 5809 148 1

Evaluation of measures for snow avalanche protection of roads

Espen Hammersland
Public Roads Administration Norway, Bergen, Norway

Harald Norem
Norwegian University of Science and Technology, Trondheim, Norway

Arnold Hustad
Public Roads Administration Norway, Ørsta, Norway

ABSTRACT: An evaluation of protection measures at 130 avalanche road sites has been carried out. The evaluation of the 97 sites protected by snow sheds and earth work protection is presented. The evaluation is based on interviews, filed trips and recorded road closures before and after the protection was installed. Experience shows that the efficiency of the snow sheds is lower than required. The low efficiency is mainly a result of too short sheds, poor adjustment between the terrain and the shed and open areas in front of the sheds. The earth works in most cases have efficiency above 70%. The protections made of earth works are most effective for moist and wet snow avalanches, less for dry snow and slush avalanches

1 INTRODUCTION

The last 25 years have been an active period for avalanche protection of the Norwegian road network. The most severe types of avalanches are snow and slush avalanches and rock falls. To a lesser degree roads have also been protected against debris flows and sediment transport in rivers.

Traditionally, the main protective measures have been construction of tunnels or snow sheds made of concrete. During the last two decades several alternative measures have been used, including bridges, earth works, supporting structures, detection of avalanches and blasting techniques. The latter measures are generally less expensive and are only realistic alternatives for specific types of avalanches and terrain features.

The last year a comprehensive study has been carried out on evaluating the efficiency of the constructed measures. The main scope of the study was to assemble experiences with the different types to improve the design of further avalanche protection of the Norwegian roads exposed to avalanches.

2 TYPE OF PROTECTIVE MEASURES INVESTIGATED

The methods used in Norway for snow avalanche protection of roads are presented by Norem (1995) This textbooks gives recommendations for the design and use of the various protective measures based on the experience assembled at that time.

The present investigation is based on interviews with the local maintenance crews and supervisors, field trips to the areas and recorded avalanche activity both before and after the construction of the protective measures. The accuracy of the recordings varies, but it is assumed that the combination of both recordings and interviews gives the necessary accuracy to use the results statistically.

The following types of protective measures have been evaluated:

Type	No	Description and use
Snow sheds	60	All types of avalanches and terrain
Earth works	37	Diverting and collecting dams. Earth mounds. Gentle terrain and small and medium-sized avalanches
Culvert-type sheds	5	Gentle terrain and all type of avalanches
Wide road ditches	3	Small avalanches only
Supporting structures	13	Made of net and mainly for falling rock. Have been used for smaller snow avalanches
Road bridges	2	Smaller and medium sized avalanches where the road crosses creeks
Relocation of roads	2	Mainly in gentle sloping terrain where the relocation may have limited length
Blasting	5	Three types of blasting techniques, radio- and cable-controlled detonation of pre-planted charges and use of cableways
Detection of avalanches	3	Avalanches are detected by geophones which switch on red lights at the road

The report presents only the experiences assembled for the two types that have been most extensively used, snow sheds and earth works.

3 EFFICIENCY OF PROTECTIVE MEASURES

3.1 *Definition of the efficiency*

The construction of protective measures has two main purposes, to improve the safety for the road-users, and to reduce the closure time of the road.

Avalanche accidents fall in three groups:
- Cars or pedestrians moving on the road are hit by an avalanche.
- Cars or pedestrians waiting for the road to be opened are hit by a new avalanche.
- Maintenance crew is hit by an avalanche during opening of the road closed by a previous avalanche.

Experience from Norway indicates that the relative number in each group is approx. 20, 40 and 40% respectively. When protecting an avalanche exposed road it is thus important to protect slopes that are running several times during an avalanche period or where the slopes are located close to each other.

In a similar way, the reduction of the closure time is dependent on the fact that the road is protected for all avalanches during a certain weather situation. The effect on the closure time is thus only limited if only two of three avalanche sites are protected for one avalanche period.

The very best method for evaluating the efficiency of the avalanche protection is to take both the improved safety and the reduced closure time for a road section into account. Unfortunately the accuracy of the recordings are too uncertain to define such efficiency. The only practical way to define the efficiency was to base the evaluation the reduction of closures caused by the protection. The efficiency, E, is then defined by:

$$E=(1-\frac{\text{Number of closures after protection}}{\text{Number of closures before protection}})100 \text{ [%]}$$

3.2 *Requirements for the efficiency*

A main question when both designing and evaluating protective measures is "What is the required efficiency"? So far, there are no specific national guidelines to answer that question. The required efficiency may be both related to the frequency of closures or to the ratio of avalanches still closing the road after the road protection

The accepted frequency of avalanches is usually dependent on the traffic volume and the importance of the road. In any case 100% safety due to avalanches is a too high requirement. In that case, too expensive solutions have to be selected, causing the protection never to be constructed or to leave other sections with poor traffic safety unprotected.

The required efficiency for the present evaluation is defined dependent on the cost of the protective measure. Costly measures like tunnels and snow sheds are usually permanent constructions and should thus have a high efficiency. In our case an efficiency of 90% has been selected.

Earth works are generally much less expensive than snow sheds. Generally the cost per meter road protection is only 5-20% of the cost of the snow sheds. One may thus accept some lower efficiency due to the lower investment. On the other hand, the road users are only interested in the number of closures, independent of the protective measures used. The accepted efficiency should thus not be selected too low. The present evaluation is based on 70% efficiency for a successful protection.

4 EVALUATION OF THE SNOW SHEDS

4.1 *Snow sheds evaluated*

The start of extensive use of snow sheds started in the 1960's, and the average age of the 60 evaluated snow sheds is 22 years. The sheds are mostly used on low-traffic roads, where the average daily traffic varies between 75 and 2000. The cost of a two-lane snow shed today is approx. 12000 US$ per meter.

4.2 *Recorded efficiency*

Fig 1 shows that a high percentage does not fulfill the requirements to a successful design. Only 40% have efficiency higher or equal 90%, and as many as 23% have less than 70% efficiency. This result is unsatisfactory and clearly indicates the need for more detailed studies to find systematic weaknesses in our practice of designing snow sheds.

The efficiency is further analyzed due to the following parameters (Figs 2,3 and 4):
The length of the snow sheds.
The avalanche frequency.
The age of the construction.

Fig 2 shows that the efficiency is too low for snow sheds less than 30 m, approx. 50%, compared to more than 80% for the longer sheds. The major parts of the snow sheds having low efficiency is thus found among the shortest sheds.

A low efficiency is also found for snow sheds protecting avalanche sites with high frequency, fig 3. The most frequent avalanche sites are usually found in distinct creeks with small and well-defined avalanche width. The results presented in figs 2 and 3 are thus correlated.

The need for carrying out evaluations is demonstrated in fig 4, showing that there has been no increase in efficiency with time since the 1960's. The low efficiency for the 90/94 snow sheds is the result of the construction of short sheds on low-traffic roads in that period.

4.3 Evaluation of the results

The main reasons for the general low efficiency for the snow sheds have been found to be:

The snow sheds are only designed for the most frequent avalanches. The less frequent ones then close the road at one or both ends. The reason for constructing too short shed is a result of the high investment costs.

Poor adjustment between the terrain and the snow shed. In practice this means that deflecting dams and walls are missing or are too low

A snow shed causes the avalanches to pass the road 6 m higher than before the construction. There will thus be a gentle section on the roof where most avalanches are retarded and increased in width. This widening is not taken into account by the design of the shed

Especially the short sheds do not have enough storage capacity to handle more than one or two avalanches.

Snow sheds constructed with open pillars are often filled with snow entering from the open side.

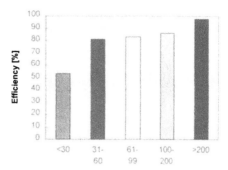

Figure 2. Efficiency versus length of the snow shed

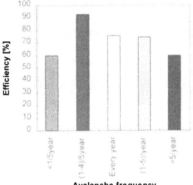

Figure 3. Efficiency versus the avalanche frequency

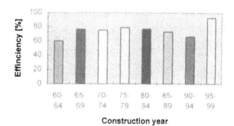

Figure 4. Efficiency versus the construction year.

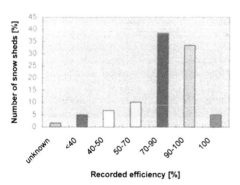

Figure 1. Recorded efficiency of the evaluated snow sheds

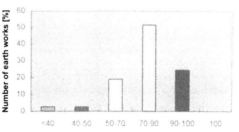

Figure 5. Recorded efficiency of the evaluated earth works

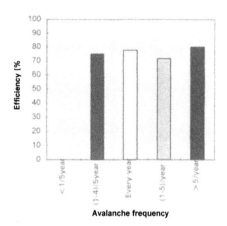

Figure 6. Efficiency versus the avalanche frequency.

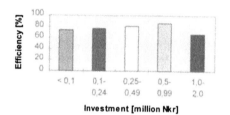

Figure 7. Efficiency versus the investment

Table 1. Efficiency of earth works against diffrent types of avalanche.

Type of avalanche	Number	Road closures	Efficiency
Dry	23	6	74
Wet	39	2	95
Slush	7	3	57

The main conclusion is that snow sheds at least shuld be more than 30 m long. There should also be room for storing at least 2-3 avalanches on the roof. The terrain upstream of the snow shed should further be channeled and combined with deflecting dams and walls.

Due to the low efficiency there has been a need to reconstruct 45% of the snow sheds. The reconstruction have consisted of:

- Extensions.
- Reconstruction of dams and concrete walls.
- Making a dense wall on the air side
- Making drainage channels on the roof of the sheds to avoid clogged pipes below the sheds.

The reconstruction of the snow sheds have improved the efficiency with more than 20% in 90% of the cases, and in 50% of the cases the reconstructed sheds are recorded to have improved the efficiency to 90% or more. Generally the reconstruction is assumed to be successful.

5 EVALUATION OF THE EARTH WORKS

5.1 Earth works evaluated

The use of earth works as a separate protection measure was first introduced in Norway in 1976. This protection included diverting dams, collecting dams and earth mounds. These measures are still the main types of earth works used. Since that time earth works as the only protective measure protects several avalanche sites. This presentation is based on evaluating 37 different sites. The sites cover all regions in Norway and are found in different climatic conditions, varying from sea level to mountainous areas.

The average age of the protective measures is 12 years. The type of the measures are deflecting and collecting dams, earth mounds and magazines. The latter is excavations of wide ditches and may be combined with collecting dams to a lesser degree.

5.2 Recorded efficiency

The evaluation indicates that 75% of the earth works fulfill the criteria of 70% efficiency, and more than 20% have more than 90% efficiency, fig 5.

Considering the low investments and the limited time for gaining experience we assume this result to be fairly satisfactory.

The efficiency was investigated due to the following parameters:
Figure 6. Efficiency versus the avalanche frequency.
Figure 7. Efficiency versus the investment

Fig 6 shows that the average efficiency is above 70% within all groups of frequencies, and it is not possible to find any trend in the efficiency versus the frequency. One should, maybe, expect some lower efficiency for high frequencies due to the fact that several avalanches gradually reduce the storage capacity of the measures. The reason for the result of the investigation is probably because sites with high frequencies usually have small avalanches.

There is also hard to find any systematic trend in the efficiency versus the cost of the measures. Fig 7 indicates a lower efficiency for investments above one million NOK (120000 US$). This is probably because there are very few sites evaluated within this group, but may also reflect that earth works should not be used to protect bigger avalanches.

5.3 Evaluation of the results

There are found some systematic weaknesses of the evaluated earth works, of which the most important are:

Earth works are most effective to stop dense flowing avalanches, which in practice means moist and wet avalanches. For one road section investigated in detail, 69 avalanches were surveyed. All these avalanches were big enough to close an unprotected road. They were classified in three groups and the recorded effects of the protection were:

The table shows that avalanches having some cohesion are most effectively stopped by earth works, while the non-cohesive avalanches consisting of dry snow or slush pass the dams and mounds more easily.

Earth works implemented in mountainous terrain have generally a low efficiency. The reason is that deep snow covers the earth works and thus reduces the effect. Up in the mountains the percentage of dry snow avalanches is also generally high.

Especially the first generation of earth works were generally too small to stop several avalanches a year. 22% of the evaluated measures have been reconstructed and 90% of them have improved the efficiency with more than 20%

6 GENERAL REMARKS

The main benefit of the snow sheds is that these protect the road for all kinds of avalanches and may be used in all kinds of terrain. The shed in addition offers full protection to the road-users when they are inside the sheds. The snow sheds have also shown a long durability. 40 years old sheds are still in proper condition.

There are, however, some disadvantages with the use of snow sheds. The main one is the high cost, which too often has resulted in making too short sheds. Another disadvantage is the reduced traffic safety at the entrances and inside the sheds. The light conditions at the entrances are some times extremely difficult, and in springtime the road inside the tunnels may be slippery. In some cases it has been difficult to find a good design that both takes care of the requirement for proper road design and avalanche safety.

The maintenance costs are recorded to be limited. Some sheds have, however, faced damages due to falling rocks and avalanches more powerful than expected.

Despite the disadvantages snow sheds will still be important for avalanche protection of roads in the future. However the present practice for designing the sheds needs to be adjusted. The short sheds should be avoided, and one should be more concerned about the adjustment between the terrain and the shed.

The experiences assembled about earth works indicate that these have been fairly successful. The relative high efficiency combined with small investments should favor the use of these protective structures in the future. In addition, earth works have no negative influence on the driving conditions and on the traffic safety.

There are so far few recorded intensive maintenance difficulties with the earth works. The main threat for damages are the eroding effects of slush avalanches. The effect of the earth works may also be reduced with time due to sedimentation of debris in front of them.

The present evaluation will go on at least for one more year. The main workload will be on analyzing the assembled experiences to also cover the other alternative measures and to go into more detail concerning snow sheds and earth works. It is thus our hope that the evaluation may make it possible to improve our practice in snow avalanche protection of roads in the future.

7 REFERENCES

Norem, Harald (1995)
Snow Engineering for Roads
Norwegian Public Roads Adm., Oslo

Snow Engineering, Hjorth-Hansen, Holand, Løset & Norem (eds) © 2000 Balkema, Rotterdam, ISBN 90 5809 148 1

Wind effects on snow drifts around two-dimensional fence

Takeshi Hongo & Manabu Tsuchiya
Kajima Technical Research Institute, Japan

Tsukasa Tomabechi
Hokkaido Institute of Technology, Sapporo, Japan

Hiroshi Ueda
Chiba Institute of Technology, Japan

ABSTRACT: The purpose of this study is to clarify the relation between snow accumulation and wind flow by wind tunnel tests, snow wind tunnel tests and field measurements. This paper investigates snow accumulation and wind flow around a two-dimensional type fence with opening at the bottom.

First, the snow accumulation around the fence was measured in a snow wind tunnel. Then the wind flow around the models which simulated the snow accumulation around the fence, was measured in detail by means of a triple-split probe anemometer in a boundary layer wind tunnel.

The test results yielded the snow accumulation patterns around the fence with duration of snowfall and clarified the relationship between the snow accumulation patterns and vortices formed front and behind the fence. The results were verified by field measurements of the full-scale model.

1. INTRODUCTION

A snow wind tunnel test using model snow is a useful method for investigating snow accumulation problems at the building design stage. However, the similarity law for this wind tunnel test has not yet been established. Snow accumulation is closely related to the wind flow around the fence. The authors have been investigating the characteristics of both air flow and snow accumulation patterns around a two-dimensional fence which possesses an opening at the bottom (Author et al. 1997). This was done in order to clarify the relationship between the flow and the snow accumulation patterns, which is closely related to the similarity law of snow wind tunnel tests. This paper describes the snow accumulation patterns obtained from wind tunnel tests and the characteristics of the flow under the condition that these patterns are discernible. Comparison of the test results with the snow accumulation patterns obtained from field tests is also shown.

2. INVESTIGATION OUTLINE

Fig.1 shows the investigation process. In general, a hot-wire anemometer used for air flow measurement cannot be applied in a snow wind tunnel for investigation into snow accumulation.

Therefore, it is very difficult to measure both snow accumulation and flow characteristics simultaneously. Therefore, model snow accumulation patterns were first investigated using a snow wind tunnel. Wind directions and wind velocity distribution were then measured in a boundary-layer wind tunnel in which snow accumulation patterns were simulated. Furthermore, the validity of the wind tunnel test results was investigated by comparing them with the field measurement data. Finally, the relationship between the flow characteristics and the snow accumulation pattern was examined.

2.1 Snow Wind Tunnel Test

A snow wind tunnel test was carried out using the large snow wind tunnel belonging to the Hokkaido Institute of Technology (Toyoda1992). Fig.2 shows the outline of this closed-circuit wind tunnel, which has a working section 1000mm wide and 800mm high. This test used as model snow, activated clay with a moisture content of 8.5% by weight and an angle of repose and viscosity closely resembling the snow in Hokkaido (Tomabechi 1985, 1986). This activated clay was jetted into the wind tunnel through a snow effusion nozzle fitted to the upper part of the testing section by compressed air, producing conditions of snowfall and a snow storm. The windward side of the working section was equipped with three screens with 10mm grids in order to approximate the distribution of the mean wind velocity and turbulence intensity for the flow measurement test. A model of a fence with a 5mm opening at the bottom was installed on a turntable. This model was 30mm high at its highest point and 400mm wide. The oncoming flow was set at about 5m/sec. Model snow was supplied during a period of

0 ~240 min. The accumulation pattern measurements for the model snow was carried out at the central section of the model at 1mm intervals using a laser displacement gauge.

2.2 Flow Wind Tunnel Test

The boundary-layer wind tunnel used for the air flow measurements was the Eiffel type wind tunnel belonging to the Kajima Technical Research Institute. In order to simulate the air flow around the fence when it snows in this wind tunnel, the model snow accumulation pattern obtained from the snow wind tests was modeled. The scale of this model was three times (H=90mm) that for the snow wind tunnel test. This was decided by comparing the characteristics of the flows in the two wind tunnels, which will be described later. On the far windward and leeward side of the fence, the wind tunnel floor was regarded as a place where the depth of snow was constant. Snow accumulation was simulated from the windward area extending to a point of 8.3 times the fence height a point of 4 times the fence height in the leeward area. The model was 500mm wide, and at both of its ends, a side board (1,200mm high, 3,250mm long, 10mm thick) was installed in order to ensure two-dimensional flow characteristics. It is very important in this test to make the flow characteristics at the inflow part in a snow wind tunnel test similar to those in a flow measurement test. First, the flow in the snow wind tunnel test was measured as a fluctuating wind using a special Pitot tube. Then, the vertical profile of the mean wind velocity and the turbulence intensity were obtained by digitally processing the measurement results. Next, a flow was created in the Eiffel type wind tunnel and the profile of this flow was compared with that in the snow wind tunnel. It was thus clarified that the results obtained from the Eiffel type wind tunnel agree closely with those from the snow wind tunnel when the scale ratio is set at three times. Fig.3 shows the vertical profile of the flow for each of the two kinds of wind tunnel tests. Spires and a saw were used to produce the flow in the Eiffel type wind tunnel. An I-type hot-wire anemometer (KANOMAX4186) was used to take the measurements.

It is difficult to conduct accurate measurements of a flow with high turbulence and reverse flow using I-type and X-type hot-wire anemometers for air flow measurements. This is so because these types of anemometer cannot measure wind direction or can only measure limited wind direction. It is important in this study to have an accurate understanding of the profiles of wind direction and velocity as well as the turbulence intensity in the surrounding the fence. Since accurate measurements of the wind direction and velocity profiles in turbulent and reversed flows are indispensable to do this study, a triple-split fiber anemometer was used. This

anemometer was equipped with three sensors installed parallel to each other on a cylindrical quartz fiber with a diameter of 0.4mm, and can accurately measure the wind direction and velocity of two-dimensional flows. Fig.4 illustrates this anemometer. The wind velocity used in the test was set at about 10m/sec at a height of 90mm (H) of 480mm (x/H= ⁻ 5.3) on the windward side of the fence, considering the performance of the wind tunnel and the anemometer. AD transformation was carried out at each measurement point at a sampling interval of Dt = 0.002sec (500Hz) and the sampling number of 15,000 (30sec). Moreover, the velocity pressure, flow temperature and atmospheric pressure outside the boundary-layer were measured simultaneously in order to correct the results obtained from the air flow measurement in the test. The measurement interval in the flow direction was set at x/H = 1/9 (10mm) in the proximity of the fence (-1.33~2.0 x/H, 0~1.67 z/H) and at x/H = 2/3 (60mm) in the other areas. The measurement interval near the ground surface (0~1.67 z/H) was set at z/H = 1/9 (10mm) and at z/H = 1/39 (30mm) above this level (1.67<z/H).

2.3 Field Observations

Field observations were carried out on the ground of the Hokkaido Institute of Technology. The fence was erected at right angles to the NW wind direction, which is the main wind direction in the winter in this district. The area on both the windward and leeward sides of the fence is flat and there were no high-rise buildings anywhere near. It is thus considered that the flow characteristics in this field observation area closely resemble those of the flow used in the two kinds of wind tunnel tests.

During the winter months, snow accumulation patterns at the central section of the fence were measured at set times every day using a snow scale. At the same time, observations of reference wind direction, velocity and snowfall were carried out. Photo.1 shows an overall view of the fence.

Photo.1. Full Scale Fence for Field Test

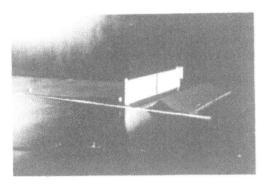

Photo. 2. Model of Accumulation Pattern after 240 minutes Supply of Model Snow for Velocity Measurements

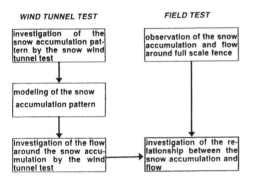

Figure 1. Process of the Investigation

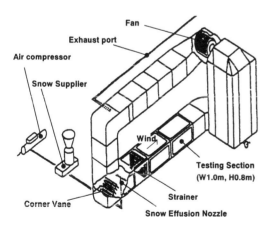

Figure 2. Large Snow Wind Tunnel

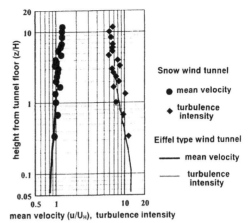

Figure 3. Comparison of Wind Profiles Used for Two Kinds of Wind Tunnel Tests.

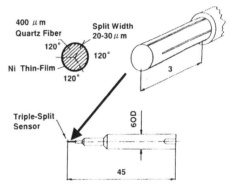

Figure 4. Outline of Triple-Split Fiber Probe

3. INVESTIGATION RESULTS AND CONSIDERATIONS

3.1 *Investigations by Snow Wind Tunnel Tests into Snow Accumulation Patterns*

Fig.5 shows the difference in model snow accumulation patterns according to the model snow supply period. The values for a distance of x/H in this case were obtained by transferring the fence height H into a non-dimensional form.

In the front and rear of the fence, two snow accumulation peaks are discernible. On the windward side of the fence, the depth of snow gently increases toward the fence and reaches a maximum at about $x/H=^-0.7$. On the leeward side of the fence, the snow accumulation reaches its peak at $x/H=1.7$. The heights of the two snow accumulation peaks of the in the front and rear of the fence are almost the same when the model snow supply period is short. However, with an increase in the supply period, the height of the peak rises on the leeward side.

3.2 Investigations by Wind Tunnel Tests into Flow in Case of Snow Accumulation

The model snow accumulation pattern after supplying snow for 240 minutes, which is thought to be the final accumulation state, was modeled for velocity measurements (Photo.2). The mean vector profile as well as the mean streamline around the fence, the mean velocity profile for horizontal components, and the mean velocity profile for vertical components are illustrated in Figs.6(a), 6(b) and 6(c) respectively. The mean velocity was obtained by transferring the mean velocity (10m/sec) at the height H of the fence into a non-dimensional form assuming that there is no fence. For comparison, distributions of mean flow vector and stream line for the fence with no snow accumulation and the fence without opening at the bottom are shown in Figs.7 and 8.

On the windward side of the fence, the wind velocity near the ground surface decreases toward the fence. With this velocity decrease, the snow depth increases. Weak separation, which occurs from the peak position of snow accumulation, produces a reverse flow streaming toward the peak position away from the fence. It can be thought that the snow accumulation peak on the windward side rises due to these two kinds of flows which stream toward this peak position.

A strong separated flow from the upper end of the fence reattaches in the vicinity of $x/H=7$ behind the fence and forms a vortex area. In the rear of the fence, an ascending flow is produced along it. However, a relatively strong flow, which comes through the opening in the lower part of the fence, streams toward the leeward side parallel to the ground surface. This flow meets with the reverse flow caused by the vortex area which has been formed in the rear of the fence and is pushed back toward the fence. It then rejoins the flow separated from the fence. It can

be deduced that the snow accumulation peak on the leeward side is formed due to this flow coming through the lower part of the fence and the reverse flow induced by the vortex area.

3.3 Investigations by Field Tests into Snow Accumulation Patterns

Fig.9 shows the results of the investigations into snow accumulation patterns obtained from field tests. The values for a distance x/H and snow depth S/H were obtained by transferring the fence height H into a non-dimensional form.

Two snow accumulation peaks are discernible in the front and rear of the fence in the same way as that seen in the snow wind test. The peak position on the windward side of the fence is about $x/H = -0.7$. This agrees with the result obtained from the wind tunnel test. However, the peak on the leeward side of the fence is positioned at about $x/H =1.0$, which does not agree with the result of the wind tunnel test. The accumulation peak in the wind tunnel test is higher on the leeward side than on the windward side. However, the peak is higher on the windward side in the field test.

It can be deduced that the difference in the peak position on the leeward side is due to the following reasons.

(1) Difference between the wind profile and turbulence intensity of the wind tunnel test and the field test

(2) Difference between the dimensional characteristics of the wind tunnel test fence and the field test fence, that is, in the field test, three-dimensional effects occur because the width of the fence is not sufficient in comparison to its height

(3) Difference in the flow characteristics, that is, in the field test, the strength of the separated flow and the flow characteristics on the leeward side of the fence fluctuate because the direction of the wind shifts constantly

SNOW DEPTH (mm)

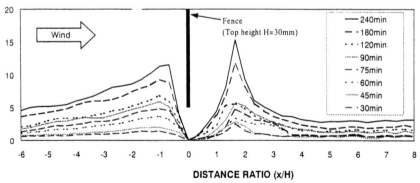

Figure 5. Difference of Model Snow Accumulation around Fence by Supply Time of Material.

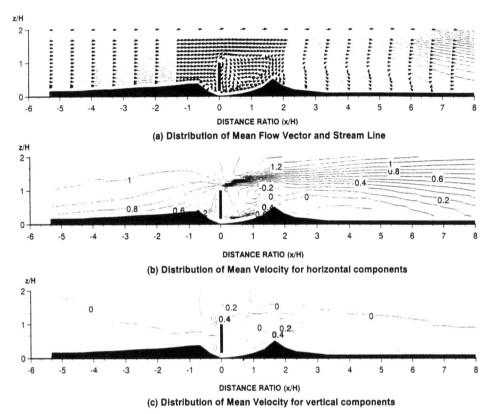

(a) Distribution of Mean Flow Vector and Stream Line

(b) Distribution of Mean Velocity for horizontal components

(c) Distribution of Mean Velocity for vertical components

Figure. 6. Distribution of Mean Flow Vector, Stream Line, Horizontal Velocity and Vertical Velocity around of an Aperture Fence and Snow Accumulation.

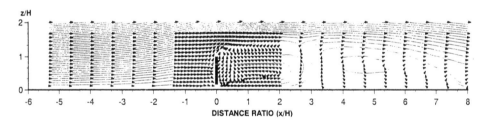

Figure.7   Distribution of Mean Flow Vector and Stream Line around a fence without snow accumulation.

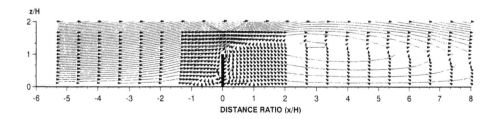

Figure 8.   Distribution of Mean Flow Vector and Stream Line around a fence without opening at the bottom.

401

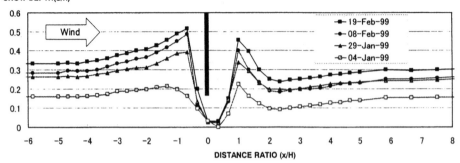

Figure. 9. Snow Accumulation of Field Tests

However, the cause of the higher accumulation peak on the leeward side cannot be fully explained.

4. CONCLUSION

This paper has described snow accumulation patterns as well as the characteristics of the flow in the case of the snow accumulation around a two-dimensional fence with an opening in its lower part, which were obtained from wind tunnel tests. The characteristics of the flow around the fence and the area where the snow accumulates were made clear. Furthermore, the test results were compared with the snow accumulation patterns obtained from field tests. As a result, the snow accumulation patterns on the windward side of the fence could be fairly well reproduced. However, some differences were found between the wind tunnel test results and those from the field test. The relationship between snow accumulation patterns and characteristics of flow need to be further clarified by carrying out wind tunnel tests under consideration of the causes of these differences.

5.REFERENCES

Hongo, T., Tsuchiya, M., Tomabechi, T. & Ueda, H., 1997, Wind effects on snow accumulation around a two-dimensional fence, *Annual Report of Kajima Technical Research Institute, Kajima Corporation*, Vol. 45, 57-62 (in Japanese).

Toyoda, T. & Tomabechi T. 1992. Development of a wind tunnel for the study of snowdrifting. *Second International Conference in Snow Engineering* :207-214.

Tomabechi, T. & Endo A. 1985. The relation between the roof inclination and the forming conditions of the snow depth. *Journal of Structural and Construction Engineering, Trans. of AIJ.* 357:20-28 (in Japanese).

Tomabechi, T. & Endo A. 1986. Snowdrift formation around buildings. *J. of Snow Engineering of Japan.* 1:1-8(in Japanese).

Snow Engineering, Hjorth-Hansen, Holand, Løset & Norem (eds) © 2000 Balkema, Rotterdam, ISBN 90 5809 148 1

Forecast model of road surface temperature in snowy areas using neural network

M. Horii
Department of Civil Engineering, College of Engineering, Nihon University, Koriyama, Japan

T. Fukuda
Faculty of Project Design, Miyagi University, Japan

ABSTRACT: A forecast model using a neural network to predict what the winter road surface temperature will be three hours later is proposed to increase the efficiency of anti-icing programs. The results of the present study revealed that the road surface temperature in three hours can be predicted with considerable precision by having the model learnt the time series change of the road surface temperature.

1 INTRODUCTION

In snowy areas, highway maintenance authorities often spread chemicals such as sodium chloride on the road surface in order to increase the safety of winter traffic. However, these chemicals damage soil, plants, vehicles and roadside structures. Furthermore, the amount of salt used increases yearly, increasing the financial burden for these services. Therefore, a forecast model of road surface temperatures that would enable salt to be spread before the road surface freezes would be beneficial. The model must be simple and easy to apply in the daily maintenance of roads. However, such a practical forecast model has not yet been established.

In the present paper, a forecast model that uses a neural network to predict what the winter road surface temperature will be three hours later is proposed in an attempt to increase the efficiency of anti-icing programs.

2 NEURAL NETWORK

Artificial neural networks are computer models that are based on the operation of components of the human brain. These are powerful tools that can be used to solve complex problems. The neural network used in the present study is a multilayered network, as shown in Figure 1. The neural network consists of three layers: an input layer, a hidden layer and an output layer. The input neurons receive signals from outside the model, and send their outputs to the hidden neurons. The output neurons receive signals from the hidden neurons, and produce the final outputs.

If we denote the outputs of the hidden neurons and the output neurons by the variables o_j, y_k, respec-

tively, these are given by

$$o_j = f(u_j) = f\left(\sum_{i=1}^{I} w_{ji}^l x_i\right) \tag{1}$$

$$y_k = f(z_k) = f\left(\sum_{j=1}^{J} w_{kj}^u o_j\right) \tag{2}$$

where u_j are inputs to the jth neuron in the hidden layer; z_k are inputs to the kth neuron in the output layer; w_{ji}^l are the connection weights between the jth neuron in the hidden layer and the ith neuron in the input layer; w_{kj}^u are the connection weights between the kth neuron in the output layer and the jth neuron in the hidden layer; and x_i are inputs to the ith neuron in the input layer. The most commonly used activation function is a sigmoid function:

$$f(x) = \frac{1}{1 + e^{-x}} \tag{3}$$

Neural networks can be trained by adjusting the connection weights in order to abstract the relationships between the presented input data and the output data. This learning procedure can be achieved by minimizing the error between the desired outputs and actual outputs:

$$E = \frac{1}{2}\sum_{k=1}^{K}(y_k - y_{dk})^2 \tag{4}$$

where y_{dk} are the desired output signals of the kth neuron in the output layer. In general, the error backpropagation algorithm is applied in order to minimize this error function. However, the backpropagation learning algorithm has some problems; the backpropagation process requires too many learning cycles to converge to the correct weights; the learning process sticks to a local minimum of the error.

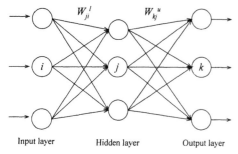

Input layer   Hidden layer   Output layer

Figure 1. Multilayered neural network

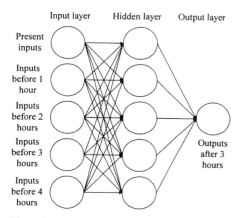

Figure 2. Basic architecture for forecasting road surface temperatures

In the present paper, a learning algorithm that is based on an extended Kalman filter has been introduced (Murase et al. 1994). We assume the following nonlinear state model (Katayama 1983):

$$w_{t+1} = F_t(w_t) + G_t v_t \qquad (5)$$

$$y_t = h(w_t) + n_t \qquad (6)$$

where w_t is the n-dimensional state vector at time t; y_t is the p-dimensional observation vector; F_t is the $n \times n$ state transition matrix; G_t is the $n \times m$ driving matrix; v_t is the zero-mean m-dimensional plant noise; and n_t is the zero-mean p-dimensional observation noise with covariance matrices

$$E\left\{ \begin{pmatrix} v_t \\ n_t \end{pmatrix} \begin{pmatrix} v_\tau^T & n_\tau^T \end{pmatrix} \right\} = \begin{bmatrix} Q_t & 0 \\ 0 & R_t \end{bmatrix} \delta_{t\tau} \qquad (7)$$

where Q_t is the matrix of size $m \times m$; R_t is the matrix of size $p \times p$; and $\delta_{t\tau}$ is the Kronecker delta.

Here, the connection weights are the state vector and are assumed to be time invariant. Therefore, the state equation can be expressed in place of Equation 5 as follows:

$$w_{t+1} = I w_t \qquad (8)$$

where I is the identity matrix and the state transition matrix F_t in Equation 5 reduces to I.

Since $h_t(w_t)$ is a nonlinear function, this function is linearized around the estimate of the state. Then, an extended Kalman filter algorithm can be applied to the estimates of the states. By linearizing the observation equation, we can derive the following expression to replace Equation 6:

$$\eta_t = H_t w_t + n_t \qquad (9)$$

where

$$\eta_t = y_t - h_t(\hat{w}_{t/t-1}) + H_t \hat{w}_{t/t-1} \qquad (10)$$

and H_t is the gradient matrix that results from linearizing the network and is defined by

$$H_t = \left(\frac{\partial h_t}{\partial w_t} \right)_{w=\hat{w}_{t/t-1}} \qquad (11)$$

Using the equations of the state and the observations, a method for updating the estimation of the system can be derived. The update equations are

$$\hat{w}_{t/t} = \hat{w}_{t/t-1} + K_t[y_t - h_t(\hat{w}_{t/t-1})] \qquad (12)$$

$$K_t = P_{t/t-1} H_t^T (H_t P_{t/t-1} H_t^T + R_t)^{-1} \qquad (13)$$

$$P_{t/t} = P_{t/t-1} - K_t H_t P_{t/t-1} \qquad (14)$$

where $\hat{w}_{t/t}$ is the Kalman filter estimate of the state at time t based on its one-step prediction $\hat{w}_{t/t-1}$; K_t is the Kalman gain; and P_{tt} is an approximate error covariance matrix.

3 APPLICATION

3.1 Learning model of road surface temperatures

We applied the neural network model based on the extended Kalman filter in order to predict road surface temperatures at two sites of the Tohoku expressway in Fukushima prefecture; one site is located in the embankment section (A site), the other is located in the cut section (B site). The survey data from January to February in 1996 and 1997 were provided by Tohoku Bureau of Japan Highway Public Corporation (Japan Highway Public Corporation 1996, 1997). These data consist of air temperatures, road surface temperatures, temperatures at 5 and 10 centimeters underground, wind velocities, and the amounts of rainfall and snowfall per hour (Horii et al. 1999).

The model structure is a three-layer network, as shown in Figure 2. We used the time series change of data as inputs to the input neurons and the road surface temperatures after three hours as outputs to the output neurons. Three hours correspond to the amount of time period required for a salt spreader to start from a snow removal station and complete the spreading operation.

The selection of the model was based on the ability of learning and testing. Here, we used a data set for learning for a period of seven days, from the 16th to the 22nd of February in 1996, and a data set for testing for the following seven-day period, from the 23rd to the 29th of February in 1996. First, we constructed the learning model by varying the input neurons and hidden neurons for each input variable, then we predicted the outputs of these models using the testing data set, which had not been previously presented to the models. The iteration times for learning were fixed as 100 iterations, taking into account practicability.

Tables 1 and 2 show the best results of the correlation coefficient at A site and B site, respectively, which were obtained by changing input variables and the number of neurons in the input and the hidden layers. The models that dealt with the road surface temperatures as the input variable were found to have the best learning and testing abilities.

Figures 3 and 4 show the learning and testing results that were obtained using the road surface temperatures as the input variable for A site. Figures 5 and 6 show the same results for B site. In every case, the predicted values follow the observed values, and we can obtain models having high-precision learning and testing.

Table 1. Results of correlation coefficients according to input variables (A site)

Input Variables	Number of input neurons	Number of hidden neurons	Correlation coefficient Learning data	Testing data
Air temperatures	5	9	0.929	0.749
Road surface temperatures	5	10	0.969	0.875
Temperatures at 5 cm underground	5	10	0.956	0.771
Temperatures at 10 cm underground	5	8	0.948	0.686

Table 2. Results of correlation coefficients according to input variables (B site)

Input Variables	Number of input neurons	Number of hidden neurons	Correlation coefficient Learning data	Testing data
Air temperatures	5	5	0.900	0.825
Road surface temperatures	5	7	0.905	0.862
Temperatures at 5 cm underground	5	6	0.912	0.750
Temperatures at 10 cm underground	5	7	0.882	0.830

Table 3. Classification results of the neural network model (A site)

	Prediction Below zero	Over zero
Below zero (observation)	28	11
Over zero (observation)	2	60
Ratio of correct classification	(28+60)/101=0.871	

Table 4. Classification results of the model obtained by multiple regression analysis (A site)

	Prediction Below zero	Over zero
Below zero (observation)	34	5
Over zero (observation)	26	36
Ratio of correct classification	(34+36)/101=0.693	

Table 5. Classification results of the neural network model (B site)

	Prediction Below zero	Over zero
Below zero (observation)	64	1
Over zero (observation)	10	26
Ratio of correct classification	(64+26)/101=0.891	

Table 6. Classification results of the model obtained by multiple regression analysis (B site)

	Prediction Below zero	Over zero
Below zero (observation)	63	2
Over zero (observation)	6	30
Ratio of correct classification	(63+30)/101=0.921	

3.2 Evaluation of the generalization of the model

Next, we evaluated the generalization of the neural network model, which is defined as the probability of outputting the correct outputs for the forecast of the road surface temperatures for an additional time period. For this purpose, we used the network mentioned in Tables 1 and 2, in which the road surface temperatures for five hours were input to the input layer, and forecasted the road surface temperatures three hours later for a period of seven days from the 16th to the 22nd of February in 1997. Comparing these results with the results obtained by multiple regression analysis, which are applied to forecast the field data now (Suzuki et al. 1993), we evaluated the generalization of the network model.

The procedure for the evaluation of the generalization is as follows: First, the data for night (from 19:00 to 9:00 the next morning) is examined. Secondly, the observed and the predicted data for three hours later are divided into four groups as follows: both data are over 0℃; both data are below 0℃; the observed data are over 0℃ whereas the predicted data are below 0℃; the observed data are below 0℃ whereas the predicted data are over 0℃. Finally, we calculate the ratio of correct classification.

Figures 7 and 8 show the prediction results for the road surface temperatures after three hours for A site and B site from the 16th to the 22nd of February in 1997. These results indicate similar shape of observed and predicted road surface temperature curves. In particular, these results are successfully fit in the range near zero. Tables 3 and 4 show the classification tables for A site obtained by the neural network model and by the model of multiple regression analysis, respectively. Tables 5 and 6 are the same classification tables for B site. The neural network model has the same forecast precision as the multiple regression analysis model, which is currently used to forecast road surface temperatures. The ratios of correct classification for the proposed neural network model, 0.871 and 0.891, are comparatively high. The total classification shows good fit, indicating that these models have the generalization for prediction over an additional time period. Therefore, the proposed neural network model can easily be applied in the daily maintenance of roads.

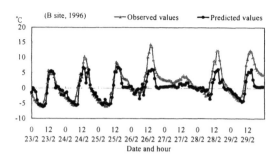

Figure 6. Testing results of road surface temperatures

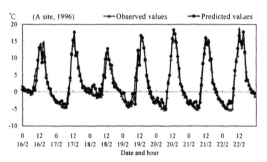

Figure 3. Learning results of road surface temperatures

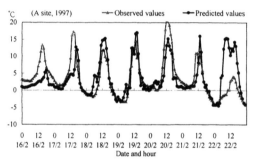

Figure 7. Forecast results of road surface temperatures

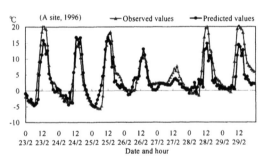

Figure 4. Testing results of road surface temperatures

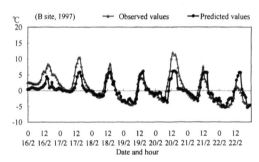

Figure 8. Forecast results of road surface temperatures

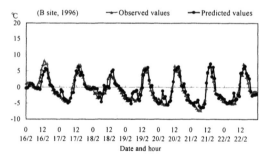

Figure 5. Learning results of road surface temperatures

4 CONCLUSION

In the present study, using a neural network, we attempted to establish the forecast model of the road surface temperatures after three hours which would enable anti-icing work before the road surface freezes.

The results are as follows:

1 The neural network model could be applied to forecast road surface temperatures.
2 This model has the ability to forecast road surface temperatures with high precision.
3 The generalization of this model was examined.

We are studying to establish a forecast model of road surface conditions by combining the proposed model and the detection model of water on road surfaces.

406

REFERENCES

Horii, M., Kato, K. & Fukuda, T. 1999. Pavement temperature forecast model using neural network for maintaining winter roads. *Journal of materials, concrete structures and pavements 620.* Japan society of civil engineers: 271-278.

Japan Highway Public Corporation 1996. Data of meteorological observation.

Japan Highway Public Corporation 1997. Data of meteorological observation.

Katayama, T. 1983. Applied Kalman filter: Asakura.

Murase, H., Koyama, S. & Ishida, R. 1994. Kalman neuro computing by personal computer: Morikita.

Suzuki, T., Amano, K. & Hirama, S. 1993. Research on new-system for predicting freezing on road surfaces. *Technical report of Research Institute of Japan Highway Public Corporation* 30: 179-190.

Snow Engineering, Hjorth-Hansen, Holand, Løset & Norem (eds) © 2000 Balkema, Rotterdam, ISBN 90 5809 148 1

Pavement snow-melting system utilizing shallow layer geothermal energy

S. Kamimura
Oyama National College of Technology, Japan

K. Kuwabara
Kowa Company Limited, Niigata, Japan

T. Umemura
Nagaoka University of Technology, Japan

ABSTRACT: Several ten meter deep ground have almost constant temperature equal to annual average air temperature at the place, such as 13 °C at Nagaoka, Japan. Region of relatively warm climate but having much snow, such like north-west coast of Japan, geothermal energy is promising as a heat source for pavement snow melting. A practical snow-melting system utilizing shallow layer geothermal energy for a pedestrian walkway was constructed in Nagaoka City in 1998. Performance test of the system in two winter season revealed that it has possibility to have enough ability to keep clear on the pavement .

1 INTRODUCTION

In the northwest coast of Japan, almost all roads are cleared by snowplowing vehicles or sprinkle snow-melting system in these days. Meanwhile the necessity of pavement snow-melting is enhancing in many cases such as exit and entrance of tunnels of highway, pedestrian walkway in urbanized area to prevent slipping, and so on.

Several kinds of snow-melting system utilizing natural and/or exhausted energy as heat source, such as geothermal energy, sea water, wind power, exhaust heat of vehicles and so on, are trying to be adopt as heat source of the system . Especially geothermal energy systems already have several practical execution cases because of following advantages:

i. Several ten-meter deep ground has almost constant temperature equal to annual average air temperature at the place.
ii. Possible to use it at anyplace.
iii. Inexhaustible natural resource and not to make anything pollute.

Morita (1997) introduced a technique of the Downhole Coaxial Heat Exchanger (DCHE), which had developed for a power generation utilizing geothermal hot dry rock sources, to snow-melting. A practical system combined with heat-pump for 266 m² roadway snow-melting constructed in Ninohe, Aomori in 1997. Fukuhara et al. (2000) developed a new system without heat-pump system, named the Borehole Heat Exchanger System (BHES), and constructed a practical system in Muraoka, Hyogo in 1998. They adopted polyethylene outer-pipe for heat exchanger. Both systems proved these have enough ability of snow-melting for practical use. However, in these cases, temperature of the ground is exceptionally high comparing ordinary place. Summary of these systems are listed in Table 1.

The aim of this research is to investigate basic performance of the system in the area of thermally strict condition, then to develop effective operating method.

2 PRINCIPLE OF THE SYSTEM

Schematic figure of heat exchange pipe is shown in Figure 1. Coaxial pipes are put down to a bore hole of 200 mm diameter. Cold water is fed into the space between outer pipe and inner pipe, then exchanges heat with ground soil during running through the pipe toward the bottom of the pipe. Warmed water return to the ground surface through the inner pipe, then it deliver to snow-melting panels. Heat carrying water is circulating in close-circuit, therefore it does not pumping any ground water but only ground heat. So we named the system as the Geothermal Heat Exchange Well (GHEW). Specifications of the system is shown in Table 1 comparing with two other systems mentioned above.

3 PRELIMINARY EXPERIMENT

3.1 *Facilities*

Experimental facility of GHEW was constructed in Sekihara, Nagaoka. It consists of two types of GHEW having steel and polyethylene outer pipe, and snow-melting panels of 63.6 m² area. Specifications of the system are shown in Table 1. Layout of the system is shown in Figure 2.

Flow rate is controlled by an inverter-controlled pump, and is measured by a propeller-type flow meter. Two GHEWs are independent, whereas these are switched by a three- way- valve.

There are two ground soil layers in the planning area from the ground line to 25 m deep in the ground: clay layer from 0 to 22 m deep and gravel layer soaked with ground water from 22 to 25 m deep.

3.2 Experimental conditions

Preliminary experiments were done in 6 flow rate ranged from 10 to 55 l/min for two GHEWs. All the experiments started around 6 p.m. in order to omit the effect of solar radiation, and complete about 15 hours after the starting time, These experiments were carried out in January 31 through March 3, 1999. There was enough amount of snow cover on the snow melting panels, i.e., it could be assume the surface temperature of snow-melting panels as 0 °C during the experiment period.

3.3 Experimental results

Experimental results are shown in Figure 3 and 4. Circulating water temperature at GHEW outlet exponentially decreases with the time. Initial temperature of more than 10 °C fell down to less than 5 °C in every flow rate for steel outer pipe system. Polyethylene outer-pipe system, temperature falling is more faster comparing with steel outer-pipe. It becomes almost constant value of

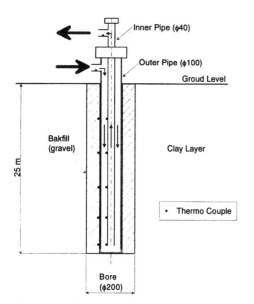

Figure 1. Heat Exchange Pipe (GHEW).

around 3 °C in every flow rate condition. Circulation water temperature at GHEW inlet almost constant value of about 1.5 °C, It means the radiating pipe running through snow- melting panel is long enough.

Figure 5 shows cumulative heat calculated with outlet and inlet temperature, T_i, using following equation:

$$Q = \Sigma C_p \rho V (T_o - T_i) \qquad (1)$$

where C_p (= 3.5 kJ/kg/°C) and ρ (= 1050 kg/m³) is specific heat and density circulating antifreeze water, and V (m³/min) is flow rate. Here T_i is assumed constant value of 1.5 °C. Steel outer-pipe system have approximately twice capacity of heat as well as polyethylene outer-pipe system except flow rate of 55 l/min. Heat flux (gradient of the curves) of steel outer-pipe system gradually decreases, while it of polyethylene outer-pipe system keeps constant value. It should be noted steel outer-pipe system have an advantage of high heat capacity in short term, but it is needed to take recovery time after heat gathering operation. While the polyethylene outer-pipe system has the low heat capacity but it could use continuously.

4 PRACTICAL SYSTEM

4.1 Facilities

A practical system of GHEW for pedestrian walkway snow melting was constructed in the Echigo Hillside Park, Nagaoka, Japan in 1998. Temperature of soil at 25 m deep in the ground is about 13 °C, which is equal to annual mean air temperature at the place. There is only clay layer in 25 m deep ground, which is generally low heat conductivity compared with gravel layer soak with ground water.

Figure 6 shows plan view of practical GHEW snow-melting system constructed in the Echigo Hillside Park, Nagaoka. Four GHEWs have installed in 6 meter interval. Inlet and outlet pipe of these pull in to header box, which has function to switches active GHEW. Warmed water from GHEWs delivers to 8 snow-melting panels. After radiating heat in snow melting panes, it returns to header box.

Figure 7 shows the cross-section view of the pavement. Radiating pipes having 15 mm diameter placed on under layer concrete. Surface asphalt of 30 mm covered over the radiating pipes.

In the planning area, as there is only clay layer and no ground water flow down to the 25 meter-deep ground which is heat source of the system, as mentioned above, therefore resin-coated steel outer-pipe system was adopted considering high initial performance of heat collecting and corrosion resistance in long-life-term. In addition switching operation using plural GHEWs (Multi-GHEW) is adopted for the system.

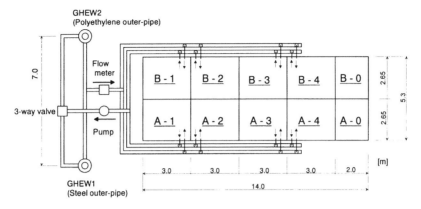

Figure 2. Schematic figure of experimental facility of the Geothermal Heat Exchange Well (GHEW) for pavement snow- melting constructed in Sekihara, Nagaoka.

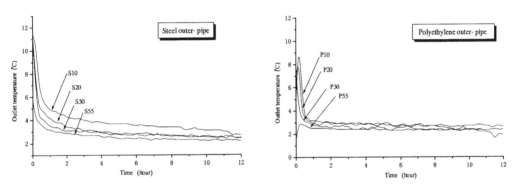

Figure 3. Outlet water temperature in prelimina ry experiments.

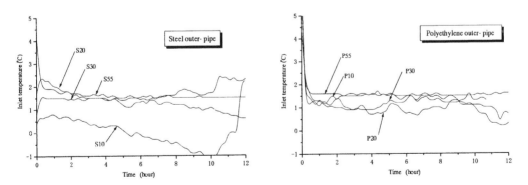

Figure 4. Inlet water temperature in preliminal experiments.

411

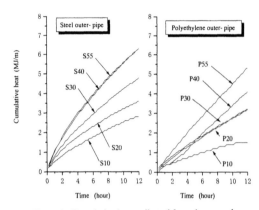

Figure 5. Cumulative heat collected from the ground.

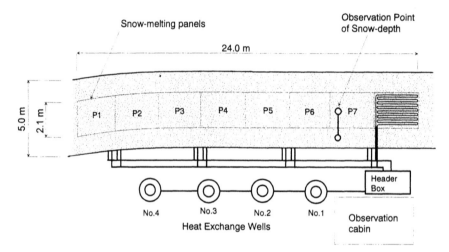

Figure 6. Plan view of a practical system of GHEW for pavement sniw- melting in the Echigo Hillside Park in Nagaoka. Japan.

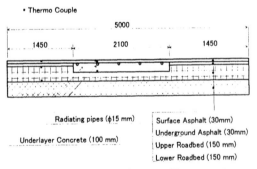

Figure 7. Cross- section view of snow- melting pavement.

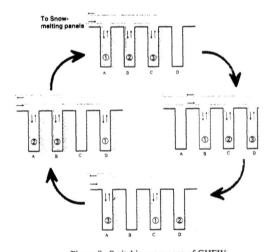

Figure 8. Switching sequence of GHEWs.

412

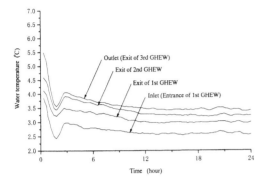

Figure 9. Circulating water temperature at exit point of three GHEW sin series.

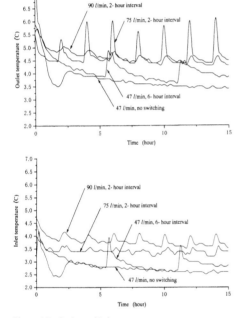

Figure 10. Outlet and Inlet temperature of the Practical System

4.2 Switching operation

Switching sequence is shown in Figure 8. In the first stage, GHEW of A, B and C connect in series. First 25 m, A, must be exhausted among three GHEWs after heat-collecting operation. So A-GHEW is bypassed and makes ground temperature recover in next stage. D-GHEW, already get recovery, put into the end of the series, i.e., in next stage, B, C and D connect in series. This switching rule is applied to 4 stages shown in Figure 8.

4.3 Experimental conditions

Effect of switching interval and flow rate is investigated by four experiments shown in Table 3. Flow rate of Exp. 0 and 1 is 47 l/min and it of Exp. 2 and 3 is higher flow rate. Exp. 0 has no switching operation. Exp. 1 has 6-hour switching interval and other two experiments has 2-hour switching interval. Shorter interval is taken with expectation to use initial high performance in steel outer-pipe system according to the result of preliminary experiments.

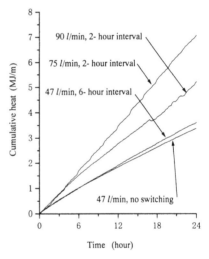

Figure 11. Cumulative amount of heat c ollected from GHEWs.

Table 1. Specifications of the GHEW sys tem comparing ot her systems.

		GHEW		DCHE	BHES
Place		Nagaoka (1998)	Nagaoka (1998)	Ninohe (1995)	Hyogo (1997)
Pavement		(only for experiment)	Pedestrian way in the Echigo Hillside Park	Roadway	Pedestrian way in the Hachikita Parking
Flow rate	l/min	11~54	47 ~95	unknown	475
Snow melting Area	m^2	63.6	50.4	266	310
Ground Temp.	°C	13	13	13~22	18
Diameter of pipes		100 + 40	100 + 40	89+unknown	90+56
Outer- pipe material		Steel / Polyethylene	Resin coated steel	Stainless Steel	Polyethylene
Depth × Number		25m / 25m	25m×4	150m×3	100m ×12
remarks		-	-	combined with heat pump	-

413

Table 2. Experimental condition.

Exp.name	Outer pipe material	Average flow rate (l/min)	Starting time
S10	steel	11.0	'99/2/19 20:45
S20	steel	20.8	'99/3/2 18:15
S30	steel	30.9	'99/2/24 18:15
S40	steel	43.4	'99/2/17 18:15
S50	steel	50.5	'99/2/3 18:45
S55	steel	53.7	'99/2/15 18:15
P10	polyethylene	10.8	'99/1/31 18:37
P20	polyethylene	20.8	'99/2/25 18:14
P30	polyethylene	28.0	'99/2/16 18:15
P40	polyethylene	41.7	'99/2/2 18:25
P50	polyethylene	52.3	'99/2/14 18:25
P55	polyethylene	53.8	'99/2/20 18:05

Table 3. Experiment condition for the practical GHEW system.

	Flow rate (l/min)	Switching Interval (hour)	Starting time	Days
Exp. 0	47	N/A	'99/1/08 9:00	2
Exp. 1	47	6	'99/1/13 16:30	55
Exp. 2	75	2	'99/12/21 13:00	8
Exp. 3	90	2	'00/01/21 11:10	5

4.4 Experimental results

4.4.1 Effect of series connection of GHEW

For effective collection of heat, a strategy of series connection of three GHEWs was taken. Figure 9 shows the result of Exp. 0. Water temperature of the entrance of 1st GHEW is initially around 3 °C, then gradually decreases with time. Water temperature at the exit of 3rd GHEW is initially around 4 °C, then also gradually decreases with time. Therefore temperature difference between inlet and outlet keeps almost constant value of 1.0 °C. Concerning bout contribution of each of 3 GHEWs, values of temperature deference at each exit are approximately 0.5, 0.4 and 0.1 °C for 1st, 2nd and 3rd GHEW until 8 hours. After that they becomes approximately 0.4, 0.3 and 0.2 °C, respectively. It is to say series connection works well as to be expected.

4.4.2 Effect of switching interval

Figure 10 shows inlet and outlet water temperature of these four experiments. In Exp. 1, outlet water temperature falls down to 3.7 °C after 6-hour continuous operation.

Table 4 Amount of heat gained from the ground in first 12- hour in all the experiments

	flow rate (l/min)	switching interval (h)	amount of heat collected 0 - 25 m (MJ)	25 - 50 m (MJ)	50 - 75 m (MJ)	total (MJ)	per pipe (W/m)	per panel (W/m²)
Steel outer- pipe system								
S10	11.0	N/A	70.4	-	-	70.4	65.1	32.3
S20	20.8	N/A	90.0	-	-	90	83.3	41.3
S30	30.9	N/A	119.5	-	-	119.5	110.6	54.9
S40	43.4	N/A	157.0	-	-	157	145.4	72.1
S50	50.5	N/A	153.7	-	-	153.7	142.3	70.6
S60	53.7	N/A	157.6	-	-	157.6	146.0	72.4
Polyethylene outer- pipe system								
P10	10.8	N/A	38.6	-	-	38.6	35.7	17.7
P20	20.8	N/A	80.4	-	-	80.4	74.4	36.9
P30	28.0	N/A	79.0	-	-	79.0	73.1	36.3
P40	41.7	N/A	102.3	-	-	102.3	94.7	47.0
P50	52.3	N/A	90.2	-	-	90.2	83.5	41.4
P60	53.8	N/A	133.4	-	-	133.4	123.5	61.3
Resin- coated steel outer pipe with switching operation								
Exp0	47	N/A	67.9	44.4	18.3	130.5	40.3	59.9
Exp1	47	6	79.6	52.0	21.4	153.0	47.2	70.3
Exp2	75	2	135.3	88.5	36.4	260.3	80.3	119.5
Exp3	90	2	108.0	70.6	29.1	207.8	64.1	95.4

Table 5 Coefficient of performance of the system

	area (m²)	total heat (kWh)	power consumption (kWh)	COP
GHEW (Exp1, first 12- h)	50.4	42.5	4.4	9.6
GHEW (Exp2, first 12- h)	50.4	72.3	5.4	13.3
GHEW (Exp1, 55 days)	50.4	2,131	475	4.5
BHES (Yokotani facility)	1,625	176,301	20,611	8.6
DHCE (Ninohe facility)	266	(22,732)	6,686	3.4

While in the other two experiments, outlet water temperature keeps more than 4.2 °C. Seeing inlet water temperature in Figure 10, higher flow rate lead higher temperature. That is why higher flow rate makes smaller releasing heat at radiating pipe.

4.4.3 *Cumulative heat collected*

Cumulative amounts of heat collected form GHEWs in four experiments are shown in Figure 11. Higher flow rate and shorter switching interval brings greater amount of heat except Exp. 4. Especially Exp. 3, 75 l/min flow rate and 2-hour switching, shows double capacity of heat as well as Exp. 0 and 1. Seeing the gradient of the Exp. 3 curve, i.e. heat collecting rate, is almost constant while it of other experiments gradually decrease. It is obviously proof of the effectiveness of switching operation.

4.4.4 *Practical use test*

Figure 12 shows a result of 55 days continuous operation in 1999 winter season with 6-hour switching interval and 47 l/min flow rate. Upper chart shows depth of snow cover on snow-melting panel and no-snow-melting pavement. Automatic observation data in nature in the park is also shown in the chart.

Lower chart shows daily average temperature of inlet water and air temperature. Gray zone shows snow-covered period on the snow melting panels. During the winter period, maximum depth of snow cover without snow melting is 115 cm and there was snowfall in 53 days. Outlet and inlet water temperatures are almost constant value of 4 °C and 3 °C the snow-melting panels respectively when these are covered with snow. Temperature difference between inlet and outlet water temperature and flow rate with Equation 1 yield average amount of heat radiated at the panels as 57 W/m².

Assuming density of newly fallen snow as 70 kg/m³, it corresponds to 20 cm snowfall. Generally, more than 150 W/m² capacity of snow melting is required to keep clear on pavement in Nagaoka area. So it must be say current performance of the system does not have enough ability, but, as shown in Figure 12, it has fairly good performance for practical use.

Practical use test carried out only in 1999 winter because it was very scarcity snow in 2000 winter. Table 4 listed the summary of all the experiments mentioned above. Comparing the experimental facility and the practical facility, for example S50 and Exp1 having almost same flow rate, amounts of heat collected in from 0 to 25 m deep ground are 152 and 80 MJ, respectively. The reason is thought because of the effect of 3 meter gravel layer of the experimental facility. Ground water flow through the gravel layer supplies heat continuously. It should be noted that the practical facility exert a performance equal to the experimental facility by introduction of switching operation, in spite that it has serious disadvantage that there is only c lay layer.

Improved performance by introducing switching operation is still short for the heat capacity to keep clear on pavement required in Nagaoka. However it is shown its possibilities to obtain good result without aux iliary heating.

5 DISCUSSION

Coefficient of performance (COP) of the system is shown in Table 5 comparing with other systems. In short period, it has excellent COP value especially such condition as Exp2. Unfortunately there is no data about long term experiment in condition of Exp2, however it is easily expected that the COP value would be several ten percent better than Exp0 condition. As shown in Figure 12, there is no snowcover in more than half of the period. So pump operation must control with some kind of snowfall or snowcover sensor. When do that, on-off operation should bring great amount of heat loss at the time restarting for warming up the whole system. To prevent the loss of heat and to reduce power consumption, introduction of inverter-controlled pump working with a snow sensor is a better idea.

When a snow-melting system designs, statistical maximum snowfall intensity is generally used, therefore almost all systems have excess performance for its safety. When utilizing natural energy for snow melting, there are two serious problems: heat source being relatively low temperature and difficult to keep performance. It is thought that the switching operation would be a solution for these problems. It is thought that an idea to introduce variable switching interval control system also effective to counter fluctuating snowfall intensity.

6 CONCLUDING REMARKS

Basic performance of Vertically buried coaxial heat exchange pipe is investigated. Based on obtained knowledge, a practical snow-melting system utilized shallow layer geothermal energy designed and constructed in the Echigo hillside Park for pedestrian walkway snow-melting. In spite of thermally strict condition, which there is only clay layer in the ground, it proved good performance with introducing switching operation of plural heat exchange pipes. Obtained performance is still short to keep clear on pavement, it is shown its possibility to use practically with no aux iliary heating.

ACKNOWLEDGMENT

This research project have partly supported by a grant from the Foundation of Japanese Snow Union

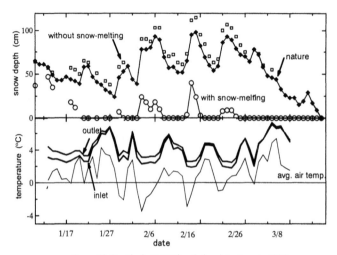

Figure 12 Practical use test in whole winter term in 1999.

(Sekisetsu-Rengo) in 1997 through 1999. The Administration office of the Echigo Hillside Park kindly offered an experimental field for our practical facilities. The authors give an appreciation for the persons whom concerned.

REFERENCES

Fukuhara, T. 2000. Performance of a Snow Melting Pavement that Couples a Bore hole Heat Exchange System with a High Thermal Conductivity Pavement. Proceedings of ICSE- 4. (in press)

Morita, K. 1997. Melting Snow with the downhole coaxial heat exchanger (DCHE), Geothermal Resources Council BULLETIN, Vol. 26 No. 3, 83- 85

Snow Engineering, Hjorth-Hansen, Holand, Løset & Norem (eds) © 2000 Balkema, Rotterdam, ISBN 90 5809 148 1

Operation of roads exposed to drifting snow in Northern Europe

Harald Norem & Skuli Thordarson
Norwegian University of Science and Technology, Trondheim, Norway

ABSTRACT: An European project (ROADEX), within the EU Northern Periphery Programme, on the operation of roads in winter climates, was started in 1999 and will last for three years . The participating countries are; Finland, Sweden, Norway, Iceland and Scotland. The aim of the study is to exchange experience and carry out a State-of–the-art study on two separate issues: (A) the management of winter roads due to traffic loads, load restrictions and how to improve road conditions, and (B) design, maintenance and operation of roads exposed to drifting snow conditions. The paper deals with the practice within the five countries on the procedures to operate roads exposed to severe climatic conditions, mainly drifting snow. The factors concentrated on are;

- safety procedures for controlling the traffic and to the extremes, to close the road
- agencies responsible for closing roads
- collecting information about weather and driving conditions, and the use of these in the daily maintenance and operation of the roads
- information on weather and driving conditions submitted to the road users

For each of these factors a comparison of the procedures for each country is presented as well as recommendations for a best practise procedure.

1 INTRODUCTION

The European Union (EU) has a program for creating effective exchange and co-operation between the northern regions of Europe, the Northern Periphery Program. As a part of this program EU is financing a project between the road districts in northern part of Finland, Sweden and Norway and the road administration of Iceland and Scotland. The project is called ROADEX. It started late 1998 and will be finished within 2001, with a total budget of 500 000 EUR

The main reason to start such a project is that the participating countries realised, that despite several differences, they also shared a lot of common problems and challenges, of which the most important are:

- Harsh winter climate with difficult driving conditions, e.g. drifting snow, poor accessibility and low friction
- Remote roads with low traffic volumes, where main part of the traffic is long distance traffic.
- The remote areas are facing reduced financial resources

- A substantial experience and a variety of engineering solutions are held locally and are seldom known outside the regions

The goal of the project is thus to establish a network and to encourage exchange of experiences as well as to establish state-of-the-art knowledge and to identify the best practice procedures.

2 DESCRIPTION OF THE PROJECT

2.1 Subprojects

The project is divided in two subprojects with the following goals:

A) Road condition management:
- Identification of measures to manage bearing capacity problems on low traffic roads in northern regions
- To establish knowledge to reduce traffic load restrictions and expensive road rehabilitation measures

B) Winter maintenance
- Identification of the-state-of-the-art maintenance techniques, traffic information systems and snow drift measures

- Improving accessibility and safety during difficult driving conditions
- Harmonising road maintenance standards

The present report presents the scope, information assembled and the preliminary conclusions of the latter sub-project.

2.2 Specific interests for the winter maintenance project

The workload in the winter maintenance part of the project has generally been concentrated on the management and maintenance of roads in remote areas exposed to harsh winter climates. The project has mainly focused on:

Road design
- Location due to wind, snow and topography
- Cross-section profiles
- Use of guard-rails
- Use of snow fences
- Interaction between road design and maintenance techniques

Operation procedures
- Co-operation with the weather service
- Safety during severe driving conditions
- Authority to close the road during extreme weather situations

Information to the traffic users
- Driving conditions
- Weather information

2.3 Climatic conditions in the Northern Periphery

The northern periphery shows a dramatic diversity concerning climatic conditions, from arid continental climate in central part of Scandinavia to pure maritime climate in the coastal areas in the northern Atlantic region. The transition from one climatic zone to another may be very distinct, and is often following watersheds.

The great variety of the climatic conditions is also reflected by the local experiences and solutions found within the different regions. For this reason a major aim of this project is not only to find best practice procedures, but also to relate them to specific climatic conditions.

The continental climatic regions are characterised by cold winters, limited amounts of snow, and few days with wind speeds above gale (12 m/s). There are thus few days with extremely difficult driving conditions. On the other hand, when strong winds occur, there is almost always sufficient dry, loose snow on the ground to be picked up and brought into transportation, causing heavy snow drifts and bad visibility. At the same wind speed the driving conditions are most often worse than in more mild climates.

The extreme opposite to the continental climate is the mild maritime climate found in Scotland, islands in Northern Norway and close to the shore in Iceland. These areas are characterised by high precipi-

tation rates and strong winds. The average temperature in the winters is close to 0^0 C, and the precipitation is either falling as snow or rain. The days with severe driving conditions have usually heavy snowfalls combined with strong winds. On such days there might be difficult to keep the roads open and the visibility may be poor.

A transition between these two climatic systems is what we have called the cold maritime climatic zones, which are mainly found in Iceland and along the coast and seafacing mountains of mainland northern Norway. These areas are extremely windy compared to other areas within Europe as well as the main part of the winter precipitation is snow. These conditions represent the very worst for keeping roads open in the wintertime, and roads exposed to snowdrifts are frequently closed every winter due to severe driving conditions.

2.4 Methods for assembling information

The data was assembled by having the participating countries filling out a questionnaire. During this work, the collected data has been available on the Internet for the different countries to review and harmonize the type of information submitted.

3 PRELIMINARY RESULTS

3.1 Design of roads in snow drifting areas

Obviously there should be an interaction between the design of the roads and the problems related to the maintenance and driving conditions. In our case those problems are mainly caused by the harsh winter climates.

To some surprise fairly minor attention has been paid to design "climatic-adapted" roads. In most cases the standard national guidelines for design of roads have been used also in areas facing both strong winds and heavy snow falls.

Norway has made a textbook for the design of roads in snow drifting areas (Norem 1995), but only the roads in the most extreme climatic regions have been specially designed for improving the maintenance and driving conditions. The most pronounced weaknesses are probably found on roads with intermediate snow problems, where no notice of the climatic conditions has been taken.

An interesting design for roads with intermediate problems is the use of "Snow blower lanes" in Norway, Figure 1. The idea of such lanes is to design the road in a way that the maintenance equipment may be used most effective. During snowstorm periods the maintenance crew is occupied with the removal of the snow from the lanes and to secure the traffic. In the calm periods there is less to do, and the roads should have a design to use the maintenance machines to prepare for the next snow storm period. The "Snow blower lanes" make it possible to use the snow blowers outside and parallel the road to create additional storage room for the snow. These lanes have shown to be very effective to delay the time for use of snow blowers during snowstorm periods as well as they improve the visibility.

Another speciality reported is the "winter roads" in Iceland, Figure 2. At road sections with heavy snow drifts an additional road with very limited standard is constructed where the snow seldom collects. This makes it possible to pass the critical section at a low speed.

The road element that probably has most negative influence on the tendency to cause snowdrifts on the road and poor visibility is the guard-rail. Any measures to avoid guard-rail and still obtain necessary safety due to driving off the road are advantageous. Experiments with more narrow guard-rails have been carried out in most of the countries. The experience indicates that narrow rails mounted along roads in continental climates create less snowdrifts on the road and improve the visibility. On roads in climate with moist snow and more snow on the ground, the snow wall collecting around the guard-rail reduces the positive effect of the narrow rail.

3.2 Operation of roads under difficult driving conditions

There seems to be great difference between the countries and regions on how roads are operated under severe driving conditions. This local practice is depending on the climatic conditions, traffic amount and traditions for responsibility developed for decades.

Roads are very seldom closed in Sweden and Finland, which both have typical continental climate. The driving conditions along those roads usually make it possible to keep the roads open with sufficient level of safety. However, to improve the safety Finland has introduced varying speed limit in the wintertime the last years. Depending on the driving conditions due to friction, snow depths and visibility the speed limit is decided and signs are adjusted to inform the road-users.

A speciality for Norway is the organised "convoy driving". When the visibility is too low to have safe and free traffic, the roads are closed by locked gates. It is then only allowed to pass the critical road section in company with a maintenance truck in front and rear. There will thus be only one-way traffic, and the maintenance crew has full control of all traffic at all times. The closures of the roads are most often a result of poor visibility, but major snow deposits on the road or high wind speeds with the risk of cars to be blown off the road may also cause closures.

Iceland, on the other hand, has developed a totally opposite approach to the operation of roads during snow storms. The roads will be signed "impassable", but the roads will never be closed by gates. This practice is probably a result of the road authorities not having the authority to close the roads, and the extensive Icelandic use of all-terrain four-wheel-drive cars, that sets reduced requirements to the removal of snow.

Scotland has less frequent snowfalls than the other participating countries in the project. However, when they have snow, the precipitation rates may be intense, and the Scottish roads then have to be closed due to too much snow on the road. In such cases the roads may be closed for several days, and not only for the storm periods, as is the general rule in the other countries.

3.3 Information of climatic and driving conditions

The operation of roads under severe winter conditions is dependent on good knowledge of the driving conditions at present and in future and effective information distribution to the road-users.

To have the best information on the driving conditions all countries report they have an organised co-operation with the national weather service. The co-operation varies from close contact to paying for detailed weather forecasts. In recent time the use of automatic weather stations has been more extensive. The data from the weather stations are usually available for the road authorities and the road information centres. In Iceland the information is also sent to the Meteorological Office and is available on the Internet. Sweden on the other hand reports that they obtain the data from the weather radar to have the best updating.

The information assembled by the automatic weather stations and video cameras has usually the purpose of:

- estimating time for icing on the road, and thus decide the use of friction control measures
- record conditions for drifting snow and poor visibility on the road
- record strong crosswinds, especially on bridges

Traditionally the information on road conditions is accessible by calling to an information centre, or when the road is closed temporarily signs have been used for information about the closure and recommended detours.

In recent times there has been an extensive use of new technology to inform the road-users. The information centres, as well as signs with varying information use local radio combined with RDS-radios. All countries have also developed Internet presentations. The minimum information is road closures and general information on the driving conditions. Some countries have in addition their weather stations and video cameras directly connected to the Internet.

Figure 1. Blower lane along a mountain pass in Norway.

Figure 2. "Winter road". Short bypass along a drifting snow exposed section.

4 CONCLUSIONS

The present paper can only give the preliminary results of the project as it will go on until 2001. The exchange of experiences and ideas has so far been fruitful. We really hope the the network established can result in further research, not only to assemble the-state-of-the-art, but to improve our practice to offer good and safe driving conditions under difficult weather situations.

ACKNOWLEDGEMENTS

Thanks are due to the EU Northern Periphery Programme, which is financing ROADEX.

REFERENCES

Norem, H. 1994. Snow Engineering for Roads, handbook no. 174. Oslo: Norwegian Public Roads Administration, Road Research Laboratory.

Snow Engineering, Hjorth-Hansen, Holand, Løset & Norem (eds) © 2000 Balkema, Rotterdam, ISBN 90 5809 148 1

Optimization of engineering decisions on road protection against snow-banks with snow properties taken into account

Ye. Prusenko, A. Dogadailo & I. Kiyashko
Kharkov State Automobile Highway Technical University, Ukraine

ABSTRACT: Consideration of regional climatic conditions influenced on snow properties, snow transferring during blizzards and forming of snow accumulations on the roads is the basic principle of engineering decisions optimization on road protection against snow-banks. The corrected procedure of snow-bringing volumes to the roads and the engineering decisions on roads snow-protection for various conditions of snow transferring are offered.

1 INTRODUCTION

Automobile transport holds a key place in a system of transport communications. The characteristics of motor transport work in a very strong degree are dependent on road's travel state in the winter period. Just in this season the roads are exposed to the strongest and adverse influence of the climatic factors. In the winter period traffic speed falls sharply and travel is being blocked sometimes.

The reduction of transport speed leads to its cost increase and to the transport losses. The average speed of cargo traffic on roads is 60-80 km/hour, but on snowy roads it is falling to several kilometers per hour. The comparison of possible transport losses and road winter maintenance cost allows to determine the profitability of the latter under those or other conditions. The calculations show that under large traffic volume the roads winter maintenance expenses in areas with intensive blizzards can be repaid within one day at the expense of transport losses averting.

The road-operational service often has some troubles with securing the travel on roads in winter period. The analysis of these troubles shows, that one of its main reasons is the insufficient efficiency of applied snow-protecting means.

On a considerable road extent the existing snow-protecting plantations work unsatisfactorily. They not only do not protect the roads against snow-banks, but often cause the formation of snow-drifts on a carriage-way because of its short distance location along a road and insufficient capacity. The cause of it is that when the engineering decisions on roads snow-protecting are worked out, the regional conditions of snow transferring and snow drifts forming are taken into account insufficiently. It results in mistakes when grounding the snow bringing volume to the roads - the basic estimated target for designing the snow-protecting measures, wrong choice of snow-retention facilities and schemes of their location. The capacity of snow-protecting plantations often is not geared to a road direction, which to a large measure determines the snow-drifting volume to the road and remains unchanged on the whole road extent. Therefore for a number of road sections the snow-protection is too much powerful, and for other sections its snow-capacity is insufficient, that results in snow-banks on roads. Whereas the extent of snow-drifted sections is highly considerable, the matter of finding the more effective and at the same time economical means of snow-protection is of very great significance.

The research laboratory of roads winter maintenance of Kharkov State Automobile Highway Technical University during already more than 30 years is concerning with resolving the basic questions of roads protection against snow-banks.

2 ELABORATION OF SNOW-DRIFTING VOLUMES TO THE ROADS

2.1 Analysis of calculated formulae

The theory of snows transferring and the mechanisms of snow-banks are a subject of studies already more than one hundred years. The modern conceptions on the blizzards behavior had been formed on the basis of classical works of A.K. Diunin, A.A. Komarov, D.M. Melnik, R.V. Bialobrhesky, A.A. Kungurtsen, N.E. Zhukovsky, S.A. Chaplygin and other scientists. The main task of snow-transferring theory is determination of blizzard's transporting capability or total snow flow rate. However, despite of considerable development of blizzard mechanics (Diunin 1963), the theoretical formulae consist of large number of the various parameters and variables, which determination with a necessary accuracy for a specific part of terrain is not possible. Therefore on the basis of an analysis of numerous blizzard measurements many authors suggest the empirical formulae for approximate determination of snow transferring intensity from the wind speed. The next formulae are most known:

Ivanov B.V. (1954)

$$Q = 0,0295 \cdot V_{1,0}^3.$$ (1)

Melnik D.M. (1952)

$$Q = 0,092 \cdot V_{1,0}^3.$$ (2)

Diunin A.K. (1954)

$$Q = 0,0334\left(1 - \frac{4}{V_{1,0}}\right) \cdot V_{1,0}^3.$$ (3)

Komarov A.A. (1954)

$$Q = 0,011 \cdot V_{1,0}^{3,5} - 0,68.$$ (4)

Akkuratov V.N. (1959)

$$Q = 21,7 \cdot V_{1,0}^2.$$ (5)

Bialobzhesky G.V. (195)

$$Q = \frac{2}{\pi} \cdot Q_{max} \cdot arctg\left(B \cdot V_{1,0}^3\right).$$ (6)

Kotliakov V.M. (1960)
a) for a combination of deflationary and upper blizzards

$$Q = 8,1 \cdot 10^{-3} \cdot \left(V_w^{2,2} - V_{w0}^{2,2}\right).$$ (7) b)

for deflationary blizzard with $\sigma_b < 1$

$$Q = 4,03 \cdot 10^{-3} \cdot \left(V_w^{2,8} - V_{w0}^{2,8}\right).$$ (8)

c) for deflationary blizzard with $1 < \sigma_b < 5$

$$Q = 6,34 \cdot 10^{-5} \cdot \left(V_w^{4,0} - V_{w0}^{4,0}\right).$$ (9)

d) for deflationary blizzard with $\sigma_b > 5$

$$Q = 1,11 \cdot 10^{-6} \cdot \left(V_w^{5,1} - V_{w0}^{5,1}\right).$$ (10)

where Q = total consumption of snow-wind flow, g/m·s; $V_{1,0}$ = horizontal component of wind speed, at height of 1 meter, m/s; Q_{max} = greatest possible solid consumption of blizzard flow, g/m·s; B = constant, determined by results of field studies; V_w = wind speed, measured at height of weathercock, m/s; and V_{w0} = wind speed by weathercock, corresponding to the beginning of snow transferring , m/s.

Data received on these formulae considerably differ for friable snow and firm snow-crust. All these formulae do not reflect maximal transporting ability of blizzard in the given conditions, i.e. its complete possible saturation with solid phase. The roughness of under-stretching surface and solid phase concentration have an influence on the kinematics of snow-

wind flow. The forces, binding surface particles with each other and with underlayer's particles, are internal forces in deflationed layer and do not affect on the wind dynamics and kinematics. They can change the speed-up length blizzard.

2.2 Regional conditions influence on blizzard

The total flow rate of blizzard reach its maximal possible value not at once. Blizzard development takes place deep into snow-collecting basin beginning from the zero flow rate at its weather side and gradually increasing further. Distance, at which total flow rate is being reached, depends on surface properties of snow cover, strength of wind and a number of other factors. This critical distance is called a blizzard speed-up length. If the road is located in a zone of blizzard speed-up, it will be subject to influence of unsaturated blizzard and volumes of snow-bringing will be less than on road that is outside this zone. In one's turn the blizzard speed-up length depends on temperature of snow and air, roughness of deflationed surface, wind action on the underlain surface, degree of sublimation hardening of snow cover surface layers, presence of surface "temperature crust', store of deflationed snow, degree of snow pollution, solar radiation, ruggedness of the terrain and presence of local obstacles.

When determinate the volumes of snow-bringing, it is necessary to take into account, that blizzards' snow-flakes, transferring by means of involving, saltation and simultaneous sublimational and mechanical grinding with final transition into steam, cannot infinitely for a long time to keep its sizes and forms. Therefore possible distance of blizzard's particles transferring by wind is limited and volumes of snow deposits caused by wind at the leeward boundaries of snow-collected basin are not proportionate to its sizes. We consider that in the formation of leeward snow accumulations takes part not the whole of snow-collecting basin, but its some part named an effective zone of snow-collection.

Blizzard speed-up length, processes of sublimation and mechanical grinding of snowy particles, the sizes of effective snow-collecting zone are the major factors that influence on the volumes of snow-bringing, largely depend on regional conditions. From the list of regional conditions only few are not functions of the given area climatic features, for example, ruggedness of the terrain and presence of local obstacles. Basic influence on parameters of blizzards is provided just by climatic factors – wind effect, precipitation, air temperature and air humidity, solar radiation. Just these factors determine the physico-mechanical properties of a snow cover. However, the flow rate method (Melnik 1954), as that has received the widest practical application, considers only wind parameter from the whole list of factors. This fact can be explained, on the one hand, by complexity of all accounted factors, and on the other hand, that the basic majority of researches had been carried out in severe conditions of Siberia, Far East, Antarctica. For these regions the thaws, wet snow-falls, winter rains and "wet blizzards" are a large rarity. At the same time in the territory of Ukraine and other countries of Central Europe such phenomena occur rather frequently and affect on the blizzard regime of the roads significantly.

The basic difference of blizzards in the regions with soft winters, that specify all the rest, is higher temperature of air during a winter period. A.K. Diunin (1963) has paid attention to some features of blizzards at high temperatures of air. He points out the distinct dependence of blizzard speed-up length against the air temperature. The higher temperature the more length of speed-up under other equal conditions. It is due to more coherency of snowy surface particles at high temperatures. Deflation conditions at high temperatures, close to 0°C, and with intensive solar radiation are becoming worse sufficiently. This effect is intensifying with snow cover pollution, which encourages an increasing of solar energy absorption in surface layer.

Blizzard speed-up length at the temperatures below -5°C, as the field studies have shown, has of the order of value 0.5 km and more. Temperature -5°C is a threshold of significant strengthening of snowflakes sticking together. Snow becomes damp or wet. A sharply increasing of snowflakes sticking together at the temperatures above -5°C is confirmed also by the fact, that under such circumstances the sieve control of snow structure becomes impossible.

Therefore blizzards at the temperature above -5°C have very large speed-up length, which reaches 1-2 km and more.

The thaws and winter rains lead in the end to decrease of snow-bringing volumes to roads. The short-term increase of air temperature +2°C does not affect essentially on a volume of snow accumulations, delayed by obstacles. But on a surface of a snow cover in snow-collecting basin, after such periods appear the various combinations of melted icy accretions.

It considerably worsens the deflation conditions and increases the blizzard speed-up length. At the higher temperatures during the thaw or with its large duration the snow particles on snow cover surface can get sluggish at all and deflation model will become impossible to using. The thaws of large duration, at high temperatures of air, can lead to complete disappearance of snow cover and also to significant decrease of snow accumulation density hear the snow fences.

The rain precipitation in a winter period can be observed both at positive and negative temperatures of air. The fall of liquid precipitation at negative air temperatures results in the forming of icy crust on the surface of snow accumulation, which completely excludes an deflating blizzard. Winter rains at positive temperatures of air considerably raise the density and humidity of snow cover and snow accumulations.

In conditions of soft winters the snow transferring usually occurs at higher temperatures than in regions of a severe climate. As was marked above, the blizzards at high temperatures of air differ essentially from that take place at low temperatures. First of all considerable increase in blizzards speed-up length and heightened sublimating losses of transferring snow are observed. In such conditions «wet blizzard» can realize its transporting ability in very rare cases even on large snow-collecting basins. And it means that «wet blizzard» will bring to a road less snow, that blizzards transferring dry snow, practically always. As a rule «wet blizzard» are general, i.e. occur with snowfall. For an opportunity of taking in to account the peculiarities of various kinds of blizzard the dependence of their basic parameters (total flow rate, speed-up length, range of snow transferring etc.) from temperature and air humidity, intensity of solar radiation, speed of wind have been established experimentally.

Besides, the snow cover state has an essentially influence on blizzard process. The snow cover's state undergoes permanent changes during winter period under action of the numerous factors. It is theoretically impossible to consider all these factors and to determine a degree of their influence on process of snow transferring. Therefore on the basis of experimental observations the major factors specifying the snow cover properties have been determined.

Such characteristic features of soft winters, as the thaws and the winter rains exercise of influence not only on the snow transferring processes and on the mechanism of snow accumulation by snow-protecting devices. Their influence has an effect and in decrease of snow accumulation volumes by their condensation basically.

In research laboratory of road winter maintenance of Kharkov State Automobile Highway Technical University the method of flow rate (Melnik 1954) for determination of snow-bringing volumes to roads has been considerably improved. This method taken into account the above-stated blizzards features in regions with soft winters and is automated completely.

3 ENGINEERING DECISIONS ON ROADS SNOW-PROTECTION

3.1 Snow-protecting forest-belts

The modern constructions of forest-belts and schemes of their location at present practically do not take into account the regional features of climate, which in many respects determine the processes of snow accumulations forming with blizzards. The main parameters of forest-belts constructions (height of trees, number of frees and bushes, their distribution on width, translucentness of forest-belt and its distribution on height) is necessary to assign depending on the most probable characteristics of blizzards.

In Ukraine and many countries of Europe the cyclonic blizzards, which frequently are accompanied by intensive snowfalls prevail. Therefore in regions with «soft winters» the occurrence probability of upper or general blizzard is higher, than lower one. With large height of trees in forest-belt the roads gets in a zone of an aerodynamic shadow of forest-belt. In this case forest-belt does not protect a road but on the contrary, promotes its snow drifting. Recommended height of forest-belt is 2-2,5 m.

Depending on regional climatic conditions it is necessary to choose density of forest-belt. Smaller volumes of snow bringing to the roads with lower blizzards in conditions of «soft winters» allow to decrease an area of the road-side forest-belt. Such forest-belt should be dense not blown through. By results of field observation the cross-section of roadside forest-belt should come nearer to rectangular. The length of a snow tail with the lee side at such forest-belt does not exceed 3-4 heights of trees. It allows to reduce distance from forest-belt up to a road and to preserve fertile soils for the agricultural needs. The recommended constructions of forest-belt are given in a figure 2.

On separate sections of roads in regions with cold winters and prevalence of lower blizzards are recommended blown forest-belts. The principle of such forest-belts action consists in snow transferring through a road with strong blizzards. At low air temperature during blizzards a snow-wind flow is saturated with dry-snow. Forest-belt owing to its aerodynamic characteristics provides the acceleration of a snow-wind flow above a road. The snow is being transferred above a road and is not being accumulated on carriage-way and on shoulders (fig. 3). The cases are known when on sections of the roads with blown forest-belts, with strong lower blizzard snow was being blown from surface of windward slope to a ground, and on the near situated road sections the powerful snow drifts were being formed.

A final choice of the optimal decision is accepted by results of the detailed analysis of regional features of snow transferring with blizzards.

3.2 Temporary snow fences

For roads protection against snow-banks the wooden latticed snow-screens, polymeric nets, snow fences from polymeric materials with the specified aerodynamic properties, snowy banks and trenches are widely practiced. In conditions of «soft winters» the snow retention with the help of snowy banks and trenches not always is effective because of frequent thaws occurring. The application peculiarities of the wooden latticed snow-screens when wet snow transferring consist in necessity of their frequent rearrangement because of screens are quickly being snow-bounded. More effective are snow fences from polymeric materials with the specified aerodynamic properties, which had been worked out in our laboratory under the direction of Professor V.M. Sidenko (fig. 4). Such constructions practically are not being drifted by snow, have large snow capacity and ensure reliable snow protection of the roads with different types of blizzards.

3.3 Complex snow protection of the roads

Complex snow protection of the roads provides an optimal combination of permanent and temporary snow fences with snow retention on the fields adjacent to a road. The objective of complex snow protection is to retain a snow in a field for increase the crop and to prevent snow accumulation forming on a road.

For snow protection in a field it is necessary to arrange the system of field-protecting forest-belts depend on regional climatic conditions too. Optimal distance between forest-belts, located parallel to a road, should make up 250-300 m for regions with primary transferring of dry snow, and 400-500 m for regions with «soft winters».

The aerodynamic properties of field-protecting forest-belts should provide an uniform forming of snow accumulations of height, that does not exceed maximum on agritechnical requirements. The blown forest-belts in a greater measure conform to these conditions.

CONCLUSIONS

The engineering decisions on roads protection should take in to account regional climatic conditions.

Predicting the volume of snow bringing to a road is the basis of roads snow-protection designing. On the basis of research of climatic conditions influence on the blizzard development the technique of determination the volume of snow bringing to a road has been specified, which allows to receive more accurate values for regions with «soft winters».

The principles of engineering decisions choice on roads snow-protection which take into account the blizzards features in region have been suggested. The rational constructions, schemes of snow-protecting forest-belts and temporary snow fences for different blizzards are determined. It is established that optimal engineering decision with all kinds of blizzards is complex snow-protection of road.

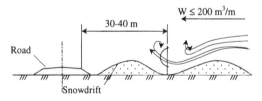

Figure 1 - Scheme of blizzard development

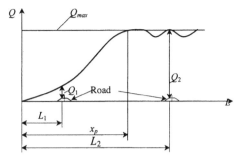

Figure 2. Recommended constructions of unblown forest-belts

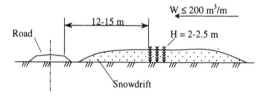

Figure 3. Principle of work of blown forest-belt

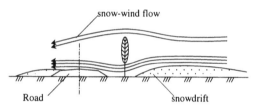

Figure 4. Snowdrifts forming near constructions from polymeric materials.

REFERENCES

Аккуратов В.Н. 1956. Прогноз наступления лавинной опасности по величинам метелевого переноса и температурного сжатия снега. *Сб. «Вопрося использования снега и борьба со снежными заносами и лавинами».* Москва: Изд-во АН СССР.

Бялобжеский Г.В., Амброс Р.А. 1956. *Повышение эффективности и экономичности снегозадерживающих устройств.* Москва: Автотрансиздат,.

Дюнин А.К. 1963. *Механика метелей.* Новосибирск: СО АН СССР.

Дюнин А.К. 1954. Твердый расход снеговетрового потока. Новосибирск: *Тр. ТЭИ ЗСФАН СССР, вып. IV.*

Иванов Б.В. 1954. Об эффективности уположения откосов мелких железнодорожных выемок как средства против образования в них снежных заносов. Новосибирск:*Тр. ТЭИ ЗСФАН СССР, вып. IV.*

Комаров А.А. 1959. *Повышение эффективности снегозащитных средств на железных дорогах Сибири.* Новосибирск: Обл. изд-во.

Котляков В.М. 1960. Метелевый перенос снега в Антарктиде и его роль в балансе питания ледника. Москва: *«География снежного покрова». Сб. ст. Изд-во АН СССР.*

Мельник Д.М. 1952. О законах переноса снега и их использовании в снегоборьбе. *«Техника железных дорог»,* №11.

Мельник Д.М. 1966. *Предупреждение снежных заносов на железных дорогах.* Москва: Транспорт

Snow Engineering, Hjorth-Hansen, Holand, Løset & Norem (eds) © 2000 Balkema, Rotterdam, ISBN 90 5809 148 1

A new retrospect of snow and ice, tribology and aircraft performance

Nirmal K. Sinha
National Research Council Canada, Ottawa, Ont., Canada

Armann Norheim
Norwegian Air Traffic and Airport Management, Oslo, Norway

ABSTRACT:

The interaction of aircraft wheel(s) on travelled winter surfaces of a movement area is described by a tribosystem with its tribological parameters. The tribological parameters of this system are structural parameters, the interfacial materials and the operational parameters. At winter-contaminated movement areas, the interfacial materials - snow and ice, cause the major operational problem with aircraft performance. Contrary to the popular thinking and belief, snow and ice exists at very high temperatures greater than $0.85\ T_m$, where T_m is the melting point in Kelvin. The central issue is how to evaluate the attainable friction force between the aircraft tyre(s) and the pavement, contaminated with a high-temperature rate sensitive material, and relate that to aircraft performance. A new retrospect of snow and ice related to such a tribosystem is described here.

1 INTRODUCTION

Runway and travelled surfaces of a movement area, contaminated with water, slush, snow and ice, are slippery depending on the nature, coverage and state of the contaminants. A slippery runway surface in an airport movement area affects the performance of an aircraft in two major ways. The most obvious effect is the increase in landing/take-off or stopping distance. The other effect is the loss of directional stability and control. In fact, a large number of landing and maneuvering accidents and incidences are caused by loss of directional control when aircraft leave the runway pavement sideways or stop on the runway with the nose pointing in the undesirable direction.

The main problem is how to characterise a surface contaminated with snow and ice and then relate it to aircraft performance in a movement area contaminated with snow and ice. In order to maintain the level of safety during ground operation, take off and landing, the operator may be forced to reduce the weight of the aircraft (e.g. fuel, number of passengers or cargo). A pragmatic approach, like this, is unavoidable in engineering problems requiring immediate working solutions. Pragmatism is useful in the sense that it makes it possible to calculate the outcome of certain tests without being worried about

details or the mechanisms of the underlying processes. Pragmatic engineering approaches do not provide answers that could be extrapolated for broad applications. The fundamental assumptions require critical examination, particularly the operating range of temperatures with snow/ice.

A critical assessment of test methods, the results obtained and the conclusions drawn from the tests require an understanding of the general background and the state-of-the-art including the human factors involved, particularly in problems involving snow/ice mechanics. It's a fact that the environmental conditions during tests, especially field tests, can be physically unbearable for both test equipment and test personnel.

Field trips are very expensive and must produce perceptible results. Serious short cuts are often made in conducting the tests for generating numbers. Even the necessary details, such as temperature, snow/ice thickness, ice type, snow cover state, etc., are often missed and the human factors are invariably not mentioned in technical reports. It is necessary to ask questions: why the tests were performed, when were they performed, how they were conducted, what tools were used, what were the test conditions, who performed the tests, what were the results and how they were analysed (Sinha, 1998)?

Because of the requirements of engineering solutions, 'strength' of snow and ice has played a central role in cold regions engineering. Considerable effort has been made for determining the strength of these materials. Both laboratory and field tests were conducted in the past for generating the numbers required in engineering calculations. Can one rely on these numbers? As the title of this paper suggests, we will address certain issues related to the analytical tools that may be useful in examining snow and ice as materials, the surfaces of winter contaminated movement area and the interaction processes.

1.1 *Traditional approach*

The traditional approach of solving the problems of contaminated surfaces has been to use tribometers (friction measuring devices) to guestimate the "friction" value. These devices fall essentially in two categories – direct measurement and indirect measurement. The direct measurement consists of an instrumented device that is integrated in or towed by a vehicle. An indirect device consists of a decelerometer mounted in a vehicle. A few examples of present day direct devices are SAAB Friction Tester, Skiddometer BV-11 and GripTester. Examples of indirect devices are Tapleymeter and ERD (Electronic Recording Decelerometer). The results obtained by these devices are given in terms of a g-force or a ratio (μ) of the tangential force (F) parallel to the surface and the force normal (N) to the pavement surface,

$$F = \mu N \qquad (1)$$

Where the constant, μ, is known as the coefficient of friction.

Equation (1) is based upon Amontons' Law of friction, after the French physicist Guillaume Amontons (Amontons, 1699) who rediscovered it; Leonardo Da Vinci, however, described it some 200 years earlier.

The substantial questions are what are these μ numbers ("friction" values)? Can they be related to the interaction processes of the aircraft tyre(s)?

The physical differences are significant between the aircraft tires and speed and those of the 'tribometers' or the so-called friction measuring devices. Moreover, the simple relationship given by Equation (1) is not valid for viscoelastic pneumatic tyres. For viscoelastic materials, deformation due to an increase in temperature is equivalent to decreasing sliding velocities, and vice versa. This equivalence of the time and temperature effects can be used to interpret the frictional behaviour. Thus a better understanding of the interaction processes between an aircraft tyre and the contaminated pavement is needed. In this respect, ideas developed in the field of 'Tribology' may be applied to elucidate these interaction processes and to link them with basic properties of the materials involved.

2 TRIBOLOGY

2.1 *Tribological definition*

Tribology is derived from the Greek word *tribos* meaning rubbing and was first reported in a landmark report by Jost (1966). Dictionaries define it as the science and technology of interacting surfaces in relative motion and of related subjects and practices. Tribology can also be understood as a single word to embrace the inter-linked, multidisciplinary topics of subjects separately labelled as friction, wear or lubrication science.

2.2 *Tribosystem*

In most friction, wear or lubrication situations four tribocomponents are involved (Czichos, 1978):

Triboelement 1
Triboelement 2
Interfacial medium
Environmental medium

For a tribosystem involving aircraft tyre and snow/ice contaminated movement area, the four tribocomponents mentioned above are respectively represented by:

Aircraft tyre (Triboelement 1)
Surface (Triboelement 2)
Snow/ice (Interfacial medium)
Atmosphere (Environmental medium)

The tribosystem of tire/surface interacting embodies mechanisms and details that are described and understood in different ways by experts within the disciplines involved. The complete tribological process in a contact in relative motion is very complex because it involves simultaneously friction, wear and deformation mechanisms at different scale levels and of different types. A complete description of the tribosystem under discussion should take account of all the components involved. To describe problems with any of the components missing is to miss most of the possible solutions to the problem. It is therefore essential to standardise terminology. Good solutions to problems are more likely achieved when everyone uses the same terms with the same meanings. The complexity of the tribosystem requires the seeking of multi-disciplinary approaches, to understand these phenomena and to solve the tribological problems.

2.3   Three fundamental aspects

According to Suh (1982) there are three fundamental aspects to the tribological science that must be understood to deal with the technological problems:

(a) The effect of environment on surface characteristics through physicochemical interactions.
(b) The interactions of the surfaces in contact which result in force generation and transmission between the surfaces.
(c) The behaviour of the material near the surface in response to the external force acting on the surface.

Those interested in studying the frictional behaviour of materials must comprehend the details of the chemical and physical interactions at the interface. This requires understanding of all three aspects of tribology. The contribution of each aspect of tribology to friction changes as a function of sliding distance (or time) and the environment. Therefore, the coefficient of friction of the same pair of materials differs depending on the specific application.

2.4   Triboelement 1 – the tyre

The tyre is made out of natural or synthetic rubber, which is a viscoelastic high-frictional material. Of all the major variables that may influence the deformation modes, time and temperature are the most significant (see section 4). In indentation, major variations in elastic and plastic behaviour are apparent. Increasing the scratching velocity often has the same effect as increasing the ambient temperature (Briscoe, 1997). For sliding of rubber against a rough (abrasive) surface there is a gradual change in friction with sliding rate. This is understood in terms of a combination of effects including adhesion, ploughing hysteresis and tearing.   The frictional force is determined by contributions from four different sources (Alliston-Grainer, 1997):

a)   Interfacial adhesive component
b)   Hysteric component arising from energy losses associated with bulk deformation
c)   Viscous component arising from dissipation in any lubricant
d)   Wear component

Sliding involves generation of so-called Schallamach waves. See (Johnson, 1981). These waves are like giant dislocations that form on the compression side of the rubber member and then move through the contact zone toward tension side. The rubber peels from the counterface and becomes reattached. The reattachment energy is much smaller than the detachment energy (work of adhesion), so the latter dominates the behaviour.

Since the low thermal conductivity of rubber can result in a very high temperature at the interface, it is important to investigate the effects of frictional heating on the sliding friction of rubbers. The temperature of the band of tread rubber 0.15 mm inside the tread surface is only 1% of the surface temperature (Schallamach, 1967). The surface temperature is therefore the main factor affecting the rubber properties, and viscoelastic properties corresponding to the surface temperature should be included in the analysis to predict hysteretic sliding friction. When a surface contact is exposed to frictional heating, an unsteady situation ensues because the temperature increase becomes a function of time as well as position. The size of the source as well as the thermal properties and speed (time) of the respective materials determine the transient behaviour. The surface receives thermal energy only for the time that the heat source exists.

2.5   Triboelement 2 - Surface

Here the surface means the exposed top surface of the pavement of the movement area. The pavements are usually made of asphalt concrete. Asphalt concrete is a nonlinear viscoelastic material and the bearing capacity of a pavement depends on a number factors such as the thickness of the pavement and the mechanical properties of the foundation supporting the pavement. The surface of a pavement is textured, depending on the size and distributions of the aggregate and the binding agent (bitumen) used, and the method of compaction applied in making the pavement. The age and usage of the pavement also affect the conditions of the surface. Moreover, the texture of the surface could vary significantly within the interaction area. A detailed description of runway pavement is outside the scope of this paper.

2.6   Interfacial Medium and the Environmental Medium

The other two tribocomponents of the present tribosystem are the snow/ice as the interfacial medium and the atmospheric or local weather conditions as the environmental medium. These two components are often linked with each other in a complex manner. The local weather conditions (air temperature, relative humidity, wind conditions, solar radiation etc.) not only influence the response of snow and ice, but also determine the outcome of the interaction processes. There is a need to discuss the material aspects of snow and ice.

3 SNOW AND ICE

Snow consists of ice particles that form in the atmosphere by complex joining processes of water molecules. These particles encompass a large variety of crystal habits and sizes. After the deposition of snow on the runway pavements, a process known as "metamorphism" modifies the geometrical features of the particles. The rate of changes of these features depends on the temperatures of the air and the pavement, and the atmospheric conditions including solar radiation. The metamorphic processes that causes these changes in shape and size are still poorly understood.

A field classification of snow, based on the physical properties of snow cover, using information on grain size, hardness and liquid water content, was proposed by Bader et al. (1939). The classification of snow on ground made by the International Commission on Snow and Ice (ICSI 1954) was used almost as a standard for many years. A new committee recently updated the ICSI (1954) classification of snow by giving considerations to the knowledge gained in the last few decades (Colbeck et al., 1990). Michel and Ramseier (1971) made river and lake ice classification and this could also be applied to describe ice on pavements.

Snow on a pavement is a mixture of ice particles, air and free water, depending on its temperature and atmospheric conditions. Freshly fallen snow can have densities in the range of 30 kg.m^{-3} to 100 kg.m^{-3}. Free spaces between the particles are known as pores. In addition to morphological processes inducing changes in the densities, a snow cover in an airport movement area could be mechanically compacted due to the vehicular movements. For further details on compacted snow and snow roads, refer to Adam (1978).

When a compacted snow mass may be called ice? This distinction is often qualitative and descriptive. In the field of glacier studies, a distinction between snow and ice is made when the density is less or more than 830 kg.m^{-3}, but this definition may not be applicable to snow or ice masses on pavements. Snow may be referred to as ice when the pores are not interconnected and when it becomes impermeable. However, this raises the question whether a solid ice cover be called snow when its density is decreased by the action of chemicals, particularly in the form of beads, making vertically oriented, but isolated holes. The authors have found

it simpler to describe the state of the cover material as clearly as possible (Sinha, 1998).

3.1 Snow/ice – a high temperature material.

Because the melting point of pure ice is considered as zero (0) degree in Celsius scale, temperatures of ice and snow (excepting very wet snow or pure slush) are always in negative numbers. These temperatures are certainly cold and uncomfortable for humans. It is natural, therefore, to treat snow/ice as a material at low temperatures and snow/ice engineering - a synonym for 'cold regions engineering'.

Question: is snow/ice a low temperature material?

This crucial question must be answered before a tribosystem involving snow/ice can be examined in a rational manner.

Consider the bottom surface of ice covers floating on water bodies, like lakes and rivers. It is always at or very close to the freezing point of water or the melting point of ice. Even the temperatures of the top surfaces are not far from the melting point. Materials at temperatures close their melting point are called hot. Thus, natural snow/ice is indeed a hot stuff. A rational basis for establishing this fact in a quantitative manner can be developed by adopting the Kelvin (K) scale for reporting purposes and Homologous (H) scale for carrying out the analysis (Figure 2).

If the temperature is expressed in Kelvin (K), the ratio of the operating temperature, T, and the melting temperature, T_m, may be defined as the Homologous (H) temperature:

$$H = T/T_m \qquad (2)$$

Where H = Homologous Temperature; T = Temperature of the material, K; and T_m = Melting Point of the material, K

Excepting the winter months in central Antarctica or Arctic regions, it would be safe to say that in nature snow/ice exists at temperatures greater than about -40°C or 233 K. This is equivalent to homologous temperatures greater than 0.85 (= 233/273), assuming the equivalent point at which snow/ice melts as 273 K, leaving aside the decimal amount. As illustrated in Fig. 2, a temperature of -27°C (246 K) is equivalent to a homologous temperature of 0.90 T_m or a temperature only 10% below the melting point on the homologous scale for ice. For a commonly used stainless steel used for high temperature applications, a temperature of 0.90 T_m is equivalent to a temperatures of 1800 K or 1527°C (Fig. 2). This is certainly a very high temperature.

Figure 1. Microstructure of snow that has been mechanically compacted to a density of 400 kg.m^{-3} and subjected to equitemperature (ET) metamorphism.

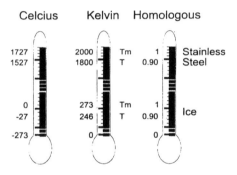

Figure 2. A comparison of operating temperatures on three temperature scales.

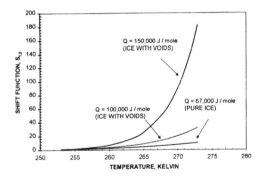

Figure 3. Dependence of shift function for pure ice, $S_{1,2}$, using $T_1 = 253K$ as the base temperature

Thus, naturally occurring snow and ice masses, indeed, are extremely hot materials. At high temperatures, the mechanical responses of all materials are significantly different than their response at low temperatures. This is why the conventional wisdom of treating snow/ice as a low-temperature material mis-

led many investigators including 'snow/ice experts'. Furthermore, this also explains why to an uninitiated engineer, snow/ice appears to be a strange material.

3.2 Certain aspects of snow/ice system

Snow/ice actually belongs to a "unique family" of materials; it is lighter than its melt. In this regard, snow/ice is like Silicon, a non-metallic element, Germanium, a rare metallic element, and a number of bismuth-based alloys. Silicon is commonly found as silica (sand, silt, quartz) and silicates, and these materials make up to about 70 per cent by volume of the rock in the earth's crust. In comparison, snow and ice cover about 30 per cent of the global surface. This "unique family", therefore, may not have a large membership, but is certainly most common around us.

Snow/ice is a crystalline solid and natural ice is a polycrystalline aggregate very similar to the familiar materials like rocks, ceramics, concrete, metals and alloys. Granular, isotropic snow-ice is similar to many ceramics. Lake-ice and sea-ice are similar to directionally solidified (DS) materials like present day nickel-based superalloys used in gas turbine engines.

Snow/ice belongs to the family of strong high-temperature materials. Ice has been found to be among the strongest and hardest materials when its 'strength' (normalized with respect to its elastic modulus) is compared with that of other solids at similar high homologous temperatures. Low diffusivity, a consequence of the lattice structure and the hydrogen bonding, govern the rate controlling processes in an ice crystal.

Drop an ice cube on the floor; it shatters like glass. A picture of a glacier makes one think that ice flows like butter. Photographs of an avalanche in action make one think that snow moves like a fluid. Therefore, depending upon the rate of loading, snow/ice could behave as a brittle material or it could behave as a ductile material. To alleviate this difficult problem and to apply the conventional wisdom (gained from low homologous temperature experience), a simple concept of brittle/ductile transition had been introduced. Snow/ice is assumed to behave as a brittle material in the 'brittle regime' above a certain rate of loading. Brittle-ductile transition, on the other hand, has been found to depend on the microstructure, the state of the material and, worst of all, the test conditions and the loading history. As the temperature increases, ice creeps more and the ductility increases (Barnes, et al., 1971). However, as the temperature increases, ice also becomes more brittle and cracks at lower stresses (Gold, 1972). The increase in ductility (and volume dilatation) with the increase in brittleness and the increase in temperature has also been noted in sea ice (Sinha, et al., 1992). Ice, therefore, seems to defy conventional wisdom and, as a consequence, makes it even more difficult to grasp the complexities of failures in snow and ice masses.

Is the mechanical behaviour of snow/ice really peculiar? Not really. An assessment of the snow/ice properties, obtained from conventionally accepted test methods, raises several questions. To answer these questions, one has to understand snow/ice as a material at extremely high temperatures and the environmental conditions to which it is subjected. This requires knowledge of the tribosystem and the associated operating range of stresses, strains and temperatures, and their rate of changes.

3.3 High temperature materials science – a new frontier

Being a geological material, it is not rare that snow/ice has been treated analytically in terms of methods developed in soil mechanics or rock mechanics. The snow and ice literature contains information on the failure envelopes not unlike that used in soils, rocks and concrete. The fundamental basis of most of the analytical techniques is the assumption that particles of snow are like soil particles and ice is a solid and it must behave like other solids. These assumptions are further supported by the macroscopically observed phenomenon that the compressive strength of ice, like rocks and concrete, is significantly higher than its tensile strength. This wisdom, however, is questionable. Can a direct comparison be made between the high-temperature mechanical behaviour of snow and ice and the known low-temperature macroscopic response of other materials? Concrete also creeps like ice but are the micromechanisms of deformation the same in both materials? The tensile strength of ice is less than its compressive strength, but at what rate of loading? At low rates of loading there are no difference between compressive strength and tensile strength. Can the phenomenological viscoelasticity be extended unequivocally to all materials without giving consideration to the microstructure-based deformation processes? Certainly not.

Most polycrystalline materials exhibit intergranular embrittlement at temperatures greater than about 0.3 T_m. Temperatures greater than about 0.3 T_m may, therefore, be classified as 'High Temperature' regime. Like other materials at high temperatures, snow and ice masses also show intergranular fracturing processes and rate sensitivities in its mechanical response. Their mechanical behaviours are not unlike steel or ceramics (Sinha, 1989a). However, one has to recognise that the similarities are with the high temperature response of other materials. The dichotomy in the behaviour of ice - ductility increase and increase in brittleness - is typical at high temperatures, greater than about 0.3 T_m, for most metallic and non metallic polycrystalline materials (Bressers, 1981). It is a result of high temperature, grain-boundary shearing (sliding) induced creep/embrittlement processes in polycrystalline

solids. Microcracks, comparable to the lengths of the grain facets and usually called wedge type cracks, occur in high-density polycrystalline ice under uniaxial loading conditions when the stress (global) exceeds a critical value of about 300 to 500 kPa (Sinha, 1989a,b), depending on ice type, or

$$\sigma_c > 3 \times 10^{-5} E \qquad (3)$$

where E is the Young's modulus (about 9.5×10^3 MPa). Many problems in ice involve stresses greater than the critical stresses, σ_c. The critical macroscopic stress for breaking bonds between snow crystals is expected to be significantly lower (depending on porosity and microstructure) than that given above, but the elastic modulus for snow is also to be expected to be lower than that of pure ice. A general statement can, therefore, be made:

Snow/Ice mechanics, relevant to snow/ice engineering, is a high temperature (> 0.3 T_m) and high stress (> 3×10^{-5} E for ice) problem.

4 TIME-TEMPERATURE DEPENDENCE – CONCEPT OF SHIFT FUNCTION

One of the high-temperature manifestations of mechanical response of polycrystalline solids is exhibition of elastic after effect or strictly speaking the phenomenon of "delayed elasticity". As a consequence of this mechanism, both elastic and plastic properties become time dependent and transient creep often dominates the measurements of mechanical response. It was shown by Sinha (1979) that the delayed elastic effect was primarily controlled by sliding or shearing and resistance to shearing brought about by the interaction between grain boundaries, accommodation at triple points, and grain deformation and re-arrangement. All these processes are thermally activated and thereby depend significantly on temperature. Sinha (1978) introduced the ideas of polymer and high-temperature glass rheology to the field of glaciology. He introduced the concept of "shift function", $S_{1,2}$ given by

$$S_{1,2} = \exp [Q (1/T_1 - 1/T_2) / R] \qquad (4)$$

where Q and R (= 8.32 J mole^{-1} K^{-1}) are the activation energy and gas constant respectively, and T_1 and T_2 are temperatures in degrees Kelvin.

If a certain deformation process takes a time, t_1 to develop at T_1, then the same process would take time, t_2 to develop at T_2, such that

$$t_2 = t_1 / S_{1,2} = t_1 / \exp [Q (1/T_1 - 1/T_2) / R] \qquad (5)$$

For pure ice with no porosity and cracks, the activation energy, Q, for both intragranular deformation and delayed elasticity was found to be 67 kJ mole^{-1} (Sinha, 1978,1979). The apparent activation energies for materials with voids are significantly higher and depend on porosity. Q values up to 150 kJ mole^{-1} are found for ice containing voids.

Examples of the dependence of the shift function on temperature and activation energy, based on Equation (4) and a base temperature, $T_1 = 253K$ are shown in Fig.3. It shows that at 272.5K, the shift function, $S_{1,2}$ could vary from 10 for pure ice with Q = 67 kJ mole^{-1} to 165 for porous ice with Q = 150 kJ mole^{-1}. It means that a time of 1 second at 253K is equivalent to a time of 0.1 seconds at 272.5K for pure ice and 0.06 seconds for porous ice. This means, for example, that a vehicle sliding with a speed of 1 km/h on a pavement covered with a layer of pure ice at 253K is equivalent to a speed of 10 km/h at 272.5K. The equivalent speed increases to 165 km/h in the case of ice with voids.

5 MATERIAL VARIATIONS IN THE FIELD

In addition to the multitude of structural features, natural snow and ice include many trapped impurities. For engineering purposes, naturally fallen snow may be considered as a pure material with large amount of air pockets unless contaminated by the materials at the ground level. In case of ice, however, the purity of the material depends on the purity of the water or the melt from which the ice grew. Freshwater ice contains impurities usually in the form of air bubbles. Refrozen runway pavement ice, on the other hand, may contain a significant amount of liquid inclusions in the form of pockets trapped between the grains and the subgrains.

6 STRENGTH OF SNOW AND ICE

Engineering designs are based on operational lifetime and a measure of the extreme values of the properties of materials involved in the design. Knowledge of the local and the global pressure that a component can exert on a structure is essential for design purposes. Since the simplest solution methods are based on the concept of strength of materials, most design equations involving snow/ice are also based on the strength properties of snow/ice. These materials have been treated as low-temperature materials like other geotechnical materials, and the approaches and methodologies of traditional engineering schools (not specialised in high-temperature materials) were followed in developing the field of snow/ice engineering. As a consequence, considerable effort was made in obtaining the numerical values of tensile and compressive strength and the failure envelope for snow/ice.

What is strength? Conventionally it is defined as the maximum load that a material can support. 'Strength' is a low-temperature concept. Because of high temperatures, strength of snow/ice is rate sensitive and is not a unique property. The strength numbers are, at best, index values and are useful only for certain limited practical uses. In the field of snow engineering, it is common to determine hardness of snow using Rammsonde cone penetrometer (see Adam,1978). The ram hardness number has units of kilograms, the magnitude of hardness has no real significance except in relative terms as index values.

Laboratory tests, for determining strength, are primarily divided into two methods: the uniaxial or multi-axial constant strain or deformation rate (CD) tests and the uniaxial constant stress (CS) deformation tests. Ideally a displacement rate is suddenly imposed to a cylindrical or prismatic specimen and held constant in the CD tests; in the CS tests, a stress is suddenly imposed and held constant. CS tests are known as 'creep tests'; CD tests are known as 'strength test'.

CD tests on snow/ice are performed by making use of commercial testing machines (Sinha, 1989b). Usually the displacement rate of the crosshead or the actuator is maintained constant during a CD test. Stress is recorded against displacement and stress-strain diagrams are drawn. The initial slope of the stress-strain diagram is used to calculate the elastic modulus and the peak stress is used as the strength.

In the CS tests, it is usually the load that is maintained constant. The usual practice is to record strain-time curves from which instantaneous strain rate is determined as a function of time.

In the field of snow/ice engineering, CD tests are preferred because these tests readily provide a means for determining both 'elastic modulus' and 'strengths'. Such usage of the stress-strain diagram is a low-temperature concept. It is based on the 'inherent assumption' that the material response is independent of 'time'. Because of this, snow literature contains elastic modulus data that are highly scattered and varies over several orders of magnitude.

Laboratory results are limited to small and selected volume of materials. There is, therefore, a requirement for field tests and field equipment. A number of portable instruments have been developed for the determination of in situ strengths of snow and ice. A few theoretical frameworks were also developed in order to interpret the field results. A dialogue, however, is necessary to clarify, at the outset, some apparent peculiarities of snow/ice as a material and its strength properties.

Mechanical properties of snow and ice vary significantly due to the natural variations in the material. The structure, the texture and the fabric of the aggregate under consideration determine its mechanical properties and the mode of failure. Consequently, empirical equations have been developed for describing the strength as a function of density or porosity such as air or brine volume. What baffles engineers is the rate sensitivity and load path dependence of the material properties. Both the strength of ice and the mode of failure depend on the loading rate and the history of loading. Even the slope of the stress-strain diagram, conventionally used for determining the elastic modulus, depends on the loading rate. There is no choice but to report strength and modulus in terms of strain rate or, more commonly, in terms of nominal strain rates, estimated from the cross-head rate and the specimen length. Yet vital information on stiffness characteristics of test systems and histories of stress, time of failure, three-dimensional specimen strain, cracking activities, etc., necessary for quantitative analyses of test results are not presented. The observed strain-rate sensitivity has, however, led to the questions: what strain rate is applicable for an indentation test or what strain rate applies to an ice-structure interaction process? These questions, on the other hand, led to the absurd methods of geometrical normalization such as the ratio of the velocity of an ice cover and some multiples of the dimension of the structure. When we focus our effort on the wrong target, in this case strain rate, we depart from our ability to solve a problem.

7 CONCLUDING REMARKS

The tribological parameters of a tribosystem, describing the interaction of aircraft wheel(s) on travelled winter surfaces of a movement area, are structural parameters, the interfacial materials and the operational parameters. Here the interfacial material is snow or ice – a crystalline material. In nature, snow/ice exists at temperatures very close to its melting point. These factors have steered us in rethinking about snow/ice and its deformation and interaction processes in a tribosystem consisting of aircraft tyre(s) and runway pavement. Snow/ice mechanics is part of a broad new subject: creep, compaction, structural damage and failure of crystalline materials at high temperatures and high stresses. The elevated-temperature micromechanical processes are manifest in rate-sensitive, loading path-dependent properties of snow and ice. The interaction processes in a tribosystem could be examined in terms of geometrical factors under discussion, and the failure processes in snow/ice as interfacial material. Progress made on the microstructure-based approaches has helped us in understanding the kinetics and the complexities of microcrack-induced failures in ice. These analytical methods will also be the determining factors in the interpretation of environmentally sensitive, in situ, three-dimensionally loaded, aircraft or ground vehicle tire interactions with winter contaminated runway surfaces.

8 ACKNOWLEDGEMENTS

The authors would like to convey thanks to the Transportation Development Centre and Aerodrome Safety Branch of Transport Canada for supporting this work. They are indebted to P. Satyawali of Snow and Avalanche Study Establishment, Manali, India for his assistance in obtaining the micrograph of snow illustrated in Figure 1.

9 REFERENCES

Adam, K.M. 1978. *Building and operating winter roads in Canada and Alaska.Environmental Studies No.4*, Environment Division, Northern Environmental Protection and Renewable Resources Branch, Department of Indian and Northern Affairs, Ottawa, Canada: 221.

Alliston-Grainer, A.F. 1997. Test methods in tribology. In *New Directions in Tribology; Proc. First World Tribology Congress, London, September 1997*: 85-93. London: Mechanical Engineering Publications.

Amontons, G. 1699. De la resistance caus'ee dans les machines. *Mémoires de l'Académie Royale.* A, (Chez Gerhard Keuper, Amsterdam, 1706): 257-282.

Barnes, P., Tabor, D., and Walker, J. C. F. 1971. The friction and creep of polycrystalline ice. In *Proc. Royal Soc. London, Ser. A, Vol. 324, No. 1557*: 127-155.

Bressers, J. 1981. *Creep and fracture in high temperature alloys*. Applied Science Publishers, London.

Briscoe, B.J. 1997. Isolated contact stress deformations of polymers; the basis for interpreting polymer tribology. In *New Directions in Tribology; Proc. First World Tribology Congress, London, September 1997: 191-196*. London: Mechanical Engineering Publications.

Colbeck, S.C., Akitaya, E., Armstrong, R., Gubler, H., Lafeuille, J., Lied, K., McClung, D., and Morris, E. 1990. *The international classification for seasonal snow on the ground*. International Association of Scientific Hydrology, International Commission on Snow and Ice (IAHS), Wallingford, Oxfordshire: 23.

Gold, L.W. 1972. *The failure process in columnar-grained ice*. National Research council of Canada, Division of Building Research, Ottawa, 1972.

ICSI. 1954. *A classification for snow on the ground*. Associate Committee on Soil and Snow Mechanics, National Research Council, Ottawa, Canada, Technical Memorandum, No. 31.

Johnson, K.L. 1981. Aspects of Friction. *Friction and Traction; Proc. 7th Leeds-Lyon Symposium on Tribology*. London: Westbury House: 3-12.

Jost, P. 1966. *Lubrication (tribology)-A Report on the Present Position and Industry's Needs*. London: Dept. of Education and Science, HMSO.

Michel, B., and Ramseier, R. 1971. Classification of river and lake ice. *Canadian Geotechnical Journal*, Vol. 8 (1): 38-45.

Schallamach, A. 1967. *J. Institution of Rubber Industry, Vol. 1.*

Sinha, N.K. 1978. Short-term rheology of polycrystalline ice. In *J. Glaciology, Vol.21, No.85*: 457-473.

Sinha, N.K. 1979. Grain boundary sliding in polycrystalline materials. *Philosophical Magazine A*, Vol. 40, No. 6: 825-842.

Sinha, N.K. 1989a. Ice and steel- a comparison of creep and failure. In A.C.F. Cocks & A.R.S. Ponter (eds), *Proc. of Euromech - 239: 'Mechanics of Creep Brittle Materials - 1'*: 201-212. Elsevier Applied Science Publishers, London.

Sinha, N.K. 1989b. Microcrack-Enhanced Creep in Polycrystalline Materials at Elevated Temperature. *Acta Metallurgica*, Vol.37, No.11: 3107-3118.

Sinha, N.K., Zhan, C., and Evgin, E. 1992. Creep of sea ice. *Proc. of 11th. Int. Conf. Offshore Mechanics and Arctic Engineering (OMAE), Calgary, Alberta, June 7-11, 1992*, American Society of Mechanical Engineers (ASME), New York, Vol. 4.

Sinha, N.K. 1998. *Characteristics of Winter Contaminants on Runway Surfaces in North Bay – January and February-March 1997 Tests*. Institute for Aerospace Research, National Research Council Canada, Ottawa, Ontario, Canada, Report No. LTR-ST-2159, Transport Canada Report No. TP 13060E, September 1998: 80.

Suh, N.P. 1982. Surface Interactions. *Tribological Technology; Proc. NATO Advanced Study Institute on Tribological Technology, Mareatea, 13-16 September 1981*. The Hague: Martinus Nijhoff.

Snow Engineering, Hjorth-Hansen, Holand, Løset & Norem (eds) © 2000 Balkema, Rotterdam, ISBN 90 5809 148 1

Simulation of two-dimensional wind flow and snow drifting application for roads: Part I

S. Thordarson & H. Norem
Department of Road and Railway Engineering, Norwegian University of Science and Technology, Trondheim, Norway

ABSTRACT: Snowdrift sedimentation on leeward facing mountain slopes and terrain ridges makes road planning in certain areas very difficult. Problem areas are frequently found on leeward facing slopes where the wind blows perpendicular over the road. We use the commercial computational fluid dynamics code Flow 3D to simulate two-dimensional wind flow, in order to quantify the snow drifting conditions for roads. To evaluate the numerical model, the study includes a comparison to Askervein Hill data. The main idea of the study is to simulate the wind flow over original terrain profiles and the flow over actual observed snowdrifts, and compare the resulting friction velocity profiles. We use this comparison to estimate the snow drifting rate, and the effect and capacity of natural snow sedimentation areas. We propose a relationship between the slope of the terrain upwind from the snowdrift area, the slope of the drift and the horizontal rate of change of the snow drifting rate.

1 INTRODUCTION

Road planning in snow drifting areas is a challenging task, especially where complex terrain influences the sedimentation of drifted snow on the ground. Wind flow in the lowest part of the atmospheric boundary layer is modified by landscape and terrain features, which explains why snow drifting conditions are subject to large spatial variations, even locally. Computational fluid dynamics (CFD) is used in this study to analyse the wind flow around roads with respect to snow drifting conditions.

Several investigators have used CFD to predict snowdrifts around snow fences (e.g. Sundsbø 1997) and buildings (e.g. Thiis 2000, Waechter et al. 1997) or made general models for snow drifting (e.g. Shao & Li 1999, Pomeroy & Gray 1990, Sato et al. 1997, Uematsu et al. 1989). Significantly less has been done with CFD in the area of road and highway engineering and more knowledge and guidelines are needed in this area of work.

The purpose of the first of the two papers (Part I) is to test and evaluate the CFD code Flow 3D as a tool for simulation of atmospheric boundary layer wind flow over complex terrain, and to simulate wind flow over several two-dimensional terrain profiles from a study site. The second part of the report deals with snow drifting application of the results from the study site.

The present part of the report includes a simulation of wind flow over Askervein Hill, and a comparison with measured and simulated results from other authors. For details on the Askervein Hill Project, see Taylor & Teunissen (1987). We conclude that this comparison confirms that the numerical procedure in use will simulate the characteristics of the mean wind flow on downslopes of the scale in question up to an acceptable accuracy. The simulations from the study site were done over original ground profiles and over measured snow cover in the same profiles. The results confirm that a plain wind simulation can be an aid to identify areas for large snowdrift sedimentation.

2 METHODS

2.1 Wind field simulation

The simulated area and the terrain height difference on the study site is on such a large scale, that the use of a plain engineering numerical flow model for the wind flow becomes questionable. Wind flow over complex terrain and hills is the subject of boundary layer meteorology and has been simulated numerically by many investigators (e.g Ying 1994, Wood 1995, Davies et al. 1995, Beljaars et al. 1987, Raithby & Stubley 1987).

The code used for the wind field simulation in this study is Flow3D (Flow Science, Los Alamos NM, USA). For post processing of the results we used

Field View (Intelligent Light, USA). The code has been customized to account for logarithmic inlet and initial conditions for the wind flow over complex terrain and a feature to trace calculated values along a surface at an arbitrary elevation above the ground has also been added. Flow3D applies a rectangular calculation mesh and uses the FAVOR technique (Flow Science, 1997) to adapt the mesh to the flow geometry, which in this case is two-dimensional terrain profiles. This is done by integrating a volume fraction into the momentum equations to account for partially blocked computational cells at the boundary to geometric objects in the flow field. The FAVOR method makes grid generation a straight forward task and enables complex geometry to be resolved without time-consuming grid generation. Although the FAVOR method is very handy for the integration of terrain in the flow field, one drawback has to be dealt with when tracing computed values along the boundary to a complex geometric feature; the rectangular discretization results in spatially fluctuating gradients in the flow field very close to the boundary. Although these fluctuations may not have serious impact on the flow physics, they make the curve of extracted values along the boundary highly fluctuating and difficult to interpret. As a mitigation, we found that choosing cell aspect ratio, dz_{cell}/dx_{cell}, that follows the slope on the ground smoothens out these fluctuations to a certain degree.

Even when optimizing the grid spacing according to terrain slope, abrupt change in slope and the appearance of minor terrain perturbations gives unphysical fluctuations as described above, and hence since the shear stress along the ground is of main interest for this study, we had to employ a different method on estimating the surface shear stress. In stead of tracing the shear stress along the surface boundary, we extract the velocity at height dz_{cell} along the surface, where dz_{cell} is the vertical spacing between computational points in the finite difference discretization. Since the velocity calculated at height dz_{cell} is in the vicinity of the surface we find it reasonable to assume a logarithmic decrease in velocity from that point and down to the surface. Hence, we can solve for the friction velocity directly from Equation (1). The resulting curve is relatively smooth and any fluctuations can be explained by the nature of the underlying terrain.

The code solves the Navier-Stokes equations for the mean wind speed together with the mass continuity equation, and turbulence closure is achieved through a standard k-ε model. For further information on the features of Flow3D we refer to Sundsbø (1997) or Flow Science (1997).

2.2 Roughness modelling

The roughness height, z_0, can not be modelled directly in Flow3D. In general, solid boundaries are treated as smooth surfaces, and here with a no-slip condition. However, object roughness can be applied by adjusting the fluid viscosity in the first cell row enclosing the object, but we did not get control over this feature and therefore the terrain surface in our model is practically smooth.

In nature, when wind with a certain velocity distribution describing a boundary layer developed over terrain of a certain roughness, travels over to a new surface which has another roughness, an inner boundary layer based on the new surface roughness will develop inside the former (Stull 1988). The information about the new roughness is transported vertically up into the flow by diffusion and the inner boundary layer will grow higher as the wind moves further down the new surface. This will also happen in a numerical model, when the inlet boundary describes a logarithmic boundary layer of certain roughness, but the underlying terrain has another roughness. The growth rate of the inner boundary layer depends on the efficiency of diffusion in the model, which depends on how fine the computational mesh is. High vertical gradients near the surface will not be resolved by a coarse grid, thus disabling diffusion. Choosing a coarser grid will therefore help to sustain the initial roughness through the simulation domain and minimize the disturbing effect of the virtually smooth underlying surface in the model.

2.3 Boundary conditions and domain dimensions

The extension of the two-dimensional profiles at the study site is up to 1000 m and the terrain height difference is about 150 m. This large height difference and the inevitable high simulation domain is on such a scale that the effect from possible thermal stratification has to be discussed. However, assuming neutral stability in the atmospheric boundary layer and a flat upwind terrain, the logarithmic wind profile can be applied (Stull 1988):

$$u(z) = \frac{u_*}{\kappa} \ln\left(\frac{z}{z_0}\right) \qquad (1)$$

where

$u(z)$ = mean wind speed at height z
u_* = friction velocity
κ = von Karman's constant
z_0 = aerodynamic roughness length

It is well documented that atmospheric shear stress on the ground is the governing factor concerning whether snow particles are eroded from the ground or deposited to form drifts. The goal for the wind field simulation is to quantify the shear stress variations along the two-dimensional terrain profiles during strong winds. The friction velocity, u_*, is per definition representing the surface shear stress:

$$u_* = \sqrt{\frac{\tau_0}{\rho}} \qquad (2)$$

where τ_0 is the shear stress at ground level and ρ is the density of air.

The bottom boundary condition applied in our model has been described in the previous section. Probably the most difficult boundary to treat in this simulation is the inlet velocity. Because the terrain upwind from the simulated area is complex, the actual shape of the inlet velocity profile will be different for different locations and we did not have the opportunity to measure the shape of the velocity profile for the individual profiles investigated here. Since the upwind terrain is relatively flat, although complex in nature, a logarithmic inlet boundary based on Equation (1) is chosen. The outflow boundary has been chosen as continuative, that is, forcing all horizontal derivatives to zero at the boundary. This outflow condition is the most common one found in previous studies.

Although located several hundred meters above the bottom of the computational domain, the choice for a boundary condition at the top was found to be critical. When applying a solid lid with free slip, the recirculation zone formed below the steep incline of the road embankment in some of the simulated terrain profiles would gradually move up the hill during the calculation until large portion of the simulation domain would lie in a recirculation zone. By applying a constant pressure boundary condition at the top, this problem was avoided and the point of separation remained at a fixed location in those profiles where separation occurred. The top boundary pressure is set to 990 mbar in all simulations.

The terrain profiles from the test site are about 1000 m long, from top of the hill down to the valley bottom. The computational domain is stretched all the way down to the flat ground to capture the whole height difference in the model. Also when choosing a suitable height for the domain, the total height difference along the terrain profile has to be considered. In their Askervein simulation, Raithby & Stubly (1987) used 700 m domain height, which corresponds to roughly 7 times the terrain height difference. We tested the domain height in our model to find that the velocity near the ground did not change remarkably when the domain height was in excess of 4 to 5 times the downslope height. This proportion being lower than the one found necessary for isolated hills can probably be explained by the nature of the flow geom-

etry when only a downslope hill is simulated instead of an isolated symmetrical hill.

The computational grid used in our model has 2 m spacing in the vertical up to a level 10 m above the highest portion of the slope and is increased to 10-15 m at the top boundary to save computational costs. The horizontal spacing is varying according to terrain slope, such that dz_{cell}/dx_{cell} roughly equals the terrain slope, dz/dx. However on the upper and flatter portion of the terrain profile the spacing is set more dense than the slope indicates.

3 RESULTS

3.1 Model comparison to Askervein Hill data

To test the numerical code against other atmospheric studies, a simulation was done over a two-dimensional sinusoidal hill. The Askervein Hill has been chosen for this comparison because the topographic dimensions are almost similar to that in the current study. It can be seen on Figure 1 that a sinusoidal hill of 120 m height and 1000 m long approximately follows a slice through the Askervein hill-top at location HT, at least with such an accuracy that main features of the flow can be compared. A description of other models referred to on Figure 1 and Figure 2 can be found in Beljaars et al. (1987) and Walmsley & Taylor (1996). Note that the other simulations are three-dimensional but the current is two-dimensional.

The model was run with z_0=3 cm and u_*=0.60 m/s, the same values as reported for many of the measurement series from the Askervein Hill field project. Fractional speed-up ratio used in this presentation is defined as:

$$FSR = \frac{U}{U_{ref}} - 1 \qquad (3)$$

where U is the measured or simulated wind speed and U_{ref} is the speed at an upwind reference location. On Figure 1, FSR at 10 m height above the ground is compared. The vertical FSR profile at the hill crest, location HT, is shown on Figure 2. We note that our model is failing from 3 m height and down to the ground at the hill top, predicting to high values. The value at 2 m height is 15% higher than the measured one, but the two meter wind speed is the basis for our shear stress estimation for the terrain profiles from the study site. At 1 m height the calculated FSR value is 27% higher than the measured value. This is probably the result of the roughness modelling problem discussed in chapter 2.2. However, the hill summit, which Figure 2 refers to, is located 1000 m downstream from the inlet boundary which is twice as far as the interesting portions of the real terrain simulation from our study site. Even if a slightly progressive

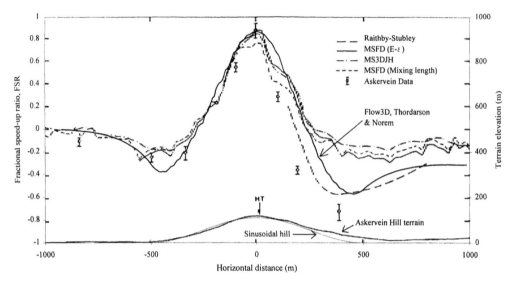

Figure 1. Model comparison to Askervein Hill data, measurements and simulations by other investigators. Number of computational cells is 180 in the horizontal direction and 180 in the vertical. Fractional speed-up ratio at 10 m height, Equation (3).

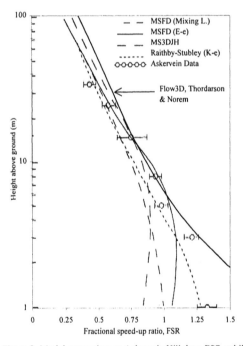

Figure 2. Model comparison to Askervein Hill data. FSR at hill crest.

overestimation of velocity happens along the simulated profiles, the local variations and horizontal velocity gradient will probably be realistic. We will thus conclude that the model is precise enough for the work in this study.

3.2 Study site simulations

The numerical model described in the previous chapter was run on several two-dimensional profiles from the test site. Characteristics of the test site are described in part two of this report. Two configurations were run for every terrain profile, the first one with original terrain and the second including the late-winter snow cover from the snow survey.

We chose a relatively high input velocity for the model, $u_* = 0.8$ m/s, and a low roughness height, $z_0=0.001$ m. According to Equation (1), this gives a wind speed of 18.4 m/s at 10 m height. The model was also run for lower velocity, $u_* = 0.6$ m/s and $z_0=0.01$ m, resulting in 10.4 m/s at 10 m height. When the two resulting curves for friction velocity along the terrain surface are normalized by the input friction velocity, the curves fall into each other. This important result indicates that the shape of the friction velocity profile along the terrain is unaffected by the input velocity. This will however not be true if a high velocity suddenly triggers a separation in the flow, but none of the profiles simulated here resulted in separated flow, at least not until downwind from the road.

Figure 3 shows the results for the first profile, numbered 1040. It can be seen that the drift starts to form where an abrupt change in terrain slope occurs about 80 m upstream from the road and is terminated by a sharp 5 m high edge at the road. The sharp edge

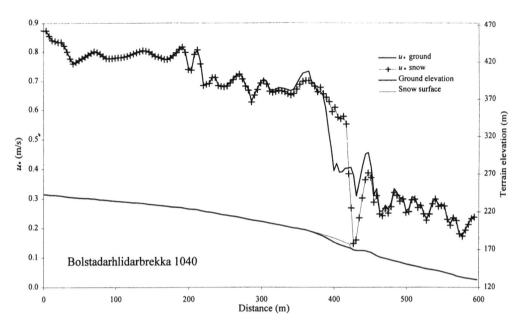

Figure 3. Calculated friction velocity, u_* along profile 1040. Simulation over original terrain and simulation over measured snow cover. The main road is located at distance 430 m. The old road is located at distance 210 m and is providing a moderate snowdrift area downwind. This drift is however not included in the snow cover simulation since only the large snowdrift adjacent to the main road was included in the snow survey.

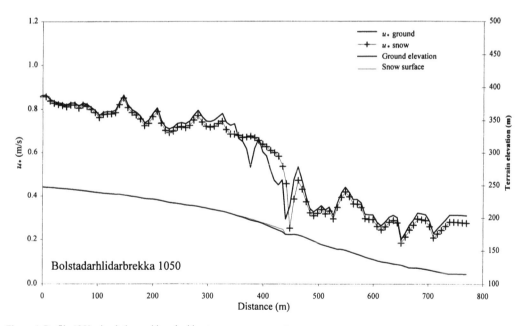

Figure 4. Profile 1050, simulations with and without snow cover present.

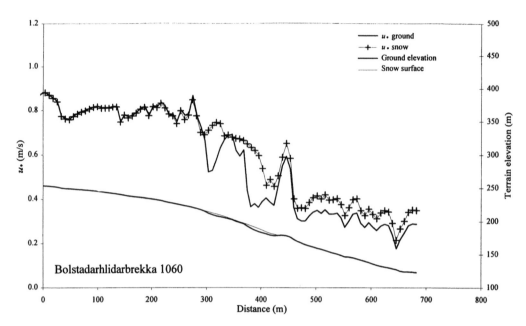

Figure 5. Profile 1060, simulations with and without snow cover present.

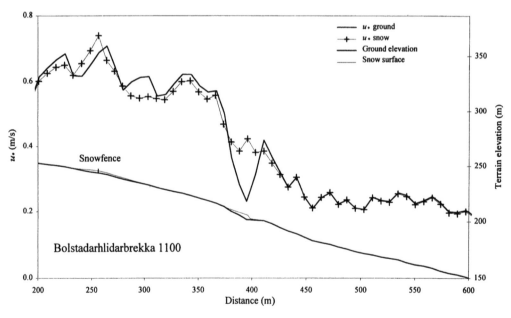

Figure 6. Profile 1100. Note that the snow fence structure is not included in the model, the calculation only takes account for the snow surface and not the 1 m high portion of the fence perturbing through the drift. The drop in u_* across the fence should thus be underestimated here.

was made by rotary blowers when re-opening the road after the latest storm. The shear stress curve for the original terrain, u_* ground, falls dramatically where the drift starts to form and maintains almost constant value towards the road. Looking at the curve from the simulation including the snow cover, u_* snow, it can be seen that some of the upstream shear stress is maintained, however the curve has a constant fall towards the drift edge where a sharp drop occurs. Not to any surprise, the simulation reveals a recirculating flow over the road beneath the sharp edge of the drift though the shear stress curve fails to indicate this, since it is based on the velocity at 2 m height. The difference in shear stress along the length of the drift between the two simulations is representing the flow energy gained by the wind by building up the drift.

Figures 4 to 6 show the results for other topographic profiles from the test site. Also here we note that the snowdrift starts where a sudden drop in friction velocity is calculated. These results are treated in part two of this report.

4 CONCLUSIONS

Comparison with Askervein Hill data indicates that the model is capable of simulating unseparated atmospheric boundary layer flow over hills. The results show that the velocity at 10 m height is well representative and the vertical velocity profile is quite accurate, at least outside of the inner boundary layer developed adjacent to the ground as a result of the virtually smooth surface applied in the model. For the estimation of shear stress on the ground this inner boundary layer will cause some inaccuracy. Therefore, we see the roughness modelling as a weakness in the model, which in worst case could result in the model not detecting separated flow where it should happen. This factor should be paid attention to in further studies.

In some of the simulated profiles from the study site, the snowdrift area starts where abrupt change in terrain slope occurs, and hence the start point could easily be predicted from looking at the terrain profile alone, but in other profiles the snowdrift start point is found on a gradually increasing slope, where no sharp changes in slope find place. Hence, the study confirms that a plain wind simulation can be an aid to identify areas for large snowdrift sedimentation in two-dimensional profiles.

ACKNOWLEDGEMENTS

We are grateful for the financial support given by the Public Road Administration of Iceland and Norway and the Nordic Road Association. This work was carried out as a part of the EU project ROADEX, which is financed within the EU Northern Periphery programme. We also thank Mr. Ernst W.M. Hansen of Sintef Energy Research for his assistance to establish the use of FLOW3D at our institute.

REFERENCES

Beljaars, A.C.M., Walmsley, J.L. & Taylor, P.A. 1987. A mixed spectral finite-difference model for neutrally stratified boundary-layer flow over roughness change and topography. *Boundary-Layer Meteorology* 38: 273-303.

Davies, T.D., Palutikof, J.D., Guo, X., Berkofsky, L. & Halliday, J. 1995. Development and testing of a two-dimensional downslope model. *Boundary-Layer Meteorology* 73: 279-297.

Flow Science, Inc. 1997. *Flow3D, Users Manual.* Los Alamos, NM: Flow Science Inc.

Pomeroy, J.W. & Gray, D.M. 1990. Saltation of Snow. *Water Resources Research* 26: 1583-1594.

Raithby, G.D. & Stubley, G.D. 1987. The Askervein Hill Project: A finite control volume prediction of three-dimensional flows over the hill. *Boundary-Layer Meteorology* 39: 247-267.

Sato, T., Uematsu, T. & Kaneda, Y. 1997. Application of a random walk model to blowing snow. In Izumi, Nakamura & Sack (ed.), *Snow Engineering: Recent Advances*: 133-138. Rotterdam: Balkema.

Shao, Y. & Li, A. 1999. Numerical modelling of saltation in the atmospheric surface layer. *Boundary-Layer Meteorology* 91: 199-225.

Stull, R.B. 1988. *An Introduction to Boundary-Layer Meteorology.* Dordrecht: Kluwer Academic Publishers.

Sundsbø, P.A. 1997. *Numerical modelling and simulation of snow drift.* Ph.D. thesis. Narvik: Norwegian University of Science and Technology.

Taylor, P.A. & Teunissen, H.W. 1987. The Askervein Hill Project: Overview and background data. *Boundary-Layer Meteorology* 39: 15-39.

Thiis, T.K. 2000. A comparison of numerical simulations and full-scale measurements of snowdrifts around buildings. *Wind and Structures* (in press).

Uematsu, T., Kaneda, Y., Takeuchi, K., Nakata, T. & Yukumi, M. 1989. Numerical simulation of snowdrift development. *Annals of Glaciology* 13: 265-268.

Waechter, B.F., Sinclair, R.J., Schuyler, G.D. & Williams, C.J. 1997. Snowdrift control design: Application of CFD simulation techniques. In Izumi, Nakamura & Sack (ed.), *Snow Engineering: Recent Advances*: 511-516. Rotterdam: Balkema.

Walmsley, J.L. & Taylor, P.A. 1996. Boundary-layer flow over topography: Impacts of the Askervein study. *Boundary-Layer Meteorology* 78: 291-320.

Wood, N. 1995. The onset of separation in neutral, turbulent flow over hills. *Boundary-Layer Meteorology* 76: 137-164.

Ying, R., Canuto, V.M. & Ypma, R.M. 1994. Numerical simulation of flow data over two-dimensional hills. *Boundary-Layer-Meteorology* 70: 401-427.

Snow Engineering, Hjorth-Hansen, Holand, Løset & Norem (eds) © 2000 Balkema, Rotterdam, ISBN 90 5809 148 1

Simulation of two-dimensional wind flow and snow drifting application for roads: Part II

S. Thordarson & H. Norem
Department of Road and Railway Engineering, Norwegian University of Science and Technology, Trondheim, Norway

ABSTRACT: Snowdrift simulation on leeward facing mountain slopes and terrain ridges makes road planning in certain areas very difficult. Problem areas are frequently found on leeward facing slopes where the wind blows perpendicular over the road. We use the commercial computational fluid dynamics code Flow 3D to simulate two-dimensional wind flow, in order to quantify the snow drifting conditions for roads. To evaluate the numerical model, the study includes a comparison to Askervein Hill data. This comparison is presented in Part I. The present report presents field observations of wind and snow from Iceland and numerical simulations of snow drift simulations. The simulations are compared with Tabler's equation for equilibrium snowdrift and the field observations.

1 INTRODUCTION

In the area of road design and drifting snow, Tabler and Norem have written engineering guidelines (Norem 1975, 1994, Tabler 1988, 1994). However, any guidelines developed with the help of CFD have not yet found it's way into design codes and standards. It is therefore of interest to road planning that the effect of terrain characteristics on the wind flow and snow drifting conditions is investigated in order to develop guidelines for designers. In the present study, the commercial fluid dynamics code Flow3D (Flow Science, Los Alamos, NM USA) is used to calculate two-dimensional wind field in several profiles through a roadway section. In a two-dimensional application, terrain characteristics refer to terrain slope and change in slope. The study is supported by wind measurements and snow surveys from the same road section.

The goal in this latter part of the report is to use the results from the wind field simulations from the study site, described in part one of this report to evaluate the drifting snow conditions.

On roads in mountainous areas, the most difficult sites with respect to heavy snowdrift sedimentation are frequently found where prevailing wind direction blows downslope, resulting in retarded wind speed and thus a lower atmospheric shear stress at the surface. Typical problem areas of this kind are where the road is aligned perpendicular to the wind on the lee-

ward side of terrain ridges and mountain sides. Under these conditions there is a great risk for the road being placed in natural deposit areas for snow. On the other hand, this may also lead to a very favourable road align in case that the road is situated downwind from the area of equilibrium snowdrift development.

Regardless of turbulence that is present in all high Reynolds number flows and always has a three-dimensional nature, fluid flow in all applications can be categorized as either flow with two-dimensional nature or flow featuring three dimensional interactions in the mean flow. With respect to this, we suggest that a roadway exposed to drifting snow can be spatially divided into sections of either two-dimensional or three-dimensional character. This paper deals with road sections where the wind flow can be described two-dimensionally.

Our simulations were done over original ground profiles and over measured snow cover in the same profiles. The difference between the two simulations is used to establish a relationship between terrain slope, drift slope, drifting snow transport rates and snow trapping efficiency of terrain features.

The study confirms that a plain wind simulation can be an aid to identify areas for large snowdrift sedimentation, and to evaluate the snow-trapping efficiency and capacity of these. We propose a relationship between the terrain slope upwind of the drift and the drift slope itself, and the horizontal gradient of the snow drifting rate. This can be considered when choosing road path near leeward sloping terrain features and when choosing appropriate cross sectional profile for the road. The model developed in the study is a potential addition to the tools already known for road and highway engineering in snow drifting areas, although further testing and generalization is necessary.

2 METHODS

2.1 *Field observations*

The study site, national road no.1 at Bolstadarhlidarbrekka in Iceland is situated on a mountain slope where strong winds and heavy snowfall occur during the winter, Figure 1. The data collected and used as a support for numerical work in this paper consists of measurement series from an automatic weather station (AWS) and a snow survey. Additionally, information from maintenance personnel on road closures and sight distance along the road during storm periods is available. The data from the test site was collected during the winter 1998-1999.

Norem (1975, 1994) has classified weather conditions for snow drifting problems on mountain roads.

According to this the data from the AWS was used to identify the prevailing direction for strong winds.

The snow survey was done at the end of the winter. Snow depths were measured in profiles parallel to the prevailing wind direction, the same profiles as chosen for the numerical simulation, Figure 1. The measured snow profiles extend up to the erosion area where no snow was registered on the ground, and are superimposed on topographic profiles taken from a digital map, Figure 2.

2.2 *Wind field simulation*

The method and tools for the wind field simulation is documented in Part I of this report.

3 RESULTS

3.1 *Field observation*

The road site chosen for the field observations has the typical characteristics of maritime climate. The worst drifting snow problems on the road are usually encountered when strong northerly winds blow from the ocean with simultaneous precipitation, which results in enormous quantities of drifted snow. A typical

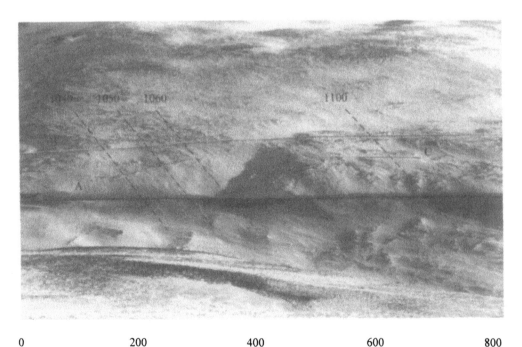

| 0 | 200 | 400 | 600 | 800 |

Distance along the road (m)

Figure 1. Photograph showing the study site at Bolstadarhlidarbrekka. The marked profiles are investigated in the study. Other symbols: A, the main road. B, an older generation road. C, a snowfence. The snowfence is 3 m high and 100 m long.

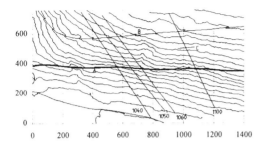

Figure 2. Topographic map of the study site, 10 m equidistance curves. Scale along the edges is in meters. The valley bottom is at the lower edge, governing wind direction from the top along the marked profiles. Profiles are aligned roughly 10° clockwise from true north. A, B and C are the same as in Figure 1.

Bólstaðarhlíðarbrekka 30.9.98 - 20.3.99

wind speed >= 10.8 m/s

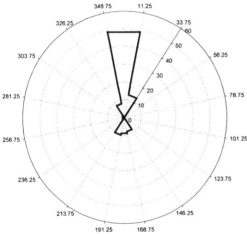

Figure 3. Wind rose of strong winds only for the winter months at Bolstadarhlidarbrekka. Note the dominating frequency of strong winds from the north, 50%. Total frequency of strong winds for the period was 17%.

storm may last from few hours to several days and will occasionally lead to closure of the road. Between storms, periods of higher temperatures accompanied by rain or wet snowfall harden the snow cover and consequently few events of drifting snow are registered between the events described above. These cycles of snowstorms and mild weather result in large spatial variations of the snow cover thickness and even leave the wind eroded areas completely snow-free throughout the winter, making the deposition areas for drifted snow easy to identify. This fact together with a monotone wind direction for drifting snow due to the surrounding landscape, makes the site very favourable for this kind of investigation.

Wind rose for wind speeds exceeding 10.8 m/s during the winter months shows that the strong winds are appearing in a narrow sector from north, Figure 3. The wind speed registered during a snow storm is generally between 15 m/s and 25 m/s and the highest 10 minute mean wind speed registered during the winter was 35 m/s, including a 48 m/s gust.

3.2 Model results and snow drifting application

3.2.1 Shear stress and snow transport

We refer to the wind field simulations for the study site in part one. A closer view of the results for profile 1040 is found on Figure 4. Figure 5 shows the shear stress difference, $(u_*$ snow) - $(u_*$ ground) for profile 1040, plotted together with the measured snow cover thickness at running intervals along the drift. Apparently, there is a strong correlation between these two curves. We use this correlation to create a relationship between the shear stress gain and the deposited snow masses in order to quantify the snowdrift rate and estimate the trapping efficiency of the terrain depression. Many authors have reported on the relationship between the friction velocity, u_* or wind speed at a certain height above the ground and the snow transport capacity of the wind. The simplest form for these models is:

$$Q = a(U-b)^{\alpha} \qquad (1)$$

where a, b and α are constants, a and α found by regression analysis on experimental data and b denoting a threshold wind speed for drifting snow. U (m/s) is the wind speed or friction velocity and Q (kg/s m) is the snowdrift rate per unit width. The constant a is usually replaced by $(\rho/g\,C)$, where ρ is the density of the fluid, g is gravitational acceleration and C is an experimental constant (Pomeroy & Gray 1990, Schmidt 1986). In general, this kind of model indicates that a change in wind speed in horizontal direction results in a change in the observed snowdrift rate. When integrated into a fluid flow model, this relationship can be used to calculate drift sedimentation and erosion (Uematsu et al. 1989):

$$\frac{\partial h}{\partial t} = -\frac{1}{\rho_s}\frac{\partial q}{\partial x} \qquad (2)$$

This model adjusts the snow surface elevation, h, in a computational cell according to change in drift rate, q, along the surface. ρ_S is the density of sedimented snow on ground.

We use this relationship in the vertical direction, and calculate the total change in drift rate caused by elevating the surface from ground level to the observed drift surface. The change in transport rate, ΔQ, is a result of drift growth, ΔV (m³/m), when divided by estimated drift density ρ_S (kg/m³), and multiplied by a time ΔT (s) for the duration of the drift build-up. For simplicity the threshold velocity for snow transport, b, is neglected here. This should not result in large error since the threshold value for snow transport is very low for a fresh loose snow cover (Kind 1981) and the snow drifting at the observation site is mainly occurring with simultaneous precipitation. The snow drifting rate formula is:

$$Q = C \frac{\rho}{g} u_*^{\alpha} \qquad (3)$$

For conservation of snow-mass in the model we get:

$$\frac{\Delta V}{\Delta T} = \Delta Q \frac{1}{\rho_S} \qquad (4)$$

where $\Delta Q = (Q \text{ snow} - Q \text{ ground})$, according to Equation (3). This model was applied in stepwise increments, dx, along the drift length to calculate the drift volume from the gain in u_*. Note that this calculation is not integrated in the fluid flow model but is done after the wind flow has been simulated over the two surfaces, ground and snow cover. Figure 6 shows the relationship between the accumulated measured drift volume on the ground for profile 1040 against the calculated drift volume from Equation (3) and Equation (4) in running intervals along the drift. The unknown parameters, ρ_S, ΔT, C and α have to be determined.

We use profile 1040 to calibrate the equation, since it has the most smooth ground surface. The drift density is chosen as $\rho_S = 400$ kg/m³ and the build-up duration, ΔT, is set to two days. The values for ρ_S and ΔT are actually arbitrary chosen, as a part of scaling the equation. The exponent α is adjusted to find the least squares difference for a linear fit between the calculated and measured drift volumes, and finally, the constant C is modified until the slope of the line equals one. This gives the constants $C = 2.30$ and $\alpha = 2.85$, and the transport rate equation becomes:

$$Q = 2.30 \frac{\rho}{g} u_*^{2.85} \qquad (5)$$

Of course, since we neither know the actual ρ_S nor ΔT, the absolute drifting rate calculated here is not likely to be correct. Anyhow, this will not matter in this application, since the exponent α, is the parameter that really describes the changes in drifting rate due to changes in friction velocity. The exponential relationship found here, $Q \sim u_*^{2.85}$ is close to the most commonly reported relationship of about 3 (Pomeroy 1989, Kind 1981, Kobayashi 1972). Takeuchi (1980) reported on $\alpha = 2.7$ for old firm snow and $\alpha = 4.16$ for settled dry snow. This model was applied to profiles 1050 and 1060 with the same value for all parameters found for profile 1040. The resulting diagrams are displayed on Figure 7 and Figure 8.

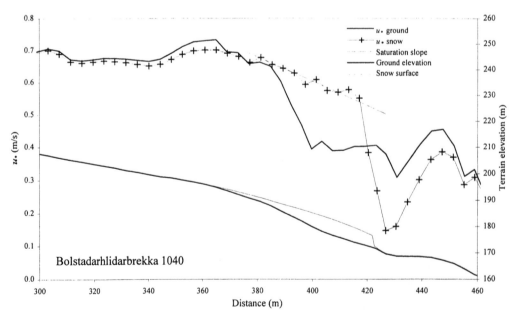

Figure 4. Closer view over the calculated friction velocity, u_* along profile 1040. The friction velocity keeps a constant drop along the drift surface. Further we note that the friction velocity is lowered on a section upstream from the drift.

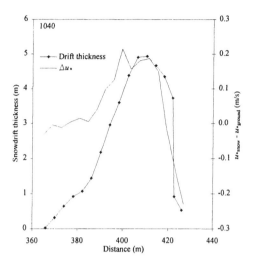

Figure 5. Profile 1040. Difference between calculated friction velocity for ground and snow cover, and measured snowdrift thickness.

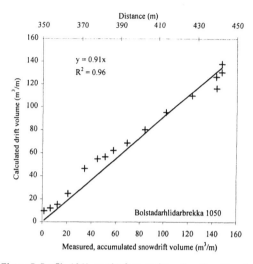

Figure 7. Profile 1050, results from applying Equation (4) and Equation (5) on the u_* difference between ground and snowcover.

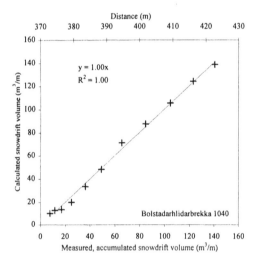

Figure 6. Profile 1040. Measured drift volume, accumulated in stepwise increments along the drift, and computed drift volume according to Equation (3) and Equation (4). Best linear fit results in $\alpha = 2.85$ and C = 2.30.

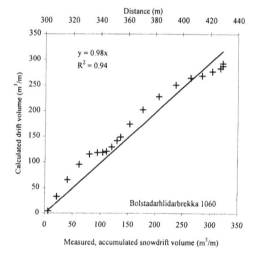

Figure 8. Profile 1060, same as for Figure 7.

3.2.2 *Terrain slope and snow transport*

The friction velocity diagrams displayed in part one indicate that there is a tendency towards a constant u_* value along terrain sections of constant slope on the moderate sloping sections. This is reasonable concerning that in nature, we don't observe snowdrifts starting suddenly on a constant slope, but we relate the starting point to a change in slope. However, when the terrain slope exceeds a certain value, there is a

sudden drop in the u_* curve.

To test this effect, we built an ideal terrain profile from constant slope segments and imported into the wind flow model. Figure 9 shows that u_* indeed keeps a constant value along each section, until the terrain slope changes from 30% to 40% where a steep descend in the friction velocity profile occurs, at distance 300 m. This slope could be in the vicinity of the critical terrain slope for flow separation, but this critical slope can be related to the proportion between horizontally length scale of the hill and the roughness height, λ/z_0 (Wood 1995). Separation is a sufficient

449

condition for snow drift sedimentation in two-dimensional flow, but according to our results, not necessary for sedimentation. Diagrams on the shear stress variations along the profiles from the study site show that the large snowdrifts observed start where a sudden and continuous decrease in shear stress is calculated. Hence, this must be a feasible start point for a large snowdrift sedimentation on the ideal terrain profile and we applied Tabler's model for equilibrium drift growth here. Tabler (1994) has suggested equilibrium shape for snowdrifts, given that the start point of the drift is known. His model is based on regression analysis of measured snowdrifts in the field and uses the terrain slope upwind and downwind of the prediction point as input parameters:

$$Y_S = 0.25\,X_1 + 0.55\,X_2 + 0.15\,X_3 + 0.05\,X_4 \quad (6)$$

where

Y_S = snow slope (%) over the main portion of the drift

X_1 = average ground slope over a distance of 45 m upwind of the catchment lip

X_2 = ground slope from 0 to 15 m downwind of the trap lip

X_3 = ground slope from 15 to 30 m downwind of the trap lip

X_4 = ground slope from 30 to 45 m downwind of the trap lip

The resulting drift starts at distance 300 m and has a 35.5% slope. The wind flow model was now run again over this new artificial snow cover and the u_* curve plotted on Figure 9. The u_* curve now has a constant slope and lies above the curve for the ground simulation as one will expect. However, we see that the relationship of constant terrain slope and constant friction velocity is not valid here. Either this could mean that the relationship is not valid for steeper slopes than about 30%, or that the distance required to establish a constant friction velocity is longer, the steeper the slope is. We note that this is also the case for the profiles from the study site.

The relationship between upwind terrain slope and drift slope, and the constant slope of the u_* curve was found worth a further study. This is because according to snow transport theory, the drift will still be growing as long as the shear stress along the drift surface is decreasing.

For the next diagram plotted, the friction velocity was replaced by calculated drift rate according to Equation (5). Considering drift rates calculated from the u_* values, these must be considered as maximum possible drift rate and not the actual drift rate, at least

until the point of sedimentation is reached. We know that at the point where snow sedimentation starts, the transport capacity of the wind is fully exerted and no further decrease can occur without snow particles being deposited on the ground. We can thus normalize the drift rate curve, such that the drift rate at the point where the sedimentation starts is set to 100%. The slope of the curve will hence give a reduction in percent per meter travelled from the initial point. On Figure 10, the difference between the upwind terrain slope and the surface slope of the drift is plotted against the slope of the friction velocity curve along the drift surface which has been modified to percent reduction per unit length as written above. The terrain slopes are average slopes along a 50 m long path. The data used to derive this relationship comes from profiles 1040, 1050, 1060 and the ideal slope profile with Tabler's equilibrium drift.

We find a linear relationship passing through the origin, Figure 10, which is in accordance with the rule that no sedimentation can be initiated if no drop in friction velocity and no change in slope happens. Besides we see that the larger the difference between upwind terrain slope and drift slope, the higher is the rate of drop in transport capacity along the drift surface and hence, the faster the drift growth or the more efficient is the natural snow trap.

For the model derived here, it is assumed that a decrease in u_* or maximum possible transport rate immediately results in a drift growth on the spot. This is really a crude simplification, since it will take some distance for the snow particles to be deposited on the ground. As a simplified example, considering a terminal fall velocity of 0.5 m/s for a snow particle transported by the wind at 15 m/s. This particle will be transported 15 m downwind if initially at 0.5 m height when the drop in u_* encourages this particle to be sedimented. Despite this, it is tempting to state that a theoretical equilibrium drift surface will actually have the same slope as the upwind terrain, though it will seldom reach it because how slow the last stages of drift growth towards equilibrium happen. Supporting this, we can refer to small scale terrain depressions in nature that are easily filled with drifting snow and smeared out in the landscape, but larger depressions take longer time to fill. In engineering application, the definition of an equilibrium snowdrift surface, which still allows a certain fall in friction velocity along it is thus feasible. Norem (1994) has used a 10% to 15% slope difference as a reference value for equilibrium drift surfaces in road planning applications.

450

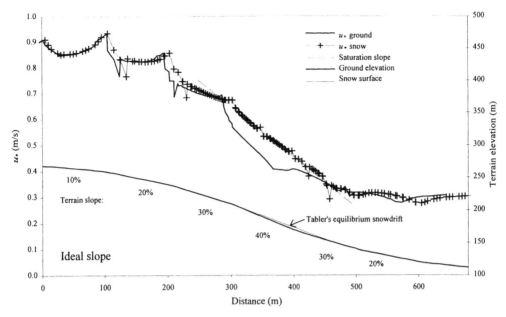

Figure 9. Model run over ideal terrain of constant sloping segments. Tabler's equilibrium drift formula is applied at distance 300 m, where a sharp drop in the u_* curve occurs, and the model was run again over the new snow cover.

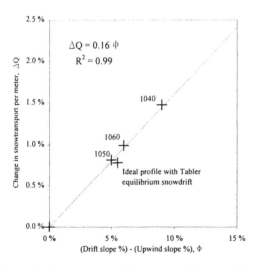

Figure 10. Calculated relationship for upwind slope and drift slope, and relative drop in transport rates per meter along the drift.

3.2.3 Profile 1100, snow fence

The last terrain profile from the study site is taken through an existing snow fence. This snow fence is 3 m high and is made of fabrics of about 50% porosity. By the time of the snow survey, the windward- and leeward drifts had grown together, leaving only one meter of the fence standing above the surface.

The following example is an effort to evaluate the location of the snow fence and its effect to reduce snow problems on the road. First, the model was run on the original terrain, without considering the snowfence, then the model was run with the measured snowdrifts included. The part of the fence still active, the topmost meter, is not included in the simulation, and hence the effect of the fence is underestimated here. The two resulting u_* curves are plotted on Figure 6 in part one.

The u_* curve for the snow surface falls steeply across the drift around the fence, the equivalent drop in drifting rate is about 29% or 1% per meter, according to Equation (5). Figure 10 suggests that 1% drop per meter is close to being a practical equilibrium state for the drift development. In the meanwhile, the fall of the u_* curve along the drift adjacent to the road corresponds to a change in drift rate from 71% of the original flux down to 24%. This equals that about 3% of the flux at the start of the drift is sedimented per meter of length along the drift. Compared to the other profiles investigated, where the drop in drift rate was from 0.8% to 1.5% along the snow cover, it is evident that the fence has delayed the development of the drift next to the road significantly.

4 CONCLUSIONS

From part one of this report we recall that the normalized shape of the friction velocity curve along the two-dimensional profiles is unaffected by the input velocity. This is important for the general application of the model developed here.

The study confirms that a plain wind simulation can be an aid to identify areas for large snowdrift sedimentation in two-dimensional profiles, and to evaluate the snow-trapping efficiency and capacity of these. We use a new method to establish a relationship between the friction velocity and the snow drifting capacity of the wind, by using the difference in calculated friction velocity between simulations with and without the snow cover present. We propose a relationship between the terrain slope upwind of the drift and the drift slope itself, and the horizontal gradient of the snow drifting rate. This can be considered when choosing road path near leeward sloping terrain features and when choosing appropriate cross sectional profile for the road.

The model developed in the study is a potential addition to the tools already known for road and highway engineering in snow drifting areas.

ACKNOWLEDGEMENTS

We are grateful for the financial support given by the Public Road Administration of Iceland and Norway and the Nordic Road Association. This work was carried out as a part of the EU project ROADEX, which is financed within the EU Northern Periphery programme. We also thank the staff of Saudarkrokur road district in Iceland for great help with the field work.

REFERENCES

Flow Science, Inc. 1997. *Flow3D, Users Manual*. Los Alamos, NM: Flow Science Inc.

Kind, R.J. 1981. Snow drifting. In Gray, D.M. & Male, D.H. (ed.), *Handbook of snow*: 338-360. Toronto: Pergamon Press.

Kobayashi, D. 1972. *Studies of snow transport in low level drifting snow*. Contributions from the Institute of Low Temperature Science. Sapporo: Hokkaido University.

Norem, H. 1975. *Designing highways situated in areas of drifting snow*. Draft translation 503. Hanover, New Hampshire: CRREL.

Norem, H. 1994. *Snow Engineering for Roads*. Handbook no. 174. Oslo: Norwegian Public Road Administration, Road Research Laboratory.

Pomeroy, J.W. 1989. A process-based model of snow drifting. *Annals of Glaciology* 13: 237-240.

Pomeroy, J.W. & Gray, D.M. 1990. Salutation of Snow. *Water Resources Research* 26: 1583-1594.

Schmidt, R.A. 1986. Transport rate of drifting snow and the mean wind speed profile. *Boundary-Layer Meteorology* 34: 213-241.

Tabler, R.D. 1988. *Snow Fence Handbook*. Laramie, Wyoming: Tabler & Associates.

Tabler, R.D. 1994. *Design Guidelines for the Control of Blowing and Drifting Snow*. SHRP-H-381. Washington, DC: National Research Council.

Takeuchi, M. 1980. Vertical profile and horizontal increase of drift-snow transport. *Journal of Glaciology* 26: 481-492.

Wood, N. 1995. The onset of separation in neutral, turbulent flow over hills. *Boundary-Layer Meteorology* 76: 137-164.

Snow Engineering, Hjorth-Hansen, Holand, Løset & Norem (eds) © 2000 Balkema, Rotterdam, ISBN 90 5809 148 1

Subject index

Snow Engineering, Hjorth-Hansen, Holand, Løset & Norem (eds) © 2000 Balkema, Rotterdam, ISBN 90 5809 148 1

Author index